DIETARY REFERENCE INTAKES

FOR

Thiamin, Riboflavin, Niacin, Vitamin B$_6$, Folate, Vitamin B$_{12}$, Pantothenic Acid, Biotin, and Choline

A Report of the
Standing Committee on the Scientific Evaluation
of Dietary Reference Intakes and its
Panel on Folate, Other B Vitamins, and Choline and
Subcommittee on Upper Reference Levels of Nutrients
Food and Nutrition Board
Institute of Medicine

NATIONAL ACADEMY PRESS
Washington, D.C.

NATIONAL ACADEMY PRESS • 2101 Constitution Avenue, N.W. • Washington, DC 20418

NOTICE: The project that is the subject of this report was approved by the Governing Board of the National Research Council, whose members are drawn from the councils of the National Academy of Sciences, the National Academy of Engineering, and the Institute of Medicine. The members of the committee responsible for the report were chosen for their special competences and with regard for appropriate balance.

This project was funded by the U.S. Department of Health and Human Services Office of Disease Prevention and Health Promotion, Contract No. 282-96-0033, T01; the National Institutes of Health Office of Nutrition Supplements, Contract No. N01-OD-4-2139, T024, the Centers for Disease Control and Prevention, National Center for Chronic Disease Prevention and Health Promotion, Division of Nutrition and Physical Activity; Health Canada; the Institute of Medicine; and the Dietary Reference Intakes Corporate Donors' Fund. Contributors to the Fund include Roche Vitamins Inc, Mead Johnson Nutrition Group, Daiichi Fine Chemicals, Inc, Kemin Foods, Inc, M&M Mars, Weider Nutrition Group, and Natural Source Vitamin E Association. The opinions or conclusions expressed herein do not necessarily reflect those of the funders.

Library of Congress Cataloging-in-Publication Data

Dietary reference intakes for thiamin, riboflavin, niacin, vitamin B6, folate, vitamin B12, pantothenic acid, biotin, and choline / a report of the Standing Committee on the Scientific Evaluation of Dietary Reference Intakes and its Panel on Folate, Other B Vitamins, and Choline and Subcommittee on Upper Reference Levels of Nutrients, Food and Nutrition Board, Institute of Medicine.
    p. cm.
Includes bibliographical references and index.
ISBN 0-309-06554-2 (pbk.) – ISBN 0-309-06411-2 (case)
   1. Vitamin B in human nutrition. 2. Reference values (Medicine) I. Institute of Medicine (U.S.). Standing Committee on the Scientific Evaluation of Dietary Reference Intakes. II. Institute of Medicine (U.S.). Panel on Folate, Other B Vitamins, and Choline. III. Institute of Medicine (U.S.). Subcommittee on Upper Reference Levels of Nutrients.

QP772.V52 D53 2000
612.3'99–dc21   00-028380

Additional copies of this report are available from National Academy Press, 2101 Constitution Avenue, N.W., Lock Box 285, Washington, DC 20055. Call (800) 624-6242 or (202) 334-3313 (in the Washington metropolitan area), or visit the NAP's on-line bookstore at **http:/ www.nap.edu**.

For more information about the Institute of Medicine or the Food and Nutrition Board, visit the IOM home page at **http://www.nas.edu/iom.**

The serpent has been a symbol of long life, healing, and knowledge among almost all cultures and religions since the beginning of recorded history. The image adopted as a logotype by the Institute of Medicine is based on a relief carving from ancient Greece, now held by the Staatliche Museen in Berlin.

"Knowing is not enough; we must apply.
Willing is not enough; we must do."
—Goethe

INSTITUTE OF MEDICINE

Shaping the Future for Health

THE NATIONAL ACADEMIES

National Academy of Sciences
National Academy of Engineering
Institute of Medicine
National Research Council

The **National Academy of Sciences** is a private, nonprofit, self-perpetuating society of distinguished scholars engaged in scientific and engineering research, dedicated to the furtherance of science and technology and to their use for the general welfare. Upon the authority of the charter granted to it by the Congress in 1863, the Academy has a mandate that requires it to advise the federal government on scientific and technical matters. Dr. Bruce M. Alberts is president of the National Academy of Sciences.

The **National Academy of Engineering** was established in 1964, under the charter of the National Academy of Sciences, as a parallel organization of outstanding engineers. It is autonomous in its administration and in the selection of its members, sharing with the National Academy of Sciences the responsibility for advising the federal government. The National Academy of Engineering also sponsors engineering programs aimed at meeting national needs, encourages education and research, and recognizes the superior achievements of engineers. Dr. William A. Wulf is president of the National Academy of Engineering.

The **Institute of Medicine** was established in 1970 by the National Academy of Sciences to secure the services of eminent members of appropriate professions in the examination of policy matters pertaining to the health of the public. The Institute acts under the responsibility given to the National Academy of Sciences by its congressional charter to be an adviser to the federal government and, upon its own initiative, to identify issues of medical care, research, and education. Dr. Kenneth I. Shine is president of the Institute of Medicine.

The **National Research Council** was organized by the National Academy of Sciences in 1916 to associate the broad community of science and technology with the Academy's purposes of furthering knowledge and advising the federal government. Functioning in accordance with general policies determined by the Academy, the Council has become the principal operating agency of both the National Academy of Sciences and the National Academy of Engineering in providing services to the government, the public, and the scientific and engineering communities. The Council is administered jointly by both Academies and the Institute of Medicine. Dr. Bruce M. Alberts and Dr. William A. Wulf are chairman and vice chairman, respectively, of the National Research Council.

PANEL ON FOLATE, OTHER B VITAMINS, AND CHOLINE

ROY M. PITKIN (*Chair*), Department of Obstetrics and Gynecology, University of California, Los Angeles (Professor Emeritus) and *Obstetrics & Gynecology* (Editor), Los Angeles

LINDSAY H. ALLEN, Department of Nutrition, University of California, Davis

LYNN B. BAILEY, Food Science and Human Nutrition Department, University of Florida, Gainesville

MERTON BERNFIELD, Department of Pediatrics, Harvard Medical School, Boston

PHILLIPE De WALS, Department of Community Health Sciences, University of Sherbrooke, Quebec

RALPH GREEN, Department of Pathology, University of California at Davis Medical Center, Sacramento

DONALD B. McCORMICK, Department of Biochemistry, Emory University School of Medicine, Atlanta

ROBERT M. RUSSELL, Department of Medicine and Nutrition at the Jean Mayer U.S. Department of Agriculture Human Nutrition Research Center on Aging, Tufts University, Boston

BARRY SHANE, Department of Nutritional Sciences, University of California, Berkeley

STEVEN H. ZEISEL, Department of Nutrition, University of North Carolina School of Public Health and School of Medicine, Chapel Hill

IRWIN H. ROSENBERG, Clinical Nutrition Division, the Jean Mayer U.S. Department of Agriculture Human Nutrition Research Center on Aging, Tufts University and New England Medical Center, Boston, *Liaison to the Panel from the Subcommittee on Upper Reference Levels of Nutrients*

Staff

CAROL W. SUITOR, Study Director
ELISABETH A. REESE, Research Associate
ALICE L. KULIK, Research Assistant
MICHELE RAMSEY, Project Assistant

SUBCOMMITTEE ON UPPER REFERENCE LEVELS OF NUTRIENTS

IAN C. MUNRO (*Chair*), CanTox, Inc., Mississauga, Ontario
WALTER MERTZ, Retired, U.S. Department of Agriculture Human Nutrition Research Center, Rockville, Maryland
RITA B. MESSING, Division of Environmental Health, Minnesota Department of Health, St. Paul
SANFORD A. MILLER, Graduate School of Biomedical Sciences, University of Texas Health Sciences Center, San Antonio
SUZANNE P. MURPHY, Department of Nutritional Sciences, University of California, Berkeley
JOSEPH V. RODRICKS, ENVIRON Corporation, Arlington, Virginia
IRWIN H. ROSENBERG, Clinical Nutrition Division, the Jean Mayer U.S. Department of Agriculture Human Nutrition Research Center on Aging, Tufts University and New England Medical Center, Boston
STEVE L. TAYLOR, Department of Food Science and Technology and Food Processing Center, University of Nebraska, Lincoln
ROBERT H. WASSERMAN, Department of Physiology, College of Veterinary Medicine, Cornell University, Ithaca

Consultants

SHEILA DUBOIS, Food Directorate, Health Canada, Ottawa
HERBERT BLUMENTHAL, Retired, Food and Drug Administration, Washington, D.C.

Staff

SANDRA SCHLICKER, Study Director
ELISABETH A. REESE, Research Associate
GERALDINE KENNEDO, Project Assistant

Preface

This report is the second in a series that presents a comprehensive set of reference values for nutrient intakes for healthy U.S and Canadian populations. It is a product of the Food and Nutrition Board of the Institute of Medicine (IOM) working in cooperation with scientists from Canada.

The report establishes a set of reference values for the B vitamins and choline to replace previously published Recommended Dietary Allowances (RDAs) for the United States and Recommended Nutrient Intakes (RNIs) for Canada. It considers evidence concerning the prevention of disease and developmental disorders along with more traditional evidence of sufficient nutrient intake; and examines data about choline, a food component that in the past has not been considered essential in the human diet. Although the reference values are based on data, the data were often scanty or drawn from studies that had limitations in addressing the question. Thus, scientific judgment was required in setting the reference values. The reasoning used is described for each nutrient in Chapters 4 through 12. Evidence concerning the use of these nutrients for the amelioration or cure of disease or disability was not considered because that was beyond the project's scope of work.

The B vitamins appear second in the series largely because recommendations for folate intake have been a subject of controversy for many years. The RDA for folate has shifted up and down. Recently, low folate intake has been linked with vascular disease and other chronic conditions as well as risk of neural tube defects and other congenital malformations in the offspring of women of reproductive age. However, high folate intake has also been implicated in delaying the diagnosis of pernicious anemia until after irreversible neurological damage has occurred. A major task of the Panel on Folate, Other B Vitamins, and Choline; the Subcommittee on Upper Reference Levels of Nutrients (UL Subcommittee); and the Standing Committee on the Scientific Evaluation of Dietary Reference Intakes (DRI Committee) was to analyze the evidence on beneficial and adverse effects of different folate intakes—in the context of setting Dietary Reference Intakes (DRIs) for all the B vitamins and choline.

Many of the questions raised about requirements for and recommended intakes of B vitamins and choline cannot be answered fully because of inadequacies in the present database. Apart from studies of overt deficiency disease, there is a dearth of studies that address specific effects of inadequate B vitamin intakes on health status. For most of the B vitamins, there is no direct information that permits estimating the amounts required by children and adolescents. For five of the B vitamins, data useful for the setting of Tolerable Upper Intake Levels (ULs) are sparse, precluding reliable estimates of how much can be ingested safely. For some of these nutrients, there are questions about how much is contained in the food North Americans eat. Thus, another major task of the report was to outline a research agenda to provide a basis for public policy decisions related to recommended intakes of the B vitamins and choline and ways to achieve those intakes. The process for establishing DRIs is an iterative process and is thus evolving as the conceptual framework is applied to new nutrients and food components. With more experience, the proposed models for establishing reference intakes for use with nutrients and food components that play a role in health will be refined and, as new information or new methods of analysis are adopted, these reference values will be reassessed. The DRI Committee and its UL Subcommittee are developing plans to explore ways to address the safety of high nutrient intakes in other age groups or situations where data are lacking. For example, although the panel chose to use metabolic body weight ($kg^{0.75}$) as the basis for adjusting all DRIs, including ULs for children for establishing intakes of the vitamins reviewed in this report, the proposed risk assessment model of the UL Subcommittee uses body weight directly as the default for extrapolation to children because of its more conservative result.

Considerations of bioavailability and nutrient-nutrient interactions played a key role in the decision-making process for several B vitamins. For example, the concept of dietary folate equivalents is introduced to help estimate folate requirements, and limitations on the absorption of vitamin B_{12} were considered when recommending B_{12} intake for the elderly.

Because the project is ongoing as indicated above, and many comments were solicited and have been received on the first report in the series (*Dietary Reference Intakes for Calcium, Phosphorus, Magnesium, Vitamin D, and Fluoride*), it has been possible to introduce refinements in introductory material (Chapters 1 through 3) and in the discussion of uses of DRIs (Chapter 13 in this report). For example, it is now clearly stated that a detrimental nutrient-nutrient interaction could be used as the critical adverse effect in setting a UL for a nutrient. Among the comments have been requests for additional guidance in the practical application of DRIs. The newly formed Subcommittee on the Interpretation and Uses of Dietary Reference Intakes will work toward that end.

This report reflects the work of the Food and Nutrition Board's DRI Committee; the expert Panel on Folate, Other B Vitamins, and Choline; and the UL Subcommittee. The support of the government of Canada and Canadian scientists in this initiative represents a pioneering first step in the standardization of nutrient reference intakes at least within one continent. A brief description of the overall project of the DRI Committee and of the panel's task are given in Appendix A. It is hoped that the critical, comprehensive analyses of available information and of knowledge gaps in this initial series of reports will greatly assist the private sector, foundations, universities, government laboratories, and other institutions with their research interests and with the development of a productive research agenda for the next decade.

The DRI Committee; the Panel on Folate, Other B Vitamins, and Choline; and the UL Subcommittee wish to extend sincere thanks to the many experts who have assisted with this report by giving presentations, providing written materials, participating in discussions, analyzing data, and other means. Many, but far from all, of these people are named in Appendix B. Special thanks go to Robert A. Jacob and Donald M. Mock, who made major contributions to chapters on niacin and biotin, respectively, and to staff at the National Center for Health Statistics, the Food Surveys Research Group of the Agricultural Research Service, and the Department of Statistics at Iowa State University for extensive analyses of survey data.

The respective chairs and members of the panel and subcommittee have performed their work under great time pressure. Their dedication made the completion of this report possible. All gave of their time willingly and without financial reward; both the science and practice of nutrition are major beneficiaries.

This report has been reviewed by individuals chosen for their diverse perspectives and technical expertise, in accordance with procedures approved by the National Research Council's Report Review Committee. The purpose of this independent review is to provide candid and critical comments to assist the authors and the IOM in making the published report as sound as possible and to ensure that the report meets institutional standards for objectivity, evidence, and responsiveness to the study charge. The content of the review comments and draft manuscript remain confidential to protect the integrity of the deliberative process. We wish to thank the following individuals for their participation in the review of this report: Frederick C. Battaglia, M.D., University of Colorado Health Sciences Center; Enriqueta C. Bond, Ph.D., Burroughs Wellcome Fund; Patricia K. Crumrine, M.D., Children's Hospital; Krishnamurti Dakshinamurti, Ph.D., University of Manitoba; Gary Flamm, Ph.D., Flamm Associates; Theresa Glanville, Ph.D., Mount Saint Vincent University; John Hathcock, Ph.D., Council for Responsible Nutrition; James Marshall, Ph.D., Arizona Cancer Center; Deborah O'Connor, Ph.D., Ross Laboratories; Claire Regan, M.S., R.D., Grocery Manufacturers of America; Eric Rimm, Sc.D., Harvard School of Public Health; Killian Robinson, M.D., Cleveland Clinic Foundation; Robert Rucker, Ph.D., University of California-Davis; Robert F. Schilling, M.D., University of Wisconsin; John Scott, Ph.D., Sc.D., M.A., University of Dublin, Trinity College.

Although the individuals listed above have provided many constructive comments and suggestions, responsibility for the final content of this report rests solely with the authoring committee and the IOM.

The DRI Committee wishes to acknowledge, in particular, the commitment shown by Roy Pitkin, chair of the panel, who steered this difficult project through what at times seemed to some of us like dangerous and uncharted waters. His ability to keep the effort and our various biases moving in a positive direction is very much appreciated.

Special thanks go to the staff of the Food and Nutrition Board and foremost to Carol Suitor, who was the study director for the panel and without whose assistance, both intellectual and managerial, this report would neither have been as polished nor as timely in its initial release. She now moves on to enjoy the peace of her new home in beautiful Vermont and we wish her well. It is, of course those at the Food and Nutrition Board who get the real work completed and so the committee wishes to thank Allison Yates, Director of the Food and Nutrition Board, for constant assistance and it also recognizes, with appreciation, the contributions of Sandra Schlicker, Elisabeth Reese, Kimberly Brewer, Alice Kulik, Sheila Moats, Gail Spears, Diane Johnson, Michele Ramsey, and Geraldine Kennedo. We also thank Judith Grumstrup-Scott and Judith Dickson for editing the manuscript and Mike Edington and Claudia Carl for assistance with publication.

Vernon Young
Chair, Standing Committee on the Scientific
  Evaluation of Dietary Reference Intakes

Cutberto Garza
Chair, Food and Nutrition Board

Contents

DIETARY REFERENCE INTAKES

FOR

Thiamin, Riboflavin,
Niacin, Vitamin B$_6$,
Folate, Vitamin B$_{12}$,
Pantothenic Acid,
Biotin, and Choline

Summary

This report on folate, other B vitamins, and choline is one of a series that presents dietary reference values for the intake of nutrients by Americans and Canadians. The overall project is a comprehensive effort undertaken by the Standing Committee on the Scientific Evaluation of Dietary Reference Intakes (DRI Committee) of the Food and Nutrition Board, Institute of Medicine, National Academy of Sciences, with active involvement of Health Canada. (See Appendix A for a description of the overall process and its origins.) This study was requested by the U.S. Federal Steering Committee for Dietary Reference Intakes and coordinated by the U.S. Department of Health and Human Services Office of Disease Prevention and Health Promotion; with funding from the Office of Dietary Supplements, National Institutes of Health; Division of Nutrition and Physical Activity, National Center for Chronic Disease Prevention and Health Promotion, Centers for Disease Control and Prevention; the Agricultural Research Service, U.S. Department of Agriculture; and Health Canada. Additional funding was provided by contributors to the DRI Corporate Donors Fund.

Major new recommendations in this report include the following:

• The use of dietary folate equivalents (DFEs) for estimating folate requirements, recommending daily folate intake, and assessing intake. This adjusts for the greater degree of absorption of folic acid (free form) compared with folate naturally found in foods (1 μg of food folate equals 0.6 μg of folate added to foods or taken with food or 0.5 μg of folate supplements taken on an empty stomach).

1

• A Recommended Dietary Allowance (RDA) for folate that is the same for men and women; however, for women who are capable of becoming pregnant, a special recommendation is made for intake of folate from fortified food or supplements.

• For adults over age 50, the use of food fortified with vitamin B_{12}, or B_{12} supplements to meet most of the RDA for B_{12}.

• Recommendations for choline intake (recommendations for all the B vitamins had been made previously).

• Suggested maximum intakes of niacin, vitamin B_6, folate from supplements and fortified foods, and choline.

• A research agenda that focuses primarily on folate and vitamin B_{12}.

WHAT ARE DIETARY REFERENCE INTAKES?

Dietary Reference Intakes (DRIs) are reference values that are quantitative estimates of nutrient intakes to be used for planning and assessing diets for healthy people. They include RDAs but also three other types of reference values (see Box S-1). Although the reference values are based on data, the data were often scanty or drawn from studies that had limitations in addressing the question. Thus, scientific judgment was required in setting the reference values.

BOX S-1   Dietary Reference Intakes

Recommended Dietary Allowance (RDA): the average daily dietary intake level that is sufficient to meet the nutrient requirement of nearly all (97 to 98 percent) healthy individuals in a particular life stage and gender group.

Adequate Intake (AI): a recommended daily intake value based on observed or experimentally determined approximations of nutrient intake by a group (or groups) of healthy people that are assumed to be adequate—used when an RDA cannot be determined.

Tolerable Upper Intake Level (UL): the highest level of daily nutrient intake that is likely to pose no risk of adverse health effects to almost all individuals in the general population. As intake increases above the UL, the risk of adverse effects increases.

Estimated Average Requirement (EAR): a daily nutrient intake value that is estimated to meet the requirement of half the healthy individuals in a group.

The reasoning used is described for each nutrient in Chapters 4 through 12.

The development of DRIs expands on the periodic reports called *Recommended Dietary Allowances,* which have been published since 1941 by the National Academy of Sciences. The chart on the inside back cover gives the recommended intake levels, whether RDAs or AIs, for the B vitamins and choline by life stage and gender group. Uses of DRIs appear in Box S-2. The transition from using RDAs alone to using all the DRIs appropriately will require time and effort by health professionals and others.

Recommended Dietary Allowances

The Recommended Dietary Allowance (RDA) is the average daily dietary intake level that is sufficient to meet the nutrient requirement of nearly all (97 to 98 percent) healthy individuals in a particular life stage (life stage considers age and, when applicable, pregnancy or lactation) and gender group.

Process for Setting the RDA

The process for setting the RDA depends on being able to set an *Estimated Average Requirement* (EAR). That is, the RDA is derived from the nutrient requirement; therefore, if an EAR cannot be set, no RDA will be set. The EAR is the daily intake value of a nutrient that is estimated to meet the nutrient requirement of half the healthy individuals in a life stage and gender group. Before setting the EAR, a specific criterion of adequacy is selected based on a careful review of the literature. When selecting the criterion, contemporary concepts of the reduction of disease risk are considered along with many other health parameters.

If the standard deviation (SD) of the EAR is available and the requirement for the nutrient is normally distributed, the RDA is set at 2 SDs above the EAR:

$$RDA = EAR + 2\ SD_{EAR}.$$

If data about variability in requirements are insufficient to calculate an SD, a coefficient of variation for the EAR of 10 percent is ordinarily assumed in this report.

The resulting equation for the RDA is then

$$RDA = 1.2 \times EAR.$$

BOX S-2 Uses of Dietary Reference Intakes for Healthy Individuals and Groups

Type of Use	For the Individual	For a Group
Planning	**RDA**: aim for this intake.	**EAR**: use in conjunction with a measure of variability of the group's intake to set goals for the mean intake of a specific population.
	AI: aim for this intake. **UL**: use as a guide to limit intake; chronic intake of higher amounts may increase risk of adverse effects.	
Assessment[a]	**EAR**: use to examine the possibility of inadequacy; evaluation of true status requires clinical, biochemical, and/or anthropometric data.	**EAR**: use in the assessment of the prevalence of inadequate intakes within a group.
	UL: use to examine the possibility of overcon-sumption; evaluation of true status requires clinical, biochemical, and/or anthropometric data.	

RDA = Recommended Dietary Allowance
EAR = Estimated Average Requirement
AI = Adequate Intake
UL = Tolerable Upper Intake Level

[a] Requires statistically valid approximation of usual intake.

If the estimated coefficient of variation is 15 percent, the formula would be

$$RDA = 1.3 \times EAR.$$

If the nutrient requirement is known to be skewed for a population, other approaches are used to find the ninety-seventh to ninety-eighth percentile to set the RDA.

The RDA for a nutrient is a value to be used as a goal for dietary intake by healthy individuals. As discussed in Chapter 13 of the report, the RDA is not intended to be used to assess the diets of either individuals or groups or to plan diets for groups.

Adequate Intakes

The *Adequate Intake* (AI) is set instead of an RDA if sufficient scientific evidence is not available for calculating an EAR. The AI is based on observed or experimentally determined estimates of nutrient intake by a group (or groups) of healthy people. For example, the AI for young infants, for whom human milk is the recommended sole source of food for the first 4 to 6 months, is based on the daily mean nutrient intake supplied by human milk for healthy, full-term infants who are exclusively breastfed. The main intended use of the AI is as a goal for the nutrient intake of individuals. Other possible uses of AIs will be considered by another expert group.

Tolerable Upper Intake Levels

The *Tolerable Upper Intake Level* (UL) is the highest level of daily nutrient intake that is likely to pose no risk of adverse health effects to almost all individuals in the general population. As intake increases above the UL, the risk of adverse effects increases. The term *tolerable intake* was chosen to avoid implying a possible beneficial effect. Instead, the term is intended to connote a level of intake that can, with high probability, be tolerated biologically. The UL is not intended to be a recommended level of intake. There is no established benefit for healthy individuals if they consume nutrient intakes above the RDA or AI.

ULs are useful because of the increased interest in and availability of fortified foods and the increased use of dietary supplements. ULs are based on total intake of a nutrient from food, water, and supplements if adverse effects have been associated with total intake. However, if adverse effects have been associated with intake from supplements or food fortificants only, the UL is based on nutrient intake from those sources only, not on total intake. The UL applies to chronic daily use.

For some nutrients, there are insufficient data on which to develop a UL. This does not mean that there is no potential for adverse effects resulting from high intake. When data about adverse effects are extremely limited, extra caution may be warranted.

COMPARISON OF RECOMMENDED DIETARY ALLOWANCES AND ADEQUATE INTAKES

Although the Recommended Dietary Allowance (RDA) and Adequate Intake (AI) are used for the same purpose—setting goals for intake by individuals—the RDA differs from the AI. Intake of the RDA for a nutrient is expected to meet the needs of 97 to 98 percent of the individuals in a life stage and gender group. If the EAR is not known, as is the case when an AI is set, it is not known what percentage of the individuals are covered by the AI. The AI for a nutrient is expected to exceed the average requirement for that nutrient, and it might even cover the needs of more than 98 percent of the individuals, but it might cover the needs of far fewer (see Figure S-1). The degree to which an AI exceeds the average requirement is likely to differ among nutrients and population groups.

For people with diseases that increase requirements or who have other special health needs, the RDA and AI may each serve as the basis for adjusting individual recommendations; qualified health professionals should adapt the recommended intake to cover higher or lower needs.

In this report, AIs rather than RDAs are being proposed for all nutrients for infants to age 1 year and for pantothenic acid, biotin, and choline for persons of all ages (see Table S-1).

APPROACH FOR SETTING DIETARY REFERENCE INTAKES

The scientific data used to develop DRIs have come from observational and experimental studies. Studies published in peer-reviewed journals were the principal source of data. Life stage and gender were considered to the extent possible, but for some nutrients the data did not provide a basis for proposing different requirements for men and women or for adults in different age groups.

Three of the categories of reference values (Estimated Average Requirement [EAR], Recommended Dietary Allowance [RDA], and

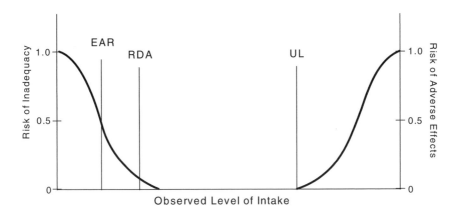

FIGURE S-1 Dietary reference intakes. This figure shows that the Estimated Average Requirement (EAR) is the intake at which the risk of inadequacy is 0.5 (50%) to an individual. The Recommended Dietary Allowance (RDA) is the intake at which the risk of inadequacy is very small—only 0.02 to 0.03 (2% to 3%). The Adequate Intake (AI) does not bear a consistent relationship to the EAR or the RDA because it is set without being able to estimate the average requirement. It is assumed that the AI is at or above the RDA if one could be calculated. At intakes between the RDA and the Tolerable Upper Intake Level (UL), the risks of inadequacy and of excess are both close to 0. At intakes above the UL, the risk of adverse effects may increase.

Adequate Intake [AI]) are defined by specific criteria of nutrient adequacy; the fourth (Tolerable Upper Intake Level [UL]) is defined by a specific indicator of excess if one is available. In all cases, data are examined closely to determine whether reduction of risk of a chronic degenerative disease or developmental abnormality could be used as a criterion of adequacy. The quality of studies was examined by considering study design; methods used for measuring intake and indicators of adequacy; and biases, interactions, and confounding factors. After careful review and analysis of the evidence, including examination of the extent of congruence of findings, scientific judgment was used to determine the basis for establishing the values. In this report, the scientific evidence was judged to be too weak to use the prevention of chronic degenerative disease as the basis for setting any of the recommended levels of intake. Thus, for the B vitamins and choline, EARs and RDAs, or AIs if applicable, are based on criteria related to their general func-

TABLE S-1 Estimated Average Requirements (EARs) and Reported Dietary Intakes of Six B Complex Vitamins by Gender for Young (19–30 years) and Older (> 70 years) Adults

Life Stage Group	Thiamin (mg/d)	Riboflavin (mg/d)
Males		
19–30 y		
EAR	**1.0**	**1.1**
CSFII Median Dietary Intake[b]	1.95	2.33
Range (5th–95th percentiles)	1.16–3.14	1.32–4.00
NHANES III Median Dietary Intake[c]	1.78	2.09
Range (5th–95th percentiles)	1.07–3.41	1.18–3.90
>70 y		
EAR	**1.0**	**1.1**
CSFII Median Dietary Intake	1.64	1.97
Range (5th–95th percentiles)	0.97–2.62	1.09–3.30
NHANES III Median Dietary Intake	1.56	1.84
Range (5th–95th percentiles)	1.03–2.68	1.13–3.28
Females		
19–30 y		
EAR	**0.9**	**0.9**
CSFII Median Dietary Intake	1.22	1.49
Range (5th–95th percentiles)	0.80–1.99	0.80–2.55
NHANES III Median Dietary Intake	1.45	1.63
Range (5th–95th percentiles)	0.94–2.49	0.99–2.85
>70 y		
EAR	**0.9**	**0.9**
CSFII Median Dietary Intake	1.18	1.40
Range (5th–95th percentiles)	0.68–1.86	0.83–2.34
NHANES III Median Dietary Intake	1.38	1.60
Range (5th–95th percentiles)	0.94–2.21	1.01–2.71

NOTE: The EAR can be used to assess the adequacy of nutrient intakes by groups. To do this, one determines the percentage of individuals whose usual intakes are less than the EAR. From this table it can be seen that less than 5 percent of young men have thiamin intakes less than the EAR, but more than half of young women have reported folate intakes less than the EAR. Appendixes G and H allow more accurate estimates of percentages for all age groups than does this excerpted table.

[a] Dietary folate equivalents for the EAR but not for reported dietary intakes. Reported intakes are likely to underestimate true intakes because of limitations of the methods used to analyze the folate content of food (see Chapter 8) and because adjustment has not been made for the higher bioavailability of the folic acid consumed in fortified

Niacin (mg/d)	B$_6$ (mg/d)	Folate (µg/d)[a]	B$_{12}$ (µg/d)
12	**1.1**	**320**	**2.0**
30.5	2.31	297	5.60
17.60–50.60	1.25–4.01	148–584	2.90–13.10
25.30	2.02	277	5.22
15.00–45.60	1.16–3.91	163–564	4.42–7.56
12	**1.4**	**320**	**2.0**
21.7	1.89	276	5.10
12.60–35.30	1.01–3.29	137–527	2.40–10.30
20.8	1.72	269	4.99
13.84–35.67	1.02–3.22	163–542	4.45–6.81
11	**1.1**	**320**	**2.0**
17.5	1.38	200	3.45
9.50–29.10	0.76–2.31	100–374	1.67–6.47
19.69	1.54	223	4.77
13.23–33.56	0.93–2.77	145–497	4.27–6.23
11	**1.3**	**320**	**2.0**
16.8	1.41	212	3.32
9.70–26.60	0.76–2.35	105–383	1.49–11.63
18.78	1.53	252	4.74
12.74–30.30	0.92–2.76	152–474	4.37–5.99

foods and supplements: 1 dietary folate equivalent = 1 µg food folate = 0.6 µg of folate from fortified food or as a supplement consumed with food = 0.5 µg of a supplement taken on an empty stomach.

[b] SOURCE: Continuing Survey of Food Intakes by Individuals (CSFII) data on B vitamin intake from food, unpublished data, A. Carriquiry, Iowa State University, 1997.

[c] SOURCE: Third National Health and Nutrition Examination Survey (NHANES III), 1988–1994, unpublished data on B vitamin intake from food, C.L. Johnson and J.D. Wright, National Center for Health Statistics, Centers for Disease Control and Prevention, 1997.

tions. For the B vitamins the EAR is somewhat higher than the amount needed to prevent deficiency disease (allowing a moderate safety margin) and there is laboratory evidence of sufficiency, but there is no observable health benefit beyond the prevention of signs and symptoms of deficiency. However, a special recommendation is included to address reduction of the risk of neural tube defects. The indicators used in deriving the RDAs and AIs are described below.

Nutrient Functions and the Indicators Used to Estimate the Requirements for the B Vitamins

Thiamin functions as a coenzyme in the metabolism of carbohydrates and branched-chain amino acids. Estimations of the requirement are based on the amount of thiamin needed to achieve and maintain normal erythrocyte transketolase activity while avoiding excessive thiamin excretion.

Riboflavin functions as a coenzyme in numerous oxidation-reduction reactions. Intake in relation to a combination of indicators is used to estimate the requirement for riboflavin. These indicators include the excretion of riboflavin and its metabolites, blood values for riboflavin, and the erythrocyte glutathione reductase activity coefficient.

Niacin functions as a cosubstrate or coenzyme with numerous dehydrogenases for the transfer of the hydride ion. The primary method used to estimate the requirement for niacin relates intake to the urinary excretion of niacin metabolites. The requirement is expressed in niacin equivalents, allowing for some conversion of the amino acid tryptophan to niacin.

Vitamin B_6 functions as a coenzyme in the metabolism of amino acids, glycogen, and sphingoid bases. To estimate the requirement, many types of biochemical data were examined; however, when possible, priority was given to the amount of B_6 consistent with maintenance of an adequate plasma pyridoxal phosphate concentration.

Folate functions as a coenzyme in single-carbon transfers in the metabolism of nucleic and amino acids. Folate is a generic term used to cover both the naturally occurring form of the vitamin (food folate) and the monoglutamate form (folic acid), which is used in fortified foods and supplements. To estimate the requirement the primary focus for all adults was on the amount of Dietary Folate Equivalents (DFEs) (values adjusted for differences in the absorption of food folate and folic acid) needed to maintain erythrocyte folate, but ancillary data on plasma homocysteine and plasma folate concentrations were also considered.

Vitamin B_{12} functions as a coenzyme in the metabolism of fatty acids of odd-chain length and in methyl transfer. To estimate the requirement, the primary focus was on the amount of B_{12} needed for the maintenance of hematological status and serum B_{12} values.

Pantothenic acid functions as a component of coenzyme A and phosphopantetheine, which are involved in fatty acid metabolism. The AI is based on data on pantothenic acid intake sufficient to replace urinary excretion.

Biotin functions as a coenzyme in bicarbonate-dependent carboxylations. The AI is based on limited intake data.

Choline functions as a precursor for acetylcholine, phospholipids, and the methyl donor betaine. The AI is based on the intake required to maintain liver function as assessed by measuring serum alanine aminotransferase levels. Although AIs have been set for choline, there are few data to assess whether a dietary supply of choline is needed at all stages of the life cycle, and it may be that the choline requirement can be met by endogenous synthesis at some of these stages.

Consideration of the Risk of Developmental Abnormalities and Chronic Degenerative Disease

Close attention was given to evidence relating intake of B vitamins and choline to reduction of the risk of developmental disability and chronic disease. Conclusions on four of these relationships follow.

Neural Tube Defects

Because pregnancy affected by a neural tube defect (NTD) occurs in only a very small fraction of the population of women in their childbearing years, reduction of the risk of NTDs is not considered compatible with the setting of the RDA for folate. That is, by definition, the EAR would need to prevent fetal NTD in 50 percent of all women in the age group and the RDA would need to prevent it in 97 to 98 percent of the women, but NTD occurrence is already much lower than this—less than 1 percent of all pregnancies.

The RDA for folate recommended in this report for women ages 19 through 50 years (400 µg/day of dietary folate equivalents) is consistent with some recommendations for the prevention of NTDs. However, the amount and form of folate demonstrated in currently available studies to minimize NTD risk is 400 µg/day of folic acid in addition to food folate. Therefore, the recommendation for women capable of becoming pregnant is to take 400 µg/day of folate from fortified foods and/or a supplement as well as food folate from a varied diet. It is not known whether the same level of protection could be achieved by using food that is naturally rich in folate. Neither is it known whether lower intakes would be protective or whether there is a threshold below which no protection occurs.

Vascular Disease and Thrombosis

Elevated homocysteine values have been associated with increased risk of vascular disease, and intakes of folate and vitamins B_6 and B_{12} have been inversely related to homocysteine values. However, conflicting evidence exists and it is premature to conclude that increasing the intake of these B vitamins could reduce the risk of vascular disease and thrombosis. Randomized trials among high-risk, healthy individuals and among patients with vascular disease are expected to provide evidence useful in resolving this matter.

Cancer

Many studies have investigated relationships between folate status and carcinogenesis. The data suggesting an inverse relationship between folate status and the occurrence of colorectal cancer are stronger than for other forms of cancer (e.g., cancer of the cervix, esophagus, stomach, and lung) but are not conclusive.

Neuropsychiatric Disorders

Although available information suggests that a link may exist between folate deficiency and abnormal mental function, more than three decades of research have not produced a definitive connection. Other than for relatively rare inborn errors of metabolism, it is not known whether low folate or vitamin B_6 status increases the risk of neuropsychiatric disorders or results from them. Neither is it known definitively how vitamin B_{12} status above that usually presumed to be adequate relates to psychiatric disturbances.

CRITERIA AND PROPOSED VALUES FOR TOLERABLE UPPER INTAKE LEVELS

A risk assessment model is used to derive the Tolerable Upper Intake Levels (ULs). The model consists of a systematic series of scientific considerations and judgments. The hallmark of the risk assessment model is the requirement to be explicit in all the evaluations and judgments made. Primarily as a result of database limitations, ULs are set for very broad age groups.

The UL values in Table S-2 for niacin, vitamin B_6, folate, and choline were set to protect the most sensitive individuals in the general population (such as those light in weight).

The ULs for folate and niacin apply to forms obtained from supplements, fortified foods, or a combination of the two. As described in Chapter 8, the UL for folate is based on examination of case studies of progression of neurological effects in vitamin B_{12}-deficient patients taking folate supplements.

Because of lack of suitable data, ULs could not be established for infants or for thiamin, riboflavin, vitamin B_{12}, pantothenic acid, or biotin. This signifies a need for data. It does not signify that people can tolerate chronic intakes of these vitamins at levels exceeding the Recommended Dietary Allowance (RDA) or Adequate Intake

TABLE S-2 Tolerable Upper Intake Levels (ULs[a]), by Life Stage Group

Life Stage Group	Niacin[b] (mg/d)	Vitamin B_6 (mg/d)	Folate[b] (µg/d)	Choline (g/d)
0 through 12 mo	ND[c]	ND	ND	ND
1 through 3 y	10	30	300	1.0
4 through 8 y	15	40	400	1.0
9 through 13 y	20	60	600	2.0
14 through 18 y	30	80	800	3.0
≥ 19 years	35	100	1,000	3.5
Pregnancy, ≤ 18 y	30	80	800	3.0
Pregnancy, ≥ 19 y	35	100	1,000	3.5
Lactation, ≤ 18 y	30	80	800	3.0
Lactation, ≥ 19 y	35	100	1,000	3.5

[a] UL = maximum level of daily nutrient intake that is likely to pose no risk of adverse effects. Unless otherwise specified, the UL represents total intake from food, water, and supplements. Because of the lack of suitable data, ULs could not be established for thiamin, riboflavin, vitamin B_{12}, pantothenic acid, or biotin. In the absence of ULs, extra caution may be warranted in consuming levels above recommended intakes.

[b] The ULs for niacin and folate apply to forms obtained from supplements, fortified foods, or a combination of the two.

[c] ND: Not determinable because of lack of data for adverse effects in this age group and concern with regard to lack of ability to handle excess amounts. Source of intake to prevent high levels of intake should be from formula and food only.

(AI). Like all chemical agents, nutrients can produce adverse effects if intakes are excessive. Therefore, when data are extremely limited, extra caution may be warranted.

USING DIETARY REFERENCE INTAKES TO ASSESS THE NUTRIENT INTAKE OF GROUPS

For statistical reasons that will be addressed in a future report, the Estimated Average Requirement (EAR) is greatly preferred over the Recommended Dietary Allowance (RDA) for use in assessing the nutrient intake of groups. Table S-1 shows, for example, that fewer than 5 percent of young men have dietary intakes of thiamin, riboflavin, niacin, or vitamin B_{12} that are less than the EARs for these nutrients. This indicates that dietary intake of these five B vitamins by young men has a high probability of being sufficient to meet their needs. A large proportion of the individuals in the population, especially women, reportedly has a total folate intake less than the EAR. However, because the reported folate content of foods is considered to be substantially underestimated (in part because of methodological problems, content not being reported in dietary folate equivalents, and data being obtained before the fortification of cereal grains was required), it is not known to what extent this discrepancy between the EAR and intake represents a problem.

The determination of ways to increase dietary intake of a nutrient should include examination of the foods that are the major contributors of the nutrient to the U.S. or Canadian diet and the specific foods that are rich in the nutrient. U.S. data on both are provided in this report.

HOW TO MEET RECOMMENDED DIETARY ALLOWANCES OR ADEQUATE INTAKES

A primary question that must be answered is How can individuals consume the Recommended Dietary Allowance (RDA) or Adequate Intake (AI) if surveys indicate that typical diets contain lower amounts? This becomes a policy issue with regard to choosing methods to increase consumption of that nutrient in order to decrease the number of individuals at risk because of inadequate dietary intakes. Such methods include educating consumers to change their food consumption behavior, fortifying foodstuffs with the nutrient, providing dietary supplements, or a combination of the three methods. It is not the function of this report, given the scope of work outlined, to provide an analysis of the impact of using these three methods.

Obtaining recommended intakes from unfortified foodstuffs has the advantage of providing intakes of other beneficial nutrients and of food components for which RDAs and AIs may not be determined. Another advantage is the potential enhancement of nutrient utilization through simultaneous interactions with other nutrients. It is recognized, however, that the low energy intakes reported in recent national surveys may mean that it would be unusual to see changes in food habits to the extent necessary to maintain intakes by all individuals at levels recommended in this report. Eating fortified food products represents one method by which individuals can increase or maintain intakes without major changes in food habits. For some individuals at higher risk, use of nutrient supplements may be desirable in order to meet reference intakes.

It is not the function of this report (see Appendix A, Origin and Framework of the Development of Dietary Reference Intakes) to address in detail the applications of the DRIs, including considerations necessary for the assessment of adequacy of intakes of various population groups and for planning for intakes of populations or for groups with special needs. However, some uses for the different types of DRIs are described briefly in Chapter 13. A subsequent report is expected to focus on the uses of DRIs in various settings.

RECOMMENDATIONS

Reporting Data

Because of the difference in the bioavailability of food folate and the monoglutamate form of folate, it is recommended that both food folate and folic acid be included in tables of food composition and in reports of intake. That is, the content or intake of naturally occurring food folate should be reported separately from that of folate provided by fortified foods and supplements.

Research

Four major types of information gaps were noted: (1) a dearth of studies designed specifically to estimate average requirements; (2) a nearly complete lack of usable data on the nutrient needs of infants, children, and adolescents; (3) a lack of appropriately designed studies to determine the role of selected B vitamins and choline in reducing the risk of certain chronic diseases; and (4) a lack of studies designed to detect adverse effects of chronic high intakes of some B vitamins.

In the judgment of the DRI Committee and its panel and subcommittees, highest priority should be given to research that has potential to prevent or retard human disease processes and to prevent deficiencies with functional consequences, as follows:

• Studies to provide the basic data for constructing risk curves and benefit curves across the exposures to food folate and folic acid. Such studies would provide estimates of the risk of developing neural tube defects, vascular disease, and neurological complications in susceptible individuals consuming different amounts of folate.

• Investigations of the size of the effect of folate, vitamin B_6, vitamin B_{12}, and related nutrients for preventing vascular disease and of possible mechanisms for the influence of genetic variation.

• Studies to overcome the methodological problems in the analysis of folate, including the development of sensitive and specific deficiency indicators and of practical, improved methods for analyzing the folate content of foods and determining its bioavailability.

• Studies to develop economical, sensitive, and specific methods to assess the prevalence, causes, and consequences of vitamin B_{12} malabsorption and deficiency and to prevent and treat these conditions.

• Investigation of how folate and related nutrients influence normal cellular differentiation and development, including embryogenesis and neoplastic transformation.

The requirements that appear to be the most productive to study are vitamin B_{12} requirements of the elderly and how they may be met; folate requirements, by trimester of pregnancy; and indicators on which to base vitamin B_6 requirements.

1

Introduction to Dietary Reference Intakes

Dietary Reference Intakes (DRIs) comprise a set of at least four nutrient-based reference values, each of which has special uses. The development of DRIs expands on the periodic reports called *Recommended Dietary Allowances,* which have been published since 1941 by the National Academy of Sciences. This comprehensive effort is being undertaken by the Standing Committee on the Scientific Evaluation of Dietary Reference Intakes of the Food and Nutrition Board, Institute of Medicine, National Academy of Sciences, with the involvement of Health Canada. See Appendix A for a description of the overall process and its origins.

WHAT ARE DIETARY REFERENCE INTAKES?

The reference values, collectively called the Dietary Reference Intakes (DRIs), include the Recommended Dietary Allowance (RDA), Adequate Intake (AI), Tolerable Upper Intake Level (UL), and Estimated Average Requirement (EAR).

A requirement is defined as the lowest continuing intake level of a nutrient that will maintain a defined level of nutriture in an individual. The chosen criterion of nutritional adequacy is identified in each chapter; note that the criterion may differ for individuals at different life stages.

Unless otherwise stated, all values given for RDAs, AIs, and EARs represent the quantity of the nutrient or food component to be supplied by foods from a diet similar to those consumed in Canada and the United States. If the degree of absorption of the nutrient is unusually low on a chronic basis (e.g., because of very high fiber intake), a higher intake may be needed. If the primary source of a B vitamin is a supplement (e.g., B_{12} for the elderly), a higher percentage of the vitamin may be absorbed and so a smaller intake may be required.

The DRIs apply to the healthy population. RDAs and AIs are levels of intake recommended for individuals. Meeting the recommended intake for the B vitamins and choline would not necessarily be sufficient for individuals who are already malnourished. People with diseases that result in malabsorption syndrome or who are receiving treatments such as hemodialysis or peritoneal dialysis may have increased requirements. Special guidance should be provided for those with greatly increased nutrient requirements. Although the RDA or AI may serve as the basis for such guidance, qualified medical and nutrition personnel should make necessary adaptations for specific situations.

CATEGORIES OF DIETARY REFERENCE INTAKES

Each type of Dietary Reference Intake (DRI) refers to average daily nutrient intake of individuals over time. The amount taken may vary substantially from day to day without ill effect in most cases.

Recommended Dietary Allowance

The Recommended Dietary Allowance (RDA) is the average daily dietary intake level that is sufficient to meet the nutrient requirement of nearly all (97 to 98 percent) healthy individuals in a particular life stage and gender group (see Figure 1-1). The RDA is intended to be used as a goal for daily intake by individuals. The process for setting the RDA is described below; it depends on being able to set an Estimated Average Requirement (EAR). That is, if an EAR cannot be set, no RDA will be set.

Estimated Average Requirement [1]

The EAR is the daily intake value that is estimated to meet the requirement—as defined by the specified indicator of adequacy—

---

[1] The definition of EAR implies a median as opposed to a mean, or average. The median and average would be the same if the distribution of requirements followed a symmetrical distribution and would diverge as a distribution became skewed.

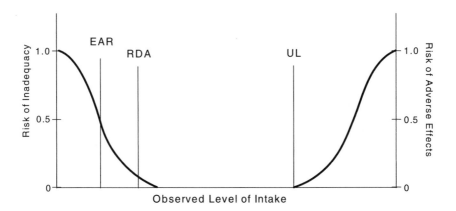

FIGURE 1-1 Dietary reference intakes. This figure shows that the Estimated Average Requirement (EAR) is the intake at which the risk of inadequacy is 0.5 (50%) to an individual. The Recommended Dietary Allowance (RDA) is the intake at which the risk of inadequacy is very small—only 0.02 to 0.03 (2% to 3%). The Adequate Intake (AI) does not bear a consistent relationship to the EAR or the RDA because it is set without being able to estimate the average requirement. It is assumed that the AI is at or above the RDA if one could be calculated. At intakes between the RDA and the Tolerable Upper Intake Level (UL), the risks of inadequacy and of excess are both close to 0. At intakes above the UL, the risk of adverse effects may increase.

in half of the healthy individuals in a life stage or gender group (see Figure 1-1). At this level of intake, the other half of a specified group would not have its nutritional needs met. The general method used to set the EAR is the same for all the B vitamins. The details, which are provided in Chapters 4 through 9, differ because of the different types of data available.

Method for Setting the RDA

The EAR is used in setting the RDA as follows. If the standard deviation (SD) of the EAR is available and the requirement for the

Three considerations prompted the choice of the term *EAR*: data are rarely adequate to determine the distribution of requirements, precedent has been set by other countries that have used EAR for reference values similarly derived (COMA, 1991), and the type of data evaluated makes the determination of a median impossible or inappropriate.

nutrient is normally distributed, the EAR plus 2 SDs of the EAR equals the RDA:

$$RDA = EAR + 2\ SD_{EAR}.$$

If data about variability in requirements are insufficient to calculate an SD, a coefficient of variation (CV_{EAR}) of 10 percent will be ordinarily assumed. Because

$$CV_{EAR} = SD_{EAR}/EAR,$$

and

$$SD = (EAR \times CV_{EAR}),$$

the resulting equation for the RDA is

$$RDA = EAR + 2\ (0.1 \times EAR)$$

or

$$RDA = 1.2 \times EAR.$$

The assumption of a 10 percent CV is based on extensive data on the variation in basal metabolic rate (FAO/WHO/UNA, 1985; Garby and Lammert, 1984), which contributes about two-thirds of the daily energy expenditure of many individuals residing in Canada and the United States (Elia, 1992) and on the similar CV of 12.5 percent estimated for the protein requirements in adults (FAO/WHO/UNA, 1985). If there is evidence of greater variation, a larger CV will be assumed. If the distribution of the nutrient requirement is known to be skewed for a population, other approaches may be used to find the ninety-seventh percentile to set the RDA. In all cases the method used to derive the RDA from the EAR is stated.

For the B vitamins there are few direct data on the requirements of children. Thus, EARs and RDAs for children are based on extrapolations from adult values. The method is described in Chapter 2.

Other Uses of the EAR

Together with an estimate of the variance of intake, the EAR may also be used in the assessment of the intake of groups or in planning for the intake of groups (Beaton, 1994) (see Chapter 13).

Adequate Intake

If sufficient scientific evidence is not available to calculate an EAR, a reference intake called an Adequate Intake (AI) is used instead of an RDA. The AI is a value based on experimentally derived intake levels or approximations of observed mean nutrient intakes by a group (or groups) of healthy people. In the opinion of the committee, the AI for children and adults is expected to meet or exceed the amount needed to maintain a defined nutritional state or criterion of adequacy in essentially all members of a specific healthy population. Examples of defined nutritional states include normal growth, maintenance of normal circulating nutrient values, or other aspects of nutritional well-being or general health.

The AI is set when data are considered to be insufficient or inadequate to establish an EAR on which an RDA would be based. For example, for young infants for whom human milk is the recommended sole source of food for most nutrients for the first 4 to 6 months, the AI is based on the daily mean nutrient intake supplied by human milk for healthy, full-term infants who are exclusively breastfed. For adults the AI may be based on data from a single experiment (e.g., choline), on estimated dietary intakes in apparently healthy population groups (e.g., biotin and pantothenic acid), or on a review of data from different approaches that considered alone do not permit a reasonably confident estimate of an EAR (e.g., dietary and experimental intakes of calcium).

The issuance of an AI indicates that more research is needed to determine with some degree of confidence the mean and distribution of requirements for a specific nutrient. When this research is completed, it should be possible to replace AI estimates with EARs and RDAs.

Comparison of the AI with the RDA

Similarities. Both the AI and RDA are to be used as a goal for individual intake. In general the values are intended to cover the needs of nearly all persons in a life stage group. (For infants the AI is the mean intake when infants in the age group are consuming human milk. Larger infants may have greater needs, which they meet by consuming more milk.) As with RDAs, AIs for children and adolescents may be extrapolated from adult values if no other usable data are available.

Differences. There is much less certainty about the AI value than about the RDA value. Because AIs depend on a greater degree of judgment than is applied in estimating the EAR and subsequently an RDA, the AI might deviate significantly from and be numerically higher than the RDA if it could be determined. For this reason, AIs must be used with greater care than is the case for RDAs. Also, the RDA is always calculated from the EAR by using a formula that takes into account the expected variation in the requirement for the nutrient (see previous section).

Tolerable Upper Intake Level

The Tolerable Upper Intake Level (UL) is the highest level of daily nutrient intake that is likely to pose no risk of adverse health effects in almost all individuals in the specified life stage group (see Figure 1-1). As intake increases above the UL, the risk of adverse effects increases. The term *tolerable* was chosen to avoid implying a possible beneficial effect; the term is intended to connote a level of intake that can, with high probability, be tolerated biologically. The UL is not intended to be a recommended level of intake, and there is no established benefit for healthy individuals if they consume a nutrient in amounts exceeding the recommended intake (the RDA or AI).

The UL is based on an evaluation conducted by using the methodology for risk assessment of nutrients (see Chapter 3). The need for setting ULs grew out of the increased fortification of foods with nutrients and the increased use of dietary supplements. For vitamin B_6 and choline, the UL refers to total intakes—from food, fortified food, and nutrient supplements. In other instances (i.e., for niacin and folate) it may refer only to intakes from supplements or fortificants or a combination of the two. The UL applies to chronic daily use. Details are given for each nutrient.

For some nutrients, data may not be sufficient for deriving a UL. This indicates the need for caution in consuming amounts greater than the recommended intakes; it does not mean that high intakes pose no risk of adverse effects.

Determination of Adequacy

In the derivation of the EAR or AI, close attention has been paid to the determination of the most appropriate indicators of adequacy. A key question is, Adequate for what? In many cases a continuum of benefits may be ascribed to various levels of intake of the same nutrient. One criterion may be deemed the most appropriate to determine the risk that an individual will become deficient in the nutrient whereas another may relate to reducing the risk of chronic degenerative disease, such as certain dementias, cardiovascular disease, cancer, diabetes mellitus, some forms of renal disease, or degenerative arthritis.

Each EAR and AI is described in terms of the selected criterion. For example, the dietary intake set for the RDA for folate for women in the childbearing years is based on a combination of biochemical indicators, but a separate recommendation is made for women capable of becoming pregnant to reduce the risk of a neural tube defect in the offspring if pregnancy occurs.

The potential role of B vitamins and choline in the reduction of disease risk was considered in developing the EARs and AIs for this group of nutrients. The types of evidence considered are described in Chapter 2.

For many of the B vitamins, the use of a single indicator of adequacy was deemed inappropriate. For any one B vitamin, several biochemical values provide information about nutrient status, but adverse effects of inadequate intake may not be observable if only one (or possibly more) of the values is outside its normal range. With the acquisition of new data, such as data relating intake to chronic disease or disability, the choice of the criterion for setting the EAR may change.

PARAMETERS FOR DIETARY REFERENCE INTAKES

Life Stage Groups

Reference nutrient intakes are expressed for 16 life stage groups,[2] as listed in Table 1-1 and described in more detail in the first report in this series (IOM, 1997). If data are too sparse to distinguish differences in requirements by life stage or gender group, the analysis may be presented for a larger grouping. Differences will be indicated by gender when warranted by the data.

---

[2] As with all quantitative estimates, mathematically derived reference intakes may appear to provide much greater precision than the data used to derive them. Conventional rules for rounding of reference intakes are followed in most cases. However, because of the number of assumptions made, particularly in extrapolations, values may have been rounded up to provide a more generous recommendation (or down to provide a more conservative UL).

TABLE 1-1 Life Stage Groups

Infants	Females
0 through 6 mo	9 through 13 y
7 through 12 mo	14 through 18 y
	19 through 30 y
Children	31 through 50 y
1 through 3 y	51 through 70 y
4 through 8 y	> 70 y
Males	Pregnancy
9 through 13 y	≤ 14 through 18 y
14 through 18 y	19 through 30 y
19 through 30 y	31 through 50 y
31 through 50 y	
51 through 70 y	Lactation
> 70 y	≤ 14 through 18 y
	19 through 30 y
	31 through 50 y

Reference Weights and Heights

The reference weights and heights selected for adults and children are shown in Table 1-2. The values are based on anthropometric data collected from 1988 to 1994 as part of the Third National Health and Nutrition Examination Survey (NHANES III) in the United States.

The median heights for the life stage and gender groups through age 30 years were identified, and the median weights for those heights were based on median body mass index for the same individuals. Because there is no evidence that weight should change as adults age if activity is maintained, the reference weights for adults aged 19 through 30 years are applied to all adult age groups.

The most recent nationally representative data available for Canadians (from the 1970–1972 Nutrition Canada Survey [Demirjian, 1980]) were reviewed. In general, median heights of children from 1 year of age in the United States were greater by 3 to 8 cm (1 to 2.5 inches) than those of children of the same age in Canada measured two decades earlier (Demirjian, 1980). This could be partly explained by approximations necessary to compare the two data sets but more possibly by a continuation of the secular trend of increased heights for age noted in the Nutrition Canada Survey when it compared data from that survey with an earlier (1953) national Canadian survey (Pett and Ogilvie, 1956).

TABLE 1-2 Reference Heights and Weights for Children and Adults in the United States[a]

Gender	Age	Median Body Mass Index[b]	Reference Height (cm [in])	Reference Weight[c] (kg [lb])
Male, female	2–6 mo	—	64 (25)	7 (16)
	7–11 mo	—	72 (28)	9 (20)
	1–3 y	—	91 (36)	13 (29)
	4–8 y	15.8	118 (46)	22 (48)
Male	9–13 y	18.5	147 (58)	40 (88)
	14–18 y	21.3	174 (68)	64 (142)
	19–30 y	24.4	176 (69)	76 (166)
Female	9–13 y	18.3	148 (58)	40 (88)
	14–18 y	21.3	163 (64)	57 (125)
	19–30 y	22.8	163 (64)	61 (133)

[a] Adapted from Third National Health and Nutrition Examination Survey (NHANES III), 1988–1994.
[b] In kg/m^2.
[c] Calculated from body mass index and height for ages 4 through 8 years and older.

Similarly, median weights beyond age 1 year derived from the recent survey in the United States (NHANES III, 1988 to 1994) were also greater than those obtained from the older Canadian survey (Demirjian, 1980). Differences were greatest during adolescence, ranging from 10 to 17 percent higher. The differences probably reflect the secular trend of earlier onset of puberty (Herman-Giddens et al., 1997) rather than differences in populations. Calculations of body mass index for young adults (e.g., a median of 22.6 for Canadian women compared with 22.8 for U.S. women) resulted in similar values, indicating greater concordance between the two surveys by adulthood.

The reference weights chosen for this report were based on the most recent data set available from either country, recognizing that earlier surveys in Canada indicated shorter stature and lower weights during adolescence than did surveys in the United States.

Reference weights are used primarily when setting the Estimated Average Requirements (EARs), Adequate Intakes (AIs), or Tolerable Upper Intake Levels (ULs) for children or when relating the nutrient needs of adults to body weight. For the 4- to 8-year-old age group, it can be assumed that a small 4-year-old child will require less than the EAR and that a large 8-year-old will require more than the EAR. However, the Recommended Dietary Allowance (RDA) or AI should meet the needs of both.

SUMMARY

Dietary Reference Intakes (DRIs) is a generic term for a set of nutrient reference values that includes the Recommended Dietary Allowance (RDA), Adequate Intake (AI), Tolerable Upper Intake Level (UL), and Estimated Average Requirement (EAR). These reference values are being developed for life stage and gender groups in a joint U.S.-Canadian activity. This report, which is the second in a series, covers the DRIs for folate, other B vitamins (thiamin, riboflavin, niacin, vitamin B_6, vitamin B_{12}, pantothenic acid, and biotin), and choline.

REFERENCES

Beaton GH. 1994. Criteria of an adequate diet. In: Shils ME, Olson JA, Shike M, eds. *Modern Nutrition in Health and Disease,* 8th ed. Philadelphia: Lea & Febiger. Pp. 1491–1505.

COMA (Committee on Medical Aspects of Food Policy). 1991. *Dietary Reference Values for Food Energy and Nutrients for the United Kingdom.* Report on Health and Social Subjects, No. 41. London: HMSO.

Demirjian A. 1980. *Anthropometry Report. Height, Weight, and Body Dimensions: A Report from Nutrition Canada.* Ottawa: Minister of National Health and Welfare, Health and Promotion Directorate, Health Services and Promotion Branch.

Elia M. 1992. Energy expenditure and the whole body. In: Kinney JM, Tucker HN, eds. *Energy Metabolism: Tissue Determinants and Cellular Corollaries.* New York: Raven Press. Pp. 19–59.

FAO/WHO/UNA (Food and Agriculture Organization of the United Nations/World Health Organization/United Nations). 1985. *Energy and Protein Requirements Report of a Joint FAO/WHO/UNA Expert Consultation.* Technical Report Series. No. 724. Geneva: World Health Organization.

Garby L, Lammert O. 1984. Within-subjects between-days-and-weeks variation in energy expenditure at rest. *Hum Nutr Clin Nutr* 38:395–397.

Herman-Giddens ME, Slora EJ, Wasserman RC, Bourdony CJ, Bhapkar MV, Koch GG, Hasemeier CM. 1997. Secondary sexual characteristics and menses in young girls seen in office practice: A study from the Pediatric Research in Office Settings Network. *Pediatrics* 99:505–512.

IOM (Institute of Medicine). 1997. *Dietary Reference Intakes for Calcium, Phosphorus, Magnesium, Vitamin D, and Fluoride.* Washington, DC: National Academy Press.

Pett LB, Ogilvie GH. 1956. The Canadian Weight-Height Survey. *Hum Biol* 28:177–188.

2

The B Vitamins and Choline: Overview and Methods

OVERVIEW

This report focuses on the eight B complex vitamins—thiamin, riboflavin, niacin, vitamin B_6, folate, vitamin B_{12}, pantothenic acid, and biotin—and choline. These water-soluble nutrients fall into two categories: those involved in the reactions of intermediary metabolism related to energy production and redox status and those involved in the transfer of single-carbon units.

Thiamin, riboflavin, niacin, vitamin B_6, and pantothenic acid are required for decarboxylation, transamination, acylation, oxidation, and reduction of substrates that ultimately are used for energy utilization. One or more of these also are important for amino acid, fatty acid, cholesterol, steroid, and glucose synthesis.

Biotin is required for carbon dioxide fixation by four carboxylases. Folate, vitamin B_{12}, choline, and riboflavin are needed for methyl-group transfer. Their metabolism intermingles at the pathway for conversion of homocysteine to methionine. Folate is also important for the supply of single-carbon units for deoxyribonucleic acid (DNA) synthesis.

Both the Système International d´Unités (SI units) and traditional units are used in this report, as described in Appendix C.

METHODOLOGICAL CONSIDERATIONS

Types of Data Used

The scientific data for developing the Dietary Reference Intakes (DRIs) have essentially come from observational and experimental studies in humans. Observational studies include single-case and case-series reports, epidemiological cohort studies, and case-control studies. Experimental studies include randomized and nonrandomized therapeutic or prevention trials and controlled dose-response, balance, turnover, and depletion-repletion physiological studies. Results from animal experiments are generally not applicable to nutritional deficiencies, chronic diseases, and toxic effects in humans, but selected animal studies are considered in the absence of human data. The strategies used for identifying potentially relevant studies are summarized in Appendix D.

As a principle, only studies published in peer-reviewed journals have been used. However, studies published in other scientific journals or readily available reports were considered if they appeared to provide important information not documented elsewhere. To the extent possible, original scientific studies and quantitative meta-analyses have been used to derive the DRIs. A thorough review of the scientific literature resulted in the identification of clinical and functional indicators of nutritional adequacy for each nutrient for each life stage and gender group. Anything that might affect dietary requirements, such as an interaction with other nutrients and the bioavailability of the nutrient, was considered when relevant. For example, the effect of energy intake was considered for thiamin, riboflavin, and niacin; the effect of protein intake was considered for vitamin B_6.

Because of the growing evidence that some B vitamins may prevent the occurrence of developmental abnormalities and chronic degenerative and neoplastic diseases, special consideration was given to the possible use of such indicators as criteria of adequacy. It was beyond the scope of the report to consider the use of nutrients in the treatment of disease or other disorders.

The quality of studies was considered in weighing the evidence. The characteristics examined included the study design and the representativeness of the study population; the validity, reliability, and precision of the methods used for measuring intake and indicators of adequacy; the control of biases and of confounders; and the power of the study to demonstrate a given difference or correlation. When applicable, greatest weight was given to randomized controlled trials and less to nonrandomized trials; prospective cohort, retrospective cohort, and case-control studies; case-series; and single-case reports. Publications solely expressing opinions were not used in setting DRIs.

Statistical association does not imply causation, and this is especially true for relationships between nutrient intake and developmental abnormalities or chronic disease risk reduction as well as for toxic effects. The criteria proposed by Hill (1971) were considered when examining the evidence that a relationship might be causal:

- strength of the association, usually expressed as a relative risk or a correlation coefficient;
- dose-response relationship;
- temporally plausible association, with exposure preceding the effect;
- consistency of association in time and place;
- specificity of cause and effect; and
- biological plausibility.

For example, biological plausibility would not be sufficient in the presence of a weak association and lack of evidence that exposure preceded the effect.

Data were examined to determine whether similar estimates of the requirement resulted from the use of different indicators and different types of studies. For a single nutrient the criterion for setting the Estimated Average Requirement (EAR) may differ from one life stage group to another because the critical function or the risk of disease may be different. When no or very poor data were available for a given life stage group, extrapolation was made from the EAR or Adequate Intake (AI) set for another group based on explicit assumptions on relative requirements.

Method to Determine the Adequate Intake for Infants

The AI for young infants is generally taken to be the average intake by full-term infants who are born to healthy, well-nourished mothers and who are exclusively fed human milk. The extent to which intake of a nutrient from human milk may exceed the actual requirements of infants is not known, and ethics of experimentation preclude testing the levels known to be potentially inadequate. Using the breastfed infant as a model is in keeping with the basis for earlier recommendations for intake (e.g., Health Canada, 1990; IOM, 1991). It also supports the recommendation that exclusive breastfeeding is the preferred method of feeding for normal full-term infants for the first 4 to 6 months of life, even though most U.S. babies are no longer breastfed by age 6 months. This recommendation has been made by the Canadian Paediatric Society (Health Canada, 1990), the American Academy of Pediatrics (AAP, 1997), the Institute of Medicine (IOM, 1991), and many other expert groups.

In general, this report does not cover possible variations in physiological need during the first month after birth or the variations in intake of nutrients from human milk that result from differences in milk volume and nutrient concentration during early lactation.

In keeping with the decision made by the Standing Committee on the Scientific Evaluation of Dietary Reference Intakes, specific DRIs to meet the needs of formula-fed infants have not been proposed in this report. The use of formula introduces a large number of complex issues, one of which is the bioavailability of different forms of the nutrient in different formula types. However, in the section "Special Considerations," issues related to bioavailability and use of different milk sources are discussed when appropriate.

Ages 0 through 6 Months

To derive the AI value for infants ages 0 through 6 months, the mean intake of a nutrient was calculated based on (1) the average concentration of the nutrient from 2 to 6 months of lactation using consensus values from several reported studies, if possible, and (2) an average volume of milk intake of 780 mL/day. This volume was reported from studies that used test weighing of full-term infants. In this procedure, the infant is weighed before and after each feeding (Allen et al., 1991; Butte et al., 1984; Chandra, 1984; Hofvander et al., 1982; Neville et al., 1988). Because there is variation in both the composition of milk and the volume consumed, the computed value represents the mean. It is expected that infants will consume increased volumes of human milk during growth spurts.

Ages 7 through 12 Months

During the period of infant growth and gradual weaning to a mixed diet of human milk and solid foods from ages 7 through 12 months, there is no evidence for markedly different nutrient needs. The basis of the AI values derived for this age category could be the sum of (1) the specific nutrient provided by 600 mL/day of human milk, which is the average volume of milk reported from studies of breastfed infants in this age category (Heinig et al., 1993), and (2) that provided by the usual intakes of complementary weaning foods consumed by infants in this age category. Such an approach would be in keeping with the current recommendations of the Canadian Paediatric Society (Health Canada, 1990), American Academy of Pediatrics (AAP, 1997), and Institute of Medicine (IOM, 1991) for continued breastfeeding of infants through 9 to 12 months of age with appropriate introduction of solid foods.

Only one relatively recent published source of information about B vitamin intake from solid foods for infants aged 7 through 12 months was found (Montalto et al., 1985), and it covered only three B vitamins: thiamin, riboflavin, and niacin. These researchers' estimates are based on data from 24-hour dietary intakes from the 1976–1980 National Health and Nutrition Examination Survey (NHANES II) for infants aged 7 to 12 months. The infants were consuming formula; intake from solid food was reported separately.

For the B vitamins and choline, two other approaches were considered as well: (1) extrapolation upward from the AI for infants ages 0 through 6 months by using the metabolic weight ratio and (2) extrapolation downward from the EAR for young adults by adjusting for metabolic body size and growth and adding a factor for variability or from the AI if the recommended intake for adults was an AI. Both of these methods are described below. The results of these methods are compared in the process of setting the AI.

Method for Extrapolating Data from Adults to Infants and Children

Setting the EAR or AI

For the B vitamins and choline, if data were not available to set the EAR and Recommended Dietary Allowance (RDA) or an AI for children ages 1 year and older and for adolescents, the EAR or AI has been extrapolated down by using a consistent basic method. The method relies on at least four assumptions:

1. Maintenance needs for the B vitamins and choline expressed with respect to body weight ([kilogram of body weight]$^{0.75}$) are the same for adults and children. Scaling requirements as the 0.75 power of body mass adjusts for metabolic differences demonstrated to be related to body weight, as described by Kleiber (1947) and explored further by West and colleagues (1997). By this scaling a child weighing 22 kg would require 42 percent of what an adult

weighing 70 kg would require—a higher percentage than that represented by actual weight.

2. The EAR for adults is an estimate of maintenance needs.

3. The percentage of extra B vitamins and choline needed for growth is comparable with the percentage of extra protein needed for growth.

4. On average, total needs do not differ substantially for males and females until age 14, when reference weights differ.

The formula for the extrapolation is

$$EAR_{child} = EAR_{adult} (F),$$

where $F = (Weight_{child}/Weight_{adult})^{0.75} (1 + growth factor)$. Reference weights from Table 1-2 are used. If the EAR differs for men and women, the reference weight used for adults differs by gender; otherwise, the average for men and women is used unless the value for women is derived from data on men. The approximate proportional increase in protein requirements for growth (FAO/WHO/UNA, 1985) is used as an estimate of the growth factor as shown in Table 2-1. If only an AI has been set for adults, it is substituted for the EAR in the above formula and an AI is calculated; no RDA will be set.

Setting the RDA

To account for variability in requirements because of growth rates and other factors, a 10 percent coefficient of variation (CV) for the

TABLE 2-1 Growth Factors Used to Extrapolate DRIs

Age Group	Growth Factor
7 mo–3 y	0.30
4–8 y	0.15
9–13 y	0.15
14–18 y, Males	0.15
14–18 y, Females	0.00

NOTE: Growth beyond age 13 for females is assumed to represent a negligible increased requirement for vitamins.

SOURCE: The proportional increase in protein requirements for growth from FAO/WHO/UNA (1985) was used to estimate the growth factor indicated.

requirement is assumed unless data are available to support another value, as described in Chapter 1.

Method for Extrapolating Data from Young to Older Infants

This adjustment, the metabolic weight ratio method, involves metabolic scaling but does not adjust for growth because it is based on a value for a growing infant. To extrapolate from the AI for infants ages 0 through 6 months to an AI for infants ages 7 through 12 months, the following formula is used:

$$AI_{7-12 \text{ mo}} = AI_{0-6 \text{ mo}} \text{ (F)},$$

where $F = (\text{Weight}_{7-12 \text{ mo}}/\text{Weight}_{0-6 \text{ mo}})^{0.75}$.

Methods for Determining Increased Needs for Pregnancy

The placenta actively transports water-soluble vitamins (except biotin) and choline from the mother to the fetus against a concentration gradient (Hytten and Leitch, 1971; Zempleni et al., 1992). Placental transport of biotin is a passive process (Hu et al., 1994; Karl and Fisher, 1992; Schenker et al., 1993). For many of the B vitamins, experimental data that could be used to set an EAR and RDA or an AI for pregnancy are lacking. In these cases the potential for increased need for these nutrients during pregnancy is based on theoretical considerations, including obligatory fetal transfer, if data are available, and increased maternal needs related to increases in energy or protein metabolism, as applicable. With the possible exception of B_{12}, vitamin absorption does not appear to improve substantially during pregnancy. For choline, the AI is based on the increase in maternal weight.

Methods to Determine Increased Needs for Lactation

For the B vitamins and choline, it is assumed that the total requirement of lactating women equals the requirement for the nonpregnant, nonlactating woman of similar age plus an increment to cover the amount of the nutrient needed for milk production. To allow for inefficiencies in use of certain B vitamins, the increment may be somewhat greater than the amount of the nutrient contained in the milk produced. Details are provided in each nutrient chapter.

ESTIMATES OF LABORATORY VALUES

Analytic Considerations

Substantial changes in methods have occurred during the 50 years of vitamin B studies considered in this report. Appendix E summarizes information about methods used to determine laboratory values related to B vitamin and choline status. Methodological problems have been documented for folate (see Chapter 8) and other nutrients (see Appendix E).

Interpretation of Results

Caution was used in the interpretation of study results. Some examples of points that were considered follow:

• The method of measurement of urinary excretion of the vitamin (e.g., fasting, random, or 24-hour specimens) introduces different types of errors. The use of creatinine corrections to allow for assay of random fasting urine samples rather than 24-hour collections may need to include considerations of differences in creatinine excretion by age.

• Depletion-repletion studies assess requirements by identifying intakes that return status indicators to the prestudy baseline values. Baseline values have been those of motivated healthy individuals on self-selected diets or on diets containing a recommended level. The assessed requirements based on this approach of returning values to baseline are invariably similar or higher than the baseline vitamin intake. This is addressed in more detail in Chapters 7 and 14.

• Some studies are too short to determine whether a tested level of nutrient intake will be sufficient to stabilize the laboratory test result at a lower but satisfactory value.

• Many laboratory values change during pregnancy (NRC, 1978) and the postpartum period (sometimes with differences between lactating and nonlactating women) (IOM, 1991). A decreased value does not necessarily mean that intake was inadequate.

Sensitivity and Specificity

The terms *sensitivity* and *specificity* each have different meanings when applied to laboratory tests as compared with public health applications. *Analytic sensitivity* is defined as the amount of a nutrient that results in a doubling of background blank in an assay. *Analytic specificity* is the ability of the assay to discriminate the nutrient of interest from nutrients that might give false positive readings in the assay. In public health usage, sensitivity refers to the ability of a criterion (e.g., a laboratory test and its cutoff point) to identify the individuals who have a particular problem. Specificity, in turn, refers to the ability of a criterion to identify those who do not have the problem. In this report, the terms sensitivity and specificity reflect the public health usage unless preceded by the word *analytic.*

ESTIMATES OF NUTRIENT INTAKE

Reliable and valid methods of food composition analysis are crucial in determining the intake of a nutrient needed to meet a requirement. For several B vitamins and choline, analytic methods to determine the content of the nutrient in food have serious limitations (see Appendix E).

Methodological Considerations

The quality of nutrient intake data varies widely across studies. The most valid intake data are those collected from the metabolic study protocols in which all food is provided by the researchers, amounts consumed are accurately measured, and the nutrient composition of the food is determined by reliable and valid laboratory analyses. Such protocols are usually possible with only a small number of subjects. Thus, in many studies, intake data are self-reported (e.g., through 24-hour recalls of food intake, diet records, or food frequency questionnaires). Potential sources of error in self-reported intake data include over- or underreporting of portion sizes and frequency of intake, omission of foods, and inaccuracies related to the use of food composition tables. Therefore, the values reported by nationwide surveys or studies that rely on self-report may be somewhat inaccurate and possibly biased.

Food composition databases that are used to calculate nutrient intake from self-reported and observed intake data introduce errors due to random variability, genetic variation in the nutrient content, analytical errors, and missing or imputed data. In general, when nutrient intakes for groups are estimated, the effect of errors in the composition data is probably considerably smaller than the effect of errors in the self-reported intake data (NRC, 1986). However, it is not known to what extent this is true for folate, biotin, pantothenic acid, or choline (see Appendix E).

The accuracy of the food composition data for folate and vitamin B_{12} is described as "conflicting" (LSRO/FASEB, 1995). Food composition data are not even routinely reported for pantothenic acid, biotin, and choline. Moreover, wide variation in the B vitamin content of similar cooked foods is likely because of susceptibility to cooking losses, especially losses when liquids in which food is cooked are not also consumed.

Adjusting for Day-to-Day Variation

Because of day-to-day variation in dietary intakes, the distribution of 1-day (or 2-day) intakes for a group is wider than the distribution of usual intakes even though the mean of the intakes may be the same (for further elaboration, see Chapter 13). To reduce this problem, statistical adjustments were developed (NRC, 1986; Nusser et al., 1996) that require at least 2 days of dietary data from a representative subsample of the population of interest. However, no accepted method is available to adjust for the underreporting of intake, which may average as much as 20 percent for energy (Mertz et al., 1991).

DIETARY INTAKES IN THE UNITED STATES AND CANADA

Sources of Dietary Intake Data

The major sources of current dietary intake data for the U.S. population are the Third National Health and Nutrition Examination Survey (NHANES III), which was conducted from 1988 to 1994 by the U.S. Department of Health and Human Services, and the Continuing Survey of Food Intakes by Individuals (CSFII), which was conducted by the U.S. Department of Agriculture (USDA). NHANES III examined 30,000 subjects aged 2 months and older. A single 24-hour diet recall was collected for all subjects. A second recall was collected for a 5 percent nonrandom subsample to allow adjustment of intake estimates for day-to-day variation. The 1994 to 1995 CSFII collected two nonconsecutive 24-hour recalls from approximately 5,600 subjects of all ages. Both surveys used the food composition database developed by USDA to calculate nutrient intakes (Perloff et al., 1990). National survey data for Canada are not currently available, but data have been collected from Québec and Nova Scotia. The extent to which these data are applicable nationwide is not known.

Additional data on nearly 700 free-living, elderly persons from the

Boston Nutritional Status Survey are provided in Appendix F. In this survey, 3-day diet records were kept by participants and subsequently checked by a qualified nutritionist.

Appendix G gives the mean and the first through ninety-ninth percentiles of dietary intakes of six of the B vitamins by age from the first phase of the CSFII, adjusted for day-to-day variation by the method of Nusser et al. (1996). Appendix H provides comparable information from NHANES III, adjusted by methods described by the National Research Council (NRC, 1986) and by Feinleib and colleagues (1993) for persons aged 6 years and older. (There were too few second dietary recalls to do the adjustment for the younger children.) Because food composition data are not readily available for pantothenic acid, biotin, and choline, neither of the U.S. national surveys has estimated intakes for these nutrients. Appendix I provides means and selected percentiles of dietary intakes of seven B vitamins for men and women in Québec and mean daily intake of thiamin, riboflavin, niacin, and folate for men and women in Nova Scotia.

Sources of Data on Supplement Intake

Although subjects in the CSFII were asked about the use of dietary supplements, quantitative information was not collected. Data on supplement intake obtained from NHANES III were reported as a part of total nutrient intake (Appendix H). NHANES III data on overall prevalence of supplement use are also available (LSRO/FASEB, 1995). In 1986, the National Health Interview Survey queried 11,558 adults and 1,877 children on their intake of supplements during the previous 2 weeks (Moss et al., 1989). The composition of the supplement was obtained directly from the product label whenever possible. Table 2-2 shows the percentage of adults, by age, taking at least one of the B vitamins.

Food Sources of Folate and Other B Vitamins

For six of the B vitamins, two types of information are provided about food sources of nutrients: identification of the foods that are the major contributors of the vitamin to diets in the United States and food sources of the nutrient. The determination of foods that are major contributors depends on both nutrient content of a food and the total consumption of the food (amount and frequency). Therefore, a food that has a relatively low concentration of the nutrient might still be a large contributor to total intake if that food

TABLE 2-2 Percentage of Persons Taking Vitamin Supplements, by Sex, Age, and Type of Vitamin Used: National Health Interview Survey, United States, 1986

Vitamin Supplement Taken	Females			
	≤18 y	18–44 y	45–64 y	≥65 y
Thiamin	29.5	29.9	30.4	27.2
Riboflavin	29.5	30.0	30.3	26.7
Niacin	29.3	29.7	30.3	26.4
Vitamin B_6	29.7	30.2	30.5	27.4
Folate	26.0	27.3	25.8	22.2
Vitamin B_{12}	29.3	29.8	29.9	26.9
Pantothenic Acid	24.9	25.4	25.3	22.9
Biotin	18.6	19.7	18.5	15.6

NOTE: The high use of supplements by pregnant women is not reflected in this table.
SOURCE: Moss et al. (1989).

is consumed in relatively large amounts. Data from the 1995 CSFII were used to identify major contributors. In contrast, the food sources listed are those with the high concentrations of the nutrient; no consideration is given to the amount consumed. Both types of data were provided for this report by USDA (A. Moshfegh, Agricultural Research Service, USDA, personal communication, 1997).

SUMMARY

General methods for examining and interpreting the evidence on requirements for B vitamins and choline are presented in this chapter, with special attention given to infants, children, and pregnant and lactating women; methodological problems; and dietary intake data. Relevant detail is provided in the nutrient chapters.

REFERENCES

AAP (American Academy of Pediatrics). 1997. Breastfeeding and the use of human milk. *Pediatrics* 100:1035–1039.

Allen JC, Keller RP, Archer P, Neville MC. 1991. Studies in human lactation: Milk composition and daily secretion rates of macronutrients in the first year of lactation. *Am J Clin Nutr* 54:69–80.

Butte NF, Garza C, Smith EO, Nichols BL. 1984. Human milk intake and growth in exclusively breast-fed infants. *J Pediatr* 104:187–195.

Males			
≤18 y	18–44 y	45–64 y	≥65 y
23.2	23.2	23.7	22.1
23.0	23.1	23.3	21.9
22.9	23.0	23.1	22.0
23.1	23.1	23.6	21.8
20.6	21.2	20.2	18.9
22.8	22.9	23.3	21.8
19.3	19.5	19.1	19.1
15.8	16.4	14.7	15.1

Chandra RK. 1984. Physical growth of exclusively breast-fed infants. *Nutr Res* 2:275–276.

FAO/WHO/UNA (Food and Agriculture Organization of the United Nations/World Health Organization/United Nations). 1985. *Energy and Protein Requirements Report of a Joint FAO/WHO/UNA Expert Consultation.* Technical Report Series. No. 724. Geneva: World Health Organization.

Feinleib M, Rifkind B, Sempos C, Johnson C, Bachorik P, Lippel K, Carroll M, Ingster-Moore L, Murphy R. 1993. Methodological issues in the measurement of cardiovascular risk factors: Within-person variability in selected serum lipid measures—results from the Third National Health and Nutrition Survey (NHANES III). *Can J Cardiol* 9:87D–88D.

Health Canada. 1990. *Nutrition Recommendations. The Report of the Scientific Review Committee.* Ottawa: Canadian Government Publishing Centre.

Heinig MJ, Nommsen LA, Peerson JM, Lonnerdal B, Dewey KG. 1993. Energy and protein intakes of breast-fed and formula-fed infants during the first year of life and their association with growth velocity: The DARLING Study. *Am J Clin Nutr* 58:152–161.

Hill AB. 1971. *Principles of Medical Statistics,* 9th ed. New York: Oxford University Press.

Hofvander Y, Hagman U, Hillervik C, Sjolin S. 1982. The amount of milk consumed by 1–3 months old breast- or bottle-fed infants. *Acta Pediatr Scand* 71:953–958.

Hu Z-Q, Henderson GI, Mock DM, Schenker S. 1994. Biotin uptake by basolateral membrane vesicles of human placenta: Normal characteristics and role of ethanol. *Proc Soc Exp Biol Med* 206:404–408.

Hytten FE, Leitch I. 1971. *The Physiology of Human Pregnancy,* 2nd ed. Oxford: Blackwell Scientific Publications.

IOM (Institute of Medicine). 1991. *Nutrition During Lactation.* Washington, DC: National Academy Press.

Karl PI, Fisher SE. 1992. Biotin transport in microvillous membrane vesicles, cultured trophoblasts and isolated perfused human placenta. *Am J Physiol* 262:C302–C308.

Kleiber M. 1947. Body size and metabolic rate. *Physiol Rev* 27:511–541.

LSRO/FASEB (Life Sciences Research Office/Federation of American Societies for Experimental Biology). 1995. *Third Report on Nutrition Monitoring in the United States.* Washington DC: U.S. Government Printing Office.

Mertz W, Tsui JC, Judd JT, Reiser S, Hallfrisch J, Morris ER, Steele PD, Lashley E. 1991. What are people really eating? The relation between energy intake derived from estimated diet records and intake determined to maintain body weight. *Am J Clin Nutr* 54:291–295.

Montalto MB, Benson JD, Martinez GA. 1985. Nutrient intake of formula-fed infants and infants fed cow's milk. *Pediatrics* 75:343–351.

Moss AJ, Levy AS, Kim I, Park YK. 1989. *Use of Vitamin and Mineral Supplements in the United States: Current Users, Types of Products, and Nutrients.* Advance Data, Vital and Health Statistics of the National Center for Health Statistics, No. 174. Hyattsville, MD: National Center for Health Statistics.

Neville MC, Keller R, Seacat J, Lutes V, Neifert M, Casey C, Allen J, Archer P. 1988. Studies in human lactation: Milk volumes in lactating women during the onset of lactation and full lactation. *Am J Clin Nutr* 48:1375–1386.

NRC (National Research Council). 1978. *Laboratory Indices of Nutritional Status in Pregnancy.* Report of the Committee on Nutrition of the Mother and Preschool Child, Food and Nutrition Board. Washington, DC: National Academy Press.

NRC (National Research Council). 1986. *Nutrient Adequacy. Assessment Using Food Consumption Surveys.* Washington, DC: National Academy Press.

Nusser SM, Carriquiry AL, Dodd KW, Fuller WA. 1996. A semiparametric transformation approach to estimating usual daily intake distributions. *J Am Stat Assoc* 91:1440–1449.

Perloff BP, Rizek RL, Haytowitz DB, Reid PR. 1990. Dietary intake methodology. II. USDA's Nutrient Data Base for Nationwide Dietary Intake Surveys. *J Nutr* 120:1530–1534.

Schenker S, Hu Z, Johnson RF, Yang Y, Frosto T, Elliott BD, Henderson GI, Mock DM. 1993. Human placental biotin transport: Normal characteristics and effect of ethanol. *Alcohol Clin Exp Res* 17:566–575.

West GB, Brown JH, Enquist BJ. 1997. A general model for the origin of allometric scaling laws in biology. *Science* 276:122–126.

Zempleni J, Link G, Kubler W. 1992. The transport of thiamine, riboflavin and pyridoxal 5'-phosphate by human placenta. *Int J Vitam Nutr Res* 62:165–172.

3

A Model for the Development of Tolerable Upper Intake Levels

BACKGROUND

The Tolerable Upper Intake Level (UL) refers to the highest level of daily nutrient intake that is likely to pose no risk of adverse health effects to almost all individuals in the general population. As intake increases above the UL, the risk of adverse effects increases. The term *tolerable* is chosen because it connotes a level of intake that can, with high probability, be tolerated biologically by individuals; it does not imply acceptability of that level in any other sense. The setting of a UL does not indicate that nutrient intakes greater than the Recommended Dietary Allowance (RDA) or Adequate Intake (AI) are recommended as being beneficial to an individual. Many individuals are self-medicating with nutrients for curative or treatment purposes. It is beyond the scope of this report to address the possible therapeutic benefits of higher nutrient intakes that may offset the risk of adverse effects. The UL is not meant to apply to individuals who are treated with the nutrient under medical supervision. It is designed to be applied to almost all individuals in the general healthy population.

The term *adverse effect* is defined as any significant alteration in the structure or function of the human organism (Klaassen et al., 1986) or any impairment of a physiologically important function, in accordance with the definition set by the joint World Health Organization, Food and Agriculture Organization of the United Nations, and International Atomic Energy Agency Expert Consultation in *Trace Elements in Human Nutrition and Health* (WHO, 1996). In the

41

case of nutrients, it is exceedingly important to consider the possibility that the intake of one nutrient may alter in detrimental ways the health benefits conferred by another nutrient. Any such alteration (referred to as an adverse nutrient-nutrient interaction) is considered an adverse health effect. When evidence for such adverse interactions is available, it is considered in establishing a nutrient's UL.

As is true for all chemical agents, adverse health effects can result if the intake of nutrients from a combination of food, water, nutrient supplements, and pharmacological agents is excessive. Some lower level of nutrient intake will ordinarily pose no likelihood (or risk) of adverse health effects in normal individuals even if the level is above that associated with any benefit. It is not possible to identify a single risk-free intake level for a nutrient that can be applied with certainty to all members of a population. However, it is possible to develop intake levels that are likely to pose no risk of adverse health effects to most members of the general population, including sensitive individuals. For some nutrients these intake levels may still pose a risk to subpopulations with extreme or distinct vulnerabilities.

A MODEL FOR THE DERIVATION OF TOLERABLE UPPER INTAKE LEVELS

The development of a mathematical model for deriving the Tolerable Upper Intake Level (UL) was rejected for reasons described elsewhere (IOM, 1997). Instead, the model for the derivation of ULs consists of a set of scientific factors that always should be considered explicitly. The framework under which these factors are organized is called risk assessment. Risk assessment (NRC, 1983, 1994) is a systematic means of evaluating the probability of occurrence of adverse health effects in humans from excess exposure to an environmental agent (in this case, a nutrient) (FAO/WHO, 1995; Health Canada, 1993). The hallmark of risk assessment is the requirement to be explicit in all the evaluations and judgments that must be made to document conclusions.

RISK ASSESSMENT AND FOOD SAFETY

Basic Concepts

Risk assessment is a scientific undertaking having as its objective a characterization of the nature and likelihood of harm resulting from human exposure to agents in the environment. The characterization of risk typically contains both qualitative and quantitative information and includes a discussion of the scientific uncertainties in that information. In the present context the agents of interest are nutrients and the environmental media are food, water, and nonfood sources such as nutrient supplements and pharmacological preparations.

Performing a risk assessment results in a characterization of the relationship between exposure to an agent and the likelihood that adverse health effects will occur in members of exposed populations. Scientific uncertainties are an inherent part of the risk assessment process and are discussed below. Deciding whether the magnitude of exposure is *acceptable* or *tolerable* in specific circumstances is not a component of risk assessment; this activity falls within the domain of risk management. Risk management decisions depend on the results of risk assessments but may also involve the public health significance of the risk, the technical feasibility of achieving various degrees of risk control, and the economic and social costs of this control. Because there is no single, scientifically definable distinction between *safe* and *unsafe* exposures, risk management necessarily incorporates components of sound, practical decision making that are not addressed by the risk assessment process (NRC, 1983, 1994).

A risk assessment requires that information be organized in rather specific ways but does not require any specific scientific evaluation methods. Rather, risk assessors must evaluate scientific information using what they judge to be appropriate methods; must make explicit the basis for their judgments about the uncertainties in risk estimates; and, when appropriate, must include alternative scientifically plausible interpretations of the available data (NRC, 1994; OTA, 1993).

Risk assessment is subject to two types of scientific uncertainties: those related to data and those associated with inferences that are required when directly applicable data are not available (NRC, 1994). Data uncertainties arise during the evaluation of information obtained from the epidemiological and toxicological studies of nutrient intake levels that are the basis for risk assessments. Examples of inferences include the use of data from experimental animals to estimate responses in humans and the selection of uncertainty factors to estimate inter- and intraspecies variabilities in response to toxic substances. Uncertainties arise whenever estimates of adverse health effects in humans are based on extrapolations of data obtained under dissimilar conditions (e.g., from experimental animal studies). Options for dealing with uncertainties are discussed below and in detail in Appendix J.

Steps in the Risk Assessment Process

In this report the organization of risk assessment is based on a model proposed by the NRC (1983, 1994) that is widely used in public health and regulatory decision making. The steps of risk assessment as applied to nutrients are as follows (see also Figure 3-1):

- Step 1. Hazard identification involves the collection, organization, and evaluation of all information pertaining to the adverse effects of a given nutrient. It concludes with a summary of the evidence concerning the capacity of the nutrient to cause one or more types of toxicity in humans.
- Step 2. Dose-response assessment determines the relationship between nutrient intake (dose) and adverse effect (in terms of incidence and severity). This step concludes with an estimate of the Tolerable Upper Intake Level (UL)—it identifies the highest level of daily nutrient intake that is likely to pose no risk of adverse health effects to almost all individuals in the general population. Different ULs may be developed for various life stage groups.
- Step 3. Intake assessment evaluates the distribution of usual total daily nutrient intakes for members of the general population. In cases where the UL pertains only to supplement use, the assessment is directed at intake from supplements only. It does not depend on step 1 or 2.
- Step 4. Risk characterization summarizes the conclusions from steps 1 and 2 with step 3 to determine the risk. The risk is generally expressed as the fraction of the exposed population, if any, having nutrient intakes (step 3) in excess of the estimated UL (steps 1 and 2). If possible, characterization also covers the magnitude of any such excesses. Scientific uncertainties associated with both the UL and the intake estimates are described so that risk managers understand the degree of scientific confidence they can place in the risk assessment.

The risk assessment contains no discussion of recommendations for reducing risk; these are the focus of risk management.

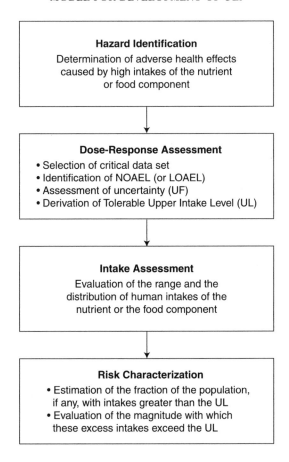

FIGURE 3-1 Risk assessment model for nutrient toxicity.

Thresholds

A principal feature of the risk assessment process for noncarcinogens is the long-standing acceptance that no risk of adverse effects is expected unless a threshold dose (or intake) is exceeded. The adverse effects that may be caused by a nutrient almost certainly occur only when the threshold dose is exceeded (NRC, 1994; WHO, 1996). The critical issues concern the methods used to identify the approximate threshold of toxicity for a large and diverse human population. Because most nutrients are not considered to be carcinogenic in humans, the approach to carcinogenic risk assessment (EPA, 1996) is not discussed here.

Thresholds vary among members of the general population (NRC, 1994). For any given adverse effect, if the distribution of thresholds in the population could be quantitatively identified, it would be possible to establish ULs by defining some point in the lower tail of the distribution of thresholds that would protect some specified fraction of the population. However, data are not sufficient to allow identification of the distribution of thresholds for the B vitamins or choline. The method for identifying thresholds for the general population described here is designed to ensure that almost all members of the population will be protected, but it is not based on an analysis of the theoretical distribution of thresholds. By using the model to derive the threshold, however, there is considerable confidence that the threshold, which becomes the UL for nutrients, lies very near the low end of the theoretical distribution and is the end representing the most sensitive members of the population. For some nutrients there may be subpopulations that are not included in the general distribution because of extreme or distinct vulnerabilities to toxicity. Such distinct groups, whose conditions warrant medical supervision, may not be protected by the UL.

When possible, the UL is based on a no-observed-adverse-effect level (NOAEL), which is the highest intake (or experimental oral dose) of a nutrient at which no adverse effects have been observed in the individuals studied. If there are no adequate data demonstrating a NOAEL, then a lowest-observed-adverse-effect level (LOAEL) may be used. A LOAEL is the lowest intake (or experimental oral dose) at which an adverse effect has been identified. The derivation of a UL from a NOAEL (or LOAEL) involves a series of choices about what factors should be used to deal with uncertainties. Uncertainty factors are applied in an attempt to deal both with gaps in data and with incomplete knowledge regarding the inferences required (e.g., the expected variability in response within the human population). The problems of both data and inference uncertainties arise in all steps of the risk assessment. A discussion of options available for dealing with these uncertainties is presented below and in greater detail in Appendix J.

A UL is not in itself a description of human risk. It is derived by application of the hazard identification and dose-response evaluation steps (steps 1 and 2) of the risk assessment model. To determine whether populations are at risk requires an intake assessment (step 3, evaluation of their intakes of the nutrient) and a determination of the fractions of those populations, if any, whose intakes exceed the UL. In the intake assessment and risk characterization steps (steps 3 and 4; described in the respective nutrient chapters), the distribution of actual intakes for the population will be used as a basis for determining whether and to what extent the population is at risk.

APPLICATION OF THE RISK ASSESSMENT MODEL TO NUTRIENTS

This section provides guidance for applying the risk assessment framework (the model) to the derivation of Tolerable Upper Intake Levels (ULs) for nutrients.

Special Problems Associated with Substances Required for Human Nutrition

In the application of accepted standards for risk assessment of environmental chemicals to risk assessment of nutrients, a fundamental difference between the two categories must be recognized: within a certain range of intakes, nutrients are essential for human well-being and usually for life itself. Nonetheless, they may share with other chemicals the production of adverse effects at excessive exposures. Because the consumption of balanced diets is consistent with the development and survival of humankind over many millennia, there is less need for the large uncertainty factors that have been used for the risk assessment of nonessential chemicals. In addition, if data on the adverse effects of nutrients are available primarily from studies in human populations, there will be less uncertainty than is associated with the types of data available on nonessential chemicals.

There is no evidence to suggest that nutrients consumed at recommended intakes (the Recommended Dietary Allowance [RDA] or Adequate Intake [AI]) present a risk of adverse effects to the general population. It is clear, however, that the addition of nutrients to a diet through the ingestion of large amounts of highly fortified food, nonfood sources such as supplements, or both may (at some level) pose a risk of adverse health effects. The UL is the highest level of daily nutrient intake that is likely to pose no risk of adverse health effects to almost all individuals in the general population. As intake increases above the UL, the risk of adverse effects increases.

If adverse effects have been associated with total intake from all sources, ULs are based on total intake of a nutrient from food, water, and supplements. For cases in which adverse effects have been associated with intake only from supplements and food fortificants, the UL is based on intake from those sources only rather than on total intake. The effects of nutrients from fortified foods or supplements may differ from those of naturally occurring constituents of foods because of the chemical form of the nutrient, the timing of the intake and amount consumed in a single bolus dose, the matrix supplied by the food, and the relation of the nutrient to the other constituents of the diet. Nutrient requirements and food intake are related to the metabolizing body mass, which is also at least an indirect measure of the space in which the nutrients are distributed. This relation between food intake and space of distribution supports homeostasis, which maintains nutrient concentrations in that space within a range compatible with health. However, excessive intake of a single nutrient from supplements or fortificants may compromise this homeostatic mechanism. Such elevations alone may pose risk of adverse effects, and imbalances among the vitamins or other nutrients may also be possible. Thus, assessment of risk from high nutrient intake includes the form and pattern of consumption, when applicable.

Consideration of Variability in Sensitivity

This risk assessment model must consider variability in the sensitivity of individuals to adverse effects of nutrients. Physiological changes and common conditions associated with growth and maturation that occur during an individual's lifespan may influence sensitivity to nutrient toxicity. For example, sensitivity increases with declines in lean body mass and with declines in renal and liver function that occur with aging; sensitivity changes in direct relation to intestinal absorption or intestinal synthesis of nutrients; in the newborn infant sensitivity is also increased because of rapid brain growth and limited ability to secrete or biotransform toxicants; and sensitivity increases with decreases in the rate of metabolism of nutrients. During pregnancy the increase in total body water and glomerular filtration results in lower blood levels of water-soluble vitamins for a given dose and therefore in reduced susceptibility to potential adverse effects. However, this effect may be offset by active placental transfer to the unborn fetus, accumulation of certain nutrients in the amniotic fluid, and rapid development of the fetal brain. There are no data to suggest increased or reduced susceptibility to adverse effects from high intake of B vitamins and choline during lactation. For the B vitamins and choline, different ULs are developed for some life stage groups, but the ULs for adults apply equally to pregnant and lactating women. For the B vitamins and choline, the ULs for infants were judged not determinable because of the lack of data on adverse effects in this age group and concern about the infant's ability to handle excess amounts of nutrients.

Even within relatively homogeneous life stage groups, there is a range of sensitivities to toxic effects. The model described below accounts for normally expected variability in sensitivity but excludes subpopulations with extreme and distinct vulnerabilities. Such subpopulations consist of individuals needing medical supervision; they are better served through the use of public health screening, product labeling, or other individualized health care strategies. The decision to treat identifiable vulnerable subgroups as distinct (not protected by the UL) is a matter of judgment and is discussed in individual nutrient chapters, as applicable.

Bioavailability

In the context of toxicity, the bioavailability of an ingested nutrient can be defined as its accessibility to normal metabolic and physiological processes. Bioavailability influences a nutrient's beneficial effects at physiological levels of intake and also may affect the nature and severity of toxicity due to excessive intakes. The concentration and chemical form of the nutrient, the nutrition and health of the individual, and excretory losses all affect bioavailability. Bioavailability data for specific nutrients must be considered and incorporated into the risk assessment process.

Certain B vitamins may be less readily absorbed when part of a meal than when taken separately. Supplemental forms of vitamins require special consideration if they have higher bioavailability and therefore may present a higher risk of producing adverse effects than does food (e.g., see Chapter 8).

Nutrient-Nutrient Interactions

A diverse array of adverse health effects can occur as a result of the interaction of nutrients. The potential risk of adverse nutrient-nutrient interactions increases when there is an imbalance in the intake of two or more nutrients. Excessive intake of one nutrient may interfere with absorption, excretion, transport, storage, function, or metabolism of a second nutrient. Possible adverse nutrient-nutrient interactions are considered as a part of setting a UL. Nutrient-nutrient interactions may be considered either as a critical endpoint on which to base a UL or as supportive evidence for a UL based on another endpoint.

STEPS IN THE DEVELOPMENT OF THE TOLERABLE UPPER INTAKE LEVEL

Hazard Identification

Based on a thorough review of the scientific literature, the hazard identification step outlines the adverse health effects that have been demonstrated to be caused by the nutrient. The primary types of data used as background for identifying nutrient hazards in humans are as follows:

- *Human studies.* Human data provide the most relevant kind of information for hazard identification and, when they are of sufficient quality and extent, are given greatest weight. However, the number of controlled human toxicity studies conducted in a clinical setting is very limited because of ethical reasons. Such studies are generally most useful for identifying very mild (and ordinarily reversible) adverse effects. Observational studies that focus on well-defined populations with clear exposures to a range of nutrient intake levels are useful for establishing a relationship between exposure and effect. Observational data in the form of case reports or anecdotal evidence are used for developing hypotheses that can lead to knowledge of causal associations. Sometimes a series of case reports, if it shows a clear and distinct pattern of effects, may be reasonably convincing on the question of causality.
- *Animal studies.* Most of the available data used in regulatory risk assessments come from controlled laboratory experiments in animals, usually mammalian species other than humans (e.g., rodents). Such data are used in part because human data on nonessential chemicals are generally very limited. Because well-conducted animal studies can be controlled, establishing causal relationships is generally not difficult. However, cross-species differences make the usefulness of animal data for establishing Tolerable Upper Intake Levels (ULs) problematic (see below).

Six key issues that are addressed in the data evaluation of human and animal studies are the following (see Box 3-1):

1. *Evidence of adverse effects in humans.* The hazard identification step involves the examination of human, animal, and in vitro published evidence addressing the likelihood of a nutrient eliciting an adverse effect in humans. Decisions regarding which observed effects are adverse are based on scientific judgments. Although toxicolo-

---

BOX 3-1   Development of Tolerable Upper Intake Levels (ULs)

Components Of Hazard Identification
- Evidence of adverse effects in humans
- Causality
- Relevance of experimental data
- Mechanisms of toxic action
- Quality and completeness of the database
- Identification of distinct and highly sensitive subpopulations

Components Of Dose-Response Assessment
- Data selection
- Identification of no-observed-adverse-effect level (NOAEL) (or lowest-observed-adverse-effect level [LOAEL]) and critical endpoint
- Uncertainty assessment
- Derivation of a UL
- Characterization of the estimate and special considerations

---

gists generally regard any demonstrable structural or functional alteration as representing an adverse effect, some alterations may be considered to be of little or self-limiting biological importance. As noted earlier, adverse nutrient-nutrient interactions are considered in the definition of an adverse effect.

2. *Causality.* As outlined in Chapter 2, the criteria of Hill (1971) are considered in judging the causal significance of an exposure-effect association indicated by epidemiological studies.

3. *Relevance of experimental data.* Consideration of the following issues can be useful in assessing the relevance of experimental data:

- *Animal data.* Animal data may be of limited utility in judging the toxicity of nutrients because of highly variable interspecies differences in nutrient requirements. Nevertheless, relevant animal data are considered in the hazard identification and dose-response assessment steps where applicable.

- *Route of exposure.* Data derived from studies involving oral exposure (rather than parenteral exposure) are most useful for evaluating nutrients. Data derived from studies involving parenteral routes of exposure may be considered relevant if the adverse effects are systemic and data are available to permit extrapolation between routes. (The terms *route of exposure* and *route of intake* refer to how a substance enters the body, for example, by ingestion, injection, or dermal absorption. These terms should not be confused with *form of intake,* which refers to the medium or vehicle used, e.g., supplements, food, and drinking water.)

• *Duration of exposure.* Consideration needs to be given to the relevance of the exposure scenario (e.g., chronic daily dietary exposure versus short-term bolus doses) to dietary intakes by human populations.

4. *Mechanisms of toxic action.* Knowledge of molecular and cellular events underlying the production of toxicity can assist in dealing with the problems of extrapolation between species and from high to low doses. It may also aid in understanding whether the mechanisms associated with toxicity are those associated with deficiency. In the case of the B vitamins, knowledge of the biochemical sequence of events resulting from toxicity and deficiency is still incomplete, and it is not yet possible to state with certainty the extent to which these sequences share a common pathway.

5. *Quality and completeness of the database.* The scientific quality and quantity of the database are evaluated. Human or animal data are reviewed for suggestions that the substances have the potential to produce additional adverse health effects. If suggestions are found, additional studies may be recommended.

6. *Identification of distinct and highly sensitive subpopulations.* The ULs are based on protecting the most sensitive members of the general population from adverse effects of high nutrient intake. For some nutrients, however, there may be distinct subgroups that have extreme sensitivities that do not fall within the range of sensitivities expected for the general population. The UL for the general population may not be protective for these subgroups. As indicated earlier, the extent to which a distinct subpopulation will be included in the derivation of a UL for the general population is an area of judgment to be addressed on a case-by-case basis.

Dose-Response Assessment

The process for deriving the UL is described in this section and outlined in Box 3-1. It includes selection of the critical data set, identification of a critical endpoint with its NOAEL (or LOAEL), and assessment of uncertainty.

Data Selection

The data evaluation process results in the selection of the most appropriate or critical data sets for deriving the UL. Selecting the critical data set includes the following considerations:

• Human data are preferable to animal data.
• In the absence of appropriate human data, information from an animal species with biological responses most like those of humans is most valuable.
• If it is not possible to identify such a species or to select such data, data from the most sensitive animal species, strain, and gender combination are given the greatest emphasis.
• The route of exposure that most resembles the route of expected human intake is preferable. This includes considering the digestive state (e.g., fed or fasted) of the subjects or experimental animals. Where this is not possible, the differences in route of exposure are noted as a source of uncertainty.
• The critical data set defines a dose-response relationship between intake and the extent of the toxic response known to be most relevant to humans. Data on bioavailability are considered and adjustments in expressions of dose-response are made to determine whether any apparent differences in response can be explained.
• The critical data set documents the route of exposure and the magnitude and duration of the intake. Furthermore, the critical data set documents the intake that does not produce adverse effects (the NOAEL), as well as the intake producing toxicity.

Identification of NOAEL (or LOAEL) and Critical Endpoint

A nutrient can produce more than one toxic effect (or endpoint), even within the same species or in studies using the same or different exposure durations. The NOAELs (and LOAELs) for these effects will differ. The critical endpoint used in this report is the adverse biological effect exhibiting the lowest NOAEL (e.g., the most sensitive indicator of a nutrient's toxicity). The derivation of a UL based on the most sensitive endpoint will ensure protection against all other adverse effects.

For some nutrients there may be inadequate data on which to develop a UL. The lack of reports of adverse effects after excess intake of a nutrient does not mean that adverse effects do not occur. As the intake of any nutrient increases, a point (see Figure 3-2) is reached at which intake begins to pose a risk. Above this point,

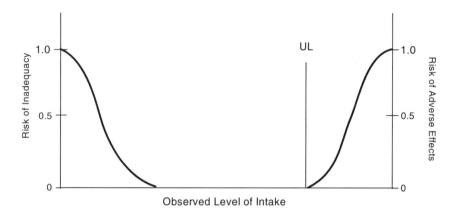

FIGURE 3-2 Theoretical description of health effects of a nutrient as a function of level of intake. The Tolerable Upper Intake Level (UL) is the highest level of daily nutrient intake that is likely to pose no risk of adverse health effects for almost all individuals in the general population. At intakes above the UL, the risk of adverse effects increases.

increased intake increases the risk of adverse effects. For some nutrients, and for various reasons, data are not adequate to identify the point where intake begins to pose a risk or even to estimate its location.

Because adverse effects are almost certain to occur for any nutrient at some level of intake, it should be assumed that such effects may occur for nutrients for which a scientifically documentable UL cannot now be derived. Until a UL is set or an alternative approach to identifying protective limits is developed, intakes greater than the Recommended Dietary Allowance (RDA) or Adequate Intake (AI) should be viewed with caution.

Uncertainty Assessment

Several judgments must be made regarding the uncertainties and thus the uncertainty factor (UF) associated with extrapolating from the observed data to the general population (see Appendix J). Applying a UF to a NOAEL (or LOAEL) results in a value for the derived UL that is less than the experimentally derived NOAEL unless the UF is 1.0. The larger the uncertainty, the larger the UF and the smaller the UL. This is consistent with the ultimate goal of the risk assessment: to provide an estimate of a level of intake that will protect the health of the healthy population (Mertz et al., 1994).

Although several reports describe the underlying basis for UFs (Dourson and Stara, 1983; Zielhuis and van der Kreek, 1979), the strength of the evidence supporting the use of a specific UF will vary. Because the imprecision of these UFs is a major limitation of risk assessment approaches, considerable leeway must be allowed for the application of scientific judgment in making the final determination. Because the data on nutrient toxicity may not be subject to the same uncertainties as are data on nonessential chemical agents, the UFs for nutrients are typically less than 10. They are lower with higher-quality data and when the adverse effects are extremely mild and reversible.

In general, when determining a UF, the following potential sources of uncertainty are considered:

• *Interindividual variation in sensitivity.* Small UFs (close to 1) are used if it is judged that little population variability is expected for the adverse effect, and larger factors (close to 10) are used if variability is expected to be great (NRC, 1994).

• *Extrapolation from experimental animals to humans.* A UF is generally applied to the NOAEL to account for the uncertainty in extrapolating animal data to humans. Larger UFs (close to 10) may be used if it is believed that the animal responses will underpredict average human responses (NRC, 1994).

• *LOAEL instead of NOAEL.* If a NOAEL is not available, a UF may be applied to account for the uncertainty in deriving a UL from the LOAEL. The size of the UF involves scientific judgment based on the severity and incidence of the observed effect at the LOAEL and the steepness (slope) of the dose response.

• *Subchronic NOAEL to predict chronic NOAEL.* When data are lacking on chronic exposures, scientific judgment is necessary to determine whether chronic exposure is likely to lead to adverse effects at lower intakes than those producing effects after subchronic exposures (exposures of shorter duration).

Derivation of a UL

The UL is derived by dividing the NOAEL (or LOAEL) by a single UF that incorporates all relevant uncertainties. For infants, ULs were not determined for any of the B vitamins or choline because of the lack of data on adverse effects in this age group and concern regarding infants' possible lack of ability to handle excess amounts.

Thus, caution is warranted; food should be the source of intake by infants.

ULs for niacin, vitamin B_{12}, and choline in children and adolescents were determined by extrapolating from the UL for adults based on body weight differences by using the formula

$$UL_{child} = (UL_{adult}) (Weight_{child}/Weight_{adult})^{0.75}.$$

See Chapter 2 for related information about extrapolation.

With the use of data from Table 1-2 (Chapter 1), the reference weight for males ages 19 through 30 years was used for adults and the reference weights for female children and adolescents were used in the formula above to obtain the UL for each age group. The use of these reference weights yields a conservative UL to protect the sensitive individuals in each age group.

The derivation of a UL involves the use of scientific judgment to select the appropriate NOAEL (or LOAEL) and UF. The risk assessment requires explicit consideration and discussion of all choices made, both regarding the data used and the uncertainties accounted for. These considerations are discussed in the chapters on nutrients. Because of lack of suitable data, ULs could not be set for infants or for thiamin, riboflavin, vitamin B_{12}, pantothenic acid, or biotin.

Characterization of the Estimate and Special Considerations

ULs are derived for various life stage groups by using relevant databases, NOAELs and LOAELs, and UFs. Where no data exist for NOAELs or LOAELs for the group under consideration, extrapolations from data in other age groups and/or animal data are made on the basis of known differences in body size, physiology, metabolism, absorption, and excretion of the nutrient.

If the data review reveals the existence of subpopulations having distinct and exceptional sensitivities to a nutrient's toxicity, these subpopulations are considered under the heading "Special Considerations."

REFERENCES

Dourson ML, Stara JF. 1983. Regulatory history and experimental support of uncertainty (safety) factors. *Regul Toxicol Pharmacol* 3:224–238.

EPA (U.S. Environmental Protection Agency). 1996. Proposed guidelines for carcinogen risk assessment; Notice. *Fed Regist* 61:17960–18011.

FAO/WHO (Food and Agriculture Organization of the United Nations/World Health Organization). 1995. *The Application of Risk Analysis to Food Standard Issues*. Recommendations to the Codex Alimentarius Commission (ALINORM 95/9, Appendix 5). Geneva: World Health Organization.

Health Canada. 1993. *Health Risk Determination—The Challenge of Health Protection*. Ottawa: Health Canada, Health Protection Branch.

Hill AB. 1971. *Principles of Medical Statistics*, 9th ed. New York: Oxford University Press.

IOM (Institute of Medicine). 1997. *Dietary Reference Intakes for Calcium, Phosphorus, Magnesium, Vitamin D, and Fluoride*. Washington, DC: National Academy Press.

Klaassen CD, Amdur MO, Doull J. 1986. *Casarett and Doull's Toxicology: The Basic Science of Poisons*, 3rd ed. New York: Macmillan.

Mertz W, Abernathy CO, Olin SS. 1994. *Risk Assessment of Essential Elements*. Washington, DC: ILSI Press.

NRC (National Research Council). 1983. *Risk Assessment in the Federal Government: Managing the Process*. Washington, DC: National Academy Press.

NRC (National Research Council). 1994. *Science and Judgment in Risk Assessment*. Washington, DC: National Academy Press.

OTA (Office of Technology Assessment). 1993. *Researching Health Risks*. Washington, DC: OTA.

WHO (World Health Organization). 1996. *Trace Elements in Human Nutrition and Health*. Prepared in collaboration with the Food and Agriculture Organization of the United Nations and the International Atomic Energy Agency. Geneva: WHO.

Zielhuis RL, van der Kreek FW. 1979. The use of a safety factor in setting health-based permissible levels for occupational exposure. *Int Arch Occup Environ Health* 42:191–201.

4
Thiamin

SUMMARY

Thiamin functions as a coenzyme in the metabolism of carbohydrates and branched-chain amino acids. The method used to estimate the Recommended Dietary Allowance (RDA) for thiamin combines erythrocyte transketolase activity, urinary thiamin excretion, and other findings. The RDA for adults is 1.2 mg/day for men and 1.1 mg/day for women. Recently, the median intake of thiamin from food in the United States was approximately 2 mg/ day, and the ninety-fifth percentile of intake from both food and supplements was approximately 6.1 mg. Intakes in two Canadian populations were slightly lower. Data concerning adverse effects are not sufficient to set a Tolerable Upper Intake Level (UL) for thiamin.

BACKGROUND INFORMATION

Thiamin (also known as vitamin B_1 and aneurin) was the first B vitamin identified. Lack of thiamin causes the deficiency disease called beriberi, which has been known since antiquity. More recently, at least in industrialized nations, thiamin deficiency has been mainly found in association with chronic alcoholism, where it presents as the Wernicke-Korsakoff syndrome.

Chemically, thiamin consists of substituted pyrimidine and thiazole rings linked by a methylene bridge. It exists mainly in various interconvertible phosphorylated forms, chiefly thiamin pyrophosphate

(TPP). TPP, the coenzymatic form of thiamin, is involved in two main types of metabolic reactions: decarboxylation of α-ketoacids (e.g., pyruvate, α-ketoglutarate, and branched-chain keto acids) and transketolation (e.g., among hexose and pentose phosphates).

Physiology of Absorption, Metabolism, and Excretion

Following ingestion, absorption of thiamin occurs mainly in the jejunum, at lower concentrations as an active, carrier-mediated system involving phosphorylation and at higher concentrations by passive diffusion. Thiamin is transported in blood both in erythrocytes and plasma.

Only a small percentage of a high dose of thiamin is absorbed, and elevated serum values result in active urinary excretion of the vitamin (Davis et al., 1984). After an oral dose of thiamin, peak excretion occurs in about 2 hours, and excretion is nearly complete after 4 hours (Levy and Hewitt, 1971; Morrison and Campbell, 1960). In a study by Davis and colleagues (1984), a 10-mg oral dose of thiamin was given in water, and the mean serum thiamin peaked at 24 nmol/L (7.2 μg/L)—42 percent above baseline. Within 6 hours the serum thiamin concentration had returned to baseline, 17 nmol/L (5.2 μg/L). Prompt urinary excretion of thiamin was also reported by Najjar and Holt (1940) and McAlpine and Hills (1941).

With higher pharmacological levels, namely repetitive 250-mg amounts taken orally and 500 mg given intramuscularly, nearly 1 week was required for steady state plasma concentrations to be reached; a mean elimination half-life of 1.8 days was estimated (Royer-Morrot et al., 1992).

Total thiamin content of the adult human has been estimated to be approximately 30 mg, and the biological half-life of the vitamin is probably in the range of 9 to 18 days (Ariaey-Nejad et al., 1970).

Clinical Effects of Inadequate Intake

Early stages of thiamin deficiency may be accompanied by nonspecific symptoms that may be overlooked or easily misinterpreted (Lonsdale and Shamberger, 1980). The clinical signs of deficiency include anorexia; weight loss; mental changes such as apathy, decrease in short-term memory, confusion, and irritability; muscle weakness; and cardiovascular effects such as an enlarged heart (Horwitt et al., 1948; Inouye and Katsura, 1965; Platt, 1967; Williams et al., 1942; Wilson, 1983). In wet beriberi, edema occurs; in dry beriberi, muscle wasting is obvious. In infants, cardiac failure may occur rather suddenly (McCormick and Greene, 1994). Severe thiamin deficiency in industrialized countries is likely to be related to heavy alcohol consumption with limited food consumption, as was noted for at least four of five Welsh cases reported by Anderson and colleagues (1985). In those cases renal and cardiovascular complications were life threatening.

SELECTION OF INDICATORS FOR ESTIMATING THE REQUIREMENT FOR THIAMIN

Biochemical changes in thiamin status occur well before the appearance of overt signs of deficiency. Thiamin status can be assessed by determining erythrocyte transketolase activity, by measuring the concentration of thiamin and its phosphorylated esters in blood or serum components using high-performance liquid chromatography, or by measuring urinary thiamin excretion under basal conditions or after thiamin loading. Commonly used reference values indicating marginal deficiency for these indicators are given in Table 4-1. Other methods have also been reported and are covered briefly below.

No currently available indicator, by itself, provides an adequate basis on which to estimate the thiamin requirement.

Urinary Thiamin Excretion

The urinary excretion of thiamin is the indicator that has been used most widely in metabolic studies of thiamin requirements and

TABLE 4-1 Reference Values for the Primary Measures of Thiamin Status

Indicator	Marginal Deficiency	Deficiency
Erythrocyte transketolase activity[a]	1.20–1.25	> 1.25
Erythrocyte thiamin (nmol/L)[a]	70–90	< 70
Thiamin pyrophosphate effect (%)[b]	15–24	≥ 25
Urinary thiamin[a]		
(nmol [µg]/g creatinine)	90–220 (27–66)	< 27
(nmol [µg]/d)	133–333 (40–100)	< 40

[a] Schrijver (1991).
[b] Stimulated value, expressed as a multiple of the basal value. Also termed the activity coefficient. Brin (1970).

was thus given careful consideration in deriving the Estimated Average Requirement (EAR). Urinary thiamin excretion decreases markedly as thiamin status declines and is also affected by recent dietary intake. Bayliss and coworkers (1984) reported a correlation of 0.86 between the oral dose of thiamin and urinary thiamin excretion. However, in doses of up to 1.05 mg there was overlap with baseline values. The use of a load test, in which thiamin excretion is measured before and after a test load of thiamin, helps differentiate between extremes of vitamin status (McCormick and Greene, 1994).

Erythrocyte Transketolase Activity

Erythrocyte transketolase activity has also been widely used and is generally regarded as the best functional test of thiamin status (McCormick and Greene, 1994), but it has some limitations for deriving the EAR and should be evaluated along with other indicators. In this test, erythrocytes are lysed and the transketolase activity is measured before and after stimulation by the addition of thiamin pyrophosphate (TPP); the basal level and the stimulated value (typically expressed as a multiple of the basal level, termed the activity coefficient or TPP effect) are measured. In thiamin-depleted individuals, basal erythrocyte transketolase typically is low and the incremental response after TPP addition is enhanced.

Although the test has long been used in assessing thiamin status, in one recent study (Bailey et al., 1994) it correlated poorly with dietary thiamin intake in English adolescents. Similarly, in a study population of 179 adult men, Gans and Harper (1991) found a wide range of TPP effect values (0 to 95 percent) associated with thiamin intakes that were all above 1.5 mg/day over a 3-day period. Similarly, they also found a TPP effect of 0 percent associated with a wide range of intakes (approximately 0.75 to 6.0 mg/day). Schrijver (1991) reported that the activity coefficient may appear normal after prolonged deficiency, making identification of the deficiency more problematic. From studies of the elderly, Pekkarinen and colleagues (1974) concluded that evaluation of thiamin status should consider other indicators along with erythrocyte transketolase activity.

Factors other than thiamin status, such as genetic defects, may influence the enzyme activity and thus the test results. Individuals and tissues both differ in their sensitivity to thiamin deficiency. This observation may be explained by the pronounced lag in the formation of active holoenzyme and the interindividual and cell type variation in the lag during thiamin deficiency (Singleton et al., 1995).

Erythrocyte Thiamin

As thiamin status declines, the concentration of TPP in erythrocytes decreases at approximately the same rate as occurs in other tissues (Brin, 1964; McCormick and Greene, 1994). The TPP effect may be noted within 2 weeks after the initiation of a thiamin-restricted diet (Brin, 1962). Baines and Davies (1988) provided evidence that, compared with erythrocyte transketolase activity, erythrocyte TPP is more stable in frozen erythrocytes, easier to standardize, and less susceptible to factors that influence enzyme activity.

Other Measurements

Because of the wide variety of signs and symptoms characteristic of thiamin deficiency, numerous other indicators of thiamin status have been reported. These include blood pyruvic acid values after exercise (Foltz et al., 1944); both pyruvic acid and lactic acid values after administration of glucose (Bueding et al., 1941; Williams et al., 1943); various indicators of work performance (e.g., maximum work test to exhaustion) (Wood et al., 1980); aerobic power, respiratory exchange ratio, and ventilatory equivalent (van der Beek et al., 1994); work output over time (Foltz et al., 1944); gross behavior changes (Williams et al., 1942); neurological changes (Wood et al., 1980); psychological changes (Wood et al., 1980); and quality of life (Wilkinson et al., 1997). None of these was judged to be a dependable criterion of thiamin status.

FACTORS AFFECTING THE THIAMIN REQUIREMENT

Bioavailability

Data on the bioavailability of thiamin in humans are extremely limited. Levy and Hewitt (1971) reported that absorption of thiamin supplements taken with breakfast does not differ from that taken on an empty stomach. No adjustments for bioavailability were judged necessary for deriving the Estimated Average Requirement (EAR).

Energy Intake

No studies were found that examined the effect of energy intake on the thiamin requirement. Some studies provided thiamin in graded doses that kept the ratio of thiamin to energy constant for those studied who had different energy requirements. Other studies provided total amounts of thiamin (and sometimes energy) that were the same for all individuals. Sauberlich and colleagues (1979) adjusted activity levels rather than energy intake to maintain weight in their subjects. Several investigators examined their data to assess whether it would be better to express thiamin as an absolute value or in relation to energy. For example, Dick and colleagues (1958) reported that the coefficient of variation of the estimated thiamin requirement for adolescent boys was 14.2 percent/person, 15.5 percent/1,000 kcal, 27.5 percent/kg body weight, 19.5 percent/m² surface area, and 19.2 percent/mg of creatinine excretion. Elsom and coworkers (1942) noted that they could not distinguish whether it was better to express thiamin in absolute values or per 1,000 kcal but that thiamin intake expressed per body weight did not discriminate between those who were deficient and those who were not. Anderson and colleagues (1986) presented evidence that expressing the thiamin requirements in absolute terms is more useful for predicting biochemical thiamin status than expressing it in relation to energy intake, and data from individuals presented by Henshaw and coworkers (1970) appear supportive.

Despite the lack of direct experimental data, the known biochemical function of thiamin as thiamin pyrophosphate (TPP) in the metabolism of carbohydrate suggests that at least a small (10 percent) adjustment to the estimated requirement to reflect differences in the average energy utilization and size of men and women, a 10 percent increase in the requirement to cover increased energy utilization during pregnancy, and a small increase to cover the energy cost of milk production during lactation may be necessary. It has been observed that during periods of starvation such as in war, larger individuals present signs of beriberi more rapidly than do those with smaller body builds, indicating their greater needs for thiamin and other energy-related nutrients (Burgess, 1946). Many studies report thiamin intake per 1,000 kcal; others report total intake. Thus, the evidence below is presented as it was done in the studies and not because the ratio is considered important.

Physical Activity

Heavy exercise under certain conditions may increase the requirement for thiamin as well as other vitamins, but the observations on the effects of physical activity on the thiamin requirement have been inconsistent, the effects small, and the experimental conditions highly variable. For example, one 14-week, double-blind, $2 \times 2 \times 2$ complete factorial experiment examined the effects of restriction of three vitamins—thiamin, riboflavin, and vitamin B_6—on physical performance in 24 healthy Dutch males (van der Beek et al., 1994). In the thiamin-restricted group, thiamin intake was 0.43 mg/day (analyzed mean value). Thiamin concentration, erythrocyte trans-ketolase activity, and urinary thiamin decreased significantly over the 11-week experimental period, and α-erythrocyte transketolase activity (or activation coefficient) increased. The decrease in thia-min status was accompanied by small but significant decrements in performance as measured during single short bouts of intense exer-cise, but these could not be attributed to any one of the three vita-mins studied.

In another double-blind study, 12 mg of thiamin (15 mg of thia-min nitrate) along with riboflavin and pyridoxine were provided to all 22 subjects in the experimental group for 5 weeks. Although the activation coefficients for transketolase (and other enzymes) de-creased in the supplemented group, no change in blood lactate was found after exercise (Fogelholm et al., 1993).

An observational study (Folgeholm et al., 1992) that found com-parable erythrocyte transketolase activation coefficients in skiers and nonskiers provided little useful information on the effect of energy expenditure on thiamin requirements. Compared with the non-skiers, the skiers had much higher energy intakes and expenditures along with much higher intakes of all reported nutrients. For both males and females, mean thiamin intakes were 0.8 mg/1,000 kcal for the skiers and 0.7 mg/1,000 kcal for the control subjects.

It was thus concluded that under normal conditions, physical activity does not appear to influence thiamin requirements to a substantial degree. However, those who are engaged in physically demanding occupations or who spend much time training for active sports may require additional thiamin.

Gender

Studies were not found that directly compare the thiamin require-ments of males and females. A small (10 percent) difference in the average thiamin requirements of men and women is assumed on the basis of mean differences in body size and energy utilization.

FINDINGS BY LIFE STAGE AND GENDER GROUP

Infants Ages 0 through 12 Months

Method Used to Set the Adequate Intake

An Adequate Intake (AI) is used as the goal for intake by infants.

Ages 0 through 6 Months. The AI reflects the observed mean thiamin intake of infants consuming human milk. Thus, the thiamin AI for young infants is based on mean intake data from infants fed human milk exclusively during their first 6 months and uses the thiamin concentration of milk produced by well-nourished mothers. There are no reports of full-term infants who were exclusively fed milk from U.S. or Canadian mothers who manifested any signs of thiamin deficiency; however, infants breastfed by mothers with beriberi have been reported to develop beriberi themselves by age 3 to 4 weeks (Hytten and Thomason, 1961). The thiamin content of human milk was similar for well-nourished mothers who received vitamin supplements and for those who did not (Nail et al., 1980; Pratt and Hamil, 1951).

The thiamin concentration is low in colostrum (approximately 0.01 µg/L). The mean concentration of thiamin in mature human milk is 0.21 ± 0.04 mg/L (mean ± standard deviation) (Committee on Nutrition, 1985). Using the mean volume for intake of human milk of 0.78 L/day (see Chapter 2) and the average thiamin content of 0.21 mg/L, the AI for thiamin is 0.16 mg/day for infants ages 0 through 6 months, which is rounded to 0.2 mg. For the reference infant weight of 7 kg, this corresponds to 0.03 mg/kg/day.

Blood concentration of total thiamin (phosphorylated and non-phosphorylated) has been shown to decrease with age: in a cross-sectional study of well-nourished individuals, blood thiamin concentrations in infants less than 3 months of age ($n = 64$) averaged 258 ± 63 nmol/L (75 ± 23 µg/L) (mean ± standard deviation), infants 3 to 12 months of age ($n = 100$) averaged 214 ± 44 nmol/L (64 ± 13 µg/L), while in children and young adults ($n = 159$) the value decreased to 187 ± 39 µmol/L (56 ± 12 µg/L) (Wyatt et al., 1991). Because total thiamin concentrations in whole blood and cerebrospinal fluid decrease in the first 12 to 18 months of life, age-specific norms should be used for determining thiamin status in infancy.

Ages 7 through 12 Months. If the reference body weight ratio method described in Chapter 2 to extrapolate from the AI for thiamin for infants ages 0 through 6 months is used, the AI for thiamin for the older infants would be 0.2 mg/day after rounding. The second method (see Chapter 2), extrapolating from the Estimated Average Requirement (EAR) for adults and adjusting for the expected variance to estimate a recommended intake, gives an AI of 0.3 mg of thiamin, a value higher than that obtained from the first method.

Alternatively, the AI for thiamin for infants ages 7 through 12 months could be calculated by using the estimated thiamin content of 0.6 L of human milk, the average volume consumed by this age group (thiamin content equals 0.13 mg), and adding the amount of thiamin provided by solid foods (0.5 mg), as estimated by Montalto et al. (1985) (see Chapter 2). The result equals approximately 0.6 mg/day. This value was judged to be unreasonably high because it is two to three times the extrapolated values given above. Thus the AI for thiamin is 0.3 mg/day for infants ages 7 through 12 months—the value extrapolated from estimates of adult requirements.

Thiamin AI Summary, Ages 0 through 12 Months

AI for Infants

0–6 months	**0.2 mg/day of thiamin**	**≈0.03 mg/kg**
7–12 months	**0.3 mg/day of thiamin**	**≈0.03 mg/kg**

Children Ages 1 through 8 Years

Method Used to Estimate the Average Requirement

No direct data were found on which to base an EAR for children ages 1 through 8 years. In the absence of additional information, EARs and Recommended Dietary Allowances (RDAs) for these age groups have been extrapolated from adult values by using the method described in Chapter 2.

Thiamin EAR and RDA Summary, Ages 1 through 8 Years

EAR for Children	**1–3 years**	**0.4 mg/day of thiamin**
	4–8 years	**0.5 mg/day of thiamin**

The RDA for thiamin is set by assuming a coefficient of variation (CV) of 10 percent (see Chapter 1) because information is not available on the standard deviation of the requirement for thiamin; the

RDA is defined as equal to the EAR plus twice the CV to cover the needs of 97 to 98 percent of the individuals in the group (therefore, for thiamin the RDA is 120 percent of the EAR).

RDA for Children 1–3 years **0.5 mg/day of thiamin**
4–8 years **0.6 mg/day of thiamin**

Children and Adolescents Ages 9 through 18 Years

Evidence Considered in Estimating the Average Requirement

Five studies were found for this age group, none of which involved children younger than 13 years. In an observational study of 19 boys and 35 girls aged 13 or 14 years, thiamin intake was calculated from 7-day food records and was also analyzed by high-performance liquid chromatography from duplicate portions (Bailey et al., 1994). The correlations of the results from the food records and the analyses of duplicate portions were significant but moderate ($r = 0.59$ for boys and 0.43 for girls). However, the indicators of thiamin status (erythrocyte transketolase, erythrocyte transketolase activity coefficient, and total erythrocyte thiamin concentration) were not correlated with each other. Moreover, none of them was correlated with thiamin intake as estimated from the food records or measured in the duplicate portions. A substantial percentage of the subjects (girls, 12 percent; boys, 17 percent) had activity coefficients that indicated a high risk of thiamin deficiency according to Brin's criterion for adults (Brin, 1970) even though estimated intakes were above 0.4 mg/1,000 kcal.

A controlled-diet, dose-response experiment was conducted with nine girls aged 16 to 18 years to examine the thiamin requirement (Hart and Reynolds, 1957). In this study, the girls were given 0.29 mg of thiamin/1,000 kcal/day (0.63 mg/day total) for the first 16-day period and 0.6 mg/1,000 kcal/day (1.3 mg/day) for the second 16-day period. The adequacy of intake was assessed by measuring total daily thiamin excretion, the percentage of consumed thiamin that was excreted, the ratio of thiamin to creatinine in the urine, and the percentage of excretion of a 5-mg oral test dose of thiamin hydrochloride. Using a modification of the thiochrome method for thiamin determination, the investigators were unable to obtain reliable measurements of the amount of thiamin excreted on the low-thiamin diet. The authors noted that the subjects became irritable and uncooperative and lost the ability to concentrate when fed the low-thiamin diet—symptoms also noted by others in the early stage of thiamin deficiency. On the diet that provided 1.3 mg/day of thiamin, 24-hour thiamin excretion ranged between 0.27 and 0.44 µmol (81 and 133 µg). These data suggest that the average thiamin requirement is less than 1.3 mg/day, especially considering the short period of repletion and the use of a generous cutoff point for urinary thiamin excretion, but they do not allow further refinement of the estimate.

In a study of eight boys aged 14 to 17 years, Dick and colleagues (1958) calculated thiamin requirements from a regression of excretion on intake at five levels of thiamin that ranged from 0.6 to 2.7 mg/day. By taking the abscissa of the intersection of two straight lines fitted to the observations on each subject, a mean requirement of 1.41 mg/day was computed. However, at this level of intake, mean urinary excretion averaged 0.618 µmol/day (186 µg/day)—a value far in excess of usual cutoffs.

In the absence of additional definitive information about requirements, EARs and RDAs for thiamin for these age groups were extrapolated from the adult values by using the method described in Chapter 2. Because only urinary excretion of thiamin was measured, the results reported by Hart and Reynolds (1957) are not considered strong enough to warrant adjustment of results from the extrapolation method.

Thiamin EAR and RDA Summary, Ages 9 through 18 Years

EAR for Boys	9–13 years	0.7 mg/day of thiamin
	14–18 years	1.0 mg/day of thiamin
EAR for Girls	9–13 years	0.7 mg/day of thiamin
	14–18 years	0.9 mg/day of thiamin

The RDA for thiamin is set by assuming a coefficient of variation (CV) of 10 percent (see Chapter 1) because information is not available on the standard deviation of the requirement for thiamin; the RDA is defined as equal to the EAR plus twice the CV to cover the needs of 97 to 98 percent of the individuals in the group (therefore, for thiamin the RDA is 120 percent of the EAR).

RDA for Boys	9–13 years	0.9 mg/day of thiamin
	14–18 years	1.2 mg/day of thiamin
RDA for Girls	9–13 years	0.9 mg/day of thiamin
	14–18 years	1.0 mg/day of thiamin

Adults Ages 19 through 50 Years

Indicators Used to Estimate the Average Requirement

It was necessary to review data from studies that used various indicators of thiamin sufficiency to derive an EAR for thiamin for adults. In reviewing the studies (Table 4-2), heavy weight was given to the carefully controlled, thiamin depletion-repletion experiment conducted with seven healthy young men (age not specified) in a metabolic unit (Sauberlich et al., 1979). The investigators concluded that thiamin at 0.30 mg/1,000 kcal (approximately 1.0 mg/day) met the minimum requirement for young men as determined by using urinary excretion of thiamin, and it appears that this value is close to the average requirement for normal erythrocyte transketolase activity. This value is slightly lower than the 1.2 and 1.0 mg total values designated by Anderson and colleagues (1986) as minimal for men and women, respectively, determined by using erythrocyte transketolase activity.

Studies by Bamji (1970) and Ziporin and coworkers (1965) support the conclusion that 0.30 mg/1,000 kcal/day is a minimum thiamin requirement to prevent overt signs and symptoms of deficiency, although at this level urinary excretion of thiamin was abnormal in some subjects. Studies by Kraut and colleagues (1966) and Reuter et al. (1967) suggest a much higher requirement. Although intakes of approximately 0.7 mg/day (0.4 mg/1,000 kcal) were found to meet the minimum requirement for thiamin based on erythrocyte transketolase activity (Reuter et al., 1967), the achievement of maximum erythrocyte transketolase activity required intakes of 2.0 to 2.5 mg/day (Kraut et al., 1966; Reuter et al., 1967). Several studies indicated that intakes of 0.075 to 0.29 mg/1,000 kcal lead to severe irritability and other symptoms and signs of deficiency (Foltz et al., 1944; Horwitt et al., 1948; Wood et al., 1980). Data are not sufficient to indicate differing requirements for adults 19 through 30 versus 31 through 50 years of age.

Thiamin EAR and RDA Summary, Ages 19 through 50 Years

Examination of the data in Table 4-2 indicates that the EAR for thiamin is at least 0.3 mg/1,000 kcal or 0.8 mg/day and that intakes greater than 1.0 mg are marginally adequate for normal erythrocyte transketolase activity and generally adequate for urinary thiamin excretion. Because of the uncertainties of the dietary intake estimates in the studies by Anderson et al. (1986) and Henshaw et al.

TABLE 4-2 Metabolic Studies Providing Evidence Used to Derive the Estimated Average Requirement (EAR) for Thiamin for Adults

Reference	Subjects	Duration of Study	Baseline Thiamin Intake
Elsom et al., 1942	9 women	28–120 d	0.8 mg/d 0.3 mg/d 0.7 mg/d NA NA NA NA NA NA
Foltz et al., 1944	4 men	1 mo 1 mo 9–12 mo	NA
Horwitt et al., 1948	24	3 y	1 mg/d
Ziporin et al., 1965	8 men	30 d depletion 12 d repletion	Mean intake during 9-d control period =1.75 mg/d
Kraut et al., 1966	4 men, 2 women	9–10 mo	NA
Reuter et al., 1967	6 obese women	NA	NA
Bamji, 1970	4 men, 4 women	2–3 wk depletion 1 wk repletion	0.1 mg/1,000 kcal (depletion level)
Henshaw et al., 1970	39 women	3 d 7 d	NA
Sauberlich et al., 1979	7 men	14 d 11 d 11 d 13 d	> 0.6 mg/1,000 kcal

Thiamin Intake During Repletion or Maintenance (mg/d)	Erythrocyte Transketolase Activity	Urinary Excretion of Thiamin	Other
0.2	NA[a]	Abnormal[b]	Abnormal[c]
0.2	NA	Abnormal	Abnormal
0.35	NA	Abnormal	Abnormal
0.41	NA	Abnormal	Marginal[d]
0.52	NA	Abnormal	Marginal
0.57	NA	Abnormal	Marginal
0.65	NA	Normal	Normal
0.70	NA	Normal	Normal
0.77	NA	Normal	Normal
0.57	NA	Abnormal[e]	Abnormal[f]
0.95	NA	50% normal	Abnormal[g]
1.44	NA	Normal	Normal
0.2	NA	Results varied	Abnormal[h]
0.4	NA	Results varied	Results varied
4.0	NA	Results varied	Normal
0.15	NA	Abnormal[i]	NA
0.58	NA	Abnormal	NA
2.0–2.5	NA	Normal[j]	NA
0.7[k]	Normal[l]	NA	NA
0.65 (men)	Abnormal[m]	Normal[n]	NA
1.3 (men)	Normal	Normal	NA
0.4 (women)	Normal	Normal	NA
0.8 (women)	Normal	Normal	NA
0.82[o]	50% abnormal[p]	50% abnormal[q]	29% abnormal[r]
1.02	29% abnormal	95% abnormal	100% normal
0.39	Abnormal[s]	Abnormal[t]	NA
0.56	Abnormal	Abnormal	NA
0.84	Normal	Marginal[u]	NA
1.08	Normal	Marginal	NA

continued

TABLE 4-2 Continued

Reference	Subjects	Duration of Study	Baseline Thiamin Intake
Wood et al., 1980	19 men	4–5 wk	NA
Anderson et al., 1986	14 women 14 men	7 d	NA

NOTE: Body weight was maintained in all studies. Thiamin intakes were measured analytically except as noted. **Bold** type is used for intakes that supported normal findings.

[a] NA = not applicable.

[b] Abnormal urinary excretion = < 133 nmol/d (40 µg/d) (no reference values given; judgment of status was based on Foltz et al., 1944).

[c] Abnormal clinical signs = appearance of unspecified manifestations of thiamin deficiency.

[d] Marginal clinical signs = appearance of some manifestations of thiamin deficiency (unspecified).

[e] Abnormal urinary excretion = < 200 nmol/d (60 µg/d) (for males).

[f] Abnormal clinical symptoms = leg pains, muscle tenderness.

[g] Abnormal clinical symptoms = decreased appetite, decreased endurance, increased irritability.

[h] Abnormal metabolism of carbohydrate and abnormal clinical signs (decreased deep reflexes, skin changes, decreased appetite, decreased blood pressure, dull vibratory sense, and edema).

[i] Abnormal urinary excretion = < 200 nmol/d (60 µg/d) (no reference values given by authors; judgment of status was based on Foltz et al., 1944).

[j] Normal urinary excretion = ≥ 67 nmol/d (20 µg/d) thiamin.

(1970), greater weight was given to the well-controlled studies of Sauberlich et al. (1979). With the assumption of a curvilinear relationship with increasing intake, it is concluded that the EAR for thiamin is 1.0 mg/day for men and 0.9 mg/day for women, which represents about a 10 percent decrease for women based on body size and energy needs.

EAR for Men	**19–30 years**	**1.0 mg/day of thiamin**
	31–50 years	**1.0 mg/day of thiamin**

Thiamin Intake During Repletion or Maintenance (mg/d)	Erythrocyte Transketolase Activity	Urinary Excretion of Thiamin	Other
0.45	Abnormal[v]	Abnormal[v]	Normal[w]
5.45	Normal	Normal	Normal
0.97[x]	Abnormal[y]	NA	NA
1.13	Abnormal	NA	NA
1.24	Normal	NA	NA
1.50	Normal	NA	NA

[k] Estimated average intake (0.42 mg/1,000 kcal × 1,745 kcal/d).

[l] Normal erythrocyte transketolase activity (ETKA) was determined by authors from a regression plot.

[m] Abnormal ETKA = > 15% thiamin pyrophosphate (TPP) effect (no reference values given by authors; judgment of status was based on Brin, 1970).

[n] Normal urinary excretion = > 90 nmol/g (27 µg/g) creatinine (no reference values given by authors; judgment of status was based on Schrijver, 1991).

[o] Intake was determined by 3-d dietary recall.

[p] Abnormal ETKA = ≥ 15% thiamin diphosphate effect.

[q] Abnormal urinary excretion = < 500 nmol/g (150 µg/g) creatinine.

[r] Abnormal erythrocyte thiamin = < 270 nmol/L (8 µg/100 mL) erythrocytes.

[s] Abnormal ETKA = ≥ 25% TPP effect.

[t] Abnormal urinary excretion = < 90 nmol/g (27 µg/g) creatinine.

[u] Marginal urinary excretion = 90–217 nmol/g (27–65 µg/g) creatinine.

[v] Abnormal ETKA and urinary excretion: based on the combination of % TPP effect (14–35%) and urinary excretion (< 90 nmol/g [27 µg/g] creatinine).

[w] Normal clinical signs (assessment based on subjective ratings, clinical examinations, psychological assessment, work performance, and neurophysiological assessment).

[x] Intake was determined by 7-d dietary recall questionnaires.

[y] Abnormal ETKA = > 15% TPP effect.

EAR for Women	**19–30 years**	**0.9 mg/day of thiamin**
	31–50 years	**0.9 mg/day of thiamin**

The RDA for thiamin is set by assuming a coefficient of variation (CV) of 10 percent (see Chapter 1) because information is not available on the standard deviation of the requirement for thiamin; the RDA is defined as equal to the EAR plus twice the CV to cover the needs of 97 to 98 percent of the individuals in the group (therefore, for thiamin the RDA is 120 percent of the EAR).

RDA for Men   19–30 years   1.2 mg/day of thiamin
                               31–50 years   1.2 mg/day of thiamin

RDA for Women   19–30 years   1.1 mg/day of thiamin
                               31–50 years   1.1 mg/day of thiamin

Adults Ages 51 Years and Older

Evidence Considered in Estimating the Average Requirement

Several studies of the thiamin status of the elderly have been conducted but provide little direct information on which to base nutrient requirements. Laboratory indicators of status suggest that a substantial percentage (20 to 30 percent) of the population has values suggestive of deficiency, but reported intake is not correlated with the laboratory results. Nichols and Basu (1994) investigated the relationship between thiamin intake and the thiamin pyrophosphate (TPP) effect in a group of 60 randomly selected, free-living elderly men and women in Alberta, Canada, aged 65 to 74 years. Thiamin intake estimated from three nonconsecutive food records was not significantly correlated with the TPP effect. In this study, only 57 percent of the subjects were described as having adequate thiamin status (TPP effect of less than 14 percent) even though the reported mean thiamin intake was 1.7 ± 0.12 mg (standard error)/day for men and 1.4 ± 0.01 mg/day for women.

O'Rourke and coworkers (1990) reported a lower TPP effect for 10 healthy elderly individuals (aged 70 to 82 years) than for 13 institutionalized elderly (aged 67 to 92 years) who were apparently free of significant gastrointestinal, hepatic, or renal disease. No differences were found in erythrocyte transketolase activity or erythrocyte thiamin content. The investigators did not report on relationships between thiamin intake and the indicators of status.

In a study of 75 elderly Finnish women and men (aged 50 to 94 years), some of whom were institutionalized, slightly over 20 percent were described as having marginal thiamin deficiency defined as greater than 15 percent TPP (Pekkarinen et al., 1974). The mean intake of thiamin was much lower than in the Nichols and Basu (1994) study mentioned above and it varied by group, ranging from 0.46 to 0.79 mg/day in women and from 0.60 to 0.84 mg/day in men. The enzyme activity of the erythrocytes was not significantly correlated with thiamin intake as determined by thiochrome analysis of collected food samples.

Similarly, Hoorn and colleagues (1975) reported that 23 percent of 153 geriatric patients aged 65 to 93 years were deficient in thiamin as determined by a transketolase activation coefficient greater than 1.27. Status became normal in all patients after the administration of 20 mg of thiamin daily for 12 days. No dietary information was provided.

A depletion-repletion study of 10 active, healthy elderly women (aged 52 to 72 years, 9 of whom were 63 years or older) measured urinary thiamin excretion after various thiamin intakes (Oldham, 1962). Eight young women aged 18 to 21 years were also studied. On a thiamin intake of less than 0.40 mg/day, the urinary thiamin excretion decreased more quickly for the older than for the younger women. By day 11 or 12, the thiamin excretion of the older women averaged only 0.153 µmol/day (0.05 µg/day), and they complained of fatigue, headaches, and irritability and canceled social engagements. Comparable results were not seen for the younger women until day 19 or 20. When thiamin intake was increased, the older women's urinary thiamin excretion did not increase as quickly as did that of the younger women. The authors concluded that the thiamin requirement of elderly women is higher than that of young women and that the ratio of thiamin to energy must be higher, but the highest thiamin intake level tested, 0.81 mg/day, showed a very wide range of urinary thiamin excretion, especially after 6 days at this intake.

In a randomized, double-blind treatment trial of either 10 mg of thiamin or placebo (Wilkinson et al., 1997), treated subjects who had low TPP concentrations twice when measured before randomization were reported to experience subjective benefits (improved quality of life) and lower blood pressure and weight; those with only one low TPP value prior to randomization did not benefit from treatment.

Although there are some data to suggest that requirements might be somewhat higher in the elderly than in younger adults (e.g., Oldham, 1962), there is also a concomitant decreased energy utilization that may offset this. Further study of this question needs to be conducted. Thus the EAR is assumed to be the same for elderly and younger adults.

Thiamin EAR and RDA Summary, Ages 51 Years and Older

The EAR for thiamin for adults ages 50 and older is set at the same level as for younger adults—1.0 mg/day for men and 0.9 mg/day for women.

EAR for Men	51–70 years	1.0 mg/day of thiamin
	> 70 years	1.0 mg/day of thiamin
EAR for Women	51–70 years	0.9 mg/day of thiamin
	> 70 years	0.9 mg/day of thiamin

The RDA for thiamin is set by assuming a coefficient of variation (CV) of 10 percent (see Chapter 1) because information is not available on the standard deviation of the requirement for thiamin; the RDA is defined as equal to the EAR plus twice the CV to cover the needs of 97 to 98 percent of the individuals in the group (therefore, for thiamin the RDA is 120 percent of the EAR).

RDA for Men	51–70 years	1.2 mg/day of thiamin
	> 70 years	1.2 mg/day of thiamin
RDA for Women	51–70 years	1.1 mg/day of thiamin
	> 70 years	1.1 mg/day of thiamin

Pregnancy

Method Used to Estimate the Average Requirement

The few studies of the thiamin need of pregnant women focus mainly on single indicators of status, usually without reference to dietary intake. For example, one measurement of transketolase activity was made in each of 556 pregnant German women at various stages of gestation (Heller et al., 1974). The mean activation coefficient was 1.13 whereas that of a reference group of 300 blood donors was 1.05; the cutoff value of normal activation coefficients, derived from data on nonpregnant adults, was 1.20. Twenty-six percent of the women with uncomplicated pregnancies and 21 percent of those with complications had activation coefficients above the cutoff and were classified as abnormal.

Regardless of the nutritional status of the mother, erythrocyte transketolase activity was higher in cord blood than in maternal blood (Tripathy, 1968). Similarly, the free thiamin concentration was higher in cord blood (Slobody et al., 1949). Transketolase activity in cord blood tended to be proportional to that in maternal blood and higher in the blood of pregnant than of nonpregnant women (Tripathy, 1968). In 103 pregnant Malaysian women whose staple diet was rice, 36 percent had a TPP effect greater than 25 percent—a larger percentage than was found for males and nonpregnant women (Chong and Ho, 1970).

Oldham and coworkers (1950) found a very strong correlation ($r = 0.98$) between total thiamin intake and excretion but no consistent decrease in thiamin excretion or in the percentage of a test dose excreted over the course of pregnancy. These investigators compared their results with those from studies of nonpregnant women of comparable ages (Daum et al., 1948; Hathaway and Strom, 1946; Oldham et al., 1946) and found that the pregnant women excreted two to three times as much thiamin as did the nonpregnant women on similar intakes (estimated at less than 1 mg) whereas their excretion of a test dose was similar. In contrast, Toverud (1940) observed no or minimal excretion of thiamin in the urine normally or after a load test in 46 percent of 114 pregnant women. Lockhart and coworkers (1943) reported that approximately three times as much thiamin, as obtained from both supplements and diet, was needed by 16 pregnant women to achieve the urinary excretion peak in the tenth lunar month as was needed by a group of nonpregnant women.

Thiamin EAR and RDA Summary, Pregnancy

For pregnancy the requirement is increased by about 30 percent based on increased growth in maternal and fetal compartments (approximately 20 percent) and a small increase in energy utilization (about 10 percent). This results in an additional requirement for pregnancy of $0.27 \cong 0.3$ mg/day of thiamin. Data from the studies cited above are equivocal about the effects of pregnancy on thiamin requirements and thus are not useful in refining this estimate. Adding 0.3 to the EAR of 0.9 mg for nonpregnant, nonlactating women gives an EAR for the second and third trimesters of pregnancy of 1.2 mg. No adjustment is made for the woman's age.

EAR for Pregnancy	14–18 years	1.2 mg/day of thiamin
	19–30 years	1.2 mg/day of thiamin
	31–50 years	1.2 mg/day of thiamin

The RDA for thiamin is set by assuming a coefficient of variation (CV) of 10 percent (see Chapter 1) because information is not available on the standard deviation of the requirement for thiamin; the RDA is defined as equal to the EAR plus twice the CV to cover the needs of 97 to 98 percent of the individuals in the group (therefore, for thiamin the RDA is 120 percent of the EAR).

RDA for Pregnancy   14–18 years   1.4 mg/day of thiamin
                         19–30 years   1.4 mg/day of thiamin
                         31–50 years   1.4 mg/day of thiamin

Lactation

Method Used to Estimate the Average Requirement

For lactating women it is assumed that 0.16 mg of thiamin is transferred in their milk each day when daily milk production is 0.78 L (see "Ages 0 through 6 Months"). To estimate the average thiamin requirement of lactating women, an additional 0.1 mg of thiamin is added to the EAR (0.9 mg/day) for the nonpregnant, nonlactating woman to cover the energy cost of milk production. Thus, the EAR for thiamin for the lactating woman is

$$0.9 + 0.16 + 0.1 = 1.16 \cong 1.2 \text{ mg/day of thiamin.}$$

Women who are breastfeeding older infants who are eating solid foods might need slightly less thiamin because of a lower volume of milk production.

Thiamin EAR and RDA Summary, Lactation

EAR for Lactation   14–18 years   1.2 mg/day of thiamin
                         19–30 years   1.2 mg/day of thiamin
                         31–50 years   1.2 mg/day of thiamin

The RDA for thiamin is set by assuming a coefficient of variation (CV) of 10 percent (see Chapter 1) because information is not available on the standard deviation of the requirement for thiamin; the RDA is defined as equal to the EAR plus twice the CV to cover the needs of 97 to 98 percent of the individuals in the group (therefore, for thiamin the RDA is 120 percent of the EAR).

RDA for Lactation   14–18 years   1.4 mg/day of thiamin
                         19–30 years   1.4 mg/day of thiamin
                         31–50 years   1.4 mg/day of thiamin

Special Considerations

Persons who may have increased needs for thiamin include those being treated with hemodialysis or peritoneal dialysis, individuals with malabsorption syndrome, women carrying more than one fetus, and lactating women who are nursing more than one infant.

INTAKE OF THIAMIN

Food Sources

Data obtained from the 1995 Continuing Survey of Food Intakes by Individuals indicate that the greatest contribution to thiamin intake of the U.S. adult population comes from the following enriched, fortified, or whole-grain products: bread and bread products, mixed foods whose main ingredient is grain, and ready-to-eat cereals (Table 4-3). Small differences are seen in the contributions of various foods to the overall thiamin intake of men and women. Other sources include pork and ham products and cereals and meat substitutes fortified with vitamins.

Dietary Intake

Data from nationally representative surveys during the past decade (Appendixes G and H) indicate that the median daily intake of thiamin in the United States by young men was approximately 2 mg and the median intake by young women was approximately 1.2 mg daily. For all life stage and gender groups except lactating females, fewer than 5 percent of the individuals had intakes that were lower than the Estimated Average Requirement (EAR). Five to 10 percent of lactating females had intakes lower than the EAR. Results from Canadian surveys indicate that thiamin intakes in two Canadian provinces were slightly lower than U.S. intakes for both men and women (Appendix I).

The Boston Nutritional Status Survey (Appendix F) indicates that this relatively advantaged group of people over age 60 had a median thiamin intake of 1.4 mg/day for men and 1.1 mg/day for women.

Intake from Supplements

Information from the Boston Nutritional Status Survey conducted on the use of thiamin supplements by a free-living elderly population is given in Appendix F. For those taking supplements, the fifti-

TABLE 4-3 Food Groups Providing Thiamin in the Diets of U.S. Men and Women Aged 19 Years and Older, CSFII, 1995[a]

Food Group	Contribution to Total Thiamin Intake[b] (%)		Foods Within the Group that Provide at Least 0.3 mg of Thiamin[c] per Serving	
	Men	Women	0.3–0.6 mg	> 0.6 mg
Food groups providing at least 5% of total thiamin intake				
Bread and bread products	17.1	17.7	—	—
Mixed foods, main ingredient is grain	9.6	8.1	NA[d]	NA
Ready-to-eat cereals	9.3	11.8	Moderately fortified	Highly fortified
Mixed foods[e]	9.1	6.5	NA	NA
Pasta, rice, and cooked cereals	6.7	7.2	Egg noodles, spinach noodles	Fortified oatmeal
Processed meats[f]	5.8	4.1	Pork sausage	—
Pork	5.6	4.9	—	Pork and ham
Thiamin from other food groups				
Finfish	0.9	1.5	Pompano, fresh tuna, catfish, and trout	—
Soy-based supplements and meal replacements	0.7	0.2	Soy milk	Soy-based meat substitutes
Seeds	0.1	0.3	Sunflower seeds	—

NOTE: Most of the grain products are enriched, whole grain, or fortified.

[a] CSFII = Continuing Survey of Food Intakes by Individuals.

[b] Contribution to total intake reflects both the concentration of the nutrient in the food and the amount of the food consumed. It refers to the percentage contribution to the American diet for both men and women, based on 1995 CSFII data.

[c] 0.3 mg = 20% of the Recommended Daily Intake (1.5 mg) of thiamin—a value set by the Food and Drug Administration.

[d] NA = not applicable. Mixed foods were not considered for this table.

[e] Includes sandwiches and other foods with meat, poultry, or fish as the main ingredient.

[f] Includes frankfurters, sausages, lunch meats, and meat spreads.

SOURCE: Unpublished data from the Food Surveys Research Group, Agricultural Research Service, U.S. Department of Agriculture, 1997.

eth percentile of supplemental thiamin intake was 2.4 mg for men and 3.2 mg for women. Approximately 27 percent of adults surveyed took a thiamin-containing supplement in 1986 (Moss et al., 1989).

TOLERABLE UPPER INTAKE LEVELS

Hazard Identification

Adverse Effects

There are no reports available of adverse effects from consumption of excess thiamin by ingestion of food and supplements. Because the data are inadequate for a quantitative risk assessment, no Tolerable Upper Intake Level (UL) can be derived for thiamin. Supplements that contain up to 50 mg/day of thiamin are widely available without prescription, but the possible occurrence of adverse effects resulting from this level or more of intake appears not to have been studied systematically. The limited evidence of adverse effects after large intakes of thiamin is summarized here.

Anaphylaxis. There have been occasional reports of serious and even fatal responses to the parenteral administration of thiamin (Stephen et al., 1992). The clinical characteristics have strongly suggested an anaphylactic reaction. Symptoms associated with thiamin-induced anaphylaxis include anxiety, pruritus, respiratory distress, nausea, abdominal pain, and shock, sometimes progressing to death (Laws, 1941; Leitner, 1943; Reingold and Webb, 1946; Schiff, 1941; Stein and Morgenstern, 1944; Stiles, 1941).

Allergic Sensitivity and Pruritus. Royer-Morrot and colleagues (1992) reported one case of pruritus after an intake of 500 mg/day of thiamin intramuscularly. Another study (Wrenn et al., 1989), which involved intravenous administration of 100 mg of thiamin hydrochloride to 989 patients, reported a burning effect at the injection site in 11 patients and pruritus in 1 patient. No reports of pruritus after thiamin ingestion were found. Because pruritus was only observed with parenteral administration and at a dosage well above the maximum that can be absorbed, it is irrelevant for setting a UL. The finding of Wrenn and coworkers (1989) supports the conclusion that even intravenous administration of high doses of thiamin is relatively safe.

The apparent lack of toxicity of supplemental thiamin may be explained by the rapid decline in absorption that occurs at intakes above 5 mg (Hayes and Hegsted, 1973; SCOGS/LSRO, 1978) and the rapid urinary excretion of thiamin (Davis et al., 1984; McAlpine and Hills, 1941; Najjar and Holt, 1940).

Dose-Response Assessment

In the absence of known toxic effects by ingestion, a lowest-observed-adverse-effect level (LOAEL) and an associated no-observed-adverse-effect level (NOAEL) cannot be determined. Supplements that contain up to 50 mg/day of thiamin are widely available without prescription, but effects of this level or more of intake do not appear to have been studied systematically.

Intake Assessment

Although no UL can be set for thiamin, an exposure assessment is provided here for possible future use. Based on data from the Third National Health and Nutrition Examination Survey, the highest mean intake of thiamin from diet and supplements for any life stage or gender group was reported for men aged 31 through 50 years: 6.7 mg/day. The highest reported intake at the ninety-fifth percentile was 11.0 mg/day in women aged 51 years and older (see Appendix H).

Risk Characterization

Although no adverse effects have been associated with excess intake of thiamin from food or supplements, this does not mean that there is no potential for adverse effects resulting from high intakes. Because data on the adverse effects of thiamin intake are extremely limited, caution may be warranted.

RESEARCH RECOMMENDATIONS FOR THIAMIN

Priority should be given to studies useful for setting Estimated Average Requirements (EARs) for thiamin for children, adolescents, pregnant and lactating women, and the elderly. Future studies should be designed around the EAR paradigm, use graded levels of thiamin intake with clearly defined cutoff values for clinical adequacy and inadequacy, and be conducted for a sufficient duration. To do this, close attention should be given to the identification of indicators on which to base thiamin requirements.

If studies are designed to test high doses of thiamin for possible beneficial effects, the design should also provide for the careful investigation of possible adverse effects.

REFERENCES

Anderson SH, Charles TJ, Nicol AD. 1985. Thiamine deficiency at a district general hospital: Report of five cases. *Q J Med* 55:15–32.

Anderson SH, Vickery CA, Nicol AD. 1986. Adult thiamine requirements and the continuing need to fortify processed cereals. *Lancet* 2:85–89.

Ariaey-Nejad MR, Balaghi M, Baker EM, Sauberlich HE. 1970. Thiamin metabolism in man. *Am J Clin Nutr* 23:764–778.

Bailey AL, Finglas PM, Wright AJ, Southon S. 1994. Thiamin intake, erythrocyte transketolase (EC 2.2.1.1) activity and total erythrocyte thiamin in adolescents. *Br J Nutr* 72:111–125.

Baines M, Davies G. 1988. The evaluation of erythrocyte thiamin diphosphate as an indicator of thiamin status in man, and its comparison with erythrocyte transketolase activity measurements. *Ann Clin Biochem* 25:698–705.

Bamji MS. 1970. Transketolase activity and urinary excretion of thiamin in the assessment of thiamin-nutrition status of Indians. *Am J Clin Nutr* 23:52–58.

Bayliss RM, Brookes R, McCulloch J, Kuyl JM, Metz J. 1984. Urinary thiamine excretion after oral physiological doses of the vitamin. *Int J Vitam Nutr Res* 54:161–164.

Brin M. 1962. Erythrocyte transketolase in early thiamine deficiency. *Ann NY Acad Sci* 98:528–541.

Brin M. 1964. Erythrocyte as a biopsy tissue for functional evaluation of thiamine adequacy. *J Am Med Assoc* 187:762–766.

Brin M. 1970. Transketolase (sedoheptulose-7-phosphate: D-glyceral-dehyde-3-phosphate dihydroxyacetonetransferase, EC 2.2.1.1) and the TPP effect in assessing thiamine adequacy. In: McCormick DB, Wright LD, eds. *Methods in Enzymology,* Vol. 18, Part A. London: Academic Press. Pp. 125–133.

Bueding E, Stein MH, Wortis H. 1941. Blood pyruvate curves following glucose ingestion in normal and thiamine-deficient subjects. *J Biol Chem* 140:697–703.

Burgess RC. 1946. Deficiency diseases in prisoners-of-war at Changi, Singapore, February 1942 to August 1945. *Lancet* 2:411–418.

Chong YH, Ho GS. 1970. Erythrocyte transketolase activity. *Am J Clin Nutr* 23:261–266.

Committee on Nutrition. 1985. Composition of human milk: Normative data. In: *Pediatric Nutrition Handbook,* 2nd ed. Elk Grove Village, IL: American Academy of Pediatrics. Pp. 363–368.

Daum K, Tuttle WW, Wilson M, Rhoads H. 1948. Influence of various levels of thiamine intake on physiologic response. 2. Urinary excretion of thiamine. *J Am Diet Assoc* 24:1049.

Davis RE, Icke GC, Thom J, Riley WJ. 1984. Intestinal absorption of thiamin in man compared with folate and pyridoxal and its subsequent urinary excretion. *J Nutr Sci Vitaminol (Tokyo)* 30:475–482.

Dick EC, Chen SD, Bert M, Smith JM. 1958. Thiamine requirement of eight adolescent boys, as estimated from urinary thiamine excretion. *J Nutr* 66:173–188.

Elsom KO, Reinhold JG, Nicholson JT, Chornock C. 1942. Studies of the B vitamins in the human subject. 5. The normal requirement for thiamine; some factors influencing its utilization and excretion. *Am J Med Sci* 203:569–577.

Fogelholm M, Rehunen S, Gref CG, Laakso JT, Lehto J, Ruokonen I, Himberg JJ. 1992. Dietary intake and thiamin, iron, and zinc status in elite Nordic skiers during different training periods. *Int J Sport Nutr* 2:351–365.

Fogelholm M, Ruokonen I, Laakso JT, Vuorimaa T, Himberg JJ. 1993. Lack of association between indices of vitamin B_1, B_2, and B_6 status and exercise-induced blood lactate in young adults. *Int J Sport Nutr* 3:165–176.

Foltz EE, Barborka CJ, Ivy AC. 1944. The level of vitamin B-complex in the diet at which detectable symptoms of deficiency occur in man. *Gastroenterology* 2:323–344.

Gans DA, Harper AE. 1991. Thiamin status of incarcerated and nonincarcerated adolescent males: Dietary intake and thiamin pyrophosphate response. *Am J Clin Nutr* 53:1471–1475.

Hart M, Reynolds MS. 1957. Thiamine requirement of adolescent girls. *J Home Econ* 49:35–37.

Hathaway ML, Strom JE. 1946. A comparison of thiamine synthesis and excretion in human subjects on synthetic and natural diets. *J Nutr* 32:1.

Hayes KC, Hegsted DM. 1973. Toxicity of the vitamins. In: *Toxicants Occurring Naturally in Foods*. Washington, DC: National Academy Press. Pp. 235–253.

Heller S, Salkeld RM, Korner WF. 1974. Vitamin B_1 status in pregnancy. *Am J Clin Nutr* 27:1221–1224.

Henshaw JL, Noakes G, Morris SO, Bennion M, Gubler CJ. 1970. Method for evaluating thiamine adequacy in college women. *J Am Diet Assoc* 57:436–441.

Hoorn RK, Flikweert JP, Westerink D. 1975. Vitamin B_1, B_2 and B_6 deficiencies in geriatric patients, measured by coenzyme stimulation of enzyme activities. *Clin Chim Acta* 61:151–162.

Horwitt MK, Liebert E, Kreisler O, Wittman P. 1948. *Investigations of Human Requirements for B-Complex Vitamins*. Bulletin of the National Research Council No. 116. Report of the Committee on Nutritional Aspects of Ageing, Food and Nutrition Board, Division of Biology and Agriculture. Washington, DC: National Academy of Sciences.

Hytten FE, Thomason AM. 1961. Nutrition of the lactating women. In: Kon SK, Cowie AT, eds. *Milk: The Mammary Gland and Its Secretion*. New York: Academic Press. Pp. 3–46.

Inouye K, Katsura E. 1965. Etiology and pathology of beriberi. In: Shimazono N, Katsura E, eds. *Review of Japanese Literature on Beriberi and Thiamine*. Igaku Shoin, Tokyo: Vitamin B Research Committee of Japan. Pp. 1–28.

Kraut H, Wildemann L, Böhm M. 1966. Human thiamine requirements. *Int Z Vitaminforsch* 36:157–193.

Laws CL. 1941. Sensitization to thiamine hydrochloride. *J Am Med Assoc* 117:146.

Leitner ZA. 1943. Untoward effects of vitamin B_1. *Lancet* 2:474–475.

Levy G, Hewitt RR. 1971. Evidence in man for different specialized intestinal transport mechanisms for riboflavin and thiamin. *Am J Clin Nutr* 24:401–404.

Lockhart HS, Kirkwood S, Harris RS. 1943. The effect of pregnancy and puerperium on the thiamine status of women. *Am J Obstet Gynecol* 46:358–365.

Lonsdale D, Shamberger RJ. 1980. Red cell transketolase as an indicator of nutritional deficiency. *Am J Clin Nutr* 33:205–211.

McAlpine D, Hills GM. 1941. The clinical value of the thiochrome test for aneurin (vitamin B_1) in urine. *Q J Med* 34:31–39.

McCormick DB, Greene HL. 1994. Vitamins. In: Burtis CA, Ashwood ER, eds. *Tietz Textbook of Clinical Chemistry*. Philadelphia: Saunders. Pp. 1275–1316.

Montalto MB, Benson JD, Martinez GA. 1985. Nutrient intake of formula-fed infants and infants fed cow's milk. *Pediatrics* 75:343–351.

Morrison AB, Campbell JA. 1960. Vitamin absorption studies. 1. Factors influencing the excretion of oral test doses of thiamine and riboflavin by human subjects. *J Nutr* 72:435–440.

Moss AJ, Levy AS, Kim I, Park YK. 1989. *Use of Vitamin and Mineral Supplements in the United States: Current Users, Types of Products, and Nutrients*. Advance Data, Vital and Health Statistics of the National Center for Health Statistics, No. 174. Hyattsville, MD: National Center for Health Statistics.

Nail PA, Thomas MR, Eakin R. 1980. The effect of thiamin and riboflavin supplementation on the level of those vitamins in human breast milk and urine. *Am J Clin Nutr* 33:198–204.

Najjar VA, Holt LE Jr. 1940. Studies in thiamin excretion. *Bull Johns Hopkins Hosp* 67:107–124.

Nichols HK, Basu TK. 1994. Thiamin status of the elderly: Dietary intake and thiamin pyrophosphate response. *J Am Coll Nutr* 13:57–61.

O'Rourke NP, Bunker VW, Thomas AJ, Finglas PM, Bailey AL, Clayton BE. 1990. Thiamine status of healthy and institutionalized elderly subjects: Analysis of dietary intake and biochemical indices. *Age Ageing* 19:325–329.

Oldham H. 1962. Thiamine requirements of women. *Ann NY Acad Sci* 98:542–549.

Oldham HG, Davis MV, Roberts LJ. 1946. Thiamine excretions and blood levels of young women on diets containing varying levels of the B vitamins, with some observations on niacin and pantothenic acid. *J Nutr* 32:163–180.

Oldham H, Sheft BB, Porter T. 1950. Thiamine and riboflavin intakes and excretions during pregnancy. *J Nutr* 41:231–245.

Pekkarinen M, Koivula L, Rissanen A. 1974. Thiamine intake and evaluation of thiamine status among aged people in Finland. *Int J Vitam Nutr Res* 44:435–442.

Platt BS. 1967. Thiamine deficiency in human beriberi and in Wernicke's encephalopathy. In: Wolstenholme GEW, O'Connor M, eds. *Thiamine Deficiency: Biochemical Lesions and their Clinical Significance*. Ciba Foundation Study Group No. 28. London: Churchill Livingstone. Pp. 135–143.

Pratt JB, Hamil BM. 1951. Metabolism of women during the reproductive cycle. 18. The effect of multivitamin supplements on the secretion of B vitamins in human milk. *J Nutr* 44:141–157.

Reingold IM, Webb FR. 1946. Sudden death following intravenous administration of thiamine hydrochloride. *J Am Med Assoc* 130:491–492.

Reuter H, Gassmann B, Erhardt V. 1967. Contribution to the question of the human thiamine requirement. *Int Z Vitaminforsch* 37:315–328.

Royer-Morrot MJ, Zhiri A, Paille F, Royer RJ. 1992. Plasma thiamine concentrations after intramuscular and oral multiple dosage regimens in healthy men. *Eur J Clin Pharmacol* 42:219–222.

Sauberlich HE, Herman YF, Stevens CO, Herman RH. 1979. Thiamin requirement of the adult human. *Am J Clin Nutr* 32:2237–2248.

Schiff L. 1941. Collapse following parenteral administration of solution of thiamine hydrochloride. *J Am Med Assoc* 117:609.

Schrijver J. 1991. Biochemical markers for micronutrient status and their interpretation. In: Pietrzik K, ed. *Modern Lifestyles, Lower Energy Intake and Micronutrient Status*. London: Springer-Verlag. Pp. 55–85.

SCOGS/LSRO (Select Committee on GRAS Substances, Life Sciences Research Office). 1978. *Evaluation of the Health Aspects of Thiamin Hydrochloride and Thiamin Mononitrate as Food Ingredients.* Bethesda, MD: LSRO/FASEB.

Singleton CK, Pekovich SR, McCool BA, Martin, PR. 1995. The thiamine-dependent hysteretic behavior of human transketolase: Implications for thiamine deficiency. *J Nutr* 125:189–194.

Slobody LB, Willner MM, Mestern J. 1949. Comparison of vitamin B_1 levels in mothers and their newborn infants. *Am J Dis Child* 77:736.

Stein W, Morgenstern M. 1944. Sensitization to thiamine hydrochloride: Report of a case. *Ann Intern Med* 70:826–828.

Stephen JM, Grant R, Yeh CS. 1992. Anaphylaxis from administration of intravenous thiamine. *Am J Emerg Med* 10:61–63.

Stiles MH. 1941. Hypersensitivity to thiamine chloride, with a note on sensitivity to pyridoxine hydrochloride. *J Allergy* 12:507–509.

Toverud KU. 1940. The excretion of aneurin in pregnant and lactating women and infants. *Z Vitaminforsch* 10:255–267.

Tripathy K. 1968. Erythrocyte transketolase activity and thiamine transfer across human placenta. *Am J Clin Nutr* 21:739–742.

van der Beek EJ, van Dokkum W, Wedel M, Schrijver J, van den Berg H. 1994. Thiamin, riboflavin and vitamin B_6: Impact of restricted intake on physical performance in man. *J Am Coll Nutr* 13:629–640.

Wilkinson TJ, Hanger HC, Elmslie J, George PM, Sainsbury R. 1997. The response to treatment of subclinical thiamine deficiency in the elderly. *Am J Clin Nutr* 66:925–928.

Williams RD, Mason HL, Smith BF, Wilder RM. 1942. Induced thiamin (vitamin B_1) deficiency and the thiamine requirement of man: Further observations. *Arch Intern Med* 69:721–738.

Williams RD, Mason HL, Wilder RM. 1943. The minimum daily requirement of thiamine in man. *J Nutr* 25:71–97.

Wilson JA. 1983. Disorders of vitamins: Deficiency, excess and errors of metabolism. In: Petersdorf RG, Harrison TR, eds. *Harrison's Principles of Internal Medicine,* 10th ed. New York: McGraw-Hill. Pp. 461–470.

Wood B, Gijsbers A, Goode A, Davis S, Mulholland J, Breen K. 1980. A study of partial thiamin restriction in human volunteers. *Am J Clin Nutr* 33:848–861.

Wrenn KD, Murphy F, Slovis CM. 1989. A toxicity study of parenteral thiamine hydrochloride. *Ann Emerg Med* 18:867–870.

Wyatt DT, Nelson D, Hillman RE. 1991. Age-dependent changes in thiamin concentrations in whole blood and cerebrospinal fluid in infants and children. *Am J Clin Nutr* 53:530–536.

Ziporin ZZ, Nunes WT, Powell RC, Waring PP, Sauberlich HE. 1965. Thiamine requirement in the adult human as measured by urinary excretion of thiamine metabolites. *J Nutr* 85:297–304.

5

Riboflavin

SUMMARY

Riboflavin functions as a coenzyme in numerous redox reactions. A combination of criteria is used to estimate the Recommended Dietary Allowance (RDA) for riboflavin, including the erythrocyte glutathione reductase activity coefficient and urinary riboflavin excretion. The RDA for riboflavin for adults is 1.3 mg/day for men and 1.1 mg/day for women. Recently, the median intake of riboflavin from food in the United States and two Canadian populations was estimated to be approximately 2 mg/day for men and 1.5 mg/day for women. The ninety-fifth percentile of U.S. intake from both food and supplements ranged from 4 to 10 mg/day. The evidence on adverse effects is not sufficient to set a Tolerable Upper Intake Level (UL) for riboflavin.

BACKGROUND INFORMATION

Subsequent to the discovery of thiamin was the discovery of a more heat-stable factor that was named vitamin B_2, or riboflavin. Riboflavin is a water-soluble, yellow, fluorescent compound. The primary form of the vitamin is as an integral component of the coenzymes flavin mononucleotide (FMN) and flavin-adenine dinucleotide (FAD) (McCormick, 1994; McCormick and Greene, 1994; Merrill et al., 1981). It is in these bound coenzyme forms that riboflavin functions as a catalyst for redox reactions in numerous

87

metabolic pathways and in energy production (McCormick and Greene, 1994).

Function

The redox reactions in which flavocoenzymes participate include flavoprotein-catalyzed dehydrogenations that are both pyridine nucleotide (niacin) dependent and independent, reactions with sulfur-containing compounds, hydroxylations, oxidative decarboxylations (involving thiamin as its pyrophosphate), dioxygenations, and reduction of oxygen to hydrogen peroxide (McCormick and Greene, 1994). There are obligatory roles of flavocoenzymes in the formation of some vitamins and their coenzymes. For example, the biosynthesis of two niacin-containing coenzymes from tryptophan occurs via FAD-dependent kynurenine hydroxylase, an FMN-dependent oxidase catalyzes the conversion of the 5'-phosphates of vitamin B_6 to coenzymic pyridoxal 5'-phosphate, and an FAD-dependent dehydrogenase reduces 5,10-methylene-tetrahydrofolate to the 5'-methyl product that interfaces with the B_{12}-dependent formation of methionine from homocysteine and thus with sulfur amino acid metabolism.

Physiology of Absorption, Metabolism, and Excretion

Absorption

Most dietary riboflavin is consumed as a complex of food protein with FMN and FAD (Merrill et al., 1981; Nichoalds, 1981). In the stomach, gastric acidification releases most of the coenzyme forms of riboflavin (FAD and FMN) from the protein. The noncovalently bound coenzymes are then hydrolyzed to riboflavin by nonspecific pyrophosphatases and phosphatases in the upper gut (McCormick, 1994; Merrill et al., 1981). Primary absorption of riboflavin occurs in the proximal small intestine via a rapid, saturable transport system (McCormick, 1994; Merrill et al., 1981). The rate of absorption is proportional to intake, and it increases when riboflavin is ingested along with other foods (Jusko and Levy, 1967, 1975) and in the presence of bile salts (Jusko and Levy, 1975; Mayersohn et al., 1969). A small amount of riboflavin circulates via the enterohepatic system (McCormick, 1994).

At low intake levels most absorption of riboflavin is via an active or facilitated transport system. Although older studies in animals (Daniel et al., 1983; Meinen et al., 1977; Rivier, 1973) suggested that this transport may depend on sodium ions, more recent work in humans (Said and Ma, 1994) indicates that uptake is independent of sodium ions. A small amount of riboflavin is absorbed in the large intestine (Sorrell et al., 1971).

In plasma some riboflavin is complexed with albumin; however, a large portion of riboflavin associates with other proteins, mainly immunoglobulins, for transport (Innis et al., 1985). Pregnancy increases the level of carrier proteins available for riboflavin (Natraj et al., 1988). This results in a higher rate of riboflavin uptake at the maternal surface of the placenta (Dancis et al., 1988).

At physiological concentrations the uptake of riboflavin into the cells of organs such as the liver is facilitated and may require specific carriers. At higher levels of intake, riboflavin can be absorbed by diffusion (Bowman et al., 1989; McCormick, 1989).

Metabolism

The metabolism of riboflavin is a tightly controlled process that depends on the riboflavin status of the individual (Lee and McCormick, 1983). Riboflavin is converted to coenzymes within the cellular cytoplasm of most tissues but mainly in the small intestine, liver, heart, and kidney (Brown, 1990; Darby, 1981). The metabolism of riboflavin begins with the adenosine triphosphate (ATP)-dependent phosphorylation of the vitamin to FMN. Flavokinase, the catalyst for this conversion, is under hormonal control. FMN can then be complexed with specific apoenzymes to form a variety of flavoproteins; however, most is converted to FAD by FAD synthetase. As a result, FAD is the predominant flavocoenzyme in body tissues (McCormick and Greene, 1994). Production of FAD is controlled by product inhibition such that an excess of FAD inhibits its further production (Yamada et al., 1990).

Excretion

When riboflavin is absorbed in excess, very little is stored in the body tissues. The excess is excreted, primarily in the urine. A wide variety of flavin-related products have been identified in the urine of humans. In healthy adults consuming well-balanced diets, riboflavin accounts for 60 to 70 percent of the excreted urinary flavins (McCormick, 1989). Urinary excretion of riboflavin varies with intake, metabolic events, and age (McCormick, 1994). In newborns, urinary excretion is slow (Jusko and Levy, 1975; Jusko et al., 1970);

however, the cumulative amount excreted is similar to the amount excreted by older infants.

Clinical Effects of Inadequate Intake

Clinical Signs of Deficiency

The signs of riboflavin deficiency (ariboflavinosis) in humans are sore throat; hyperemia and edema of the pharyngeal and oral mucous membranes; cheilosis; angular stomatitis; glossitis (magenta tongue); seborrheic dermatitis; and normochromic, normocytic anemia associated with pure erythrocyte cytoplasia of the bone marrow (Wilson, 1983). Riboflavin deficiency is most often accompanied by other nutrient deficiencies. Severe riboflavin deficiency may impair the metabolism of vitamin B_6 by limiting the amount of FMN required by pyridoxine (pyridoxamine) 5-phosphate oxidase and the conversion of tryptophan to functional forms of niacin (McCormick, 1989).

Prevalence of Deficiency

Riboflavin deficiency has been documented in industrialized and developing nations and across various demographic groups (Komindr and Nichoalds, 1980; Nichoalds, 1981). Diseases such as cancer (Rivlin, 1975), cardiac disease (Steier et al., 1976), and diabetes mellitus (Cole et al., 1976; Prager et al., 1958) are known to precipitate or exacerbate riboflavin deficiency.

SELECTION OF INDICATORS FOR ESTIMATING THE REQUIREMENT FOR RIBOFLAVIN

Several indicators have been used to estimate the adequacy of riboflavin status in humans (McCormick, 1994; McCormick and Greene, 1994). Principal among them are erythrocyte glutathione reductase; erythrocyte flavin concentration; and urinary excretion of the vitamin in fasting, random, or 24-hour specimens or by load tests.

Erythrocyte Glutathione Reductase

Currently, one of the most commonly used methods for assessing riboflavin status involves the determination of erythrocyte glutathione reductase (EGR) activity, as described by Sauberlich and coworkers

(1972). The EGR value is an enzymatic and hence functional indicator that is conventionally determined with and without the addition of flavin-adenine dinucleotide (FAD)—the coenzyme required for the activity of EGR. Results are expressed as an activity coefficient (EGRAC), which is the ratio of activities in the presence of added FAD and without its addition. An EGRAC ratio of 1.0 indicates no stimulation by FAD and the presence of holoenzyme only, which means that more than adequate amounts of FAD (and riboflavin) were present in the original erythrocytes. Suggested guidelines for interpreting such coefficients are as follows: less than 1.2, acceptable; 1.2 to 1.4, low; and greater than 1.4, deficient (McCormick and Greene, 1994). However, many different cutoff values have been used by investigators. An upper limit of normality has been set at 1.34 based on the mean plus 2 standard deviations of the EGRAC value of several hundred apparently healthy elderly individuals aged 60 years and older (Sadowski, 1992).

Because FAD is a labile compound, the EGRAC must be obtained by using fresh erythrocytes that are washed, lysed, and measured promptly for enzymatic activity (McCormick and Greene, 1994). Because the glutathione reductase in the erythrocytes of individuals with glucose 6-phosphate dehydrogenase deficiency has an increased avidity for FAD (Nichoalds, 1981), this test is not valid in individuals with that condition. The dehydrogenase deficiency has been estimated to occur in 10 percent of black Americans (Frischer et al., 1973).

Erythrocyte Flavin

Erythrocyte flavin has been used as an indicator of the cellular concentration of the vitamin in its coenzyme forms because these coenzymes comprise over 90 percent of flavin (Burch et al., 1948). Because of the instability of the predominant FAD, which is rapidly hydrolyzed enzymatically when cells rupture, erythrocyte flavins are deliberately hydrolyzed and measured either microbiologically or fluorometrically as riboflavin. Values greater than 400 nmol/L (15 µg/100 mL) of cells are considered adequate (based on other observations and measurements) and values below 270 nmol/L (10 µg/100 mL) reflect deficiency (e.g., see the study by Ramsay et al. [1983] on the correlation between cord blood and maternal erythrocyte riboflavin deficiencies). Because the margin of difference between adequacy and inadequacy is rather small, there is some concern about the sensitivity and interpretation of results. After mild hydrolysis to convert FAD to the more stable flavin mononucleotide (FMN), high-performance liquid chromatography (HPLC) separation leads to a more exact determination of FMN plus traces of riboflavin. This is a useful indicator that reflects the functional, cellularly trapped forms of riboflavin.

Urinary Flavin

Urinary riboflavin has often been used in metabolic studies to estimate the riboflavin requirement. It can be measured by fluorometric HPLC methods (Chastain and McCormick, 1987; Roughhead and McCormick, 1991) as well as by microbiological procedures. Without chromatographic separation by HPLC or other means, the fluorometric assessment of the vitamin may lead to incorrectly high values, as was the case in early studies. In some cases, flavin catabolites found in food (such as the 10-formylmethyl- and 2'-hydroxyethyl-flavins found in milk) are a significant fraction of flavin catabolites that comprise as much as one-third of the total assay of urinary flavin. Such dietary flavin catabolites can also interfere with the response of bacteria or even rats used for assay of riboflavin. Additionally, in some cases flavin catabolites such as these can depress the cellular uptake of riboflavin (Aw et al., 1983) or its conversion to coenzymes (McCormick, 1962).

As HPLC techniques evolved to allow easier separation of riboflavin from other fluorescent flavin catabolites (such as 7- and 8-hydroxymethylriboflavins [McCormick, 1994]), specific quantitations of the vitamin (which contributes more than two-thirds of the total flavin) were found generally useful to relate the recent dietary intake to urinary output. Under conditions of sufficiency (that is, an average riboflavin intake of approximately 1.5 mg/day), the amount of riboflavin excreted per day exceeded 319 nmol (120 µg) riboflavin (Roughead and McCormick, 1991). The amount of riboflavin excreted per gram of creatinine was greater for children than adults.

For adults, a low urinary concentration of riboflavin is considered to be 50 to 72 nmol/g (19 to 27 µg/g) creatinine and a deficient concentration to be below 50 nmol/g (19 µg/g) creatinine. Sauberlich and colleagues (1974), reviewing the literature based on fluorometric methods without HPLC, suggested that riboflavin values less than 72 nmol/g (27 µg/g) creatinine be considered deficient, 72 to 210 nmol/g (27 to 79 µg/g) creatinine be considered low, and greater than 213 nmol/g (80 µg/g) creatinine be considered acceptable. Compared with the fluorometric method alone, the HPLC method with fluorometry tends to give lower values because riboflavin is separated from other flavins (Smith, 1980).

Load tests may be used to gauge the degree to which the body is saturated with riboflavin; the result generally agrees with tests using other indicators. Subcutaneous administration of 1 mg of riboflavin followed by assessment of urinary flavin output for a 4-hour period was found by Horwitt et al. (1950) to correspond well to riboflavin excretion over 24 hours. A break point for increased urinary excretion of riboflavin occurred with or without the load when adult men received more than 1.1 mg/day of dietary riboflavin (see Figure 5-1). Above this level, there is a sharp linear increase in the slope of urinary excretion for riboflavin intakes up to 2.5 mg/day (Sauberlich

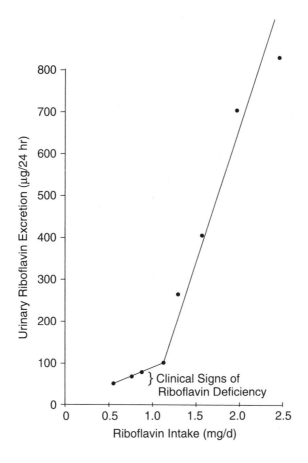

FIGURE 5-1 Relationship of riboflavin intake to urinary excretion of riboflavin as observed in the studies of Horwitt (1972) and Horwitt et al. (1950). Reprinted with permission, from Sauberlich et al. (1974). Copyright 1994 by CRC Press.

et al., 1974). Adapting the work of Lossy et al. (1951), Sauberlich and colleagues (1974) suggested a reference value of 1.4 mg or more for the normal 4-hour urinary excretion of riboflavin after a 5-mg load.

The change in slope of urinary excretion of riboflavin after a load test is an especially useful status assessment. Caution in interpretation is needed because the size of the test dose, method of administration, and method of calculating amount recovered have varied among studies. Moreover, the break point may reflect not only tissue saturation but also renal threshold and solubility (compartment) effects (Sauberlich et al., 1974). Urinary riboflavin has been shown to increase under conditions causing negative nitrogen balance and with the administration of antibiotics and certain psychotropic drugs (e.g., phenothiazine) (McCormick, 1994).

Indicators of Carbohydrate Metabolism

In an early study (Horwitt et al., 1949), indices of carbohydrate metabolism (lactic and pyruvic acid concentrations) were measured in riboflavin-depleted subjects after a short period of exercise. Because no changes were observed, these do not appear to be promising indicators of riboflavin status.

Possible Reduction of Chronic Disease Risk

Riboflavin status (low as assessed by EGRAC) has been related to certain site-specific cancers (e.g., esophageal) in areas of China (Merrill et al., 1991). However, randomized nutrition intervention trials in Linxian, China, indicated that a riboflavin and niacin combination, given for about 5 years, did not reduce total or cancer mortality (Blot et al., 1995).

Although lens opacities in humans have been associated with high glutathione reductase activity (with FAD) (Leske et al., 1995), evidence is insufficient for considering the use of risk of cataract as the basis for setting the Estimated Average Requirement (EAR).

Concurrent Analyses

Overall, greatest credence is given to status assessments that use more than one indicator, because the response variables indicate somewhat different aspects of riboflavin status. Many investigators have obtained data concurrently on several indicators of riboflavin status. Sauberlich and colleagues (1972) noted that, for the most part, "subjects with elevated EGR activity coefficients . . . had urinary excretion levels considered low or deficient." Boisvert and colleagues (1993) indicated that the EGRAC method is preferred for the assessment of riboflavin status whereas the urinary excretion method is better for determining riboflavin requirements. Both EGRAC and urinary riboflavin respond more rapidly to dietary riboflavin intake than does erythrocyte riboflavin (Bamji, 1969), but Bates (1987) showed a good relationship between EGRAC and erythrocyte riboflavin expressed as micrograms of riboflavin per gram of hemoglobin.

FACTORS AFFECTING THE RIBOFLAVIN REQUIREMENT

Bioavailability

Overall, a reasonable estimation of bioavailability is approximately 95 percent of food flavin, up to a maximum of about 27 mg absorbed per single meal or dose (Zempleni et al., 1996).

There is considerable diversity of flavins in foods, but over 90 percent of riboflavin is estimated to be in readily digested flavocoenzymes (mainly flavin-adenine dinucleotide [FAD] and to a lesser degree flavin mononucleotide [FMN]), with lesser amounts of the free vitamin and traces of glycosides and esters that are also hydrolyzed during absorption from the gut (McCormick, 1994; Merrill et al., 1981). Perhaps no more than 7 percent of food flavin is found as covalently attached 8α-FAD. In these substituted flavins, the methylene carbon at position 8 of the isoalloxazine ring is bound to heterocyclic atoms (N, S, and O) of proteins that function catalytically (e.g., mitochondrial succinate dehydrogenase and monoamine oxidase). Although some portion of the 8α-(amino acid) riboflavins are released by proteolysis of such flavoproteins, they do not have vitaminic activity (Chia et al., 1978).

Nutrient-Nutrient Interactions

Composition of the Diet

The proportions of fat and carbohydrate in the diet appear to influence the riboflavin requirements of the elderly (Boisvert et al., 1993); when fat was decreased from 31.4 to 20.0 percent of calories and carbohydrate was increased from 57.5 to 68.2 percent of calories, the riboflavin requirement was lower. Thus, a lower ratio of fat to carbohydrate decreased the requirement. This relationship has not been examined in other age groups.

Other B Vitamins

Riboflavin interrelates with other B vitamins, notably niacin, which requires FAD for its formation from tryptophan, and vitamin B_6, which requires FMN for conversion to the coenzyme pyridoxal 5'-phosphate (McCormick, 1989). These interrelationships are not known to affect the requirement for riboflavin.

Energy Intake

No studies were found that examined the effect of energy intake on the riboflavin requirement. Some studies provided riboflavin in graded doses that kept the ratio of riboflavin-to-energy constant for subjects with different energy requirements. Others provided total amounts of riboflavin (and sometimes energy) that were the same for all individuals. Five studies using 2- to 3-day complete urine collection (Belko et al., 1983, 1984, 1985; Soares et al., 1993; Winters et al., 1992) reported that the urinary excretion of riboflavin is decreased when physical activity is increased, suggesting higher utilization of riboflavin with increased energy expenditure.

Despite the lack of experimental data, the known biochemical function of riboflavin in the utilization of energy suggests at least a small (10 percent) adjustment to reflect differences in the average energy utilization and size of men and women, a small increase in the requirement to cover increased energy use during pregnancy, and a small increase to cover the inefficiencies of milk production.

Physical Activity

Riboflavin status measurements seem to be affected by physical activity. Some studies have demonstrated a moderate rise in the erythrocyte glutathione reductase activity coefficient (EGRAC) as well as a decrease in urinary riboflavin excretion with an increase in physical activity (Belko et al., 1983, 1984, 1985; Soares et al., 1993; Winters et al., 1992). For example, approximately 20 percent additional riboflavin was required to normalize EGRAC and urinary flavin values of exercising, weight-reducing women (Belko et al., 1985) and older women undergoing exercise training (Winters et al., 1992). Tucker and colleagues (1960) suggest that the decrease in riboflavin after exercise may be related to reduced renal blood flow, but in the studies cited above, all urine was collected over a period of 48 to 72 hours.

In a group of East Indian men aged 27 to 47 years with a mean EGRAC of 1.53, short periods of increased physical activity while on a diet providing riboflavin at 0.42 mg/1,000 kcal for 16 days led to an increase in EGRAC. When riboflavin status is marginal, increased physical activity may be more likely to lead to further deterioration as assessed by EGRAC, and values may not return to baseline after the extra exercise segment of the study is completed (Soares et al., 1993). However, the energy cost of the exercise and a measure of mechanical efficiency remained stable throughout.

A number of studies have failed to show an improvement in work performance (Powers et al., 1987; Prasad et al., 1990; Tremblay et al., 1984; Weight et al., 1988) and endurance (elderly subjects) (Winters et al., 1992) with riboflavin supplementation even when subjects were described as subclinically deficient (having lower-than-normal ranges of biochemical indices but no clinically observable signs of deficiency).

It is possible that the riboflavin requirement is increased for those who are ordinarily very active physically (e.g., athletes or those who carry heavy packs much of the day), but data are not available on which to quantify the adjustment that should be made.

Other Factors

Although a number of reports indicate that women taking high-dose oral contraceptives have impaired riboflavin status, no difference was seen when dietary riboflavin intake was controlled (Roe et al., 1982).

APPROACHES FOR DERIVING THE ESTIMATED AVERAGE REQUIREMENT

Primary: Maintenance or Restoration of Riboflavin Status by Using Biochemical Indicators

To derive the Estimated Average Requirement (EAR) for adults, more weight was given to experimental studies that included information on response to diets in which the source of riboflavin was food or food plus supplemental riboflavin (Table 5-1). Studies that used more than one indicator were considered most useful, especially if a functional assay (e.g., EGRAC) was conducted along with measurement of erythrocyte or urinary riboflavin. However, the

TABLE 5-1 Metabolic Studies Providing Evidence Used to Derive the Estimated Average Requirement (EAR) for Riboflavin for Adults

Reference	Duration of Study	Baseline Riboflavin Intake	Number of Subjects	Riboflavin Intake During Repletion or Maintenance (mg/d)
Women				
Sebrell et al., 1941	19–36 wk depletion 7–24 wk repletion[b]	0.5 mg/d	10 (institutionalized)	0.5 **1.56–2.05** **2.54–3.68**
Williams et al., 1943	246 d 288 d 288 d 288 d 288 d 264 d	≥ 2 mg/d	2 4 1 3 1 2	0.6 0.76 1.0 1.05 **1.2** **1.76–3.76**
Brewer et al., 1946	12 d at each intake level	NA	14	0.79 **1.04** **1.26** **1.62** **2.23** **2.73**
Davis et al., 1946	8 mo	NA	12	0.29 mg/1,000 kcal[g] 0.49 mg/1,000 kcal **0.66 mg/1,000 kcal** **4.10 mg/1,000 kcal**
Roe et al., 1982	10 wk (includes 2-wk acclimation)	1.69 ± 0.5 mg/d (based on 7-d diet records and diet history)	10	0.6 mg/1,000 kcal[g] 0.8 mg/1,000 kcal 1.0 mg/1,000 kcal
Belko et al., 1983	12 wk	1.45 ± 0.4 mg/d (based on 7-d diet records)	12 2 3 10 11 2 8	0.6 mg/1,000 kcal[g] **0.8 mg/1,000 kcal** **0.8 mg/1,000 kcal[i]** 1.0 mg/1,000 kcal 1.0 mg/1,000 kcal[i] **1.2 mg/1,000 kcal[i]** **1.4 mg/1,000 kcal[i]**

EGRAC[a]	Urinary Excretion of Riboflavin	Erythrocyte Riboflavin	Other
NA[c]	Abnormal[d]	NA	60% abnormal[e]
	Normal		Normal
	Normal		Normal
NA	Abnormal[d]	NA	Normal[f]
	75% abnormal		Normal
	Abnormal		Normal
	33% abnormal		Normal
	Normal		Normal
	Normal		Normal
NA	Abnormal[d]	NA	NA
	Normal		
	Normal		
	Normal		
	Normal		
	Normal		
NA	45% abnormal[d]	NA	NA
	17% abnormal		
	Normal		
	Normal		
50% abnormal[h]	NA	NA	NA
20% abnormal			
10% abnormal			
92% abnormal[j]	NA	NA	NA
100% normal			
100% normal			
60% abnormal			
73% abnormal			
100% normal			
100% normal			

continued

TABLE 5-1 Continued

Reference	Duration of Study	Baseline Riboflavin Intake	Number of Subjects	Riboflavin Intake During Repletion or Maintenance (mg/d)
Kuizon et al., 1992	40 d	Measured but not reported	7 (non-pregnant Filipino population)	0.25 mg/1,000 kcal[g] 0.4 mg/1,000 kcal **0.5 mg/1,000 kcal** **0.6 mg/1,000 kcal**
Women and Men				
Boisvert et al., 1993	16 wk	NA	14 (healthy, elderly Guatemalan population)	0.65 0.7 0.9 1.1 **1.3** **1.5**
Men				
Keys et al., 1944	5 mo		6	**0.99**
Horwitt et al., 1949	9–10 mo 3 mo	1.6 mg/d	39 (mental patients)	0.55 **1.1**
Horwitt et al., 1950	15 wk 12 wk 15 wk 13 wk 100 wk 10–78 wk 2–44 wk 1.5 wk	1.1 mg/d (3 mo prelim. period)	15 11 12 28 39 12 13 13	0.55 0.75 0.85 1.1 **1.6** **2.15** **2.55** **3.55**
Bessey et al., 1956	247 d 16 mo 247 d 247 d 16 mo	NA	8 10 10 23 6	0.5 0.55 1.6 **2.4** **2.55–3.55**

NOTE: Maintenance of body weight was not indicated in Davis et al. (1946), Kuizon et al. (1992), Horwitt et al. (1950), Bessey et al. (1956); body weight was maintained in the other studies. Riboflavin intake was measured analytically except as noted. Intakes given in **bold** supported normal results for all indicators.

[a] EGRAC = Erythrocyte glutathione reductase activity coefficient.

[b] During repletion, various amounts of supplements (1–15 mg/d orally) were given in addition to the basal diet. Two sample intakes for this period are given here.

EGRAC[a]	Urinary Excretion of Riboflavin	Erythrocyte Riboflavin	Other
43% abnormal[k]	NA	NA	NA
43% abnormal			
100% normal			
100% normal			
Abnormal[l]	Abnormal[m]	NA	Normal[n]
Abnormal	Abnormal		Normal
Abnormal	Abnormal		Normal
Abnormal	Normal		Normal
Normal	Normal		Normal
Normal	Normal		Normal
NA	Normal[d]	NA	Normal[o]
NA	Abnormal[d]	NA	Abnormal[p]
	Normal		Normal
NA	Abnormal[d]	NA	Abnormal[q]
	Abnormal		Normal
	Abnormal		Normal
	Abnormal		Normal
	Normal		Normal
	Normal		Normal
	Normal		Normal
	Normal		Normal
NA	NA	Abnormal[r]	50% abnormal[t]
		Abnormal	30% abnormal
		63% abnormal	Normal
		Normal[s]	Normal
		Normal	Normal

[c] NA = not applicable.

[d] Abnormal urinary excretion = < 266 nmol (100 μg)/d of riboflavin (based on Sauberlich et al., 1974).

[e] Abnormal = symptoms of cheilosis.

[f] Normal = no signs or symptoms of deficiency (i.e., no ulcers around the mouth, no changes in the eyes, normal pyruvic and lactic acid levels, no anemia).

continued

TABLE 5-1 Continued

g Intake measured as mg/1000 kcal; energy intake not provided.

h Abnormal EGRAC = > 1.2.

i During exercise period.

j Abnormal EGRAC = > 1.25.

k Abnormal EGRAC = ≥ 1.3.

l Abnormal EGRAC = > 1.4.

m Abnormal urinary excretion = values before which there is a sharp increase (breakpoint for increase in slope of urinary excretion).

n Normal = no classical signs of riboflavin deficiency.

o Normal = no signs of ariboflavinosis (assessment based on eye exams, work performance, and psychomotor tests).

p Abnormal = nonspecific symptoms (cheilosis, angular stomatitis, scrotal skin changes).

q Abnormal = severe skin lesions.

r Abnormal total erythrocyte riboflavin = < 400 nmol/L (15 µg/100 mL); authors' conclusion.

s Normal total erythrocyte riboflavin = ≥ 530 nmol/L (20 µg/100 mL); authors' conclusion.

t Abnormal = severe symptoms of ariboflavinosis.

latter occurred in only one study on elderly Guatemalans (Boisvert et al., 1993). The value selected for the EAR for riboflavin was the intake that was sufficient to maintain or restore adequate status in half the individuals in the groups studied.

Ancillary: Kinetic, Catabolic, and Clinical Reflections of Riboflavin Status

Whole-body dynamics based on pharmacokinetic analysis were used to set limits for rates and amounts of riboflavin absorption and excretion and appeared to reflect the flux of major metabolites (Zempleni et al., 1996). Such analysis assumes that the upper limits for utilization and storage have been reached if there is a rapid increase in the excretion of vitamin in the urine. The suppression or regression of clinical signs, largely dermatological, provide guideposts for lowest limits.

FINDINGS BY LIFE STAGE AND GENDER GROUP

Infants Ages 0 through 12 Months

Method Used to Set the Adequate Intake

Because there are not sufficient data that reliably reflect response to dietary riboflavin intake in infants, an Adequate Intake (AI) is used as the goal for intake by infants. In this case, the AI reflects the observed mean riboflavin intake of infants fed principally with human milk.

The riboflavin content of human milk used for calculations in this report is based on milk from healthy, well-nourished mothers who are not taking supplements. In the first few weeks postpartum, the riboflavin content of milk tends to increase with the use of riboflavin supplements (Nail et al., 1980). However, no difference in riboflavin content of human milk was found between supplemented and un-supplemented well-nourished mothers at 6 months postpartum (Thomas et al., 1980). Previously, estimates of the riboflavin content of human milk from well-nourished, unsupplemented mothers included 0.37 ± 0.13 (standard deviation) mg/L at 5 to 7 days post-partum, 0.49 ± 0.12 mg/L at 43 to 45 days postpartum, 0.24 ± 0.35 mg/L at 6 months postpartum (Thomas et al., 1980), and 0.35 ± 0.02 mg/L after 3 weeks postpartum (Committee on Nutrition, 1985).

More recent studies (Roughead and McCormick, 1990a, b) in five subjects showed that riboflavin in milk may have been previously underestimated as a result of a lack of detection of flavin-adenine dinucleotide (FAD), which by weight is 48 percent riboflavin (and accounts for 41 percent of total flavin in milk), by fluorescent measurement techniques. However, the antivitaminic potential of certain metabolites in milk, such as the 2'-hydroxyethylflavin, could suppress riboflavin efficiency by competition for cellular uptake (Aw et al., 1983) and utilization (McCormick, 1962). The net consequence is that the riboflavin equivalent of milk, either from humans or cows, may be somewhat higher than earlier estimates based on uncorrected fluorescence measurements. On the basis of this and studies of microbiologically determined riboflavin in human milk (WHO, 1965), a riboflavin concentration of 0.35 mg/L will be used for human milk consumed by infants younger than 6 months (see Box 5-1).

BOX 5-1   Riboflavin Concentration in Human Milk

Study   *Riboflavin Concentration*

Roughead and McCormick, 1990a   Approximately 0.580 mg total flavin/
                                           L with 54% in the form of FAD (which
                                           is 48% riboflavin by weight), 41% as
                                           riboflavin, and 5% other. Hence 41%
                                           riboflavin + (0.48 × 54% FAD) = 67%
                                           of the total flavin is riboflavin. Thus,
                                           67% × 0.58 mg = 0.39 mg/L

WHO, 1965   0.26 and 0.21 mg/d (for 0.75 L) =
                                           0.24 mg/0.75 L = 0.31 mg/L

                                           Average of 0.39 and 0.31 = 0.35 mg/L

Ages 0 through 6 Months. The AI for riboflavin for infants ages 0 through 6 months is based on the reported mean volume of milk consumed by this age group (0.78 L/day; see Chapter 2) and the estimate of the riboflavin concentration in human milk of 0.35 mg/L (0.78 L × 0.35 mg/L = 0.27 mg). Thus, the AI for riboflavin when rounded is 0.3 mg/day for infants ages 0 through 6 months.

In a study of 55 low-income, East Indian infants aged 0 through 6 months who were exclusively fed human milk, the mothers' EGRAC value was 1.80 ± 0.06 (standard error), whereas the infants' EGRAC value was 1.36 ± 0.04 (Bamji et al., 1991). Only 9 percent of the mothers and 36 percent of the infants had EGRAC values lower than 1.2, the standard for adequacy used in the study. The mean riboflavin concentration of the human milk the infants were consuming was 0.22 ± 0.01 (standard error) mg/L. This study suggests that when there was deficiency in the mothers, the riboflavin concentration of milk was approximately one-third lower than values reported for Western women.

Ages 7 through 12 Months. If the reference body weight ratio method described in Chapter 2 to extrapolate from the AI for riboflavin for infants ages 0 through 6 months is used, the AI for riboflavin for the older infants would be 0.35 mg/day. The second method (see Chapter 2), extrapolating from the Estimated Average Requirement (EAR) for adults and adjusting for the expected variance to estimate a recommended intake, also gives an AI of 0.35 mg of riboflavin.

Alternatively, the AI for riboflavin for infants ages 7 through 12 months could be calculated by using the estimated riboflavin content of 0.6 L of human milk, the average volume consumed by this age group (riboflavin content equals 0.21 mg), and adding the amount of riboflavin provided by solid foods (0.6 mg) as estimated by Montalto et al. (1985) (see Chapter 2). The result equals approximately 0.8 mg/day. This value was judged to be unreasonably high because it is more than twice the extrapolated value given above. Thus the AI for riboflavin after rounding is 0.4 mg/day for infants ages 7 through 12 months—the value extrapolated from younger infants and from estimates of adult requirements.

Riboflavin AI Summary, Ages 0 through 12 Months

With use of the mean value for intake of human milk of 0.78 L/day and the average riboflavin content of 0.35 mg/L, the AI for riboflavin is 0.27 mg/day for infants ages 0 through 6 months, which is rounded to 0.3 mg; the AI for riboflavin is 0.35 for infants ages 7 through 12 months, which is rounded to 0.4 mg/day.

AI for Infants
0–6 months	**0.3 mg/day of riboflavin**	**≈0.04 mg/kg**
7–12 months	**0.4 mg/day of riboflavin**	**≈0.04 mg/kg**

Children and Adolescents Ages 1 through 18 Years

Method Used to Estimate the Average Requirement

Very limited data were found concerning the riboflavin requirements of children or adolescents. An older study of two normal 5-year-old boys given different levels of intake and with urinary output measured microbiologically revealed that riboflavin at 0.53 mg/1,000 kcal was ample (Oldham et al., 1944). Riboflavin requirement was judged to be the lowest level at which daily excretions did not change. This level of riboflavin intake was comparable with the 0.5 mg/1,000 kcal found close to the "daily requirement necessary for maintenance of satisfactory tissue stores of riboflavin" in adults in whom urinary flavin was measured fluorometrically at about the same time (Williams et al., 1943). Sauberlich and coworkers (1972) reported EGRAC values for a group of 431 students aged 14 to 17 years. For an EGRAC cutoff value of 1.20, 4 percent of the white males exceeded the cutoff compared with 20 percent of the black males, 11 percent of the white females, and 38 percent of the black females. This result suggests that black adolescents had a modest insufficiency of riboflavin intake compared with whites.

In the absence of additional information, EARs and RDAs for riboflavin for children ages 1 through 18 years have been extrapolated from adult values by using the method described in Chapter 2 using a metabolic body weight ratio multiplied by a growth factor. Calculated values have been rounded where appropriate.

Riboflavin EAR and RDA Summary, Ages 1 through 18 Years

EAR for Children	**1–3 years**	**0.4 mg/day of riboflavin**
	4–8 years	**0.5 mg/day of riboflavin**
EAR for Boys	**9–13 years**	**0.8 mg/day of riboflavin**
	14–18 years	**1.1 mg/day of riboflavin**
EAR for Girls	**9–13 years**	**0.8 mg/day of riboflavin**
	14–18 years	**0.9 mg/day of riboflavin**

The RDA for riboflavin is set by assuming a coefficient of variation (CV) of 10 percent (see Chapter 1) because information is not available on the standard deviation of the requirement for riboflavin; the RDA is defined as equal to the EAR plus twice the CV to cover the needs of 97 to 98 percent of the individuals in the group (therefore, for riboflavin the RDA is approximately 120 percent of the EAR).

RDA for Children	**1–3 years**	**0.5 mg/day of riboflavin**
	4–8 years	**0.6 mg/day of riboflavin**
RDA for Boys	**9–13 years**	**0.9 mg/day of riboflavin**
	14–18 years	**1.3 mg/day of riboflavin**
RDA for Girls	**9–13 years**	**0.9 mg/day of riboflavin**
	14–18 years	**1.0 mg/day of riboflavin**

Adults Ages 19 through 70 Years

Evidence Considered in Estimating the Average Requirement

Studies of the riboflavin requirements of adults have focused primarily on the occurrence of signs of clinical deficiency and on urinary excretion of riboflavin (Table 5-1). Thus, the EAR is derived from the findings of a number of studies that addressed clinical deficiency signs and biochemical values, including EGRAC, in relation to measured dietary intake of riboflavin. Biochemical changes in riboflavin status occur well before the appearance of overt signs of deficiency. Such studies help to bracket the riboflavin requirement.

Early studies of women by Sebrell et al. (1941) and of men by Horwitt et al. (1950) helped establish that riboflavin intakes of less than 0.5 to 0.6 mg/day led to clinical signs of deficiency. The use of microbiological assays in a study of riboflavin balance in women under institutional care led to a suggestion that 3 mg/day was sufficient (Sebrell et al., 1941). However, this level was soon deemed too high by others (Davis et al., 1946; Williams et al., 1943).

The work of Horwitt et al. (1950), Keys et al. (1944), and Williams et al. (1943) indicated that riboflavin intakes of approximately 0.8 mg/day for men or women were minimally sufficient to avoid signs of clinical deficiency. For example, when young men were given 0.31 mg/1,000 kcal (equivalent to 0.77 mg/day at 2,500 kcal) for at least 5 months, urinary riboflavin excretion decreased for approximately 45 days and then stabilized at 10 to 14 percent of intake. No changes related to riboflavin intake were observed in the structure of the eyes, work performance, or psychomotor test results (Keys et al., 1944).

On the basis of estimates of urinary riboflavin, it was determined that a reserve was not maintained at riboflavin intakes below 1.1 mg/day (Horwitt et al., 1949, 1950). Nearly three times as much riboflavin was excreted by subjects on diets providing riboflavin at 1.6 mg/day compared with those providing 1.1 mg/day (see Figure 5-1) (Horwitt et al., 1950). Also, symptoms of deficiency seen with 0.55 mg/day given for 4 months could be corrected in 15 days when 6 mg/day was given, although lower repetitive doses of 2 or 4 mg/day were also effective (Horwitt et al., 1950). Bessey and colleagues (1956) showed that men required from 0.6 to 1.6 mg/day of riboflavin for maintenance of erythrocyte riboflavin. This level of intake agrees with intake to achieve satisfactory urinary flavin excretion determined by Horwitt (1972) and Horwitt and coworkers (1950).

Moreover, it adds another point between 1.1 and 1.6 mg of riboflavin intake to Figure 5-1, which indicates that a markedly increased excretion of riboflavin has been noted at intakes exceeding 0.8 to 0.9 mg/day for women and 1.0 to 1.1 mg/day for men. A similar break in the riboflavin excretion curve for young women had also been observed earlier (Brewer et al., 1946; Davis et al., 1946). The more recent work of Boisvert and coworkers (1993), studying an elderly population, pointed to approximately 1.0 mg/day as the intake at which the excretion of excess vitamin became apparent. Although some earlier fluorescence-based assays of extracts of urine gave higher values than attributable to the vitamin alone because riboflavin was not separated from other flavins, break points in urinary excretion would not have been affected.

In the Philippines, Kuizon and colleagues (1992) found that riboflavin intake at 0.4 mg/1,000 kcal would lower the activity coefficient to less than 1.3 within 20 days for five of seven nonpregnant women.

Roughead and McCormick (1991) determined the urinary riboflavin and flavin metabolite contents before and after the ingestion of a 1.7-mg riboflavin supplement. Mean dietary intake of the group, obtained from three 1-day dietary records for each subject, was at least 1.7 mg in addition to the supplement. These investigators found that most of the 1.7 mg of the supplemental riboflavin appeared in the urine without being metabolized.

Intakes of 1.8 mg/day or more were required in a group of Gambian subjects to lower the mean EGRAC to a level of 1.3 to 1.4 (Bates et al., 1989); however, this study in The Gambia was judged not relevant for estimating requirements of the U.S. and Canadian populations. The subjects were severely deficient as judged by very high initial EGRAC values, making it difficult to know what length of time and what doses would be required to achieve EGRAC values that were below 1.4. Moreover, glucose 6-dehydrogenase status was not determined even though deficiency of this enzyme may be even greater in this population than the 10 percent estimated for blacks in the U.S. population.

Finally, two studies by Belko and colleagues (1984, 1985) are not included because they were conducted with moderately obese young women on low-calorie (1,200 to 1,250 kcal) diets.

Study results suggest that no distinctions in the riboflavin requirements of men and women other than those based on general size and energy expenditure can be made, nor is there any data to indicate that requirements for younger adults differ from those of older (over 70 years) adults.

Riboflavin EAR and RDA Summary, Ages 19 through 70 Years

Because clinical signs of deficiency appear at intakes of less than 0.5 to 0.6 mg/day whereas most studies report normal EGRAC values at intakes of less than 1.3 mg/day, and because there is an expected curvilinear biological increase of values from deficient to minimally adequate, it is estimated that the EAR for riboflavin for men is 1.1 mg/day and for women is 0.9 mg/day.

EAR for Men	19–30 years	1.1 mg/day of riboflavin
	31–50 years	1.1 mg/day of riboflavin
	51–70 years	1.1 mg/day of riboflavin
EAR for Women	19–30 years	0.9 mg/day of riboflavin
	31–50 years	0.9 mg/day of riboflavin
	51–70 years	0.9 mg/day of riboflavin

The RDA for riboflavin is set by assuming a coefficient of variation (CV) of 10 percent (see Chapter 1) because information is not available on the standard deviation of the requirement for riboflavin; the RDA is defined as equal to the EAR plus twice the CV to cover the needs of 97 to 98 percent of the individuals in the group (therefore, for riboflavin the RDA is 120 percent of the EAR).

RDA for Men	19–30 years	1.3 mg/day of riboflavin
	31–50 years	1.3 mg/day of riboflavin
	51–70 years	1.3 mg/day of riboflavin
RDA for Women	19–30 years	1.1 mg/day of riboflavin
	31–50 years	1.1 mg/day of riboflavin
	51–70 years	1.1 mg/day of riboflavin

Adults Ages Older Than 70 Years

Evidence Considered in Estimating the Average Requirement

Few additional studies estimating the riboflavin requirements have been conducted in the elderly. In healthy elderly women aged 70 years or older, doubling the estimated riboflavin intake by means of a supplement containing 1.7 mg of riboflavin doubled the urinary riboflavin excretion in the supplemented group compared to the unsupplemented group, from 4.36 to 9.06 μmol/g (1.64 to 3.41 mg/g) creatinine (Alexander et al., 1984). Initially all the women had EGRAC values lower than 1.25 on a mean baseline riboflavin intake of 1.8 ± 0.1 (standard error) mg/day, and urinary excretion values in both groups were substantially above levels indicating possible risk of deficiency.

In elderly Guatemalans, normalization of EGRAC to less than 1.34 was achieved with approximately 1.3 mg/day of riboflavin (0.7 mg/day from the diet plus 0.6 mg/day from a supplement), and a sharp increase in urinary riboflavin occurred at intakes above 1.0 to 1.1 mg/day. Requirements of the elderly did not differ from those of young adults (Boisvert et al., 1993).

Riboflavin EAR and RDA Summary, Ages Older Than 70 Years

Although there is a decrease in energy expenditure with aging and the EAR for older adults would be expected to decrease, the study by Boisvert and colleagues (1993) supports the use of the same EAR for the elderly as for younger adults.

EAR for Men   **> 70 years**   **1.1 mg/day of riboflavin**
EAR for Women   **> 70 years**   **0.9 mg/day of riboflavin**

The RDA for riboflavin is set by assuming a coefficient of variation (CV) of 10 percent (see Chapter 1) because information is not available on the standard deviation of the requirement for riboflavin; the RDA is defined as equal to the EAR plus twice the CV to cover the needs of 97 to 98 percent of the individuals in the group (therefore, for riboflavin the RDA is 120 percent of the EAR).

RDA for Men   **> 70 years**   **1.3 mg/day of riboflavin**
RDA for Women   **> 70 years**   **1.1 mg/day of riboflavin**

Pregnancy

Evidence Considered in Estimating the Average Requirement

Few studies provide information about the riboflavin requirements of pregnant women. Bamji (1976) reported that infants born to women with high erythrocyte glutathione reductase (EGR) activity may also have a high EGRAC but to a lesser degree than their mothers. Bates and colleagues (1981) reported that a sample of 59 pregnant women in the United Kingdom had a mean riboflavin intake of 2.2 mg/day and a mean EGRAC of 1.19 ± 0.08 (standard deviation). Maternal riboflavin intake (estimated from a crosscheck dietary history) was positively associated with fetal growth in a study of 372 pregnant women (Badart-Smook et al., 1997), but the data are insufficient to warrant use of fetal growth as an indicator for setting the riboflavin requirement for pregnant women.

For pregnancy an additional riboflavin requirement of 0.3 mg/day is estimated based on increased growth in maternal and fetal compartments and a small increase in energy utilization. This increased need is supported by the urinary excretion of less riboflavin during the progression of pregnancy and the more frequent appearance of clinical signs of ariboflavinosis in pregnant women on low intakes (less than 0.8 mg/day) than in their nonpregnant counterparts (Brzezinski et al., 1952; Jansen and Jansen, 1954). Also, EGRAC tends to increase during pregnancy (Bates et al., 1981; Heller et al., 1974; Vir et al., 1981). In the Philippines, Kuizon and colleagues (1992) found that riboflavin intake at 0.7 mg/1,000 kcal was required to lower the activity coefficient within 20 days to less than 1.3 for four of eight pregnant women compared with only 0.41 mg/1,000 kcal for five of seven nonpregnant women.

Riboflavin EAR and RDA Summary, Pregnancy

Adding 0.3 mg/day to the EAR of 0.9 mg/day of riboflavin for the nonpregnant woman gives an EAR of 1.2 mg/day. No adjustment is made for age or stage of pregnancy.

EAR for Pregnancy	14–18 years	1.2 mg/day of riboflavin
	19–30 years	1.2 mg/day of riboflavin
	31–50 years	1.2 mg/day of riboflavin

The RDA for riboflavin is set by assuming a coefficient of variation (CV) of 10 percent (see Chapter 1) because information is not available on the standard deviation of the requirement for riboflavin; the RDA is defined as equal to the EAR plus twice the CV to cover the needs of 97 to 98 percent of the individuals in the group (therefore, for riboflavin the RDA is 120 percent of the EAR). The results have been rounded up.

RDA for Pregnancy	14–18 years	1.4 mg/day of riboflavin
	19–30 years	1.4 mg/day of riboflavin
	31–50 years	1.4 mg/day of riboflavin

Lactation

Method Used to Estimate the Average Requirement

For lactating women, it is assumed that 0.3 mg of riboflavin is transferred in their milk each day when their daily milk production is 0.78 L (during the first 6 months of lactation; see "Infants Ages 0 through 12 Months"). If the use of riboflavin for milk production by the mother is assumed to be 70 percent efficient (WHO, 1965), values are adjusted upward to 0.4 mg/day for the amount of the vitamin that should be replaced. Women who are breastfeeding older infants who are eating solid foods need slightly less, in proportion to lower volume of milk production.

Riboflavin EAR and RDA Summary, Lactation

To the EAR of 0.9 mg/day of riboflavin for the nonpregnant and nonlactating woman, 0.4 mg/day is added, giving an EAR of 1.3 mg/day.

EAR for Lactation	14–18 years	1.3 mg/day of riboflavin
	19–30 years	1.3 mg/day of riboflavin
	31–50 years	1.3 mg/day of riboflavin

The RDA for riboflavin is set by assuming a coefficient of variation (CV) of 10 percent (see Chapter 1) because information is not available on the standard deviation of the requirement for riboflavin; the RDA is defined as equal to the EAR plus twice the CV to cover the needs of 97 to 98 percent of the individuals in the group (therefore, for riboflavin the RDA is 120 percent of the EAR).

RDA for Lactation	14–18 years	1.6 mg/day of riboflavin
	19–30 years	1.6 mg/day of riboflavin
	31–50 years	1.6 mg/day of riboflavin

Special Considerations

As with other B vitamins, persons undergoing hemodialysis or peritoneal dialysis and those with severe malabsorption are likely to require extra riboflavin. Women pregnant with more than one fetus and those breastfeeding more than one infant are also likely to require more riboflavin.

INTAKE OF RIBOFLAVIN

Food Sources

Most plant and animal tissues contain at least small amounts of riboflavin. Data obtained from the 1995 Continuing Survey of Food Intakes by Individuals (CSFII) indicate that the greatest contribution to the riboflavin intake of the U.S. adult population comes from milk and milk drinks followed by bread products and fortified cereals (Table 5-2). Other sources of riboflavin are organ meats. Milk is both a rich source of riboflavin and a commonly consumed food. Riboflavin loss occurs if it is exposed to the light, for example, if milk is stored in clear glass under light.

Dietary Intake

Based on data from CSFII (Appendix G) and the Third National Health and Nutrition Examination Survey (Appendix H), the median intake of riboflavin from food in the United States is approximately 2 mg/day for men and 1.5 mg/day for women. Similarly, a group of healthy residents of rural Georgia was found to have a mean daily intake of riboflavin of 2.1 mg (Roughead and McCormick, 1991). For all life stage and gender groups, fewer than 5 percent of individuals have estimated intakes that are less than the Estimated Average Requirement (EAR) for riboflavin. Dietary riboflavin intake in two Canadian provinces was reported to be similar to U.S. intake (Appendix I).

The Boston Nutritional Status Survey (Appendix F) indicates that this relatively advantaged group of people over age 60 had an estimated median riboflavin intake of 1.9 mg/day for men and 1.5 mg/day for women.

Intake from Supplements

Information from the Boston Nutritional Status Survey on the use of riboflavin supplements by a free-living elderly population is given in Appendix F. For those taking supplements, the fiftieth percentile of supplemental riboflavin intake was estimated to be 1.9 mg for men and 2.9 mg for women. Approximately 26 percent of all adults took a riboflavin-containing supplement in 1986 (Moss et al., 1989).

TABLE 5-2 Food Groups Providing Riboflavin in the Diets of U.S. Men and Women Aged 19 Years and Older, CSFII, 1995[a]

Food Group	Contribution to Total Riboflavin Intake[b] (%)		Foods Within the Group that Provide at Least 0.3 mg of Riboflavin[c] per Serving	
	Men	Women	0.3–0.7 mg	> 0.7 mg
Food groups providing at least 5% of total riboflavin intake				
Milk and milk drinks[d]	14.5	16.0	Milk and milk products	Fortified milk drinks
Bread and bread products	10.8	11.2	—	—
Mixed foods[e]	9.1	6.7	NA[f]	NA
Ready-to-eat cereals	8.7	10.9	Moderately fortified	Highly fortified
Mixed foods, main ingredient is grain	7.9	6.6	NA	NA
Riboflavin from other food groups				
Pasta, rice, and cooked cereals	2.1	2.4	Instant oatmeal	—
Pork	2.0	1.7	Pork cutlet and spareribs	—
Finfish	0.7	0.9	Trout	—
Organ meats	0.7	0.8	—	Liver, kidney, and heart
Soy-based supplements and meal replacements	0.6	0.2	Soy-based meat replacements	—
Lamb, veal, game, and other carcass meat	0.3	0.2	Veal chop and venison	—

[a] CSFII = Continuing Survey of Food Intakes by Individuals.

[b] Contribution to total intake reflects both the concentration of the nutrient in the food and the amount of the food consumed. It refers to the percentage contribution to the American diet for both men and women based on 1995 CSFII data.

[c] 0.3 mg represents 20% of the Recommended Daily Intake (1.7 mg) of riboflavin—a value set by the Food and Drug Administration.

[d] Includes yogurt.

[e] Includes sandwiches and other foods with meat, poultry, or fish as the main ingredient.

[f] NA = not applicable. Mixed foods were not considered for this table.

SOURCE: Unpublished data from the Food Surveys Research Group, Agricultural Research Service, U.S. Department of Agriculture, 1997.

TOLERABLE UPPER INTAKE LEVELS

Hazard Identification

No adverse effects associated with riboflavin consumption from food or supplements have been reported. However, studies involving large doses of riboflavin (Schoenen et al., 1994; Stripp, 1965; Zempleni et al., 1996) have not been designed to systematically evaluate adverse effects. The limited evidence from studies involving large intakes of riboflavin is summarized here.

No adverse effects were reported in humans after single oral doses of up to 60 mg of supplemental riboflavin and 11.6 mg of riboflavin given intravenously as a single bolus dose (Zempleni et al., 1996). This study is of limited use in setting a Tolerable Upper Intake Level (UL) because it was not designed to assess adverse effects. It is possible that chronic administration of these doses would pose some risk.

A study by Schoenen and coworkers (1994) reported no short-term side effects in 49 patients treated with 400 mg/day of riboflavin taken with meals for at least 3 months. Schoenen and coworkers (1994) reported that one patient receiving riboflavin and aspirin withdrew from the study because of gastric upset. This isolated finding may be an anomaly because no side effects were reported in other patients.

The apparent lack of harm resulting from high oral doses of riboflavin may be due to its limited solubility and limited capacity for absorption in the human gastrointestinal tract (Levy and Jusko, 1966; Stripp, 1965; Zempleni et al., 1996); its rapid excretion in the urine (McCormick, 1994). Zempleni et al. (1996) showed that the maximal amount of riboflavin that was absorbed from a single oral dose was 27 mg. A study by Stripp (1965) found limited absorption of 50 to 500 mg of riboflavin with no adverse effects. The poor intestinal absorption of riboflavin is well recognized: riboflavin taken by mouth is sometimes used to mark the stool in experimental studies. There are no data from animal studies suggesting that uptake of riboflavin during pregnancy presents a potential hazard for the fetus or newborn.

The only evidence of adverse effects associated with riboflavin comes from in vitro studies showing the formation of active oxygen species on intense exposure to visible or ultraviolet light (Ali et al., 1991; Floersheim, 1994; Spector et al., 1995). However, because there are no demonstrated functional or structural adverse effects in humans or animals after excess riboflavin intake, the relevance of this evidence to human health effects in vivo is highly questionable. Nevertheless, it is theoretically plausible that riboflavin increases photosensitivity to ultraviolet irradiation. Additionally, there is a theoretical risk that excess riboflavin will increase the photosensitized oxidations of cellular compounds, such as amino acids and proteins (McCormick, 1977) in infants treated for hyperbilirubinemia, with possible undesirable consequences.

Dose-Response Assessment

The data on adverse effects from high riboflavin intake are not sufficient for a quantitative risk assessment, and a UL cannot be derived.

Special Considerations

There is some in vitro evidence that riboflavin may interfere with detoxification of chrome VI by reduction to chrome III (Sugiyama et al., 1992). This may be of concern in people who may be exposed to chrome VI, for example, workers in chrome plating. Infants treated for hyperbilirubinemia may also be sensitive to excess riboflavin, as previously mentioned.

Intake Assessment

Although no UL can be set for riboflavin, an intake assessment is provided here for possible future use. Data from the Third National Health and Nutrition Examination Survey (see Appendix H) showed that the highest mean intake of riboflavin from diet and supplements for any life stage and gender group reported was for males aged 31 through 50 years: 6.9 mg/day. The highest reported intake at the ninety-fifth percentile was 11 mg/day in females over age 70 years.

Risk Characterization

No adverse effects have been associated with excess intake of riboflavin from food or supplements. This does not mean that there is no potential for adverse effects resulting from high intakes. Because data on the adverse effects of riboflavin intake are limited, caution may be warranted.

RESEARCH RECOMMENDATIONS FOR RIBOFLAVIN

Priority should be given to studies useful for setting Estimated Average Requirements (EARs) for riboflavin for children, adolescents, pregnant and lactating women, and the elderly. Future studies should be designed specifically around the EAR paradigm, use graded levels of riboflavin intake and clearly defined cutoff values for clinical adequacy and inadequacy, and be conducted for a sufficient duration. Two specific research areas may be productive:

- development of another functional test for riboflavin status to corroborate and augment the presently used flavin-adenine dinucleotide–dependent erythrocyte glutathione reductase (e.g., a test using a flavin mononucleotide–dependent erythrocyte enzyme such as the pyridoxine [pyridoxamine] 5'-phosphate oxidase) and
- examination of the effects of physical activity on the requirement for riboflavin.

REFERENCES

Alexander M, Emanuel G, Golin T, Pinto JT, Rivlin RS. 1984. Relation of riboflavin nutriture in healthy elderly to intake of calcium and vitamin supplements: Evidence against riboflavin supplementation. *Am J Clin Nutr* 39:540–546.

Ali N, Upreti RK, Srivastava LP, Misra RB, Joshi PC, Kidwai AM. 1991. Membrane damaging potential of photosensitized riboflavin. *Indian J Exp Biol* 29:818–822.

Aw TY, Jones DP, McCormick DB. 1983. Uptake of riboflavin by isolated rat liver cells. *J Nutr* 113:1249–1254.

Badart-Smook A, van Houwelingen AC, Al MD, Kester AD, Hornstra G. 1997. Fetal growth is associated positively with maternal intake of riboflavin and negatively with maternal intake of linoleic acid. *J Am Diet Assoc* 97:867–870.

Bamji MS. 1969. Glutathione reductase activity in red blood cells and riboflavin nutritional status in humans. *Clin Chim Acta* 26:263–269.

Bamji MS. 1976. Enzymic evaluation of thiamin, riboflavin and pyridoxine status of parturient women and their newborn infants. *Br J Nutr* 35:259–265.

Bamji MS, Chowdhury N, Ramalakshmi BA, Jacob CM. 1991. Enzymatic evaluation of riboflavin status of infants. *Eur J Clin Nutr* 45:309–313.

Bates CJ. 1987. Human requirements for riboflavin. *Am J Clin Nutr* 47:122–123.

Bates CJ, Prentice AM, Paul AA, Sutcliffe BA, Watkinson M, Whitehead RG. 1981. Riboflavin status in Gambian pregnant and lactating women and its implications for Recommended Dietary Allowances. *Am J Clin Nutr* 34:928–935.

Bates CJ, Powers HJ, Downes R, Brubacher D, Sutcliffe V, Thurnhill A. 1989. Riboflavin status of adolescent vs elderly Gambian subjects before and during supplementation. *Am J Clin Nutr* 50:825–829.

Belko AZ, Obarzanek E, Kalkwarf HJ, Rotter MA, Bogusz S, Miller D, Haas JD, Roe DA. 1983. Effects of exercise on riboflavin requirements of young women. *Am J Clin Nutr* 37:509–517.

Belko AZ, Obarzanek E, Roach R, Rotter M, Urban G, Weinberg S, Roe DA. 1984. Effects of aerobic exercise and weight loss on riboflavin requirements of moderately obese, marginally deficient young women. *Am J Clin Nutr* 40:553–561.

Belko AZ, Meredith MP, Kalkwarf HJ, Obarzanek E, Weinberg S, Roach R, McKeon G, Roe DA. 1985. Effects of exercise on riboflavin requirements: Biological validation in weight reducing women. *Am J Clin Nutr* 41:270–277.

Bessey OA, Horwitt MK, Love RH. 1956. Dietary deprivation of riboflavin and blood riboflavin levels in man. *J Nutr* 58:367–383.

Blot WJ, Li JY, Taylor PR, Guo W, Dawsey SM, Li B. 1995. The Linxian trials: Mortality rates by vitamin-mineral intervention group. *Am J Clin Nutr* 62:1424S–1426S.

Boisvert WA, Mendoza I, Castañeda C, De Portocarrero L, Solomons NW, Gershoff SN, Russell RM. 1993. Riboflavin requirement of healthy elderly humans and its relationship to macronutrient composition of the diet. *J Nutr* 123:915–925.

Bowman BB, McCormick DB, Rosenberg IH. 1989. Epithelial transport of water-soluble vitamins. *Ann Rev Nutr* 9:187–199.

Brewer W, Porter T, Ingalls R, Ohlson MA. 1946. The urinary excretion of riboflavin by college women. *J Nutr* 32:583–596.

Brown ML. 1990. *Present Knowledge in Nutrition,* 6th ed. Washington, DC: International Life Sciences Institute-Nutrition Foundation.

Brzezinski A, Bromberg YM, Braun K. 1952. Riboflavin excretion during pregnancy and early lactation. *J Lab Clin Med* 39:84–90.

Burch HB, Bessey OA, Lowry OH. 1948. Fluorometric measurements of riboflavin and its natural derivatives in small quantities of blood serum and cells. *J Biol Chem* 175:457–470.

Chastain JL, McCormick DB. 1987. Flavin catabolites: Identification and quantitation in human urine. *Am J Clin Nutr* 46:830–834.

Chia CP, Addison R, McCormick DB. 1978. Absorption, metabolism, and excretion of 8α-(amino acid) riboflavins in the rat. *J Nutr* 108:373–381.

Cole HS, Lopez R, Cooperman JM. 1976. Riboflavin deficiency in children with diabetes mellitus. *Acta Diabetol Lat* 13:25–29.

Committee on Nutrition. 1985. Composition of human milk: Normative data. In: *Pediatric Nutrition Handbook,* 2nd ed. Elk Grove Village, IL: American Academy of Pediatrics. Pp. 363–368.

Dancis J, Lehanka J, Levitz M. 1988. Placental transport of riboflavin: Differential rates of uptake at the maternal and fetal surfaces of the perfused human placenta. *Am J Obstet Gynecol* 158:204–210.

Daniel H, Wille U, Rehner G. 1983. In vitro kinetics of the intestinal transport of riboflavin in rats. *J Nutr* 113:636–643.

Darby WJ. 1981. *Annual Review of Nutrition,* Vol. 1. Palo Alto, CA: Annual Reviews.

Davis MV, Oldham HG, Roberts LJ. 1946. Riboflavin excretions of young women on diets containing varying levels of the B vitamins. *J Nutr* 32:143–161.

Floersheim GL. 1994. Allopurinol, indomethacin and riboflavin enhance radiation lethality in mice. *Radiat Res* 139:240–247.

Frischer H, Bowman JE, Carson PE, Reickmann KH, Willerson D Jr, Colwell EJ. 1973. Erythrocyte glutathione reductase, glucose-6-phosphate dehydrogenase, and 6-phosphogluconic dehydrogenase deficiencies in populations of the United States, South Vietnam, Iran, and Ethiopia. *J Lab Clin Med* 81:603–612.

Heller S, Salkeld RM, Korner WF. 1974. Riboflavin status in pregnancy. *Am J Clin Nutr* 27:1225–1230.

Horwitt MK. 1972. Riboflavin. Requirements and factors influencing them. In: Sebrell WH Jr, Harris RS, eds. *The Vitamins*, 2nd ed., Vol. 5. New York: Academic Press.

Horwitt MK, Hills OW, Harvey CC, Liebert E, Steinberg DL. 1949. Effects of dietary depletion of riboflavin. *J Nutr* 39:357–373.

Horwitt MK, Harvey CC, Hills OW, Liebert E. 1950. Correlation of urinary excretion of riboflavin with dietary intake and symptoms of ariboflavinosis. *J Nutr* 41:247–264.

Innis WS, McCormick DB, Merrill AH Jr. 1985. Variations in riboflavin binding by human plasma: Identification of immunoglobulins as the major proteins responsible. *Biochem Med* 34:151–165.

Jansen AP, Jansen BC. 1954. Riboflavin-excretion with urine in pregnancy. *Int Z Vitaminforsch* 25:193–199.

Jusko WJ, Levy G. 1967. Absorption, metabolism, and excretion of riboflavin 5'-phosphate in man. *J Pharmacol Sci* 156:58–62.

Jusko WJ, Levy G. 1975. Absorption, protein binding and elimination of riboflavin. In: Rivlin RS, ed. *Riboflavin*. New York: Plenum Press. Pp. 99–152.

Jusko WJ, Khanna N, Levy G, Stern L, Yaffe SJ. 1970. Riboflavin absorption and excretion in the neonate. *Pediatrics* 45:945–949.

Keys A, Henschel AF, Mickelsen O, Brozek JM, Crawford JH. 1944. Physiological and biochemical functions in normal young men on a diet restricted in riboflavin. *J Nutr* 27:165–178.

Komindr S, Nicholalds GE. 1980. Clinical significance of riboflavin deficiency. In: Brewster MA, Naito HK, eds. *Nutritional Elements in Clinical Biochemistry*. New York: Plenum Press. Pp. 15–68.

Kuizon MD, Natera MG, Alberto SP, Perlas LA, Desnacido JA, Avena EM, Tajaon RT, Macapinlac MP. 1992. Riboflavin requirement of Filipino women. *Eur J Clin Nutr* 46:257–264.

Lee SS, McCormick DB. 1983. Effect of riboflavin status on hepatic activities of flavin-metabolizing enzymes in rats. *J Nutr* 113:2274–2279.

Leske MC, Wu SY, Hyman L, Sperduto R, Underwood B, Chylack LT, Milton RC, Srivastava S, Ansari N. 1995. Biochemical factors in the lens opacities. Case-control study. The Lens Opacities Case-Control Study Group. *Arch Ophthalmol* 113:1113–1119.

Levy G, Jusko WJ. 1966. Factors affecting the absorption of riboflavin in man. *J Pharm Sci* 55:285–289.

Lossy FT, Goldsmith GA, Sarett HP. 1951. A study of test dose excretion of five B complex vitamins in man. *J Nutr* 45:213.

Mayersohn M, Feldman S, Gribaldi M. 1969. Bile salt enhancement of riboflavin and flavin mononucleotide absorption in man. *J Nutr* 98:288–296.

McCormick DB. 1962. The intracellular localization, partial purification, and properties of flavokinase from rat liver. *J Biol Chem* 237:959–962.

McCormick DB. 1977. Interactions of flavins with amino acid residues: Assessments from spectral and photochemical studies. *Photochem Photobiol* 26:169–182.

McCormick DB. 1989. Two interconnected B vitamins: Riboflavin and pyridoxine. *Physiol Rev* 69:1170–1198.

McCormick DB. 1994. Riboflavin. In: Shils ME, Olson JE, Shike M, eds. *Modern Nutrition in Health and Disease*. Philadelphia: Lea & Febiger. Pp. 366–375.

McCormick DB, Greene HL. 1994. Vitamins. In: Burtis CA, Ashwood ER, eds. *Tietz Textbook of Clinical Chemistry*. Philadelphia: Saunders.

Meinen M, Aeppli R, Rehner G. 1977. Studies on the absorption of thiamine, riboflavin and pyridoxine in vitro. *Nutr Metab* 21:264–266.

Merrill AH Jr, Lambeth JD, Edmondson DE, McCormick DB. 1981. Formation and mode of action of flavoproteins. *Annu Rev Nutr* 1:281–317.

Merrill AH Jr, Foltz AT, McCormick DB. 1991. Vitamins and cancer. In: Alfin-Slater RB, Kritchevsky D, eds. *Cancer and Nutrition*. New York: Plenum. Pp. 261–320.

Montalto MB, Benson JD, Martinez GA. 1985. Nutrient intake of formula-fed infants and infants fed cow's milk. *Pediatrics* 75:343–351.

Moss AJ, Levy AS, Kim I, Park YK. 1989. *Use of Vitamin and Mineral Supplements in the United States: Current Users, Types of Products, and Nutrients*. Advance Data, Vital and Health Statistics of the National Center for Health Statistics, No. 174. Hyattsville, MD: National Center for Health Statistics.

Nail PA, Thomas MR, Eakin R. 1980. The effect of thiamin and riboflavin supplementation on the level of those vitamins in human breast milk and urine. *Am J Clin Nutr* 33:198–204.

Natraj U, George S, Kadam P. 1988. Isolation and partial characterisation of human riboflavin carrier protein and the estimation of its levels during human pregnancy. *J Reprod Immunol* 13:1–16.

Nichoalds GE. 1981. Riboflavin. Symposium in Laboratory Medicine. In: Labbae RF, ed. *Symposium on Laboratory Assessment of Nutritional Status*. Clinics in Laboratory Medicine Series, Vol. 1. Philadelphia: WB Saunders. Pp. 685–698.

Oldham H, Johnston F, Kleiger S, Hedderich-Arismendi H. 1944. A study of the riboflavin and thiamine requirements of children of preschool age. *J Nutr* 27:435–446.

Powers HJ, Bates CJ, Eccles M, Brown H, George E. 1987. Bicycling performance in Gambian children: Effects of supplements of riboflavin or ascorbic acid. *Hum Nutr Clin Nutr* 41:59–69.

Prager MD, Hill JM, Speer RJ. 1958. Whole blood riboflavin levels in healthy individuals and in patients manifesting various blood dyscrasias. *J Lab Clin Med* 52:206.

Prasad PA, Bamji MS, Lakshmi AV, Satyanarayana K. 1990. Functional impact of riboflavin supplementation in urban school children. *Nutr Res* 10:275–281.

Ramsay VP, Neumann C, Clark V, Swendseid ME. 1983. Vitamin cofactor saturation indices for riboflavin, thiamine, and pyridoxine in placental tissue of Kenyan women. *Am J Clin Nutr* 37:969–973.

Rivier DA. 1973. Kinetics and Na-dependence of riboflavin absorption by intestine in vivo. *Experientia* 29:1443–1446.

Rivlin RS. 1975. Riboflavin and cancer. In: Rivlin RS, ed. *Riboflavin*. New York: Plenum Press. Pp. 369–391.

Roe DA, Bogusz S, Sheu J, McCormick DB. 1982. Factors affecting riboflavin requirements of oral contraceptive users and nonusers. *Am J Clin Nutr* 35:495–501.

Roughead ZK, McCormick DB. 1990a. Flavin composition of human milk. *Am J Clin Nutr* 52:854–857.

Roughead ZK, McCormick DB. 1990b. Qualitative and quantitative assessment of flavins in cow's milk. *J Nutr* 120:382–388.

Roughead ZK, McCormick DB. 1991. Urinary riboflavin and its metabolites: Effects of riboflavin supplementation in healthy residents of rural Georgia (USA). *Eur J Clin Nutr* 45:299–307.

Sadowski JA. 1992. Riboflavin. In: Hartz SC, Russell RM, Rosenberg IH, eds. *Nutrition in the Elderly. The Boston Nutritional Status Survey.* London: Smith-Gordon. Pp. 119–125.

Said HM, Ma TY. 1994. Mechanism of riboflavine uptake by Caco-2 human intestinal epithelial cells. *Am J Physiol* 266:G15–G21.

Sauberlich HE, Judd JH Jr, Nicholalds GE, Broquist HP, Darby WJ. 1972. Application of the erythrocyte glutathione reductase assay in evaluating riboflavin nutritional status in a high school student population. *Am J Clin Nutr* 25:756–762.

Sauberlich HE, Skala JH, Dowdy RP. 1974. *Laboratory Tests for the Assessment of Nutritional Status.* Boca Raton, FL: CRC Press.

Schoenen J, Lenaerts M, Bastings E. 1994. Rapid communication: High-dose riboflavin as a prophylactic treatment of migraine: Results of an open pilot study. *Cephalalgia* 14:328–329.

Sebrell WH Jr, Butler RE, Wooley JG, Isbell H. 1941. Human riboflavin requirement estimated by urinary excretion of subjects on controlled intake. *Public Health Rep* 56:510–519.

Smith MD. 1980. Rapid method for determination of riboflavin in urine by high-performance liquid chromatography. *J Chromatogr* 182:285–291.

Soares MJ, Satyanarayana K, Bamji MS, Jacob CM, Ramana YV, Rao SS. 1993. The effect of exercise on the riboflavin status of adult men. *Br J Nutr* 69:541–551.

Sorrell MF, Frank O, Thompson AD, Aquino H, Baker H. 1971. Absorption of vitamins from the large intestine in vivo. *Nutr Rep Int* 3:143–148.

Spector A, Wang GM, Wang RR, Li WC, Kleiman NJ. 1995. A brief photochemically induced oxidative insult causes irreversible lens damage and cataracts. 2. Mechanism of action. *Exp Eye Res* 60:483–493.

Steier M, Lopez R, Cooperman JM. 1976. Riboflavin deficiency in infants and children with heart disease. *Am Heart J* 92:139–143.

Stripp B. 1965. Intestinal absorption of riboflavin by man. *Acta Pharmacol Toxicol* 22:353–362.

Sugiyama M, Tsuzuki K, Lin X, Costa M. 1992. Potentiation of sodium chromate (VI)-induced chromosomal aberrations and mutation by vitamin B_2 in Chinese hamster V79 cells. *Mutat Res* 283:211–214.

Thomas MR, Sneed SM, Wei C, Nail PA, Wilson M, Sprinkle EE 3rd. 1980. The effects of vitamin C, vitamin B_6, vitamin B_{12}, folic acid, riboflavin, and thiamin on the breast milk and maternal status of well-nourished women at 6 months postpartum. *Am J Clin Nutr* 33:2151–2156.

Tremblay A, Boilard M, Bratton MF, Bessette H, Roberge AB. 1984. The effects of a riboflavin supplementation on the nutritional status and performance of elite swimmers. *Nutr Res* 4:201–208.

Tucker RG, Mickelsen O, Keys A. 1960. The influence of sleep, work, diuresis, heat, acute starvation, thiamine intake and bed rest on human riboflavin excretion. *J Nutr* 72:251–261.

Vir SC, Love AH, Thompson W. 1981. Riboflavin status during pregnancy. *Am J Clin Nutr* 34:2699–2705.

Weight LM, Noakes TD, Labadarios D, Graves J, Jacobs P, Berman PA. 1988. Vitamin and mineral status of trained athletes including the effects of supplementation. *Am J Clin Nutr* 47:186–191.

WHO (World Health Organization). 1965. *Nutrition in Pregnancy and Lactation.* Report of a WHO Expert Committee. Technical Report Series No. 302. Geneva: World Health Organization.

Williams RD, Mason HL, Cusick PL, Wilder RM. 1943. Observations on induced riboflavin deficiency and the riboflavin requirement of man. *J Nutr* 25:361–377.

Wilson JA. 1983. Disorders of vitamins: Deficiency, excess and errors of metabolism. In: Petersdorf RG, Harrison TR, eds. *Harrison's Principles of Internal Medicine*, 10th ed. New York: McGraw-Hill. Pp. 461–470.

Winters LR, Yoon JS, Kalkwarf HJ, Davies JC, Berkowitz MG, Haas J, Roe DA. 1992. Riboflavin requirements and exercise adaptation in older women. *Am J Clin Nutr* 56:526–532.

Yamada Y, Merrill AH Jr, McCormick DB. 1990. Probable reaction mechanisms of flavokinase and FAD synthetase from rat liver. *Arch Biochem Biophys* 278:125–130.

Zempleni J, Galloway JR, McCormick DB. 1996. Pharmacokinetics of orally and intravenously administered riboflavin in healthy humans. *Am J Clin Nutr* 63:54–66.

6

Niacin

SUMMARY

Niacin functions as a cosubstrate or coenzyme for the transfer of the hydride ion with numerous dehydrogenases. The primary criterion used to estimate the Recommended Dietary Allowance (RDA) for niacin is the urinary excretion of niacin metabolites. No adjustment is made for bioavailability, but the requirement is expressed in niacin equivalents (NEs), allowing for some conversion of the amino acid tryptophan to niacin. The RDA for adults is 16 mg/day of NEs for men and 14 mg/day of NEs for women. Recently, the median intake of preformed niacin from food in the United States was approximately 28 mg for men and 18 mg for women, and the ninety-fifth percentile of intake from both food and supplements was 40 to 70 mg of NEs, depending on age. In two Canadian populations the median intake of preformed niacin was approximately 41 mg/day for men and 28 mg/day for women. The Tolerable Upper Intake Level (UL) for niacin for adults is 35 mg/day, which was based on flushing as the critical adverse effect.

BACKGROUND INFORMATION

The term *niacin* refers to nicotinamide (nicotinic acid amide), nicotinic acid (pyridine-3-carboxylic acid), and derivatives that exhibit the biological activity of nicotinamide. The nicotinamide moiety of the pyridine nucleotide coenzymes nicotinamide adenine dinucleotide (NAD) and nicotinamide adenine dinucleotide phosphate (NADP) acts as a hydride ion acceptor or donor in many biological redox reactions. NAD has also been shown to be required for important nonredox adenosine diphosphate (ADP)–ribose transfer reactions involved in deoxyribonucleic acid (DNA) repair and calcium mobilization (Kim et al., 1994; Lautier et al., 1993; Lee et al., 1989). The amino acid tryptophan is converted in part to nicotinamide and thus can contribute to meeting the requirement for niacin.

Function

In the form of the coenzymes NAD and NADP, niacin functions in many biological redox reactions. NAD functions in intracellular respiration and as a codehydrogenase with enzymes involved in the oxidation of fuel molecules such as glyceraldehyde 3-phosphate, lactate, alcohol, 3-hydroxybutyrate, pyruvate, and α-ketoglutarate. NADP functions in reductive biosyntheses such as in fatty acid and steroid syntheses and, like NAD, as a codehydrogenase—as in the oxidation of glucose 6-phosphate to ribose 5-phosphate in the pentose phosphate pathway.

Three classes of enzymes cleave the β-N-glycosylic bond of NAD to free nicotinamide and catalyze the transfer of ADP-ribose in non-redox reactions (Lautier et al., 1993). Two of the three classes catalyze ADP-ribose transfer to proteins: mono-ADP-ribosyltransferases and poly-ADP-ribose polymerase (PARP). The third class promotes the formation of cyclic ADP-ribose, which mobilizes calcium from intracellular stores in many types of cells (Kim et al., 1994).

The enzyme PARP is found in the nuclei of eukaryotic cells and catalyzes the transfer of many ADP-ribose units from NAD to an acceptor protein and also to the enzyme itself. These nuclear poly-ADP-ribose proteins seem to function in DNA replication and repair and in cell differentiation. DNA damage greatly enhances the activity of PARP (Stierum et al., 1994); PARP activity is strongly correlated with cellular apoptosis (Stierum et al., 1994).

Physiology of Absorption, Metabolism, and Excretion

Absorption and Transport

Absorption of nicotinic acid and nicotinamide from the stomach and the intestine is rapid (Bechgaard and Jespersen, 1977) and at low concentrations is mediated by sodium ion-dependent facilitated diffusion. At higher concentrations, passive diffusion predominates, with doses of 3 to 4 g of niacin almost completely absorbed (Bech-gaard and Jespersen, 1977). Glycohydrolases in the liver and intestines catalyze the release of nicotinamide from NAD (Henderson and Gross, 1979). Nicotinamide is then transported to tissues to be used in synthesis of NAD when needed. Both forms of the vitamin enter cells by simple diffusion; however, both nicotinic acid and nicotinamide also enter erythrocytes by facilitated transport (Lan and Henderson, 1968).

Metabolism and Excretion

The niacin coenzymes NAD and NADP are synthesized in all tissues of the body from nicotinic acid or nicotinamide. Tissue concentrations of NAD appear to be regulated by the concentration of extracellular nicotinamide, which in turn is under hepatic control and is hormonally influenced. Hydrolysis of hepatic NAD allows the release of nicotinamide for transport to tissues that lack the ability to synthesize the NAD and NADP coenzymes from tryptophan. In the liver some excess plasma nicotinamide is converted to storage NAD (i.e., NAD not bound to enzymes). Tryptophan and nicotinic acid also contribute to storage NAD following the biosynthetic pathway, going through NAMN, which is then reamidated to NAD. In the degradation of NAD, the nicotinamide formed can be reconverted to NAD via nicotinamide ribonucleotide. Nicotinamide can be deamidated in the intestinal tract by intestinal microflora (Bernofsky, 1980).

The body's niacin requirement is met not only by nicotinic acid and nicotinaminde present in the diet, but also by conversion from the dietary protein containing tryptophan. The relative contribution of tryptophan is estimated as follows: 60 mg of tryptophan = 1 mg of niacin = 1 mg of niacin equivalents (Horwitt et al., 1981).

Excess niacin is methylated in the liver to N^1-methyl-nicotinamide, which is excreted in the urine along with the 2- and 4-pyridone oxidation products of N^1-methyl-nicotinamide. The two major excretion products are N^1-methyl-nicotinamide and its pyridone derivative (Mrochek et al., 1976). The proportions differ somewhat depending on the amount and form of niacin ingested and the niacin status of the individual.

Clinical Effects of Inadequate Intake

Pellagra is the classic manifestation of a severe niacin deficiency. It is characterized by a pigmented rash that develops symmetrically in areas exposed to sunlight; changes in the digestive tract that are associated with vomiting, constipation or diarrhea, and a bright red tongue; and neurological symptoms including depression, apathy, headache, fatigue, and loss of memory. Pellagra was common in the United States and parts of Europe in the early twentieth century in areas in which corn or maize (which is low in both niacin and the amino acid tryptophan) was the dietary staple. Although a world-wide problem, pellagra has virtually disappeared from industrial-ized countries except for its occurrence in chronic alcoholism and in individuals with conditions that disrupt tryptophan pathways. It still appears in India and parts of China and Africa. For example, pellagra was reported in Mozambican refugees in Malawi (Malfait et al., 1993).

SELECTION OF INDICATORS FOR ESTIMATING THE REQUIREMENT FOR NIACIN

Niacin status and dietary requirement can be estimated by using biochemical or clinical endpoints of niacin deficiency. Biochemical changes occur well before the appearance of overt signs of deficien-cy. Biochemical markers include daily urinary excretion of methy-lated metabolites, the ratio of 2-pyridone to N^1-methyl-nicotinamide in urine, erythrocyte pyridine nucleotides, oral dose uptake tests, erythrocyte nicotinamide adenine dinucleotide (NAD), and plasma 2-pyridone derivative.

Urinary Excretion

The most reliable and sensitive measures of niacin status are urinary excretion of the two major methylated metabolites, N^1-methyl-nicotinamide and its 2-pyridone derivative (N^1-methyl-2-pyridone-5-carboxamide). Criteria for interpreting urinary N^1-methyl-nicotina-mide excretion amounts in adults and pregnant women indicate that for adults, 24-hour excretion rates of less than 5.8 μmol/day represent deficient niacin status and 5.8 to 17.5 μmol/day repre-sents low status (Sauberlich et al., 1974). The ratio of the 2-pyri-done to N^1-methyl-nicotinamide, although independent of age and creatinine excretion, is a measure of protein adequacy rather than niacin status (Shibata and Matsuo, 1989). It is relatively insensitive to a marginal niacin intake of 10 mg/day of niacin equivalents (NEs) and has been shown to be not totally reliable for evaluating an intake of 6 mg/day of NEs (Jacob et al., 1989). The ratio of the 6-pyridone (N^1-methyl-nicotinamide-3-carboxamide) to N^1-methyl-nicotinamide appears to be associated with the development of clinical symptoms of pellagra, principally dermatitis (Dillon et al., 1992).

Plasma Concentrations

In plasma the 2-pyridone derivative drops below detection limits after a low niacin intake (Jacob et al., 1989). With an oral niacin load (nicotinamide at 20 mg/70 kg body weight), postdose changes in 2-pyridone in both plasma and urine were more responsive to niacin status than changes seen in N^1-methyl-nicotinamide. Plasma concentrations of other niacin metabolites and of niacin are not useful markers of niacin status.

Erythrocyte Pyridine Nucleotides

Analysis of erythrocyte NAD concentration promises to be a sensitive indicator of niacin depletion. In an experimental study in which adult male subjects were fed low-niacin diets containing either 6 or 10 mg NE/day, the erythrocyte NAD concentration decreased by 70 percent, whereas the NADP concentration remained unchanged (Fu et al., 1989). An earlier study reported a similar decrease in NAD relative to NADP in fibroblasts grown in niacin-restricted culture (Jacobson et al., 1979). Erythrocyte NAD concentrations provided a marker of niacin depletion equally as sensitive and reliable as excretion of urine metabolites in a study of seven healthy young men (Fu et al., 1989) and in an experimental niacin depletion study of elderly subjects (Ribaya-Mercado et al., 1997).

Transfer of Adenosine Diphosphate Ribose

A possible functional measure for niacin status could be poly-adenosine diphosphate (ADP) ribosylation, because ADP ribosylation may contribute to gene stability (poly-ADP-ribose polymerase in the nucleus) and may function in deoxyribonucleic acid (DNA) replication and repair (Stierum et al., 1994). However, the assays have not been developed or refined well enough to be used to judge niacin status at present.

Pellagra

Clinical pellagra may represent various degrees of combined niacin and riboflavin deficiencies (Carpenter and Lewin, 1985).

Deficiencies of other micronutrients (e.g., pyridoxine and iron) required to convert tryptophan to niacin may also contribute to the appearance of pellagra. Because pellagra is a late and serious manifestation of deficiency, it was determined that the average requirement must exceed the amount required to prevent pellagra.

FACTORS AFFECTING THE NIACIN REQUIREMENT

Bioavailability

Niacin in mature cereal grains is largely bound and thus is only about 30 percent available; alkali treatment of the grain increases the percentage absorbed (Carpenter and Lewin, 1985; Carter and Carpenter, 1982). Niacin in the coenzyme nicotinamide adenine dinucleotide/nicotinamide adenine dinucleotide phosphate (NAD/NADP) form in meats appears to be much more available. Niacin added during enrichment or fortification is in the free form and thus highly available. Foods that contain niacin in the free form include beans and liver. Quantitative data are not available on which to base adjustments for the bioavailability from different types of foods.

The conversion efficiency of tryptophan to niacin, although assumed to be 60:1 (Horwitt et al., 1981), varies depending on a number of dietary and metabolic factors. The requirement for preformed niacin is increased by factors that reduce the conversion of tryptophan to niacin (McCormick, 1988), including low tryptophan intake; carcinoid syndrome in which tryptophan is preferentially oxidized to 5-hydroxytryptophan and serotonin; prolonged treatment with the drug Isoniazid, which competes with pyridoxal 5'-phosphate, a vitamin B_6–derived coenzyme required in the tryptophan-to-niacin pathway; and Hartnup's disease, an autosomal recessive disorder that interferes with the absorption of tryptophan in the intestine and kidney.

The efficiency of the conversion of dietary tryptophan to niacin is decreased by deficiencies of other nutrients (see "Nutrient-Nutrient Interactions"). Conversion efficiency may increase with some dietary restrictions because of changes in activities of pathway enzymes including tryptophan oxygenase, quinolinate phosphoribosyltransferase, and picolinate carboxylase.

The requirement for preformed niacin as a proportion of the total niacin requirement tends to be lower with higher tryptophan intakes (a greater proportion of tryptophan is available for conversion to NAD once protein synthesis needs are met) and pregnancy (the conversion of tryptophan to niacin is more efficient [Knox, 1951; Rose and Braidman, 1971; Wertz et al., 1958]). Oral contraceptives that contained high doses of estrogen have also been found to increase conversion efficiency (Rose and Braidman, 1971; Wertz et al., 1958). The priority for body tryptophan utilization is protein synthesis before NAD or NADP synthesis.

The 60 to 1 conversion value represents the mean of a wide range of individual values found in human studies that measured the conversion of tryptophan to urinary niacin metabolites (Horwitt et al., 1981). Irrespective of dietary factors, substantial individual differences (30 percent) in the conversion efficiency of tryptophan to niacin have been reported (Horwitt et al., 1981; Patterson et al., 1980).

Nutrient-Nutrient Interactions

Interactions occur between riboflavin and vitamin B_6 metabolism in which flavin mononucleotide (FMN) is required for the oxidase that forms coenzymic pyridoxal 5'-phosphate required for conversion of tryptophan to niacin (McCormick, 1989). There is also interdependence of enzymes within the tryptophan-to-niacin pathway where pyridoxal 5'-phosphate, flavin-adenine dinucleotide (FAD), and iron are functional. Hence, inadequate iron, riboflavin, or vitamin B_6 status decreases the conversion of tryptophan to niacin. Data are not available to quantitatively assess the effects of these nutrient-nutrient interactions on the niacin requirement.

Energy Intake and Expenditure

No directly relevant studies were found that examined the effect of energy intake or expenditure on the niacin requirement. Examination of individual data from four studies showed no relationship between excretion of N^1-methyl-nicotinamide and either energy intake or body weight (Goldsmith et al., 1952, 1956; Horwitt et al., 1956; Jacob et al., 1989). Studies of niacin requirements are consistently presented in relation to energy intake on the basis of the known biochemical function of niacin, as have been past recommendations for niacin intake (NRC, 1989).*

---

*It is recognized that studies of individuals in forced starvation situations have reported that individuals with the greatest initial body weights (and thus the greatest energy expenditure) developed signs of deficiency more rapidly than did others consuming similar diets but who weighed less initially. However, the panel could not quantify this difference as assessed by changes in available indicators of adequacy by those consuming typical diets in Canada and the United States.

Despite the lack of experimental data, the known biochemical function of niacin in the oxidation of fuel molecules suggests at least a small (10 percent) adjustment to reflect differences in the average energy utilization and size of men and women, a 10 percent increase in the requirement to cover increased energy use during pregnancy, and a small increase in the requirement to account for the efficiency of niacin use in milk production during lactation.

FINDINGS BY LIFE STAGE AND GENDER GROUP

Infants Ages 0 through 12 Months

Method Used to Set the Adequate Intake

As for other nutrients, the Adequate Intake (AI) level for niacin is set for infants based on the observed mean intake of infants fed principally with human milk.

Ages 0 through 6 Months. One study (Ford et al., 1983) of unsupplemented mothers estimated the niacin concentration of their milk at 1.8 mg/L. The adequate intake for niacin for infants ages 0 through 6 months is based on the reported mean volume of milk consumed by this age group (0.78 L/day; see Chapter 2) and the estimate of the niacin concentration in human milk of 1.8 mg/L (0.78 L × 1.8 mg/L = 1.4 mg). The tryptophan content of human milk is approximately 210 mg/L (Committee on Nutrition, 1985). Because of the high rate of protein turnover and net positive nitrogen retention in infancy, it is likely that the standard method for estimating niacin equivalents (NEs) would overestimate the contribution from tryptophan. Thus, the AI for niacin for infants is given in milligrams of preformed niacin only—2 mg/day after rounding up.

Ages 7 through 12 Months. No difference in human milk composition was noted between the first and second 6 months of lactation (Ford et al., 1983). If the reference body weight ratio method described in Chapter 2 to extrapolate from the AI for niacin for infants ages 0 through 6 months is used, the AI for preformed niacin for the older infants is 2 mg/day after rounding. The second method (see Chapter 2), extrapolating from the Estimated Average Requirement (EAR) for adults and adjusting for the expected variance to estimate a recommended intake, gives an AI of 4.1 mg of NEs. A 15 percent coefficient of variation (CV) is used (for discussion, see "Niacin EAR and RDA Summary, Ages 19 Years and Older"). This AI value is expected to be higher than that obtained with the first method because it reflects NEs rather than preformed niacin.

Alternatively, the AI for niacin for infants ages 7 through 12 months can be calculated by using the estimated niacin content of 0.6 L of human milk and the average volume consumed by this age group (niacin content equals 1.1 mg), and adding the amount of niacin provided by solid foods (8 mg), as estimated by Montalto et al. (1985) (see Chapter 2). The result in NEs equals more than 9 mg/day. This value was judged to be unreasonably high because it is more than twice the extrapolated values and because mean intakes of older age groups appear to be much higher than their requirements. Thus the AI for niacin given in NEs is 4 mg/day for infants ages 7 through 12 months—the value extrapolated from estimates of adult requirements.

Niacin AI Summary, Ages 0 through 12 Months

AI for Infants

0–6 months	**2 mg/day of preformed niacin**	**≈0.2 mg/kg**
7–12 months	**4 mg/day of niacin equivalents**	**≈0.4 mg/kg**

Children and Adolescents Ages 1 through 18 Years

Method Used to Estimate the Average Requirement

No data were found on which to base the EAR for niacin for children or adolescents. In the absence of additional information, EARs and RDAs for children and adolescents have been estimated by using the method described in Chapter 2, which extrapolates from adult values.

Niacin EAR and RDA Summary, Ages 1 through 18 years

EAR for Children	**1–3 years**	**5 mg/day of niacin equivalents**
	4–8 years	**6 mg/day of niacin equivalents**
EAR for Boys	**9–13 years**	**9 mg/day of niacin equivalents**
	14–18 years	**12 mg/day of niacin equivalents**
EAR for Girls	**9–13 years**	**9 mg/day of niacin equivalents**
	14–18 years	**11 mg/day of niacin equivalents**

The RDA for niacin is set by using a coefficient of variation (CV)

of 15 percent (see Chapter 1 and the discussion of adult requirements that follows) because information is not available on the standard deviation of the requirement for these age groups; the RDA is
defined as equal to the EAR plus twice the CV to cover the needs of
97 to 98 percent of the individuals in the group (therefore, for
niacin the RDA is 130 percent of the EAR).

RDA for Children	**1–3 years**	**6 mg/day of niacin equivalents**
	4–8 years	**8 mg/day of niacin equivalents**
RDA for Boys	**9–13 years**	**12 mg/day of niacin equivalents**
	14–18 years	**16 mg/day of niacin equivalents**
RDA for Girls	**9–13 years**	**12 mg/day of niacin equivalents**
	14–18 years	**14 mg/day of niacin equivalents**

Adults Ages 19 Years and Older

Method Used to Estimate the Average Requirement

The best biochemical measure for estimating the average requirement was judged to be niacin metabolite excretion data, namely the
metabolites N^1-methyl-nicotinamide and its 2-pyridone derivative.
These measures have been reported the most extensively. The niacin
intakes that result in these measures being above the levels considered barely adequate represent some degree of body pool reserve.
Niacin metabolites are not excreted until adequate tryptophan is
available to meet needs for the synthesis of protein, nicotinamide
adenine dinucleotide, and nicotinamide adenine dinucleotide phosphate (Fu et al., 1989; Vivian et al., 1958). Metabolite excretion
measures are more sensitive to niacin depletion than are other biochemical measures, such as blood levels of pyridine nucleotides or
tryptophan. Whereas pellagra is prevented with NEs of about 11
mg/day (at 2,500 kcal/day), restoration of niacin metabolite excretion beyond minimal occurs at intakes of NEs of about 12 to 16 mg/
day. Excretion of N^1-methyl-nicotinamide rather than the 2-pyridone
derivative is preferred as the target measure for estimating the
niacin requirement because this metabolite better differentiates
marginal from adequate niacin intakes (Jacob et al., 1989), inter-
pretive guidelines exist for this metabolite (ICNND, 1963), and
more data are available for this than for other metabolites.

An average niacin requirement can be estimated as the niacin
intake corresponding to an excretion of N^1-methyl-nicotinamide

that is above the minimal excretion at which pellagra symptoms occur. Urinary excretion of N^1-methyl-nicotinamide was found to be 1.2 mg/day in men on low but adequate niacin diets that resulted in no signs or symptoms of pellagra (Horwitt et al., 1956; Jacob et al., 1989). Urinary N^1-methyl-nicotinamide excretion was found to be 0.6 mg/day in three females on corn-based, niacin-deficient diets that resulted in clinical signs and symptoms of pellagra (Goldsmith et al., 1952, 1955). A urinary excretion value for N^1-methyl-nicotinamide of 1.0 mg/day has been chosen as an interpolated level of niacin excretion; it reflects a niacin intake that is above the intake that results in clinical niacin deficiency and thus is minimal or barely adequate.

Niacin intakes (expressed in NEs) that would correspond to N^1-methyl-nicotinamide excretions of 1.0 mg/day are calculated from the results of four experimental studies in which NEs of 8 to 12 mg/day were fed (Table 6-1). A study by Patterson and colleagues (1980) was not used because no individual data were provided. In calculating the niacin intakes that would result in the N^1-methyl-nicotinamide excretion level of 1 mg/day, a linear relationship was assumed. This approach is conservative; if more niacin was fed, the ratio of niacin to the excretion of N^1-methyl-nicotinamide would become somewhat lower. Because all the studies were judged satisfactory in quality, a weighted average of the results was calculated. The overall average intake equivalent to the excretion of 1 mg/day of N^1-methyl-nicotinamide is 11.6 ± 3.94 (standard deviation), with a CV of 34 percent.

From Table 6-1, it might be interpreted that women require more niacin per 1,000 kcal than do men. However, the studies are too few and not sufficiently comparable to justify such a conclusion, especially considering a lack of evidence that utilization of niacin is less efficient for women than for men. The studies varied widely in methods, including the types of foods provided, which could have marked effects on the bioavailability of niacin. It is thus assumed that women have a slightly lower requirement than do men because of their size and average energy utilization. Therefore, the EAR is estimated to be 12 mg of NEs for men and 11 mg of NEs for women. Data are not available for determining whether the niacin requirement changes with age in adults.

The results of other studies were examined (Goldsmith, 1958; Goldsmith et al., 1956; Horwitt, 1958; Leklem et al., 1975), but coexisting deficiencies, a lack of valid dietary measurements, or both made it necessary to exclude them when deriving the EAR.

TABLE 6-1 Experimental Human Studies of Niacin Intake and Urine N^1-methylnicotinamide Excretion

Reference	Subjects	Regimen	NE Intake Calculated to Result in N^1-methylnicotinamide Excretion of 1.0 mg/d, Mean ± SD (CV%)[a]
Goldsmith et al., 1952	5 females, 25–54 y	Four fed a corn-based diet low in niacin and trp (7.7 mg of NEs) and one fed a wheat-based diet (9.5 mg of NEs) for up to 135 d. Supplemented with N^1-methylnicotinamide or trp.[b]	12.6 ± 3.0 (23%) or ≈6.8 mg of NEs/1,000 kcal
Goldsmith et al., 1955	3 females, 26–60 y	Fed wheat-based diet low in niacin and trp (8.3 mg of NEs) for up to 80 d. Supplemented with N^1-methylnicotinamide.	10.9 ± 0.9 (8%) or ≈5.8 mg of NEs/1,000 kcal
Horwitt et al., 1956	14 male mental patients, 30–65 y	Fed three ordinary food diets low in niacin and trp in a series of studies for up to 87 wk. Supplemented with N^1-methylnicotinamide or trp to provide 9–12 mg/d of NEs.	11.5 ± 4.5 (39%) or ≈4.9 mg of NEs/1,000 kcal
Jacob et al., 1989	7 males, 23–39 y	Gelatin-based diet of ordinary foods fed over 11 wk with varying ratios of trp and N^1-methylnicotinamide to provide 6 or 10 mg/d of NEs (and varying leucine).	11.3 ± 4.6 (41%) or ≈4.4 mg of NEs/1,000 kcal
Average of the four studies, two with females and two with males, on low-niacin diets (6–12 mg/d of NEs) for 4–24 wk			11.6 ± 3.9 (34%) or ≈4.8 mg of NEs/1,000 kcal

[a] NE = niacin equivalent (1 NE = 1 mg niacin = 60 mg tryptophan); SD = standard deviation; CV = coefficient of variation.
[b] trp = tryptophan.

Niacin EAR and RDA Summary, Ages 19 Years and Older

On the basis of the data in Table 6-1 and with a minor (approximately 10 percent) decrease for energy in women, the EAR is set at 12 mg/day of NEs for men and 11 mg/day of NEs for women.

EAR for Men	19–30 years	12 mg/day of niacin equivalents
	31–50 years	12 mg/day of niacin equivalents
	51–70 years	12 mg/day of niacin equivalents
	> 70 years	12 mg/day of niacin equivalents
EAR for Women	19–30 years	11 mg/day of niacin equivalents
	31–50 years	11 mg/day of niacin equivalents
	51–70 years	11 mg/day of niacin equivalents
	> 70 years	11 mg/day of niacin equivalents

The data in Table 6-1 also suggest a CV of the niacin requirement that is greater than 10 percent. The wide variation in the efficiency of converting tryptophan to niacin may contribute to the large apparent variation. Thus, a CV of 15 percent is used; the RDA is defined as equal to the EAR plus twice the CV to cover the needs of 97 to 98 percent of the individuals in the group (therefore, for niacin the RDA is 130 percent of the EAR).

RDA for Men	19–30 years	16 mg/day of niacin equivalents
	31–50 years	16 mg/day of niacin equivalents
	51–70 years	16 mg/day of niacin equivalents
	> 70 years	16 mg/day of niacin equivalents
RDA for Women	19–30 years	14 mg/day of niacin equivalents
	31–50 years	14 mg/day of niacin equivalents
	51–70 years	14 mg/day of niacin equivalents
	> 70 years	14 mg/day of niacin equivalents

Pregnancy

Method Used to Estimate the Average Requirement

There is no direct evidence that would suggest a change in the niacin requirement during pregnancy. To derive the EAR for pregnant women, it is estimated that the need for niacin increases by 3 mg/day of NEs to cover increased energy utilization and growth in maternal and fetal compartments, especially during the second and third trimesters.

Niacin EAR and RDA Summary, Pregnancy

By adding 3 mg of NEs to the EAR of 11 mg of NEs for nonpregnant, nonlactating women, the EAR for pregnancy becomes 14 mg of NEs. No adjustment is made for the woman's age.

EAR for Pregnancy
14–18 years	14 mg/day of niacin equivalents
19–30 years	14 mg/day of niacin equivalents
31–50 years	14 mg/day of niacin equivalents

The data in Table 6-1 suggest a CV for the niacin requirement that is greater than 10 percent. The wide variation in the efficiency of converting tryptophan to niacin may contribute to the larger apparent variation. Thus, a CV of 15 percent is used because information is not available on the standard deviation of the requirement for pregnant women; the RDA is defined as equal to the EAR plus twice the CV to cover the needs of 97 to 98 percent of the individuals in the group (therefore, for niacin the RDA is 130 percent of the EAR).

RDA for Pregnancy
14–18 years	18 mg/day of niacin equivalents
19–30 years	18 mg/day of niacin equivalents
31–50 years	18 mg/day of niacin equivalents

Lactation

Method Used to Estimate the Average Requirement

An estimated 1.4 mg of preformed niacin is secreted daily during lactation. Added to this is a small amount (1 mg) to cover energy expenditure involved in milk production. Thus, the additional amount of niacin needed is 2.4 mg/day of NEs for women who are exclusively breastfeeding an infant.

Niacin EAR and RDA Summary, Lactation

Adding 2.4 mg of NEs to the EAR of 11 mg of NEs for nonpregnant, nonlactating women gives an EAR for niacin for lactation of 13.4 mg, rounded down to 13.

EAR for Lactation
14–18 years	**13 mg/day of niacin equivalents**
19–30 years	**13 mg/day of niacin equivalents**
31–50 years	**13 mg/day of niacin equivalents**

The data in Table 6-1 suggest a CV for the niacin requirement that is greater than 10 percent. The wide variation in the efficiency of converting tryptophan to niacin may contribute to the larger apparent variation. Thus, a CV of 15 percent is used because information is not available on the standard deviation of the requirement during lactation; the RDA is defined as equal to the EAR plus twice the CV to cover the needs of 97 to 98 percent of the individuals in the group (therefore, for niacin the RDA is 130 percent of the EAR).

RDA for Lactation
14–18 years	**17 mg/day of niacin equivalents**
19–30 years	**17 mg/day of niacin equivalents**
31–50 years	**17 mg/day of niacin equivalents**

Special Considerations

The RDAs given above are not expected to be sufficient to meet the needs of persons with Hartnup's disease, liver cirrhosis, or carcinoid syndrome or of those on long-term isoniazid treatment. As for other B vitamins, extra niacin may be required by persons treated with hemodialysis or peritoneal dialysis, those with malabsorption syndrome, pregnant women bearing multiple fetuses, and women breastfeeding more than one infant.

INTAKE OF NIACIN

Food Sources

Data obtained from the 1995 Continuing Survey of Food Intakes by Individuals indicate that the greatest contribution to the niacin intake of the U.S. adult population comes from mixed dishes high in meat, fish, or poultry; poultry as an entree; enriched and whole-grain breads and bread products; and fortified ready-to-eat cereals

(Table 6-2). In addition to the foods listed in Table 6-2, most flesh foods are sources of niacin, providing at least 2 mg per serving.

Dietary Intake

Niacin intake in the United States is generous in comparison with the Estimated Average Requirement (EAR). For example, the median intake by adult women is 17 to 20 mg of niacin (Appendixes G and H) in comparison with an EAR of 11 mg of niacin equivalents (NEs). (Survey data are reported as preformed niacin; no addition has been made for the conversion of tryptophan to niacin.) For all life stage and gender groups it appears that almost all individuals' usual niacin intakes would exceed the EAR if intake was expressed in NEs and thus the contribution of tryptophan was included. Intakes of niacin in two Canadian provinces are reported in NEs and are well above the EAR for all life stage and gender groups (Appendix I).

The Boston Nutritional Status Survey (Appendix F) indicates that this relatively advantaged group of people over age 60 has a median niacin intake of 21 mg/day for men and 17 mg/day for women, again, significantly above the EARs for adult men and women.

Intake from Supplements

Information from the Boston Nutritional Status Survey on the use of niacin supplements by a free-living elderly population is given in Appendix F. For those taking supplements, the fiftieth percentile of supplemental niacin intake was 20 mg for men and 30 mg for women. In the 1986 National Health Interview Survey, 26 percent of all adults reported use of supplements containing niacin (Moss et al., 1989). Supplements containing up to about 400 mg of niacin are available without a prescription.

TOLERABLE UPPER INTAKE LEVELS

Hazard Identification

Adverse Effects

There is no evidence of adverse effects from the consumption of naturally occurring niacin in foods. Therefore, this review is limited to evidence concerning intake of niacin as a supplement, food fortificant, or pharmacological agent.

TABLE 6-2 Food Groups Providing Niacin[a] in the Diets of U.S. Men and Women Aged 19 Years and Older, CSFII, 1995[b]

Food Group	Contribution to Total Niacin Intake[c] (%)		Foods Within the Group that Provide at Least 4 mg of Niacin[d] per Serving	
	Men	Women	4–8 mg	> 8 mg
Food groups providing at least 5% of total niacin intake				
Mixed foods[e]	14.7	12.4	NA[f]	NA
Poultry	10.8	11.9	Turkey light meat, duck, and chicken light and dark meat	Chicken breast and Cornish game hen
Bread and bread products	10.7	11.5	—	—
Ready-to-eat cereals	8.3	10.9	Moderately fortified	Highly fortified
Mixed foods, main ingredient is grain	8.0	7.1	NA	NA
Beef	6.5	4.7	—	—
Processed meats[g]	4.7	3.5	—	—
Niacin from other food groups				
Pork	2.8	2.5	Pork and ham	—
Finfish	1.7	2.7	Salmon, mackerel, mullet, croaker, and porgy	Tuna, swordfish, sturgeon, and trout
Lamb, veal, game	0.4	0.4	Lamb, venison, rabbit, and other game	Veal
Organ meats	0.2	0.3	—	Liver

[a] Preformed niacin only, not niacin equivalents.

[b] CSFII = Continuing Survey of Food Intakes by Individuals.

[c] Contribution to total intake reflects both the concentration of the nutrient in the food and the amount of the food consumed. It refers to the percentage contribution to the American diet for both men and women, based on 1995 CSFII data.

[d] 4 mg represents 20% of the Recommended Daily Intake (20.0 mg) of niacin—a value set by the Food and Drug Administration.

[e] Includes sandwiches and other foods with meat, poultry, or fish as the main ingredient.

[f] NA = not applicable. Mixed foods were not considered for this table.

[g] Includes frankfurters, sausages, lunch meats, and meat spreads.

SOURCE: Unpublished data from the Food Surveys Research Group, Agricultural Research Service, U.S. Department of Agriculture, 1997.

One report showed adverse effects after consumption of bagels to which 60 times the normal amount of niacin had been added inadvertently (CDC, 1983). Most of the data on the adverse effects of excess niacin intake are from studies and case reports involving patients with hyperlipidemia or other disorders who were treated with pharmacological preparations containing immediate-release nicotinic acid or slow- or sustained-release nicotinic acid. The Tolerable Upper Intake Level (UL) developed here applies to all forms of niacin added to foods or taken as supplements (e.g., immediate-release, slow or sustained-release nicotinic acid, and niacinamide [nicotinamide]). Adverse effects such as nausea, vomiting, and signs and symptoms of liver toxicity have been observed at nicotinamide intakes of 3,000 mg/day (Rader et al., 1992) compared with intakes of nicotinic acid of 1,500 mg/day (McKenney et al., 1994). The generic term *niacin* may be considered interchangeable with *nicotinic acid*. As described below, the critical adverse effect selected was flushing to the extent that it results in a change in the dosing pattern or withdrawal from treatment.

Vasodilatory Effects (Flushing). The term flushing covers a burning, tingling, and itching sensation as well as a reddened flush primarily on the face, arms, and chest. Flushing occurs in many patients treated with nicotinic acid therapeutically. It is often accompanied by pruritus, headaches, and increased intracranial blood flow (Miller and Hayes, 1982). Occasionally, it is accompanied by pain (Bean and Spies, 1940). Case reports and clinical trials have reported flushing effects at oral doses of 30 to 1,000 mg/day within 30 minutes to 6 weeks of the initial dose (CDC, 1983; Estep et al., 1977; Henkin et al., 1990; McKenney et al., 1994; Sebrell and Butler, 1938; Spies et al., 1938). Although flushing is a transient effect, it often results in patients deciding to withdraw from treatment.

In a study of the flushing effects of nicotinic acid and other pyridine compounds in humans, Bean and Spies (1940) suggest that pyridine compounds without a carboxyl radical in the 3 position of the pyridine ring do not produce flushing effects. Nicotinamide, which does not have a carboxyl radical in the 3 position, does not appear to be associated with flushing.

Flushing appears to be more closely related to a continuous rise in plasma nicotinic acid concentrations than to the absolute dose. Tolerance to nicotinic acid-induced flushing can develop whereby the effects are minimized when the dose is slowly increased over time (Stern et al., 1991). Flushing effects can also be reduced somewhat by taking niacin with food (Knodel and Talbert, 1987). Flushing has been shown to be prostaglandin mediated, and tolerance results from reduction in prostaglandin levels with continued administration (Morrow et al., 1989, 1992).

Gastrointestinal Effects. In addition to flushing, nonspecific gastrointestinal effects are common in patients treated with nicotinic acid (Knodel and Talbert, 1987), especially with slow-release preparations (McKenney et al., 1994; Rader et al., 1992). The gastrointestinal effects in patients taking the slow-release form have been shown to be associated with liver enzyme elevations (Gibbons et al., 1995).

Hepatotoxicity. The hepatotoxicity of niacin has been demonstrated in numerous case reports (Clementz and Holmes, 1987; Dalton and Berry, 1992; Einstein et al., 1975; Etchason et al., 1991; Ferenchick and Rovner, 1989; Frost, 1991; Goldstein, 1988; Henkin et al., 1990; Hodis, 1990; Knapp and Middleton, 1991; Knopp, 1991; McKenney et al., 1994; Mullin et al., 1989; Palumbo, 1991; Patterson et al., 1983). In the most severe cases, patients develop liver dysfunction and fulminant hepatitis and may progress to stage 3 and 4 encephalopathy requiring liver transplantation (Clementz and Holmes, 1987; Hodis, 1990; Mullin et al., 1989). The most frequently observed manifestations of niacin-induced hepatitis are jaundice and increased levels of serum transaminases (Etchason et al., 1991). These effects are typically associated with high doses (3 to 9 g/day of niacin) used to treat patients with hypercholesterolemia for periods of months to years (Clementz and Holmes, 1987; Einstein et al., 1975; Pardue, 1961; Patterson et al., 1983; Rivin, 1959; Winter and Boyer, 1973). Almost all of the patients in these case reports were taking the slow-release form. A recent double-blind comparison study suggested that the slow-release form is more hepatotoxic than the immediate-release form (McKenney et al., 1994). However, another study reported hepatotoxicity with both forms of niacin (Gray et al., 1994).

Glucose Intolerance. Large doses (3 g/day) of nicotinic acid used to treat patients with hypercholesterolemia have produced impaired glucose tolerance in otherwise apparently healthy individuals (Miettinen et al., 1969). The adverse effects on glucose tolerance have been observed during short- and long-term administration of the drug (Schwartz, 1993).

Ocular Effects. Niacin treatment may produce significant ocular effects, including blurred vision, toxic amblyopia, macular edema, and cystic maculopathy. However, there are very few cases in the literature (Fraunfelder et al., 1995). Doses of 1.5 to 5 g/day of niacin have been associated with ocular effects (Fraunfelder et al., 1995; Gass, 1973; Millay et al., 1988). Niacin-induced ocular effects appear to be reversible and dose dependent.

Summary

Flushing is the adverse effect first observed after excess niacin intake and is generally observed at lower doses than are other effects. Flushing that results in patients deciding to change the pattern of niacin intake (i.e., reduce the amount taken at a time or withdraw from treatment) was selected as the most appropriate endpoint on which to base a UL. Although nicotinamide appears not to be associated with flushing effects, a UL for nicotinic acid that is based on flushing is considered protective against potential adverse effects of nicotinamide. The data on hepatotoxicity are considered less relevant to the general population because they involve large doses taken for long periods of time for the treatment of a medical condition.

Dose-Response Assessment

Adults

Data Selection. The data sets used to identify the lowest-observed-adverse-effect level (LOAEL) for niacin included anecdotal reports and clinical trials involving oral intake of niacin by healthy individuals. Studies involving parenteral administration were not considered in the dose-response assessment. Studies involving immediate-release forms of niacin were considered more relevant to niacin intake by the general population than were studies involving sustained-release forms.

Identification of a LOAEL. The data are not adequate to identify a no-observed-adverse-effect level (NOAEL) for flushing. To identify a LOAEL, flushing reactions were considered if they resulted in a patient either changing the form or amount of niacin used or withdrawing from treatment. A LOAEL of 50 mg/day was identified based on a study by Sebrell and Butler (1938) in which four (66 percent) of six persons experienced a flushing sensation after oral intake of 50 mg/day of nicotinic acid given with meals for 92 days. In one of the four subjects who experienced flushing effects, the daily dose of 50 mg was given as 25 mg in the morning and evening. Although this study also reported a flushing reaction in one of six subjects taking 30 mg of nicotinic acid daily on day 32 of intake, this reaction was not bothersome enough to change the dosing pattern. Sebrell and Butler (1938) was selected as the critical study for identifying a LOAEL and deriving a UL because it provides the lowest effect level. A study by Spies et al. (1938) provides supportive evidence for a LOAEL of 50 mg/day. In this study, five of 100 individuals (5 percent) experienced flushing after a single oral dose of 50 mg of nicotinic acid, 50 individuals (50 percent) experienced flushing after 100 mg, and all individuals experienced flushing after 500 mg.

There is one case report showing that 14 of 69 persons (20 percent) experienced onset of rash, pruritus, and a sensation of warmth about 30 minutes after consuming one or more pumpernickel bagels to which niacin had been inadvertently added from an improperly labeled container (CDC, 1983). The bagels were found to contain an average of 190 mg of niacin.

Uncertainty Assessment. Because of the transient nature of the flushing effect, a small uncertainty factor (UF) of 1.5 was selected. A smaller UF was not appropriate because it is applied to a LOAEL rather than a NOAEL.

Derivation of a UL. A LOAEL of 50 mg/day was divided by a UF of 1.5 to obtain the UL for adults of 35 mg/day, a rounded estimate.

UL for Adults   19 years and older   35 mg/day of niacin

Other Life Stage Groups

For infants the UL was judged not determinable because of a lack of data on adverse effects in this age group and concern about the infant's ability to handle excess amounts. To prevent high levels of intake, the only source of intake for infants should be from food. No data were found to suggest that other life stage groups have increased susceptibility to flushing effects from excess niacin intake. Therefore, the UL of 35 mg/day is also set for pregnant and lactating adult women. The UL of 35 mg/day for adults was adjusted for children and adolescents on the basis of relative body weight as described in Chapter 3 and by using reference weights from Chapter 1, Table 1-2. Values have been rounded down.

UL for Infants
0–12 months   Not possible to establish; source of intake
                     should be formula and food only

UL for Children
1–3 years   10 mg/day of niacin
4–8 years   15 mg/day of niacin
9–13 years   20 mg/day of niacin

UL for Adolescents
14–18 years   30 mg/day of niacin

UL for Pregnancy
14–18 years   30 mg/day of niacin
19 years and older   35 mg/day of niacin

UL for Lactation
14–18 years   30 mg/day of niacin
19 years and older   35 mg/day of niacin

Special Considerations

A review of the literature identified individuals with the following conditions as being distinctly susceptible to the adverse effects of excess niacin intake: hepatic dysfunction or a history of liver disease, diabetes mellitus, active peptic ulcer disease, gout, cardiac arrhythmias, inflammatory bowel disease, migraine headaches, and alcoholism. Therefore, people with these conditions may not be protected by the UL for niacin for the general population.

Intake Assessment

On the basis of data from the Third National Health and Nutrition Survey, the highest mean intake of niacin from diet and supplements for any life stage and gender group was 39 mg/day. This intake was being consumed by men aged 31 through 50 years, women over age 70, and pregnant women aged 14 through 55 years. The highest reported intake at the ninety-fifth percentile was 77 mg/day in women aged 51 through 70 years (see Appendix H). Niacin is available over the counter in dosages ranging up to 100 mg or more (in the immediate-release form).

Risk Characterization

Niacin intake data indicate that a small percentage of the U.S. population is likely to exceed the UL. Individuals who take over-the-counter niacin to treat themselves, for example, for high blood cholesterol, might exceed the UL on a chronic basis. The UL is not meant to apply to individuals who are receiving niacin under medical supervision.

RESEARCH RECOMMENDATIONS FOR NIACIN

Data useful for setting the Estimated Average Requirement (EAR) for children, adolescents, pregnant women, and lactating women are scanty, but evidence suggests that niacin intake in the United States and Canada is generous relative to need. Priority should be given to studies in two areas:

• the niacin requirement to satisfy nicotinamide adenine dinucleotide (NAD) needs for increased adenosine diphosphate ribosylation resulting from oxidant-deoxyribonucleic acid damage and
• sensitive and specific blood measures of niacin status. Current assessments of niacin status and requirement are based solely on urinary metabolite measures; measurements of plasma metabolites such as the 2-pyridone derivatives may be productive. Two recent experimental studies have suggested erythrocyte NAD as a functional blood measure of niacin status (Fu et al., 1989; Ribaya-Mercado et al., 1997), but further work is needed in clinical populations.

REFERENCES

Bean WB, Spies TD. 1940. A study of the effects of nicotinic acid and related pyridine and pyrazine compounds on the temperature of the skin of human beings. *Am Heart J* 20:62–75.

Bechgaard H, Jespersen S. 1977. GI absorption of niacin in humans. *J Pharm Sci* 66:871–872.

Bernofsky C. 1980. Physiology aspects of pyridine nucleotide regulation in mammals. *Mol Cell Biochem* 33:135–143.

Carpenter KJ, Lewin WJ. 1985. A reexamination of the composition of diets associated with pellagra. *J Nutr* 115:543–552.

Carter EG, Carpenter KJ. 1982. The bioavailability for humans of bound niacin from wheat bran. *Am J Clin Nutr* 36:855–861.

CDC (Centers for Disease Control and Prevention). 1983. Niacin intoxication from pumpernickel bagels—New York. *MMWR* 32:305.

Clementz GL, Holmes AW. 1987. Nicotinic acid-induced fulminant hepatic failure. *J Clin Gastroenterol* 9:582–584.

Committee on Nutrition. 1985. Composition of human milk: Normative data. In: *Pediatric Nutrition Handbook,* 2nd ed. Elk Grove Village, IL: American Academy of Pediatrics. Pp. 363–368.

Dalton TA, Berry RS. 1992. Hepatotoxicity associated with sustained-release niacin. *Am J Med* 93:102–104.

Dillon JC, Malfait P, Demaux G, Foldihope C. 1992. Urinary metabolites of niacin during the course of pellagra. *Ann Nutr Metab* 36:181–185.

Einstein N, Baker A, Galper J, Wolfe H. 1975. Jaundice due to nicotinic acid therapy. *Am J Dig Dis* 20:282–286.

Estep DL, Gay GR, Rappolt RT Sr. 1977. Preliminary report of the effects of propranolol HCl on the discomfiture caused by niacin. *Clin Toxicol* 11:325–328.

Etchason JA, Miller TD, Squires RW, Allison TG, Gau GT, Marttila JK, Kottke BA. 1991. Niacin-induced hepatitis: A potential side effect with low-dose time-release niacin. *Mayo Clin Proc* 66:23–28.

Ferenchick G, Rovner D. 1989. Case report: Hepatitis and hematemesis complicating nicotinic acid use. *Am J Med Sci* 298:191–193.

Ford JE, Zechalko A, Murphy J, Brooke OG. 1983. Comparison of the B vitamin composition of milk from mothers of preterm and term babies. *Arch Dis Child* 58:367–372.

Fraunfelder FW, Fraunfelder FT, Illingworth DR. 1995. Adverse ocular effects associated with niacin therapy. *Br J Ophthalmol* 79:54–56.

Frost PH. 1991. All niacin is not the same. *Ann Intern Med* 114:1065.

Fu CS, Swendseid ME, Jacob RA, McKee RW. 1989. Biochemical markers for assessment of niacin status in young men: Levels of erythrocyte niacin coenzymes and plasma tryptophan. *J Nutr* 119:1949–1955.

Gass JD. 1973. Nicotinic acid maculopathy. *Am J Ophthalmol* 76:500–510.

Gibbons LW, Gonzalez V, Gordon N, Grundy S. 1995. The prevalence of side effects with regular and sustained-release nicotinic acid. *Am J Med* 99:378–385.

Goldsmith GA. 1958. Niacin-tryptophan relationships in man and niacin requirement. *Am J Clin Nutr* 6:479–486.

Goldsmith GA, Sarett HP, Register UD, Gibbens J. 1952. Studies on niacin requirement in man. 1. Experimental pellagra in subjects on corn diets low in niacin and tryptophan. *J Clin Invest* 31:533–542.

Goldsmith GA, Rosenthal HL, Gibbens J, Unglaub WG. 1955. Studies on niacin requirement in man. 2. Requirement on wheat and corn diets low in tryptophan. *J Nutr* 56:371–386.

Goldsmith GA, Gibbens J, Unglaub WG, Miller ON. 1956. Studies on niacin requirement in man. 3. Comparative effects of diets containing lime-treated and untreated corn in the production of experimental pellagra. *Am J Clin Nutr* 4:151–160.

Goldstein MR. 1988. Potential problems with the widespread use of niacin. *Am J Med* 85:881.

Gray DR, Morgan T, Chretien SD, Kashyap ML. 1994. Efficacy and safety of controlled-release niacin in dyslipoproteinemic veterans. *Ann Intern Med* 121:252–258.

Henderson LM, Gross CJ. 1979. Metabolism of niacin and niacinamide in perfused rat intestine. *J Nutr* 109:654–662.

Henkin Y, Johnson KC, Segrest JP. 1990. Rechallenge with crystalline niacin after drug-induced hepatitis from sustained-release niacin. *J Am Med Assoc* 264:241–243.

Hodis HN. 1990. Acute hepatic failure associated with the use of low-dose sustained-release niacin. *J Am Med Assoc* 264:181.

Horwitt MK. 1958. Niacin-tryptophan requirements of man. *J Am Diet Assoc* 34:914–919.

Horwitt MK, Harvey CC, Rothwell WS, Cutler JL, Haffron D. 1956. Tryptophan-niacin relationships in man: Studies with diets deficient in riboflavin and niacin, together with observations on the excretion of nitrogen and niacin metabolites. *J Nutr* 60:1–43.

Horwitt MK, Harper AE, Henderson LM. 1981. Niacin-tryptophan relationships for evaluating niacin equivalents. *Am J Clin Nutr* 34:423–427.

ICNND (Interdepartmental Committee on Nutrition for National Defense). 1963. *Manual for Nutrition Surveys*, 2nd ed. Bethesda, MD: National Institutes of Health. P. 244.

Jacob RA, Swendseid ME, McKee RW, Fu CS, Clemens RA. 1989. Biochemical markers for assessment of niacin status in young men: Urinary and blood levels of niacin metabolites. *J Nutr* 119:591–598.

Jacobson EL, Lange RA, Jacobson MK. 1979. Pyridine nucleotide synthesis in 3T3 cells. *J Cell Physiol* 99:417–425.

Kim H, Jacobson EL, Jacobson MK. 1994. NAD glycohydrolases: A possible function in calcium homeostasis. *Mol Cell Biochem* 138:237–243.

Knapp TR, Middleton RK. 1991. Adverse effects of sustained-release niacin. *Ann Pharmacother* 25:253–254.

Knodel LC, Talbert RL. 1987. Adverse effects of hypolipidaemic drugs. *Med Toxicol* 2:10–32.

Knopp RH. 1991. Niacin and hepatic failure. *Ann Intern Med* 111:769.

Knox WE. 1951. Two mechanisms which increase in vivo the liver tryptophan peroxidase activity: Specific enzyme adaptation and stimulation of the pituitary-adrenal system. *Br J Exp Pathol* 32:462–469.

Lan SJ, Henderson LM. 1968. Uptake of nicotinic acid and nicotinamide by rat erythrocytes. *J Biol Chem* 243:3388–3394.

Lautier D, Lagueux J, Thibodeau J, Menard L, Poirier GG. 1993. Molecular and biochemical features of poly (ADP-ribose) metabolism. *Mol Cell Biochem* 122:171–193.

Lee HC, Walseth TF, Bratt GT, Hayes RN, Clapper DL. 1989. Structural determination of a cyclic metabolite of NAD+ with intracellular CA2+-mobilizing activity. *J Biol Chem* 264:1608–1615.

Leklem JE, Brown RR, Rose DP, Linkswiler H, Arend RA. 1975. Metabolism of tryptophan and niacin in oral contraceptive users receiving controlled intakes of vitamin B_6. *Am J Clin Nutr* 28:146–156.

Malfait P, Moren A, Dillon JC, Brodel A, Begkoyian G, Etchegorry MG, Malenga G, Hakewill P. 1993. An outbreak of pellagra related to changes in dietary niacin among Mozambican refugees in Malawi. *Int J Epidemiol* 22:504–511.

McCormick DB. 1988. Niacin. In: Shils ME, Young VR, eds. *Modern Nutrition in Health and Disease*. Philadelphia: Lea & Febiger. Pp. 370–375.

McCormick DB. 1989. Two interconnected B vitamins: Riboflavin and pyridoxine. *Physiol Rev* 69:1170–1198.

McKenney JM, Proctor JD, Harris S, Chinchili VM. 1994. A comparison of the efficacy and toxic effects of sustained- vs immediate-release niacin in hypercholesterolemic patients. *J Am Med Assoc* 271:672–677.

Miettinen TA, Taskinen M-R, Pelkonen R, Nikkila EA. 1969. Glucose tolerance and plasma insulin in man during acute and chronic administration of nicotinic acid. *Acta Med Scand* 186:247–253.

Millay RH, Klein ML, Illingworth DR. 1988. Niacin maculopathy. *Ophthalmology* 95:930–936.

Miller DR, Hayes KC. 1982. Vitamin excess and toxicity. In: Hathcock JN, ed. *Nutritional Toxicology*, Vol. 1. New York: Academic Press. Pp. 81–133.

Montalto MB, Benson JD, Martinez GA. 1985. Nutrient intake of formula-fed infants and infants fed cow's milk. *Pediatrics* 75:343–351.

Morrow JD, Parsons WG 3rd, Roberts LJ 2nd. 1989. Release of markedly increased quantities of prostaglandin D2 in vivo in humans following the administration of nicotinic acid. *Prostaglandins* 38:263–274.

Morrow JD, Awad JA, Oates JA, Roberts LJ. 1992. Identification of skin as a major site of prostaglandin D2 release following oral administration of niacin in humans. *J Invest Dermatol* 98:812–815.

Moss AJ, Levy AS, Kim I, Park YK. 1989. *Use of Vitamin and Mineral Supplements in the United States: Current Users, Types of Products, and Nutrients.* Advance Data, Vital and Health Statistics of the National Center for Health Statistics, No. 174. Hyattsville, MD: National Center for Health Statistics.

Mrochek JE, Jolley RL, Young DS, Turner WJ. 1976. Metabolic response of humans to ingestion of nicotinic acid and nicotinamide. *Clin Chem* 22:1821–1827.

Mullin GE, Greenson JK, Mitchel MC. 1989. Fulminant hepatic failure after ingestion of sustained-release nicotinic acid. *Ann Intern Med* 111:253–255.

NRC (National Research Council). 1989. *Recommended Dietary Allowances,* 10th ed. Washington, DC: National Academy Press.

Palumbo PJ. 1991. Rediscovery of crystalline niacin. *Mayo Clin Proc* 66:112–113.

Pardue WO. 1961. Severe liver dysfunction during nicotinic acid therapy. *J Am Med Assoc* 175:137–138.

Patterson JI, Brown RR, Linkswiler H, Harper AE. 1980. Excretion of tryptophan-niacin metabolites by young men: Effects of tryptophan, leucine, and vitamin B$_6$ intakes. *Am J Clin Nutr* 33:2157–2167.

Patterson DJ, Dew EW, Gyorkey R, Graham GY. 1983. Niacin hepatitis. *South Med J* 76:239–241.

Rader JI, Calvert RJ, Hathcock JN. 1992. Hepatic toxicity of unmodified and time-release preparations of niacin. *Am J Med* 92:77–81.

Ribaya-Mercado JD, Russell RM, Rasmussen HM, Crim MC, Perrone-Petty G, Gershoff SN. 1997. Effect of niacin status on gastrointestinal function and serum lipids. *FASEB J* 11:A179.

Rivin AU. 1959. Jaundice occurring during nicotinic acid therapy for hypercholesteremia. *J Am Med Assoc* 170:2088–2089.

Rose DP, Braidman IP. 1971. Excretion of tryptophan metabolites as affected by pregnancy, contraceptive steroids, and steroid hormones. *Am J Clin Nutr* 24:673–683.

Sauberlich HE, Skala JH, Dowdy RP. 1974. *Laboratory Tests for the Assessment of Nutritional Status.* Boca Raton, FL: CRC Press.

Schwartz ML. 1993. Severe reversible hyperglycemia as a consequence of niacin therapy. *Arch Intern Med* 153:2050–2052.

Sebrell WH, Butler RE. 1938. A reaction to the oral administration of nicotinic acid. *J Am Med Assoc* 111:2286–2287.

Shibata K, Matsuo H. 1989. Effect of supplementing low protein diets with the limiting amino acids on the excretion of N^1-methylnicotinamide and its pyridones in rats. *J Nutr* 119:896–901.

Spies TD, Bean WB, Stone RE. 1938. The treatment of subclinical and classic pellagra. *J Am Med Assoc* 111:584–592.

Stern RH, Spence JD, Freeman DJ, Parbtani A. 1991. Tolerance to nicotinic acid flushing. *Clin Pharmacol Ther* 50:66–70.

Stierum RH, Vanherwijnen MH, Hageman GJ, Kleinjans JC. 1994. Increased poly (ADP-ribose) polymerase activity during repair of (+/-)-anti-benzo[a]pyrene diolepoxide-induced DNA damage in human peripheral blood lymphocytes in vitro. *Carcinogenesis* 15:745–751.

Vivian VM, Chaloupka MM, Reynolds MS. 1958. Some aspects of tryptophan metabolism in human subjects. 1. Nitrogen balances, blood pyridine nucleotides, and urinary excretion of N-methylnicotinamide and N-methyl-2-pyridone-5-carboxamide on a low-niacin diet. *J Nutr* 66:587–598.

Wertz AW, Lojkin ME, Bouchard BS, Derby MB. 1958. Tryptophan-niacin relationships in pregnancy. *J Nutr* 64:339–353.

Winter SL, Boyer JL. 1973. Hepatic toxicity from large doses of vitamin B_3 (nicotinamide). *N Engl J Med* 289:1180–1182.

7

Vitamin B$_6$

SUMMARY

Vitamin B$_6$ (pyridoxine and related compounds) functions as a coenzyme in the metabolism of amino acids, glycogen, and sphingoid bases. The primary criterion used to estimate the Recommended Dietary Allowance (RDA) for vitamin B$_6$ is a plasma 5'-pyridoxal phosphate value of at least 20 nmol/L. Bioavailability of 75 percent is assumed from a mixed diet. The RDA for young adults is 1.3 mg. Recently, the median intake of vitamin B$_6$ from food in the United States was approximately 2 mg/day for men and 1.4 mg/day for women; in one Canadian population study the median intake was approximately 1.8 mg/day for men and 1.3 mg/day for women. The ninety-fifth percentile of U.S. intake from both food and supplements has been estimated to be 6 to 10 mg/day. The critical adverse effect from high intake of the vitamin is sensory neuropathy. The data fail to demonstrate a causal association between pyridoxine intake and other endpoints (e.g., dermatological lesions and vitamin B$_6$ dependency in newborns). The Tolerable Upper Intake Level (UL) for adults is 100 mg/day of vitamin B$_6$.

BACKGROUND INFORMATION

Vitamin B$_6$ (B$_6$) comprises a group of six related compounds: pyridoxal (PL), pyridoxine (PN), pyridoxamine (PM), and their respective 5'-phosphates (PLP, PNP, and PMP). The major forms in animal tissues are PLP and PMP; plant-derived foods contain primarily PN and PNP, sometimes in the form of a glucoside. In humans, the major excretory form is 4-pyridoxic acid (4-PA).

Function

PLP is a coenzyme for more than 100 enzymes involved in amino acid metabolism, including aminotransferases, decarboxylases, racemases, and dehydratases. It is a coenzyme for δ-aminolevulinate synthase, which catalyzes the first step in heme biosynthesis, and for cystathionine β-synthase and cystathioninase, enzymes involved in the transsulfuration pathway from homocysteine to cysteine. The carbonyl group of PLP binds to proteins as a Schiff's base with the ε-amine of lysine. For practically all PLP enzymes the initial step in catalysis involves formation of a Schiff's base between an incoming amino acid, via its α-amino group, and the carbonyl group of PLP. Much of the total PLP in the body is found in muscle bound to phosphorylase. PLP is a coenzyme in the phosphorylase reaction and is also directly involved in catalysis.

Physiology of Absorption, Metabolism, and Excretion

Absorption and Transport

In animal tissue the major form of B$_6$ is PLP; next is PMP. Absorption in the gut involves phosphatase-mediated hydrolysis followed by transport of the nonphosphorylated form into the mucosal cell. Transport is by a nonsaturable passive diffusion mechanism. Even extremely large doses are well absorbed (Hamm et al., 1979). PN glucoside is absorbed less effectively than are PLP and PMP and, in humans, is deconjugated by a mucosal glucosidase (Nakano and Gregory, 1997). Some PN glucoside is absorbed intact and can be hydrolyzed in various tissues.

Metabolism

Most of the absorbed nonphosphorylated B$_6$ goes to the liver. PN, PL, and PM are converted to PNP, PLP, and PMP by PL kinase. PNP, which is normally found only at very low concentrations, and PMP are oxidized to PLP by PNP oxidase. PMP is also generated from PLP via aminotransferase reactions. PLP is bound to various proteins in tissues; this protects it from the action of phosphatases. The capacity for protein binding limits the accumulation of PLP by tissues at very high intakes of B_6 (Merrill et al., 1984). When this capacity is exceeded, free PLP is rapidly hydrolyzed and nonphosphorylated forms of B_6 are released by the liver and other tissues into circulation. At pharmacological doses of B_6, the high capacities for PLP-protein binding of muscle, plasma, and erythrocytes (hemoglobin) allow them to accumulate very high levels of PLP when other tissues are saturated (Lumeng et al., 1978).

PLP in the liver can be oxidized to 4-PA, which is released and excreted. The major PLP-binding protein in plasma is albumin. PLP is the major form of the vitamin in plasma and is derived entirely from liver as a PLP-albumin complex (Fonda et al., 1991; Leklem, 1991). Tissues and erythrocytes can transport nonphosphorylated forms of the vitamin from plasma. Some of this is derived from plasma PLP after phosphatase action. In tissues, conversion of the transported vitamin to PLP, coupled with protein binding, allows accumulation and retention of the vitamin. B_6 in tissues is found in various subcellular compartments but primarily in the mitochondria and the cytosol.

Excretion

Normally, the major excretory product is 4-PA, which accounts for about half the B_6 compounds in urine (Shultz and Leklem, 1981). Other forms of the vitamin are also found in urine. With large doses of B_6, the proportion of the other forms of the vitamin increases. At very high doses of PN, much of the dose is excreted unchanged in the urine. B_6 is also excreted in feces but probably to a limited extent (Lui et al., 1985). Microbial synthesis of B_6 in the lower gut makes it difficult to evaluate the extent of this excretion.

Body Stores

Pharmacokinetic analyses of urinary excretion of a tracer dose of labeled PN and its metabolites have suggested a two-compartment model for body B_6 stores (Johansson et al., 1966). With this approach, body stores have been estimated at 365 µmol (61 mg) or 7.7 µmol/kg in a healthy 20-year-old woman and 660 µmol (110 mg) or 8.8 µmol/kg in a 25-year-old man. Overall body half-lives were about 25 days (Shane, 1978). Intake (and excretion) was estimated to be 1.5 mg (9 µmol) for the woman and 3.4 mg (20 µmol) for the man. The two-compartment model has been questioned because muscle stores most of the body's B_6; the pool in muscle appears to turn over very slowly. This fact may have not been considered, resulting in a substantial underestimation of body stores. Coburn and colleagues (1988a) estimated the B$_6$ content of muscle biopsies and, by assuming that muscle represented 80 percent of the body B$_6$ store, calculated a total body store of about 1,000 μmol (167 mg).

Extrapolation of data from studies with experimental animals to assess the B$_6$ requirement for maintenance and growth (Coburn et al., 1987, 1988b) indicates that 1 mg of PN would be an adequate intake for the adult. Modeling of human B$_6$ pools has also led to an assessment of a minimum requirement of about 0.4 mg/day of PN (Coburn, 1990).

Clinical Effects of Inadequate Intake

The classical clinical symptoms of B$_6$ deficiency are a seborrheic dermatitis (Mueller and Vilter, 1950), microcytic anemia (Snyderman et al., 1953), epileptiform convulsions (Bessey et al., 1957; Coursin, 1954), and depression and confusion (Hawkins and Barsky, 1948). Microcytic anemia reflects decreased hemoglobin synthesis. The first enzyme and committed step in heme biosynthesis, aminolevulinate synthase, uses PLP as a coenzyme. Because PLP is also a coenzyme of decarboxylases that are involved in neurotransmitter synthesis, defects in some of these enzymes could explain the onset of convulsions in B$_6$ deficiency. Many studies have demonstrated that the levels of neurotransmitters such as dopamine, serotonin, and γ-aminobutyrate are reduced in B$_6$-depleted experimental animals, especially in extreme B$_6$ depletion (Dakshinamurti and Stephens, 1969; Dakshinamurti et al., 1991, 1993; Sharma and Dakshinamurti, 1992; Sharma et al., 1994; Stephens et al., 1971). Some of these studies were reviewed in a conference report (Dakshinamurti, 1990). However, it has not been definitely shown whether the convulsions are due to the reduced level of one of these neurotransmitters in particular. Guilarte (1993) proposed that the convulsions are caused by abnormal tryptophan metabolites that accumulate in the brain in B$_6$ deficiency.

Electroencephalogram (EEG) abnormalities have also been reported in controlled studies of B$_6$ depletion. In one depletion-repletion study (Kretsch et al., 1991) 2 of 11 young women placed on a diet containing less than 0.05 mg of B$_6$ exhibited abnormal EEG patterns within 12 days. The abnormal patterns were promptly corrected by 0.5 mg/day of PN. Similar abnormalities were reported in young men placed on a diet containing less than 0.06 mg/day of B$_6$ for 21 days (Canham et al., 1964). However, no EEG changes were detected when young men were placed on a diet containing

0.16 mg/day of B_6 for 21 days (Grabow and Linkswiler, 1969). Although a longer depletion period with 0.16 mg/day of B_6 may have eventually resulted in some abnormalities, diets containing 0.5 mg/day of B_6 have consistently failed to demonstrate abnormal EEG patterns or any hematological symptoms. Convulsions and dermatitis were not seen in these studies.

Inadequate intakes of B_6 have also been reported to impair platelet function and clotting mechanisms (Brattstrom et al., 1990; Subbarao and Kakkar, 1979), but these effects may also be due to the hyperhomocysteinemia noted in such patients (Brattstrom et al., 1990).

SELECTION OF INDICATORS FOR ESTIMATING THE REQUIREMENT FOR VITAMIN B_6

Indicators of vitamin B_6 status have traditionally been described as direct (vitamin concentrations in plasma, blood cells, or urine), indirect, or as functional (erythrocyte aminotransferase saturation by pyridoxal 5'-phosphate [PLP] or tryptophan metabolites). In most instances, the concentrations of these indicators change with increases or decreases in vitamin intake. As such, they are useful as indicators of relative B_6 status, especially in controlled depletion-repletion studies (Leklem, 1990; Sauberlich et al., 1972). However, there is little scientific information concerning which concentration of a particular indicator represents a clinical deficiency or inadequate status of the vitamin. Because of this, B_6 requirements have often been evaluated by using a combination of status indicators. However, this does not overcome the problem of establishing absolute values reflecting impaired status. The increase in methionine metabolites after a methionine load has also been used as an indicator of B_6 status (Leklem, 1994) but it has not found extensive use in B_6 requirement studies.

A review of established indicators of B_6 status suggests that plasma PLP is probably the best single indicator because it appears to reflect tissue stores (Lui et al., 1985).

Plasma Pyridoxal 5'-Phosphate

The plasma PLP concentration reflects liver PLP (Lumeng and Li, 1974) and changes fairly slowly in response to changes in vitamin intake, taking about 10 days to reach a new steady state (Lui et al., 1985). The plasma PLP concentration generally correlates with other indices of B_6 status. PLP is the major form of B_6 in tissues and is the active coenzyme species. In animals fed graded levels of pyridoxine (PN), the plasma PLP concentration correlated well with tissue B$_6$ (Lumeng et al., 1978).

Protein-bound PLP in the plasma is in equilibrium with free PLP. Binding of PLP to protein protects it from hydrolysis by alkaline phosphatase. Conditions of increased plasma phosphatase activity can lead to reduced plasma PLP. Hydrolysis of plasma PLP is required before it can be transported into tissues.

In the controlled depletion-repletion study (Kretsch et al., 1991) in which 2 of 11 young women placed on a diet containing less than 0.05 mg of B$_6$ exhibited abnormal electroencephalogram patterns, plasma PLP dropped to about 9 nmol/L. Similar PLP values were also observed in the other 9 depleted but asymptomatic subjects. This suggests that PLP concentrations of about 10 nmol/L represent a suboptimal concentration associated with clinical consequences in some subjects. Although fewer than half the subjects in this study exhibited signs of deficiency, more subjects might have shown signs if the depletion diet had been continued longer than 12 days.

Leklem (1990) has suggested a plasma PLP concentration of 30 nmol/L as the lower end of normal status. Results from a large number of studies involving various population groups (Brown et al., 1975; Driskell and Moak, 1986; Lindberg et al., 1983; Lumeng et al., 1974; Miller et al., 1975, 1985; Rose et al., 1976; Tarr et al., 1981) have shown that a substantial proportion of individuals in these populations, in some cases half, have plasma PLP concentrations below 30 nmol/L, but there are no confirming clinical or other data to suggest B$_6$ deficiency. Other investigators have proposed a cutoff of 20 nmol/L for plasma PLP as an index of adequacy (Lui et al., 1985). The more conservative cutoff of 20 nmol/L is not accompanied by observable health risks but it allows a moderate safety margin to protect against the development of signs or symptoms of deficiency. A cutoff for PLP of 20 nmol/L was selected as the basis for the average requirement (EAR) for B$_6$ although its use may overestimate the B$_6$ requirement for health maintenance of more than half the group.

A recent random sampling of the Dutch population indicated a 3 to 7 percent prevalence of plasma PLP concentrations of less than 19 nmol/L in various life stage and gender groups (Brussaard et al., 1997a, b). The prevalence was slightly higher in men aged 50 to 79 years. Although plasma PLP values in this population correlated with dietary variables, some of the fundamental tests for B$_6$ status, including the increase in homocysteine after a methionine load, did not correlate as well.

Plasma PLP concentrations decrease slightly with increased protein intake. They are very high by comparison in the fetus, decrease fairly rapidly in the first year, and then decrease more gradually throughout the lifespan (Hamfelt and Tuvemo, 1972). It is not possible to evaluate whether the higher values in newborns and infants reflect ample body stores or whether the higher concentrations reflect normal status for this age group. Because of this, it is not possible to state that a 20 nmol/L concentration in the infant reflects a status equivalent to that for a 20 nmol/L PLP concentration in the adult.

Normally, plasma PLP is measured by using an apotyrosine decarboxylase assay. This assay has been well standardized and there is usually good interlaboratory agreement with it.

Erythrocyte and Total Blood Pyridoxal 5'-Phosphate

Erythrocyte and total blood PLP concentrations have also been used as measures of B_6 status but not as extensively as plasma PLP. Erythrocyte PLP concentrations are similar to those for plasma PLP in individuals on normal diets, but they increase to much higher values than does plasma PLP in subjects taking large doses of the vitamin (Bhagavan et al., 1975). This reflects the high binding capacity of hemoglobin for PLP. Erythrocyte PLP is derived from plasma pyridoxal (PL); the erythrocyte contains PL kinase activity. Because of lower kinase activity, blacks may have lower erythrocyte PLP values than do whites. The small number of studies using erythrocyte values limits the ability to derive a concentration consistent with adequate status.

Blood Total Vitamin Concentrations

Blood concentrations of total vitamers of B_6 as well as individual concentrations of specific B_6 vitamers have been determined in some studies. These values tend to fluctuate considerably. They also fluctuate throughout the menstrual cycle, which limits their usefulness as status indicators (Contractor and Shane, 1968).

Urinary Pyridoxic Acid and Total Vitamin B_6

Urinary B_6 excretion and 4-pyridoxic acid (4-PA) excretion have been used extensively to evaluate B_6 requirements. Approximately

50 percent of the B$_6$ intake is excreted as 4-PA, but this proportion can vary somewhat. 4-PA excretion responds almost immediately to changes in dietary B$_6$ intake (Lui et al., 1985). Because it reflects recent intake, it is of essentially no value in assessing status. Leklem (1990) has suggested a value of greater than 3 μmol/day as indicative of adequate status. This is achieved with intakes of about 1 mg of B$_6$. However, the use of this cutoff value represents a circular argument; it presupposes that 1 mg/day of B$_6$ is an adequate intake.

Erythrocyte Aspartate Aminotransferase and Alanine Aminotransferase

The stimulation (activation) of erythrocyte aspartate aminotransferase (α-EAST) and erythrocyte alanine aminotransferase (α-EALT) by PLP has been used extensively to evaluate long-term B$_6$ status. These tests measure the amount of enzyme in the apoenzyme form; the ratio of the apoenzyme to total enzyme increases with B$_6$ depletion. Leklem (1990) has suggested an α-EAST of less than 1.6 and an α-EALT of less than 1.25 as indicative of adequate B$_6$ status. Variations in values reported in different studies, which may reflect blood storage conditions and time, have interfered with the setting of a well-documented cutoff point. As described in the later section "Women Ages 19 through 50 Years," aminotransferase activation factors stabilize slowly in response to changes in diet; this leads to an overestimation of the amount of B$_6$ required to return values to a preset value in depletion-repletion studies.

The absolute EALT and EAST enzyme activities, both holo- and total enzyme, have also been measured in many studies, but the large variation in values limits their usefulness as indicators of status (Raica and Sauberlich, 1964).

Tryptophan Catabolites

One of the earliest markers for B$_6$ deficiency was the urinary excretion of xanthurenic acid, which is normally a minor tryptophan catabolite. The major pathway of tryptophan catabolism proceeds via the PLP-dependent kynureninase reaction (Shane and Contractor, 1980). The xanthurenic acid pathway also involves PLP-dependent enzymes. However, under conditions of B$_6$ deficiency, this minor pathway is used to a greater extent, leading to the increased excretion of abnormal tryptophan metabolites. Mitochondrial enzymes involved in xanthurenic acid production probably retain their

PLP more effectively than does the cytosolic kynureninase under these conditions.

The evaluation of B_6 status by measuring tryptophan catabolites after a loading dose of tryptophan has been used extensively to assess B_6 status. Various challenge doses of tryptophan have been used in different studies. Xanthurenic acid excretion is responsive to B_6 intake in controlled depletion-repletion studies. However, as for many of the tests of B_6 status, it is not clear what level of excretion represents adverse B_6 status under the conditions of the tryptophan challenge dose. Leklem (1990) has suggested that a 24-hour urinary excretion of less than 65 µmol xanthurenate after a 2-g tryptophan oral dose indicates normal B_6 status.

The first enzyme in the tryptophan catabolic pathway is a dioxygenase that is induced by various steroid hormones. Consequently, the flux through this pathway and the excretion of minor tryptophan catabolites can be influenced by conditions of changed hormonal status. For example, both pregnancy and the use of high-dose oral contraceptive agents increase the excretion of these catabolites (Rose, 1978) (see "Oral Contraceptive Agents").

Plasma Homocysteine

Homocysteine catabolism proceeds via transsulfuration to cysteine and involves two PLP-dependent enzymes. Homocysteine can also be remethylated to methionine via folate and vitamin B_{12}–dependent enzymes. Thus, plasma concentrations of homocysteine are influenced by B_6 and folate and, to a lesser extent, B_{12} intakes (Selhub et al., 1993). Racial and gender differences in homocysteine values and response to vitamin intervention have been found in some studies (Ubbink et al., 1995). In a South African comparison of black and white subjects with similar lifestyles and folate and vitamin B_{12} status, plasma PLP concentrations were significantly lower in the black subjects; fasting plasma homocysteine concentration was similar. The increase in plasma homocysteine concentration after a methionine load was significantly less in the black subjects despite their lower plasma PLP values (Ubbink et al., 1995). After 6 weeks of daily supplementation with a multivitamin containing 10 mg of PN, 1 mg of folate, and 0.4 mg of vitamin B_{12}, fasting homocysteine concentration decreased in both groups. The elevation in plasma homocysteine concentration after a methionine load was unaffected by supplementation in the black subjects whereas the elevation in the white subjects decreased to about the same level as observed in the black subjects. These data suggest that despite apparently lower B$_6$ status, as indicated by plasma PLP levels, the black subjects were more efficient than the whites in catalyzing the transsulfuration of homocysteine to cysteine.

The increase in plasma homocysteine concentration after a methionine load or a meal is responsive to and primarily affected by B$_6$ status, but data are not sufficient to support using this as the criterion on which to base the EAR. Because the fasting homocysteine concentration is primarily responsive to folate status (Ubbink et al., 1996), it is not a good candidate for use in setting the EAR. Results from population-based studies using data adjusted for folate and B$_{12}$ status and for age indicate that B$_6$ status as measured by PLP is inversely correlated with nonfasting plasma homocysteine concentration (Selhub et al., 1993). At least part of the increase in plasma homocysteine concentration that occurs with aging may be due to decreased renal function (Hultberg et al., 1993) rather than B$_6$ status.

Possible Reduction of Chronic Disease Risk

Moderate hyperhomocysteinemia was identified recently as a possible risk factor for vascular disease (Selhub et al., 1995; see also Chapter 8), and vitamin intervention can be used to reduce plasma homocysteine values. A recent prospective observational study has examined the effect of self-selection for intake of folate and B$_6$ on the incidence of myocardial infarction (MI) and fatal coronary heart disease (CHD) (Rimm et al., 1998). After other risk factors for CHD were controlled for and vitamin intake was adjusted for energy intake, about a twofold reduction in MI and CHD was found for individuals in the quintile with the highest folate and B$_6$ intakes compared with those with the lowest intakes. When intakes of each of the vitamins were considered separately, the multivariate analyses suggested about a 30 percent reduction in disease incidence between individuals in the highest and lowest quintiles of intake for each of the vitamins. For B$_6$ the data are compatible with the Framingham study (Selhub et al., 1993), in which the lowest deciles of B$_6$ intake were associated with higher circulating homocysteine. However, in the current study although multivariate analysis indicated a trend in risk reduction across the quintiles of intake, the major reduction appeared to occur between the fourth and fifth quintiles of intake (median intakes 2.7 and 4.6 mg). At these high B$_6$ intakes, there is little effect of B$_6$ intake on homocysteine levels, which are mainly affected by changes in intake at much lower intakes. Although these data are intriguing and suggest that self-selection for high B_6 intake may lower CHD incidence, the highest quintile of intake was associated with increased supplement use. Some of these individuals may also exhibit other lifestyle differences that influence CHD risk, some of which were not and others that could not have been considered in the analysis. In addition, vitamin intakes were normalized to energy intake, which may have had an effect.

A study of elderly patients with coronary disease indicated a significantly elevated plasma homocysteine concentration compared with control subjects; homocysteine values were inversely correlated with plasma vitamin concentrations (Robinson et al., 1995). Plasma PLP values below 20 nmol/L were seen in 10 percent of the patients but in only 2 percent of the control subjects ($p < 0.01$).

Studies of B_6-homocysteine-vascular disease relationships were not considered in this analysis if conducted with patients with end-stage renal disease. Because homocysteine is metabolized in the kidney, this condition would exacerbate any effects of vitamin deficiency. Kidney disease may also affect B_6 metabolism and turnover.

Several ongoing randomized trials are addressing whether supplementation will decrease risks of CHD. Thus, it would be premature to establish a B_6 intake level and corresponding homocysteine value for lowest risk for disease.

Cognitive Function

The relationship of vitamin status to cognitive function was recently evaluated in the elderly (Riggs et al., 1996). B_6 status, as evaluated by plasma PLP concentrations, was related to 2 out of a battery of about 20 tests. The usefulness of these tests for evaluating B_6 status will require further validation of the putative relationships.

FACTORS AFFECTING THE VITAMIN B_6 REQUIREMENT

Bioavailability

Vitamin B_6 bioavailability was recently reviewed by Gregory (1997). B_6 in a mixed diet is about 75 percent bioavailable (Tarr et al., 1981). A mixed diet typically contains about 15 percent pyridoxine (PN) glucoside (Gregory, 1997), which is about 50 percent as bioavailable as the other B_6 vitamins. The bioavailability of nonglucoside forms of the vitamin is greater than 75 percent.

The absorption of B_6 compounds in the absence of food is comparable, even at very high doses. About 70 percent of a loading dose of 50 mg pyridoxal (PL) or the equivalent dose of pyridoxal phosphate (PLP) can be accounted for in the urine within 24 hours, demonstrating that the phosphorylated form is effectively hydrolyzed and absorbed in the gut (Shane, 1978; Snell, 1958). Under the same conditions, about 40 percent of an equivalent dose of PN can be accounted for in the urine, but PN at high doses raises the plasma PLP concentration and is retained more effectively than is PL (Shane, 1978).

Similarly, dietary pyridoxamine (PM) and PL are about 10 percent less effective than PN in raising the plasma PLP concentration, and slightly more of these vitamins is excreted in the urine as 4-pyridoxic acid (4-PA) (Wozenski et al., 1980). Most controlled B$_6$ studies have used PN as the added B$_6$ source, but requirements calculated from these studies would underestimate the B$_6$ requirement by only 5 percent or less for individuals deriving most of their B$_6$ as PLP and PMP from animal sources.

Nutrient-Nutrient Interactions

Because of PLP's role as a coenzyme for many enzymes involved in amino acid metabolism, it has been proposed that B$_6$ requirements are influenced by protein intake. Many studies have demonstrated that increased protein intake causes a relative decrease in B$_6$ status as judged by a variety of B$_6$ status indicators (Baker et al., 1964; Hansen et al., 1996b; Linkswiler, 1978; Miller et al., 1985; Sauberlich, 1964). This had led some to define B$_6$ requirements in terms of protein intake. A number of other studies, however, have failed to demonstrate an effect of protein intake on B$_6$ status parameters. A study in young and elderly men and women found little effect of dietary protein levels (12 and 21 percent of total energy) on B$_6$ status as measured by plasma PLP and erythrocyte aspartate aminotransferase (Pannemans et al., 1994).

Almost all studies investigating the effects of different protein intakes have assessed the effects of graded levels of PN on status indicators to obtain a B$_6$ requirement in milligrams. They have then adjusted this value based on the protein intake to obtain a value per gram of protein. This approach assumes a linear relationship between B$_6$ requirements and protein intake for which there is little experimental justification. This approach may also overestimate the requirement for B$_6$ because the requirement has been set by assuming a protein intake of 100 g for men. Thus, the approach cannot be used for setting an Estimated Average Requirement (EAR). Increased protein may cause a relative decrease in B$_6$ status indicators because induction of PLP-dependent enzymes may lead to tissue retention of PLP. In such a case, a decrease in a status indicator, such as plasma PLP, may not necessarily indicate a relative decrease in B_6 status. Increased excretion of tryptophan metabolites seen may also partly be due to the increased dietary tryptophan content.

The relationship between plasma PLP concentrations and the amount of B_6 per kilogram of protein intake (derived from a number of studies and compiled by J. Leklem, Oregon State University [Hansen et al., 1997; Huang et al., 1998; Kretsch et al., 1995; Ribaya-Mercado et al., 1991]), is shown in Figure 7-1A. Although a good correlation is observed ($r = 0.928$), when the same data are plotted as a function purely of B_6 intake (Figure 7-1B), the correlation appears to be equally as good ($r = 0.896$). Consequently, expressing B_6 requirements in terms of protein intake appears to add an unnecessary manipulation of the intake data that does not substantially add to the precision of assessed requirements. Moreover, this relationship is not supported by all studies. In the assessment of requirements that follows, note is taken of effects of increased protein intakes in the setting of values for B_6 requirements.

Physical Activity

A number of studies have examined the effect of physical activity on B_6 status (Crozier et al., 1994; Fogelholm, 1992), metabolism (Dreon and Butterfield, 1986; Leklem and Shultz, 1983; Manore and Leklem, 1988; Manore et al., 1987), and physical performance (van der Beek et al., 1994). Effects, if any, have been small. However, these studies were not designed to quantitate the effects of physical activity on B_6 requirements.

Other Factors

Drug Interactions

Drugs that can react with carbonyl groups have the potential to interact with PLP. Isoniazid, which is used in the treatment of tuberculosis, and L-DOPA, which is metabolized to dopamine, have been reported to reduce plasma PLP concentrations (Bhagavan, 1985; Weir et al., 1991).

Oral Contraceptive Agents

A number of studies have reported decreases in B_6 status indica-

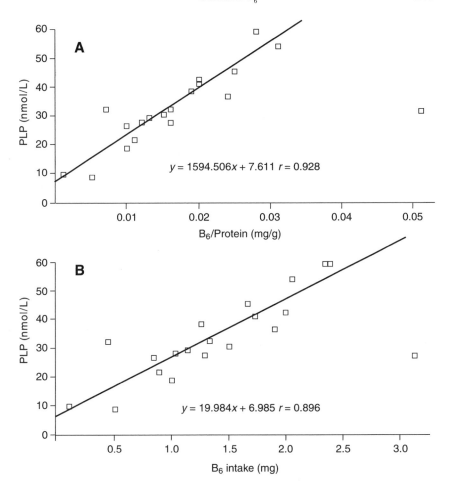

FIGURE 7-1 The relationships between (**A**) plasma pyridoxal phosphate (PLP) concentrations and vitamin B$_6$ intake per gram of dietary protein and (**B**) plasma PLP concentrations and B$_6$ intake. Data derived from a compilation by J. Leklem, Oregon State University and from Hansen et al. (1997), Huang et al. (1998), Kretsch et al. (1995), and Ribaya-Mercado et al. (1991).

tors in women receiving high-dose oral contraceptives (Rose, 1978; Shane and Contractor, 1975). Plasma PLP concentrations are decreased but the decrease is quite small. Normalization of the tryptophan load test in subjects receiving oral contraceptives requires very high levels of PN, up to 25 mg (Rose, 1978). This probably reflects hormonal stimulation of tryptophan catabolism rather than any deficiency of B$_6$ per se. These studies were conducted when the level of estrogen in oral contraceptive agents was three to five times higher than it is currently.

Alcohol

Alcoholics have low plasma PLP concentrations, and this reduced B_6 status is distinct from that caused by liver disease or poor diet. Acetaldehyde but not ethanol decreases net PLP formation by cells and is thought to compete with PLP for protein binding. This may make cellular PLP more susceptible to hydrolysis by membrane-bound phosphatase (Lumeng and Li, 1974). The extent to which this causes an increased B_6 requirement is not known.

Preeclampsia

The lowered plasma PLP concentration observed in pregnancy is lowered further in subjects with preeclampsia or eclampsia (Brophy and Siiteri, 1975; Shane and Contractor, 1980). Cord blood PLP of the newborn and placenta enzymes involved in PLP synthesis are also reduced (Gaynor and Dempsey, 1972), suggesting a potentially increased B_6 requirement in preeclampsia.

FINDINGS BY LIFE STAGE AND GENDER GROUP

In controlled studies, clinical symptoms of vitamin B_6 deficiency have only been observed during depletion with very low levels of B_6 and have never been seen at intakes of 0.5 mg/day or more. Most studies of B_6 requirements have focused on adults and have been depletion-repletion studies. Starting with the pioneering studies of the Sauberlich group (Raica and Sauberlich, 1964; Sauberlich, 1964; Sauberlich et al., 1972), studies (described below) have demonstrated the usefulness of B_6 status indicators for tracking relative vitamin status. However, many of the studies are flawed in that requirements have usually been assessed by identifying the B_6 intakes that return status indicators to the prestudy baseline values. Baseline values have been those of motivated healthy individuals on self-selected diets or on diets containing 1.5 to 2 mg of pyridoxine (PN). It is not surprising that the assessed requirements based on this approach of normalizing values are invariably similar to or higher than the baseline B_6 intake. Normalization to baseline should re-quire the same intake as in the baseline period, but studies general-ly come up with a higher value. This suggests that equilibration was not reached during the study periods for some of the indicators.

This approach of normalization to baseline is a complicated way to determine baseline vitamin intake and does not directly address the requirement for the vitamin.

In the discussion that follows, data are reanalyzed, particularly from some recent studies, to arrive at an Estimated Average Requirement (EAR) for some adult age groups. Various indicators of B$_6$ status are used, but when possible, a plasma pyridoxal phosphate (PLP) value of at least 20 nmol/L is used as the major indicator of adequacy.

Most controlled studies on B$_6$ requirements have used a liquid formula diet (which contains some food-bound B$_6$) supplemented with synthetic PN. Because PN is 95 percent bioavailable whereas food B$_6$ is only about 75 percent bioavailable, synthetic B$_6$ is 95/75 (i.e., 1.27) times more available. Thus, dietary B$_6$ equivalents are calculated as follows to determine the EAR:

$$\text{mg of dietary B}_6 \text{ equivalents provided} = \text{mg of food B}_6 + (1.27 \times \text{mg of synthetic B}_6).$$

Infants Ages 0 through 12 Months

Method Used to Set the Adequate Intake

An Adequate Intake (AI) is used as the goal for intake by infants. Limited data are available on B$_6$ requirements of the infant, but an AI can be set based on human milk B$_6$ content, which varies with maternal B$_6$ intake. West and Kirksey (1976) reported an average B$_6$ content of 0.13 mg/L of human milk at maternal B$_6$ intakes of less than 2.5 mg and 0.24 mg/L at intakes of 2.5 to 5 mg. However, milk B$_6$ content was quite variable between subjects at similar B$_6$ intakes. Borschel and colleagues (1986) reported that the milk of mothers receiving 2.5 mg/day of PN as a supplement contained 0.15 to 0.21 mg/L of B$_6$.

Ages 0 through 6 Months. The AI reflects the observed mean B$_6$ intake of infants consuming human milk. Thus, the B$_6$ AI for young infants is based on mean intake data from infants fed human milk as the principal food during their first 6 months and uses the B$_6$ concentration of milk produced by well-nourished mothers. There are no reports of full-term infants exclusively and freely fed human milk from U.S. or Canadian mothers who manifested any signs of B$_6$ deficiency. The mean concentration of B$_6$ in human milk of well-nourished but unsupplemented mothers with intakes near the Recommended Dietary Allowance (RDA) is 0.13 mg/L (West and Kirksey, 1976). Using the mean volume of milk consumed of 0.78 L/day (see Chapter 2) and the average B_6 content of 0.13 mg/L, the AI for B_6 is 0.1 mg/day for infants ages 0 through 6 months.

Little information is available on optimal plasma B_6 concentrations in the infant. Cord vein and artery PLP concentrations are very high in the fetus and in the newborn. They progressively decrease throughout the first year and then more slowly throughout life. Kang-Yoon and coworkers (1995) found lower B_6 content in milk of mothers of preterm infants even when the mothers were receiving 2 mg/day of B_6 as a supplement. Plasma PLP in infants correlated with their B_6 intake. In infants of mothers supplemented with 2 mg/day of PN, plasma PLP fell to 32 nmol/L by 1 month compared with 270 nmol/L in plasma of infants supplemented with 0.4 mg/day of B_6 directly. Formula-fed infants also had much higher plasma PLP values (about 200 nmol/L). In healthy infants, 0.3 mg of B_6 prevented abnormal tryptophan metabolites (Bessey et al., 1957).

In the 1950s convulsions were reported in infants receiving formula in which much of the B_6 content had been destroyed in processing. The convulsions were promptly treated by PN administration (Coursin, 1954); 0.26 mg was sufficient (Bessey et al., 1957) but 1 mg was required to correct abnormal tryptophan metabolites. The B_6 content of the overly processed formula has been estimated to average 0.062 mg/L or less (Borschel, 1995). Deficiency symptoms were not observed in infants receiving other formula preparations containing 0.096 mg/L of B_6, which is equivalent to an intake of 0.075 mg/day (Borschel, 1995). This level of intake is similar to the AI of 0.1 mg suggested from normal intakes and raises questions about whether an AI of 0.1 mg, which appears to be adequate on the basis of normal human milk intakes, may be slightly low. Although it is not possible to set a plasma PLP cutoff value for the infant for adequate status, one would expect it to be at least that used for the adult, if not higher. The drop in plasma PLP concentrations to 32 nmol/L, reported above, in 1-month-old infants of mothers supplemented with B_6 suggests that plasma PLP values are somewhat lower than 32 nmol/L in breastfed infants of mothers who are not supplemented. These data also suggest that the infant fed human milk does not have a large reservoir of B_6.

Other studies, however, have reported higher plasma PLP concentrations in infants of unsupplemented mothers. Andon and colleagues (1989) reported an average human milk B_6 content of 0.12

mg/L in mothers with average dietary intakes of 1.4 mg of B$_6$. The plasma PLP in the infants of these mothers averaged 54 nmol/L.

A plasma PLP concentration of 15 nmol/L was reported for a full-term infant experiencing B$_6$-responsive convulsions (Kirksey and Roepke, 1981). This is the only report of a plasma PLP concentration in an infant experiencing convulsions, and the possibility could not be excluded that other factors or a B$_6$-dependency syndrome contributed to the convulsions.

Ages 7 through 12 Months. If the reference body weight ratio method described in Chapter 2 to extrapolate from the AI for B$_6$ for infants ages 0 through 6 months is used, the AI for the older infants would be 0.2 mg/day after rounding. The second method (see Chapter 2), extrapolating from the Estimated Average Requirement (EAR) for adults and adjusting for the expected variance to estimate a recommended intake, gives an AI for B$_6$ of 0.4 mg/day, a value that is higher than that obtained from the first method. Because the AI for the young infants appears to be close to the requirement on the basis of studies mentioned in this section, the AI for B$_6$ is set at 0.3 mg/day for infants ages 7 through 12 months— as the mean obtained from the two types of extrapolation.

B$_6$ AI Summary, Ages 0 through 12 Months

AI for Infants
0–6 months   **0.1 mg/day of vitamin B$_6$**   **≈0.014 mg/kg**
7–12 months   **0.3 mg/day of vitamin B$_6$**   **≈0.033 mg/kg**

Special Considerations

Infant formula typically provides much higher levels of B$_6$ to the infant than does human milk (Borschel et al., 1986), and formula-fed infants have higher PLP concentrations and lower erythrocyte aspartate aminotransferase (α-EAST) values than do infants fed human milk (Heiskanen et al., 1994).

Children and Adolescents Ages 1 through 18 Years

Evidence Considered in Estimating the Average Requirement

Information on the B$_6$ requirements of children and of adolescents is too limited for an EAR to be established directly. For example, the few studies assessing intakes and 4-pyridoxic acid (4-PA) excretion are not useful for determining a requirement.

A recent study that tracked some B_6 status indicators from children aged 2 months to 11 years (Heiskanen et al., 1995) demonstrated a gradual decrease in erythrocyte PLP with age, similar to that previously reported for plasma PLP (Hamfelt and Tuvemo, 1972), and an increased erythrocyte α-EAST. Interestingly, there was a tendency for those at the lowest percentiles of values to remain so over this period, which could reflect a genetic component, diet, or both. The setting of normal ranges for B_6 status indicators is complicated by age-dependent changes in these indicators.

A few studies have investigated indicators of B_6 status in preschool children and adolescents. Fries and coworkers (1981) measured plasma PLP concentrations in 35 boys and girls aged 3 and 4 years. Average food intakes of B_6 ranged from 0.9 mg in the 4-year-old boys to 1.3 mg in the 3-year-old boys. Some of the children received additional supplements. The average plasma PLP concentrations in unsupplemented children ranged from 58 to 78 nmol/L in the different groups, and the supplemented children had higher concentrations. This suggests that nonsupplemented intakes were considerably above an EAR level (if it could be determined), but it is not possible to set an EAR on the basis of these data.

Kirksey and colleagues (1978) studied the nutritional status of 12- to 14-year-old female adolescents. Thirteen percent of the subjects were judged to have inadequate B_6 status as assessed by erythrocyte alanine aminotransferase (α-EALT) ratios with average dietary B_6 intakes estimated to be 1.24 mg. Driskell and colleagues (1985) studying 12-, 14-, and 16-year-old adolescent girls reported 20 percent with marginal B_6 status and 13 percent with B_6 deficiency as indicated by abnormal α-EALT ratios. Daily dietary B_6 intakes in these groups averaged about 1.2 mg. In a follow-up longitudinal assessment of 12- and 14-year-old adolescent girls, Driskell and colleagues (1987) reported a prevalence of about 20 percent B_6 deficiency and almost 30 percent marginal deficiency with average dietary intakes of 1.25 mg B_6, again as assessed by α-EALT ratios. Although these data may suggest an EAR for adolescents somewhat below 1.2 mg, a number of concerns limit the usefulness of these data for deriving an EAR. As indicated previously, cutoff α-EALT ratios for B_6 deficiency have not been correlated with any clinical indicator of deficiency, and considerable variation in α-EALT ratios has been reported (Leklem, 1990). In addition, in the Driskell et al. (1985) study, the data were stratified by supplement intake ranging from less than 1 mg to greater than 3.5 mg. In each group about 60 percent of the subjects had adequate B$_6$ status as determined by α-EALT whereas the percentages of subjects with marginal and with deficient status were essentially identical. This seriously calls into question the appropriateness of using α-EALT ratios for assessing B$_6$ status.

In the absence of additional information, EARs and RDAs for these age groups have been estimated by using the method described in Chapter 2 and then extrapolating from adult values. For comparative purposes the requirement was also estimated by assuming that there is a direct relationship between protein intake and the B$_6$ requirement. The results of applying that method to data on protein intake from the Third National Health and Nutrition Examination Survey for children were judged to be unreasonably high. (See also the previous discussion about the protein-B$_6$ relationship.) As with adults, there is no compelling evidence to suggest that requirements within any of these age groups should be adjusted based on protein intake.

B$_6$ EAR and RDA Summary, Ages 1 through 18 Years

EAR for Children	1–3 years	0.4 mg/day of vitamin B$_6$
	4–8 years	0.5 mg/day of vitamin B$_6$
EAR for Boys	9–13 years	0.8 mg/day of vitamin B$_6$
	14–18 years	1.1 mg/day of vitamin B$_6$
EAR for Girls	9–13 years	0.8 mg/day of vitamin B$_6$
	14–18 years	1.0 mg/day of vitamin B$_6$

The RDA for B$_6$ is set by assuming a coefficient of variation (CV) of 10 percent (see Chapter 1) because information is not available on the standard deviation of the requirement for B$_6$; the RDA is defined as equal to the EAR plus twice the CV to cover the needs of 97 to 98 percent of the individuals in the group (therefore, for B$_6$ the RDA is 120 percent of the EAR).

RDA for Children	1–3 years	0.5 mg/day of vitamin B$_6$
	4–8 years	0.6 mg/day of vitamin B$_6$
RDA for Boys	9–13 years	1.0 mg/day of vitamin B$_6$
	14–18 years	1.3 mg/day of vitamin B$_6$
RDA for Girls	9–13 years	1.0 mg/day of vitamin B$_6$
	14–18 years	1.2 mg/day of vitamin B$_6$

Adults Ages 19 through 50 Years

Men Ages 19 through 50 Years

Evidence Considered in Estimating the Average Requirement. The initial studies on B_6 requirements of young men are not very useful for determining an EAR because of the small number of repletion intakes. Most of these studies used the tryptophan load test to determine status. In a repletion study by Baker and colleagues (1964), eight young men on a diet containing 30 g/day of protein required 1.25 mg of PN (equivalent to 1.5 mg of food B_6) to return tryptophan metabolite excretion back to baseline after a 10-g tryptophan challenge dose, whereas subjects on a 100-g protein diet needed 1.5 mg of PN (equivalent to 1.9 mg of food B_6). This suggests an EAR considerably lower than 1.25 or 1.5 mg. Yess and co-workers (1964) placed six subjects on a diet containing 0.16 mg of PN and 100 g of protein for 55 days and investigated various tryptophan metabolites after a 2-g load. Supplementation with 0.6 or 0.9 mg of PN normalized tryptophan catabolites in nearly all the subjects, suggesting an EAR of less than 0.9 mg of PN (equivalent to less than 1.0 mg adjusted for food B_6).

Miller and Linkswiler (1967) measured tryptophan metabolites in 11 men receiving a basal diet containing 0.16 mg of PN and 54 or 150 g of protein. Over a 40-day period there was a slow increase in the excretion of tryptophan catabolites. The addition of 0.6 mg of B_6 corrected tryptophan catabolite excretion in most of the subjects, suggesting an EAR of less than 0.76 mg of PN (equivalent to less than 0.9 mg adjusted for food B_6). Complete normalization was observed with 1.5 mg (1.9 mg food B_6), which suggests that this level of intake exceeded the requirements of all subjects.

Linkswiler (1978) summarized a large number of early studies of men receiving various protein intakes. In subjects on a diet containing 100 g of protein, normalization of the load test required 1.0 to 1.5 mg, suggesting an EAR of considerably less than 1.0 to 1.5 mg. Miller and colleagues (1985) fed diets containing 1.6 mg of B_6 (from food and food fortified with PN) and 0.5, 1.0, or 2 g/kg of protein to eight young men (21 to 31 years of age). In this study, plasma PLP remained above 30 nmol/L, suggesting that 1.5 mg was substantially higher than the EAR.

In a study by Selhub and colleagues (1993), which adjusted for folate status, gender, and age, half of the subjects consuming freely selected diets containing an average of 1.3 mg of B_6 (as estimated from food frequency questionnaires and thus from food and vitamin supplements) had plasma homocysteine concentrations that were similar to those of subjects receiving much higher intakes of the vitamin. Subjects had not fasted or received a methionine load before being evaluated. Because no value for optimal homocysteine concentrations has been established and it is not clear that there is an advantage to driving homocysteine concentrations down to the lowest possible level, this suggests an EAR of less than 1.3 mg if plasma homocysteine is used as an index of adequacy.

It is difficult to derive a precise EAR for young men because most studies have identified a B$_6$ intake that restores test parameters to baseline values for all subjects, which may be considerably higher than that needed for health. Moreover, in most studies the levels of B$_6$ tested appear to have been in excess of an average requirement.

B$_6$ EAR and RDA Summary, Men Ages 19 through 50 Years. The various studies suggest an average requirement that ranges from less than 0.9 mg of food B$_6$ to considerably less than 1.9 mg. Taken together, the studies suggest an EAR of 1.1 mg/day.

EAR for Men   **19–30 years**   **1.1 mg/day of vitamin B$_6$**
                            31–50 years   **1.1 mg/day of vitamin B$_6$**

The RDA for B$_6$ is set by assuming a coefficient of variation (CV) of 10 percent (see Chapter 1) because information is not available on the standard deviation of the requirement for B$_6$; the RDA is defined as equal to the EAR plus twice the CV to cover the needs of 97 to 98 percent of the individuals in the group (therefore, for B$_6$ the RDA is 120 percent of the EAR).

RDA for Men   **19–30 years**   **1.3 mg/day of vitamin B$_6$**
                            31–50 years   **1.3 mg/day of vitamin B$_6$**

Women Ages 19 through 50 Years

Evidence Considered in Estimating the Average Requirement. Six studies provide useful evidence about the amount of B$_6$ required by young women. Analysis of these studies focuses on the amount required for a plasma PLP value of at least 20 nmol/L. Brown and colleagues (1975) depleted 10 young women (average age 22) on a diet that contained 0.16 mg of B$_6$ and 78 g of protein for 28 days. The subjects were repleted with 1.0 or 2.2 mg of B$_6$ for 28 days. During the depletion period, plasma PLP concentrations fell from about 50 to about 14 nmol/L. On the two repletion regimens they increased to about 24 and 60 nmol/L. An EAR of just under 1.0 mg (equivalent to 1.2 mg based on food B_6) can be assessed based on a PLP concentration of $\geq$ 20 nmol/L.

Kretsch and coworkers (1995), using a depletion-repletion approach, studied a group of eight women (aged 21 to 30 years) on a high-protein diet (1.55 g/kg of body weight) and also investigated the effects of animal protein compared to plant protein. A formula diet plus 2 mg of PN was supplied for the 7-day adjustment period to develop baseline values. The depletion diet contained less than 0.05 mg of PN and was continued for 11 to 28 days (two subjects developed electroencephalogram [EEG] abnormalities). Repletion was with 0.5, 1, 1.5, or 2 mg/day of PN for 14 to 21 days. Between 1 and 1.5 mg of PN restored PLP to baseline, and 1.5 to 2 mg of PN restored xanthurenate to baseline values. In this study, plasma PLP in the baseline period averaged 25 nmol/L, a very low value compared with most studies, and dropped to less than 10 nmol/L during depletion. Surprisingly, plasma PLP fell slightly further during the first repletion with 0.5 mg of B_6. Values averaged close to 20 nmol/L on intake of 1 mg of B_6, suggesting an EAR of about 1 mg (1.25 mg of food B_6). Normalization of aminotransferase activation factors occurred with intakes of about 1.5 to 2 mg of B_6 and lagged behind the diet changes. Activation factor values increased during the depletion period and then increased further during the first repletion period, indicating that they had not stabilized within the depletion period. No differences in response were observed when comparing the animal to the plant protein source.

In a depletion-repletion study, Huang et al. (1998) studied B_6 requirements in young women (aged 28 to 34 years) fed a high-protein diet (1.55 g/kg of body weight). Subjects were fed a baseline diet (1.6 mg/day of B_6 from food) for 9 days and a depletion diet (0.45 mg/day also from food) for 27 days. Repletion diets of 1.26, 1.66, and 2.06 mg of B_6 were given for 21, 21, and 14 days, respectively, the additional B_6 provided in the form of pyridoxine hydrochloride once daily. As would be expected for a baseline intake of 1.6 mg of B_6, 1.5 to 2 mg of PN normalized plasma PLP and urinary 4-PA concentration to baseline values. No hematological or EEG abnormalities were observed during depletion, and the plasma PLP concentration was above 20 nmol/L at the end of the depletion period. With longer depletion, PLP values might have dropped more. At a B_6 intake of 1.26 mg, mean PLP was 38 nmol/L. On the basis of plasma PLP, an EAR should be between the two levels tested, 0.45 and 1.26 mg, probably less than 1 mg. If suggested cutoffs for erythrocyte aminotransferase concentrations were applied to ratios for α-EAST and α-EALT observed in the study, the EAR values would be around 1.3 mg based on α-EAST or above 1.7 mg based on α-EALT. Again, these values appear to lag with supplementation and may be overestimates.

Hansen and coworkers (1996b) investigated the effects of protein intake on B$_6$ status in women aged 19 to 38 years. The subjects received a diet containing 1.25 mg of PN, and protein intake was 0.5, 1, or 2 g/kg of body weight. At the two lower levels of protein intake, plasma PLP remained above 30 nmol/L. For women on the highest protein intake, PLP concentrations dropped below 30 nmol/L but remained above 20 nmol/L. There was a slight elevation in tryptophan catabolites after a load, which may have been due to the increased dietary tryptophan intake. The decrease in plasma PLP with increased protein intake was accompanied by a decrease in urinary 4-PA and urinary B$_6$, suggesting that more B$_6$ was retained by tissues under these conditions and that a steady state may not have been reached. This also calls into question whether the decrease in plasma PLP concentrations on a high-protein intake truly reflects a decrease in B$_6$ status. If a PLP cutoff of 20 nmol/L is applied, these studies suggest an EAR of considerably less than 1.25 mg.

Hansen and coworkers (1996a) also investigated the effects of diets low and high in PN glucoside on status indicators. Ten women (mean age 29 years) were fed two food-based diets each containing about 1.5 mg of B$_6$ of which 9 or 27 percent was as the naturally occurring B$_6$ glucoside. The status indicators demonstrated lower bioavailability of the glucoside, as expected, and suggested increased fecal excretion. On the low-glucoside diet, mean PLP concentrations were above 30 nmol/L, suggesting an EAR of less than 1.5 mg.

In a recent study, Hansen and coworkers (1997) investigated the B$_6$ requirements of young women receiving 85 g/day of protein. Subjects were adapted to a basal diet containing either 1.03 or 0.84 mg of B$_6$, which was fed for 15 days. The subjects then received three or four different levels of repletion for periods of 10 or 12 days. EAR values of less than 1.18 mg and 1.15 mg could be assessed based on plasma PLP concentrations. α-EAST ratios were in the normal range even in the basal period, which suggests an EAR of less than 0.8 mg. α-EALT was also normal in all subjects receiving the basal 0.84-mg diet, but half the subjects in the group receiving the basal 1.03-mg diet had values in the abnormal range. Although this suggests an EAR of 0.8 to 1.0 mg based on aminotransferase activation factors, these data are quite different from those reported in other studies.

Driskell and colleagues (1989) investigated the B_6 status of 15 obese black adult women (21 to 51 years of age). Dietary B_6 intake of the obese women and a nonobese control group averaged 1.18 mg, as assessed by dietary recall. Plasma PLP concentrations averaged 60 nmol/L in the obese group and 63 nmol/L in the control group, indicating no apparent effect of obesity on B_6 status. These data do, however, suggest an EAR of considerably less than 1.2 mg for adult women.

B_6 EAR and RDA Summary, Women Ages 19 through 50 Years. Based on the above evidence, the average requirement of PN is less than 1.0 mg. This is adjusted to 1.1 mg for food B_6.

EAR for Women   **19–30 years**   **1.1 mg/day of vitamin B_6**
                              31–50 years   **1.1 mg/day of vitamin B_6**

The RDA for B_6 is set by assuming a coefficient of variation (CV) of 10 percent (see Chapter 1) because information is not available on the standard deviation of the requirement for B_6; the RDA is defined as equal to the EAR plus twice the CV to cover the needs of 97 to 98 percent of the individuals in the group (therefore, for B_6 the RDA is 120 percent of the EAR).

RDA for Women   **19–30 years**   **1.3 mg/day of vitamin B_6**
                              31–50 years   **1.3 mg/day of vitamin B_6**

Adults Ages 51 Years and Older

Evidence Considered in Estimating the Average Requirement. Studies of older individuals have been quite limited. Ribaya-Mercado and co-workers (1991) investigated B_6 requirements in 12 men and women over 60 years of age who received protein intakes of 0.8 or 1.2 g/kg of body weight. Depletion and repletion B_6 intakes were based on protein intake but averaged, respectively, about 0.15, 1.2, 1.8, and 2.6 mg in men and 0.1, 0.9, 1.3, and 1.9 mg in women. The source of B_6 during the repletion diets was a mixture of food and pyridoxine supplements. Individuals were kept on self-selected diets for baseline studies; these diets averaged 1.6 mg of B_6 for men and 1.4 mg for women. About 1.9 mg of B_6 was required to reach baseline values of plasma PLP and urinary 4-PA in both men and women consuming 1.2 g protein/kg of body weight. Normalization of erythrocyte aminotransferase activation factors required higher levels of B_6, but these activation factors did not stabilize during each period.

The EAR based on a plasma PLP of at least 20 nmol/L would be less than 1.3 mg (less than 1.6 mg for food B$_6$) in men and about 0.9 mg (1.1 mg for food B$_6$) in women. Xanthurenate excretion in nearly half the men and all the women was near the baseline value at the equivalent of a food B$_6$ intake of 1.6 mg/day. The four subjects on the lower protein diet (0.8 g/kg of body weight) reached baseline values of plasma PLP at a lower level, 1.33 mg/day of B$_6$, yet xanthurenate excretion did not return to baseline for some of the subjects by the end of the three repletion periods.

Selhub and colleagues (1993) investigated the relationship among plasma homocysteine, dietary B$_6$, and plasma PLP concentration in subjects from the Framingham Heart Study aged 67 to 96 years. The data were adjusted for age, gender, and folate and B$_{12}$ intakes. No clear cutoff for homocysteine has been established, but about half the subjects whose B$_6$ intakes were 1.4 mg (and who had plasma PLP concentrations of 25 nmol/L) were in the same homocysteine range as subjects consuming much higher intakes. At a plasma PLP concentration of 20 nmol/L, homocysteine concentrations averaged about 13 μmol/L, and this corresponded to a dietary intake of about 1.3 mg B$_6$.

Meydani and coworkers (1991) reported impairments in interleukin-2 and lymphocyte proliferation in eight healthy elderly subjects (four men and four women) placed on a B$_6$-deficient diet of 0.3 μg/kg body weight/day for up to 20 days (approximately 0.17 mg/day for the men and 0.10 mg/day for the women). Restoration of various parameters of cell-mediated immunity to baseline values (obtained when the subjects had been on self-selected diets) required more than 22.5 μg/kg body weight/day, the second highest repletion level tested, because most of the parameters measured returned to baseline values at the highest repletion level tested in both the men and women, which was 33.75 μg/kg body weight/day or about 2 mg/day of B$_6$ for the women and 2.88 mg/day for the men. Some measures of immunity did not return to baseline during the study period whereas others exceeded baseline values only at very high (50 mg) B$_6$ intakes. No indication was given as to which level of any of these parameters indicated a dysfunction in immune response. Although the changes in these parameters with B$_6$ intake suggest they may be of potential use as indicators of B$_6$ status, it is presently not possible to determine a requirement from these data.

B$_6$ EAR and RDA Summary, Ages 51 Years and Older. From the above studies, it appears that the EAR for B$_6$ is higher than the EAR for younger men and women and, unlike that for younger adults, is higher for men than for women. The increase due to age and gender appears to be approximately 0.2 to 0.3 mg of food B_6/day.

EAR for Men	**51–70 years**	**1.4 mg/day of vitamin B_6**
	> 70 years	**1.4 mg/day of vitamin B_6**
EAR for Women	**51–70 years**	**1.3 mg/day of vitamin B_6**
	> 70 years	**1.3 mg/day of vitamin B_6**

The RDA for B_6 is set by assuming a coefficient of variation (CV) of 10 percent (see Chapter 1) because information is not available on the standard deviation of the requirement for B_6; the RDA is defined as equal to the EAR plus twice the CV to cover the needs of 97 to 98 percent of the individuals in the group (therefore, for B_6 the RDA is 120 percent of the EAR).

RDA for Men	**51–70 years**	**1.7 mg/day of vitamin B_6**
	> 70 years	**1.7 mg/day of vitamin B_6**
RDA for Women	**51–70 years**	**1.5 mg/day of vitamin B_6**
	> 70 years	**1.5 mg/day of vitamin B_6**

Special Considerations

Note that the EARs (and RDAs) for adults were derived by using biochemical indicator cutoff values that have not been directly linked to clinical or physiological insufficiency. As described above, clinical symptoms of B_6 deficiency have been observed only in controlled studies during depletion with very low levels of B_6 and have never been seen at intakes of 0.5 mg/day or greater. This suggests that 1 mg/day is sufficient for most adults, but requirements may possibly be higher in individuals on very-high-protein diets.

Pregnancy

Evidence Considered in Estimating the Average Requirement

Concentrations of indicators of B_6 status in plasma and blood decrease throughout pregnancy, especially in the third trimester (Cleary et al., 1975; Hamfelt and Tuvemo, 1972; Shane and Contractor, 1980). Many studies have demonstrated a drop in plasma PLP to about 10 nmol/L; this is substantially more than can be accounted for by increased blood volume. Fetal blood PLP concentrations in the second and third trimester and at term are much higher than in the mother, and significant fetal sequestration of the vitamin has been demonstrated by cord vein and artery differences (Cleary et al., 1975; Contractor and Shane, 1970; Shane and Contractor, 1980). Barnard and colleagues (1987) reported that much of the decrease in plasma PLP concentration during pregnancy is offset by increases in plasma PL, but other investigators have not observed this compensation (Contractor and Shane, 1970). Studies in animals suggest direct transport of PLP to the fetus (Contractor and Shane, 1971). In the pregnant rat about 15 percent of an intraperitoneal dose of PN is initially taken up by the uterus, placenta, and fetus.

Maintenance of plasma PLP concentrations at nonpregnant values requires about 2 mg/day of supplemental PN in the first trimester and between 4 and 10 mg/day in the third trimester (Cleary et al., 1975; Hamfelt and Tuvemo, 1972; Lumeng et al., 1976). Maintenance of other status indicators such as tryptophan metabolites at nonpregnant levels after a tryptophan load requires even higher intakes, but this test may be affected by hormonal changes (Shane and Contractor, 1980). It is not clear whether these changes in status indicators during pregnancy reflect poorer vitamin status or represent normal physiological changes during pregnancy. The latter is more reasonable. There is no a priori reason to use laboratory values for nonpregnant women as controls for pregnant women. There is also no evidence of significant problems in B$_6$ status during pregnancy despite the reduced levels of status markers. Schuster and coworkers (1981) examined the relationship between α-EAST ratios and Apgar scores in low-income mothers receiving an average of 1.3 mg/day of B$_6$ in the diet. A small but significant effect of B$_6$ status was noted. However, the possibility could not be eliminated that other variables unrelated to B$_6$ may have been responsible.

For an assumed body store of 1,000 μmol and a fetal, uterine, and placental accumulation of 15 percent, the fetus and placenta would accumulate approximately 25 mg of B$_6$. This would be about 0.1 mg/day averaged over gestation. With additional allowances made for the increased metabolic needs and weight of the mother and about 75 percent bioavailability of food B$_6$, an additional average need in pregnancy of 0.25 mg can be estimated. This increased need would be concentrated more in the second half of gestation. Unlike nutrients such as iron, B$_6$ is not stored in the body to any substantial extent, so it is unlikely that a surplus in early gestation would satisfy the increased need in the latter stages of gestation.

Consequently, an extra 0.5 mg/day of B_6 can be reasonably justified to meet the need in the third trimester.

B_6 EAR and RDA Summary, Pregnancy

Although 0.5 mg/day of B_6 may overestimate the additional need in early gestation, it was considered judicious to err on the side of ensuring sufficiency and add 0.5 mg/day to the EAR for nonpregnant women throughout pregnancy. Because of the approximation involved, no additional adjustment for adolescence is included.

EAR for Pregnancy 14–18 years   1.6 mg/day of vitamin B_6
19–30 years   1.6 mg/day of vitamin B_6
31–50 years   1.6 mg/day of vitamin B_6

The RDA for B_6 is set by assuming a coefficient of variation (CV) of 10 percent (see Chapter 1) because information is not available on the standard deviation of the requirement for B_6; the RDA is defined as equal to the EAR plus twice the CV to cover the needs of 97 to 98 percent of the individuals in the group (therefore, for B_6 the RDA is 120 percent of the EAR).

RDA for Pregnancy 14–18 years   1.9 mg/day of vitamin B_6
19–30 years   1.9 mg/day of vitamin B_6
31–50 years   1.9 mg/day of vitamin B_6

Lactation

Evidence Considered in Estimating the Average Requirement

As described above, the B_6 concentration in human milk varies depending on the mother's B_6 intake, and some women consuming less than 2.5 mg/day of B_6 produce milk with a B_6 content that is not much higher than that associated with consuming formula that resulted in convulsions in infants because of low levels of B_6. There is some variation in reported human milk B_6 content, which may be due to methodological differences. Between individuals there are variations in milk B_6 content at similar B_6 intakes. Existing data previously described suggest that the amount of B_6 required to increase the milk B_6 content by a small increment is much higher than that increment and that the additional requirement for lactation is considerably in excess of that suggested by the amount secreted via lactation (Borschel et al., 1986; West and Kirksey, 1976).

To ensure a milk B$_6$ concentration of 0.13 mg/L, it is estimated that about five times that amount of B$_6$ must be consumed in addition to the EAR of 1.1 mg/day.

B$_6$ EAR and RDA Summary, Lactation

In light of the evidence that low maternal intakes could lead to compromised B$_6$ status in the infant (Borschel, 1995), it would be prudent to add 0.6 mg of B$_6$ to the EAR of 1.1 mg for nonpregnant women, giving an EAR for lactation of 1.7 mg/day of B$_6$. Because this is an approximation based on a number of assumptions, no adjustment is made for adolescent females who are lactating.

EAR for Lactation	**14–18 years**	**1.7 mg/day of vitamin B$_6$**
	19–30 years	**1.7 mg/day of vitamin B$_6$**
	31–50 years	**1.7 mg/day of vitamin B$_6$**

The RDA for B$_6$ is set by assuming a coefficient of variation (CV) of 10 percent (see Chapter 1) because information is not available on the standard deviation of the requirement for B$_6$; the RDA is defined as equal to the EAR plus twice the CV to cover the needs of 97 to 98 percent of the individuals in the group (therefore, for B$_6$ the RDA is 120 percent of the EAR).

RDA for Lactation	**14–18 years**	**2.0 mg/day of vitamin B$_6$**
	19–30 years	**2.0 mg/day of vitamin B$_6$**
	31–50 years	**2.0 mg/day of vitamin B$_6$**

INTAKE OF VITAMIN B$_6$

Food Sources

Data obtained from the 1995 Continuing Survey of Food Intakes by Individuals indicates that the greatest contribution to vitamin B$_6$ intake of the U.S. adult population comes from fortified, ready-to-eat cereals; mixed foods (including sandwiches) with meat, fish, or poultry as the main ingredient; white potatoes and other starchy vegetables; and noncitrus fruits (Table 7-1). Especially rich sources are highly fortified cereals; beef liver and other organ meats; and highly fortified, soy-based meat substitutes.

TABLE 7-1   Food Groups Providing Vitamin B$_6$ in the Diets of U.S. Men and Women Aged 19 Years and Older, CSFII, 1995[a]

Food Group	Contribution to Total Vitamin B$_6$ Intake[b] (%)		Foods Within the Group that Provide at Least 0.4 mg of Vitamin B$_6$[c] per Serving	
	Men	Women	0.4–0.8 mg	> 0.8 mg

Food groups providing at least 5% of total vitamin B$_6$ intake

Food Group	Men	Women	0.4–0.8 mg	> 0.8 mg
Ready-to-eat cereals	10.8	13.7	Moderately fortified	Highly fortified
Mixed foods[d]	10.1	8.6	NA[e]	NA
White potatoes and other starchy vegetables[f]	9.7	9.2	White potato with peel and plantain	—
Non-citrus fruits[g]	7.0	9.5	Bananas and watermelon	
Poultry[h]	6.9	7.1	Chicken breast, turkey light meat, and Cornish game hen	—
Beef	6.4	4.2	—	—
Mixed foods, main ingredient is grain	5.4	4.8	NA	NA
Other vegetables[i]	4.1	4.6	—	—

Vitamin B$_6$ from other food groups

Food Group	Men	Women	0.4–0.8 mg	> 0.8 mg
Organ meats	0.2	0.2	Calf, chicken, or pork liver and kidney	Beef liver
Soy-based supplements and meal replacements	0.7	0.2	Some soy-based meat substitutes	Some soy-based meat substitutes
Pasta, rice, and cooked cereals	3.2	3.3	—	Fortified instant oatmeal
Finfish	1.2	1.7	Fresh tuna and trout	—
Deep yellow vegetables	0.8	1.2	Sweet potatoes with peel	—

[a] CSFII = Continuing Survey of Food Intakes by Individuals.

[b] Contribution to total intake reflects both the concentration of the nutrient in the food and the amount of the food consumed. It refers to the percentage contribution to the American diet for both men and women, based on 1995 CSFII data.

[c] 0.4 mg = 20% of the Recommended Daily Intake (2.0 mg) of B$_6$—a value set by the Food and Drug Administration.

[d] Includes sandwiches and other foods with meat, poultry, or fish as the main ingredient.

[e] NA = not applicable. Mixed foods were not considered for this table.

TABLE 7-1 Continued

f Includes plantain, cassava, yam, and taro.

g Includes apple, apricot, banana, berries, cantaloupe, figs, grapes, honeydew, peach, pear, pineapple, plum, rhubarb, and watermelon.

h Includes chicken and turkey.

i Includes artichoke, asparagus, green beans, fresh lima beans, beets, Brussels sprouts, cabbage, cauliflower, corn, cucumber, eggplant, kohlrabi, lettuce, mushrooms, onions, green peas, peppers, rutabaga, snow peas, squash, turnips, vegetable salads, vegetable combinations, and vegetable soups.

SOURCE: Unpublished data from the Food Surveys Research Group, Agricultural Research Service, U.S. Department of Agriculture, 1997.

Dietary Intake

Data from nationally representative U.S. surveys (Appendixes G and H) indicate that the median daily intake of B$_6$ by men is approximately 2 mg/day and the median intake by women is approximately 1.5 mg/day. Groups having low B$_6$ intakes are shown in Table 7-2.

A survey done in one Canadian province reported that the dietary intake of B$_6$ by both men and women is slightly lower than intakes in the United States (Appendix I). The Boston Nutritional Status Survey (Appendix F) indicates that this relatively advantaged group of people over age 60 reported a median B$_6$ intake of 1.2 mg/day for men and 1.0 mg/day for women.

TABLE 7-2 Life Stage and Gender Groups in the United States with Reported Vitamin B$_6$ Intake Less than the Estimated Average Requirement (EAR)

Life Stage and Gender Group	Individuals Below the EAR (%)
Males, 51+ y	10–25
Women, 14–18 y	10–15
Women, 19–50 y	15
Women, 51+ y	25–50
Pregnant women (all ages)	25–50
Lactating women (all ages)	10–15

NOTE: See Table G-4 in Appendix D for vitamin B$_6$ intake data by category and percentile from the Continuing Survey of Food Intakes by Individuals, 1994–1995.

Intake from Supplements

Information from the Boston Nutritional Status Survey on use of B_6 supplements by a free-living elderly population is given in Appendix F. For those reporting use of supplements, the fiftieth percentile of supplemental B_6 intake was 2.2 mg for both men and women. Approximately 26 percent of all adults reported taking a B_6-containing supplement in 1986 (Moss et al., 1989).

TOLERABLE UPPER INTAKE LEVELS

Hazard Identification

Adverse Effects

No adverse effects have been associated with high intake of vitamin B_6 from food sources. This review focuses on pyridoxine, the form of B_6 that was consumed in the reports cited below. Large oral supplemental doses of pyridoxine used to treat many conditions, including carpal tunnel syndrome and premenstrual syndrome, have been associated with the development of sensory neuropathy and dermatological lesions (Cohen and Bendich, 1986; Schaumburg and Berger, 1988). The causal association between high-dose pyridoxine and neuropathy has been well documented in animals since 1940 (Unna and Antopol, 1940) and in humans since 1983 (Schaumburg et al., 1983).

Sensory Neuropathy. The first clinical report of pyridoxine-induced neurotoxicity in humans (Schaumburg et al., 1983) describes seven patients (five women and two men) with severe sensory neuropathy of the extremities after 2,000 to 6,000 mg/day of pyridoxine for 2 to 40 months. Four individuals were unable to walk. Neurological signs and symptoms were diagnosed through objective neurological assessment and improved in all patients after withdrawal of medication. Other reports of peripheral sensory neuropathy associated with high-dose pyridoxine therapy (1 to 4 g/day) appeared in the 1980s (Baer, 1984; Bredesen and Parry, 1984; De Zegher et al., 1985; Friedman et al., 1986). The pathogenesis of pyridoxine-induced peripheral sensory neuropathy and dose-response relationships have been well-described in animal models (Phillips et al., 1978; Schaeppi and Krinke, 1985). Review of the limited data involving lower pyridoxine doses (Bernstein and Lobitz, 1988; Del Tredici et al., 1985) reveals that the risk of developing sensory neuropathy decreases quite rapidly at doses below 1 g/day.

Other Adverse Effects. Painful and disfiguring dermatological lesions were reported in humans after consumption of 2 to 4 g/day of pyridoxine for more than 1 year (Friedman et al., 1986; Schaumburg and Berger, 1988). However, the limited data fail to demonstrate a relationship between this endpoint and the dose or duration of treatment. Also, the mechanism of pyridoxine dermatoses remains unclear (Schaumburg and Berger, 1988).

There are isolated, uncontrolled case reports of congenital defects (Donaldson and Bury, 1982; Gardner et al., 1985; Philpot et al., 1995) and B$_6$ dependency (Hunt et al., 1954) in newborns whose mothers were treated with pyridoxine during the first half of pregnancy. No controlled data in humans were found that confirm these findings. Observational data in women taking up to 200 mg/day of pyridoxine orally (Ellis, 1987; Weinstein et al., 1944) and up to 100 mg/day of pyridoxine parenterally (Dorsey, 1949; Hart and McConnell, 1943; Weinstein et al., 1944; Willis et al., 1942) during the first trimester of pregnancy report no adverse effects. Controlled studies in animals (Hendrickx et al., 1985; Khera, 1975; Schumacher et al., 1965) show no evidence of teratogenicity.

Very high pyridoxine doses (200 to 600 mg/day parenterally; 600 mg/day orally) after delivery have been used to successfully inhibit or suppress lactation (Foukas, 1973; Marcus, 1975; Scaglione and Vecchione, 1982). However, the results of other clinical trials on the antilactogenic effect of pyridoxine are conflicting (Scaglione and Vecchione, 1982). No evidence of antilactogenic effects at lower oral doses was found.

Summary

On the basis of considerations of causality, relevance, and the quality and completeness of the database, sensory neuropathy was selected as the critical endpoint on which to base a Tolerable Upper Intake Level (UL). The data fail to demonstrate a causal association between pyridoxine intake and other endpoints reviewed (e.g., dermatological lesions and B$_6$ dependency in newborns).

Dose-Response Assessment

Adults

Data Selection. Studies involving long-term oral administration of pyridoxine at doses of less than 1 g/day (Berger and Schaumburg, 1984; Bernstein and Lobitz, 1988; Dalton, 1985; Dalton and Dalton, 1987; Del Tredici et al., 1985; Parry and Bredesen, 1985) were considered more relevant for deriving a UL than studies involving short-term or single-dose intake, parenteral administration, or supplemental doses greater than 1 g/day.

Identification of a No-Observed-Adverse-Effect Level (NOAEL) and a Lowest-Observed-Adverse-Effect Level (LOAEL). A NOAEL of 200 mg/day can be identified by a critical evaluation of two studies (Bernstein and Lobitz, 1988; Del Tredici et al., 1985). Bernstein and Lobitz (1988) treated 70 patients with diabetic neuropathy or carpal tunnel syndrome with 100 to 150 mg/day of pyridoxine—some for up to 5 years. Despite rigorous neurological examination and testing, sensory neuropathy was not detected in any of these patients. Similarly, Del Tredici and colleagues (1985) reported on 24 patients treated for carpal tunnel syndrome with pyridoxine at doses of 100 to 300 mg/day for 4 months. A NOAEL of 200 mg/day represents the average of 100 and 300 mg/day.

Additional investigations report no neuropathy in hundreds of individuals given pyridoxine doses of 40 to 500 mg/day (Brush et al., 1988; Ellis et al., 1979; Mitwalli et al., 1984; Tolis et al., 1977). Although these studies were not as carefully executed or reported as the study by Bernstein and Lobitz (1988), they provide supportive evidence for a NOAEL of 200 mg/day.

A LOAEL of 500 mg/day was identified from a case report by Berger and Schaumburg (1984). These investigators reported that a 34-year-old woman who had been taking 200 mg/day of pyridoxine for 2 years without adverse effects developed sensory neuropathy when she increased her daily dose to 500 mg supplemented occasionally with a 300-mg dose once a week. Another report by Schaumburg and coworkers (1983) showed severe sensory neuropathy in seven adults after pyridoxine intakes that started at 50 to 100 mg/day and were steadily increased to 2 to 6 g/day over 2 to 40 months. None of these patients experienced signs or symptoms at doses of less than 2 g/day.

Several reports show sensory neuropathy at doses lower than 500 mg/day. Tabulated data in the report by Parry and Bredesen (1985) show that two patients who took 500 mg/day of pyridoxine for 8 and 36 months developed sensory neuropathy and as did one patient who took 100 to 200 mg/day for 36 months. This finding conflicts with the weight of evidence showing that daily doses in the range of 100 to 200 mg are not associated with the development of this condition. It is not clear from the report whether the neurological symptoms of the patient taking 100 to 200 mg/day were confirmed by a clinical neurological examination. The report notes that clinical neurological examinations were performed on about half of the patients. In addition, the report notes that electrophysiological studies were performed on seven patients and sural nerve biopsies were performed on two patients. It is not clear from the report which patients were examined clinically and which received additional tests. A local television report publicizing this syndrome before the study may have biased the selection of patients and reporting of neurological symptoms.

Two additional studies that report sensory neuropathy at doses of less than 200 mg/day (Dalton, 1985; Dalton and Dalton, 1987) warrant examination. In a letter to the editor, Dalton (1985) reported sensory neuropathy (characterized as burning, shooting, and tingling pains; paresthesia of limbs; clumsiness, ataxia; or perioral numbness) in 23 of 58 women (40 percent) being treated with 50 to 300 mg/day of pyridoxine for premenstrual syndrome. This case report contains methodological flaws including lack of information on the duration of treatment, use of other medications or herbal preparations, and lack of confirmatory information on actual doses consumed.

In a subsequent publication, which may have included patients from the earlier case report (Dalton, 1985), Dalton and Dalton (1987) retrospectively studied 172 women who were attending a private practice specializing in premenstrual syndrome and who were reported to have taken 50 to 500 mg/day of pyridoxine. The subjects were divided into two groups: 103 women who complained of neurological symptoms and 69 who did not. Neurological symptoms were not adequately detailed in this study. In addition, the actual doses may be underestimated because the patients were also taking vitamin supplements. In summary, the weaknesses of this study and the inconsistency of the results with the weight of evidence pertaining to the safety of higher doses of pyridoxine rule out the use of these data to determine a UL.

Uncertainty Assessment. An uncertainty factor (UF) of 2 was selected based on the limitations of the data involving pyridoxine doses of less than 500 mg/day.

Derivation of a UL. The NOAEL of 200 mg/day was divided by the UF of 2 to obtain a UL of 100 mg/day for adults:

$$\frac{NOAEL}{UF} = \frac{200 \text{ mg/day}}{2} = 100 \text{ mg/day}.$$

B_6 UL Summary, Ages 19 Years and Older

UL for Adults
19 years and older   100 mg/day of vitamin B_6 as pyridoxine

Other Life Stage Groups

Some concern for pregnant and lactating women could arise from the data available in the literature on congenital defects, B_6 dependency, and antilactogenic effects (Donaldson and Bury, 1982; Foukas, 1973; Gardner et al., 1985; Hunt et al., 1954; Marcus, 1975; Philpot et al., 1995; Scaglione and Vecchione, 1982). Scientifically based, controlled studies designed to assess the potential adverse effects of pyridoxine intake by pregnant and lactating women are lacking. As noted above, the weight of the evidence from controlled studies in animals during pregnancy reveals no adverse effects related to teratogenicity, and the evidence from humans reveals no adverse effects from intakes up to 200 mg/day. Therefore, a UL of 100 mg/day was set for pregnant and lactating women as well. The ULs for children and adolescents were calculated from the UL for adults by using the method described in Chapter 3; this method adjusts for body size.

B_6 UL Summary, Ages 1 through 18 Years, Pregnancy, Lactation

UL for Infants
0–12 months   Not possible to establish; source of
**   intake should be formula and food only**

UL for Children

1–3 years	30 mg/day of vitamin B$_6$ as pyridoxine
4–8 years	40 mg/day of vitamin B$_6$ as pyridoxine
9–13 years	60 mg/day of vitamin B$_6$ as pyridoxine

UL for Adolescents

14–18 years	80 mg/day of vitamin B$_6$ as pyridoxine

UL for Pregnancy

14–18 years	80 mg/day of vitamin B$_6$ as pyridoxine
19 years and older	100 mg/day of vitamin B$_6$ as pyridoxine

UL for Lactation

14–18 years	80 mg/day of vitamin B$_6$ as pyridoxine
19 years and older	100 mg/day of vitamin B$_6$ as pyridoxine

Special Considerations

A review of the literature failed to identify special subgroups that are distinctly susceptible to sensory neuropathy after excess pyridoxine intake.

Intake Assessment

Based on data from the Third National Health and Nutrition Survey (Appendix H), 9 mg/day was the highest mean intake of B$_6$ from food and supplements reported for any life stage and gender group; this was the reported intake of pregnant females aged 14 through 55 years. The highest reported intake at the ninety-fifth percentile was 21 mg/day in pregnant females aged 14 through 55 years, most of which is pyridoxine from supplements. B$_6$ (pyridoxine) is available over the counter in many dosages ranging up to 100 mg or more.

Risk Characterization

The risk of adverse effects resulting from excess intake of B$_6$ from food and supplements appears to be very low at the highest intakes noted above. Increased risks are likely to result from large intakes of PN used to treat conditions such as carpal tunnel syndrome, painful neuropathies, seizures, premenstrual syndrome, asthma, and sickle cell disease. The UL is not meant to apply to individuals who are being treated with PN under close medical supervision.

RESEARCH RECOMMENDATIONS FOR VITAMIN B_6

Priority should be given to studies useful for setting Estimated Average Requirements (EARs) for vitamin B_6 for children, adolescents, pregnant and lactating women, and the elderly. Future studies should be designed around the EAR paradigm, use graded levels of nutrient intake and clearly defined cutoff values for clinical adequacy and inadequacy, and be conducted for a sufficient duration. To do this, close attention should be given to the identification of indicators on which to base B_6 requirements.

REFERENCES

Andon MB, Reynolds RD, Moser-Veillon PB, Howard MP. 1989. Dietary intake of total and glycosylated vitamin B_6 and the vitamin B_6 nutritional status of unsupplemented lactating women and their infants. *Am J Clin Nutr* 50:1050–1058.

Baer RL. 1984. Cutaneous skin changes probably due to pyridoxine abuse. *J Am Acad Dermatol* 10:527–528.

Baker EM, Canham JE, Nunes WT, Sauberlich HE, McDowell ME. 1964. Vitamin B_6 requirement for adult men. *Am J Clin Nutr* 15:59–66.

Barnard HC, de Kock JJ, Vermaak WJ, Potgieter GM. 1987. A new perspective in the assessment of vitamin B_6 nutritional status during pregnancy in humans. *J Nutr* 117:1303–1306.

Berger A, Schaumburg HH. 1984. More on neuropathy from pyridoxine abuse. *N Engl J Med* 311:986–987.

Bernstein AL, Lobitz CS. 1988. A clinical and electrophysiologic study of the treatment of painful diabetic neuropathies with pyridoxine. In: Leklem JE, Reynolds RD, eds. *Clinical and Physiological Applications of Vitamin B_6. Current Topics in Nutrition and Disease.* New York: Alan R. Liss. Pp. 415–423.

Bessey OA, Adam DJ, Hansen AE. 1957. Intake of vitamin B_6 and infantile convulsions: A first approximation of requirements of pyridoxine in infants. *Pediatrics* 20:33–44.

Bhagavan HN. 1985. Interaction between vitamin B_6 and drugs. In: Reynolds RD, Leklem JE, eds. *Vitamin B_6: Its Role in Health and Disease.* New York: Liss. Pp. 401–415.

Bhagavan HN, Coleman M, Coursin DB. 1975. The effect of pyridoxine hydrochloride on blood serotonin and pyridoxal phosphate contents in hyperactive children. *Pediatrics* 55:437–441.

Borschel MW. 1995. Vitamin B_6 in infancy: Requirements and current feeding practices. In: Raiten DJ, ed. *Vitamin B_6 Metabolism in Pregnancy, Lactation and Infancy.* Boca Raton, FL: CRC Press. Pp. 109–124.

Borschel MW, Kirksey A, Hanneman RE. 1986. Effects of vitamin B_6 intake on nutriture and growth of young infants. *Am J Clin Nutr* 43:7–15.

Brattstrom LE, Israelsson B, Norrving B, Bergkvist D, Thorne J, Hultberg B, Hamfelt A. 1990. Impaired homocysteine metabolism in early-onset cerebral and peripheral occlusive arterial disease. Effects of pyridoxine and folic acid treatment. *Atherosclerosis* 81:51–60.

Bredesen E, Parry GJ. 1984. Pyridoxine neuropathy. *Neurology* 34:136.

Brophy MH, Siiteri PK. 1975. Pyridoxal phosphate and hypertensive disorders of pregnancy. *Am J Obstet Gynecol* 121:1075–1079.

Brown RR, Rose DP, Leklem JE, Linkswiler H, Anand R. 1975. Urinary 4-pyridoxic acid, plasma pyridoxal phosphate, and erythrocyte aminotransferase levels in oral contraceptive users receiving controlled intakes of vitamin B$_6$. *Am J Clin Nutr* 28:10–19.

Brush MG, Bennett T, Hansen K. 1988. Pyridoxine in the treatment of premenstrual syndrome: A retrospective survey in 630 patients. *Br J Clin Pract* 42:448–452.

Brussaard JH, Lowik MR, van den Berg H, Brants HA, Bemelmans W. 1997a. Dietary and other determinants of vitamin B$_6$ parameters. *Eur J Clin Nutr* 51:S39–S45.

Brussaard JH, Lowik MR, van den Berg H, Brants HA, Kistemaker C. 1997b. Micronutrient status, with special reference to vitamin B$_6$. *Eur J Clin Nutr* 51:S32–S38.

Canham JE, Nunes WT, Eberlin EW. 1964. Electroencephalographic and central nervous system manifestations of vitamin B$_6$ deficiency and induced vitamin B$_6$ dependency in normal human adults. In: *Proceedings of the Sixth International Congress on Nutrition.* Edinburgh: E & S Livingstone.

Cleary RE, Lumeng L, Li TK. 1975. Maternal and fetal plasma levels of pyridoxal phosphate at term: Adequacy of vitamin B$_6$ supplementation during pregnancy. *Am J Obstet Gynecol* 121:25–28.

Coburn SP. 1990. Location and turnover of vitamin B$_6$ pools and vitamin B$_6$ requirements of humans. *Ann NY Acad Sci* 585:76–85.

Coburn SP, Mahuren JD, Szadkowska Z, Schaltenbrand WE, Townsend DW. 1987. Kinetics of vitamin B$_6$ metabolism examined in minature swine by continuous administration of labelled pyridoxine. In: Canolty NL, Caine TP, eds. *Proceedings of the 1985 Conference on Mathematical Models in Experimental Nutrition.* Athens: University of Georgia. Pp. 99–111.

Coburn SP, Lewis DL, Fink WJ, Mahuren JD, Schaltenbrand WE, Costill DL. 1988a. Human vitamin B$_6$ pools estimated through muscle biopsies. *Am J Clin Nutr* 48:291–294.

Coburn SP, Mahuren JD, Kennedy MS, Schaltenbrand WE, Sampson DA, O'Connor DK, Snyder DL, Wostmann BS. 1988b. B$_6$ vitamin content of rat tissues measured by isotope tracer and chromatographic methods. *Biofactors* 1:307–312.

Cohen M, Bendich A. 1986. Safety of pyridoxine—A review of human and animal studies. *Toxicol Lett* 34:129–139.

Contractor SF, Shane B. 1968. Estimation of vitamin B$_6$ compounds in human blood and urine. *Clin Chim Acta* 21:71–77.

Contractor SF, Shane B. 1970. Blood and urine levels of vitamin B$_6$ in the mother and fetus before and after loading of the mother with vitamin B$_6$. *Am J Obstet Gynecol* 107:635–640.

Contractor SF, Shane B. 1971. Metabolism of [^{14}C] pyridoxol in the pregnant rat. *Biochim Biophys Acta* 230:127–136

Coursin DB. 1954. Convulsive seizures in infants with pyridoxine-deficient diet. *JAMA* 154:406–408.

Crozier PG, Cordain L, Sampson DA. 1994. Exercise-induced changes in plasma vitamin B-6 concentrations do not vary with exercise intensity. *Am J Clin Nutr* 60:552–558.

Dakshinamurti K, ed. 1990. Vitamin B$_6$. *Ann NY Acad Sci* 585:1–570.

Dakshinamurti K, Stephens MC. 1969. Pyridoxine deficiency in the neonatal rat. *J Neurochem* 6:1515–1522.

Dakshinamurti K, Sharma SK, Sundaram M. 1991. Domoic acid induced seizure activity in rats. *Neurosci Lett* 127:193–197.

Dakshinamurti K, Sharma SK, Sundaram M, Watanabe T. 1993. Hippocampal changes in developing postnatal mice following intrauterine exposure to domoic acid. *J Neurosci* 13:4486–4495.

Dalton K. 1985. Pyridoxine overdose in premenstrual syndrome. *Lancet* 1168–1169.

Dalton K, Dalton MJT. 1987. Characteristics of pyridoxine overdose neuropathy syndrome. *Acta Neurol Scand* 76:8–11.

Del Tredici AM, Bernstein AL, Chinn K. 1985. Carpal tunnel syndrome and vitamin B_6 therapy. In: Reynolds RD, Leklem JE, eds. *Vitamin B_6: Its Role in Health and Disease. Current Topics in Nutrition and Disease.* New York: Alan R. Liss. Pp. 459–462.

De Zegher F, Przyrembel H, Chalmers RA, Wolff ED, Huijmans JG. 1985. Successful treatment of infantile type I primary hyperoxaluria complicated by pyridoxine toxicity. *Lancet* 17:392–393.

Donaldson GL, Bury RG. 1982. Multiple congenital abnormalities in a newborn boy associated with maternal use of fluphenazine enanthate and other drugs during pregnancy. *Acta Paediatr Scand* 71:335–338.

Dorsey CW. 1949. The use of pyridoxine and suprarenal cortex combined in the treatment of the nausea and vomiting of pregnancy. *Am J Obstet Gynecol* 58:1073–1078.

Dreon DM, Butterfield GE. 1986. Vitamin B_6 utilization in active and inactive young men. *Am J Clin Nutr* 43:816–824.

Driskell JA, Moak SW. 1986. Plasma pyridoxal phosphate concentrations and coenzyme stimulation of erythrocyte alanine aminotransferase activities of white and black adolescent girls. *Am J Clin Nutr* 43:599–603.

Driskell JA, Clark AJ, Bazzarre TL, Chopin LF, McCoy H, Kenney MA, Moak SW. 1985. Vitamin B_6 status of southern adolescent girls. *J Am Diet Assoc* 85:46–49.

Driskell JA, Clark AJ, Moak SW. 1987. Longitudinal assessment of vitamin B_6 status in Southern adolescent girls. *J Am Diet Assoc* 87:307–310.

Driskell JA, McChrisley B, Reynolds LK, Moak SW. 1989. Plasma pyridoxal 5'-phosphate concentrations in obese and nonobese black women residing near Petersburg, VA. *Am J Clin Nutr* 50:37–40.

Ellis JM. 1987. Treatment of carpal tunnel syndrome with vitamin. *South Med J* 80:882–884.

Ellis J, Folkers K, Watanabe T, Kaji M, Saji S, Caldwell JW, Temple CA, Wood FS. 1979. Clinical results of a cross-over treatment with pyridoxine and placebo of the carpal tunnel syndrome. *Am J Clin Nutr* 32:2040–2046.

Fogelholm M. 1992. Micronutrient status in females during a 24-week fitness-type exercise program. *Ann Nutr Metab* 36:209–218.

Fonda ML, Trauss C, Guempel UM. 1991. The binding of pyridoxal 5'-phosphate to human serum albumin. *Arch Biochem Biophys* 288:79–86.

Foukas MD. 1973. An antilactogenic effect of pyridoxine. *J Obstet Gynaecol Br Commonw* 80:718–720.

Friedman MA, Resnick JS, Baer RL. 1986. Subepidermal vesicular dermatosis and sensory peripheral neuropathy caused by pyridoxine abuse. *J Am Acad Dermatol* 14:915–917.

Fries ME, Chrisley BM, Driskell JA. 1981. Vitamin B_6 status of a group of preschool children. *Am J Clin Nutr* 34:2706–2710.

Gardner LI, Welsh-Sloan J, Cady RB. 1985. Phocomelia in infant whose mother took large doses of pyridoxine during pregnancy. *Lancet* 1:636.

Gaynor R, Dempsey WB. 1972. Vitamin B₆ enzymes in normal and pre-eclamptic human placentae. *Clin Chim Acta* 37:411–416.

Grabow JD, Linkswiler H. 1969. Electroencephalographic and nerve-conduction studies in experimental vitamin B₆ deficiency in adults. *Am J Clin Nutr* 22:1429–1434.

Gregory JF 3rd. 1997. Bioavailability of vitamin B₆. *Eur J Clin Nutr* 51:S43–S48.

Guilarte TR. 1993. Vitamin B₆ and cognitive development: Recent research findings from human and animal studies. *Nutr Rev* 51:193–198.

Hamfelt A, Tuvemo T. 1972. Pyridoxal phosphate and folic acid concentration in blood and erythrocyte aspartate aminotransferase activity during pregnancy. *Clin Chim Acta* 41:287–298.

Hamm MW, Mehansho H, Henderson LM. 1979. Transport and metabolism of pyridoxamine and pyridoxamine phosphate in the small intestine of the rat. *J Nutr* 109:1552–1559.

Hansen CM, Leklem JE, Miller LT. 1996a. Vitamin B-6 status indicators decrease in women consuming a diet high in pyridoxine glucoside. *J Nutr* 126:2512–2518.

Hansen CM, Leklem JE, Miller LT. 1996b. Vitamin B-6 status of women with a constant intake of vitamin B-6 changes with three levels of dietary protein. *J Nutr* 126:1891–1901.

Hansen CM, Leklem JE, Miller LT. 1997. Changes in vitamin B-6 status indicators of women fed a constant protein diet with varying levels of vitamin B-6. *Am J Clin Nutr* 66:1379–1387.

Hart BF, McConnell WT. 1943. Vitamin B factors in toxic psychosis of pregnancy and the puerperium. *Am J Obstet Gynecol* 46:304.

Hawkins WW, Barsky J. 1948. An experiment on human vitamin B₆ deprivation. *Science* 108:284–286.

Heiskanen K, Salmenperä L, Perheentupa J, Siimes MA. 1994. Infant vitamin B-6 status changes with age and with formula feeding. *Am J Clin Nutr* 60:907–910.

Heiskanen K, Kallio M, Salmenperä L, Siimes MA, Ruokonen I, Perheentupa J. 1995. Vitamin B-6 status during childhood: Tracking from 2 months to 11 years of age. *J Nutr* 125:2985–2992.

Hendrickx AG, Cukierski M, Prahalada S, Janos Booher S, Nyland T. 1985. Evaluation of bendectin embryotoxicity nonhuman primates: II. Double-blind study term cynomolgus monkeys. *Teratology* 32:191–194.

Huang Y-C, Chen W, Evans MA, Mitchell ME, Shultz TD. 1998. Vitamin B₆ requirement and status assessment of young women fed a high-protein diet with various levels of vitamin B-6. *Am J Clin Nutr* 67:208–220.

Hultberg B, Andersson A, Sterner G. 1993. Plasma homocysteine in renal failure. *Clin Nephrol* 40:230–235.

Hunt AD Jr, Stokes J Jr, McCrory WW, Stroud HH. 1954. Pyridoxine dependency: Report of a case of intractable convulsions in an infant controlled by pyridoxine. *Pediatrics* 13:140–145.

Johansson S, Lindstedt S, Register U, Wadstrom L. 1966. Studies on the metabolism of labeled pyridoxine in man. *Am J Clin Nutr* 18:185–196.

Kang-Yoon SA, Kirksey A, Giacoia GP, West KD. 1995. Vitamin B₆ adequacy in neonatal nutrition: Associations with preterm delivery, type of feeding, and vitamin B-6 supplementation. *Am J Clin Nutr* 62:932–942.

Khera KS. 1975. Teratogenicity study in rats given high doses pyridoxine (vitamin B₆) during organogenesis. *Experientia* 31:469–470.

Kirksey A, Roepke JL. 1981. Vitamin B_6 nutriture of mothers of three breast-fed neonates with central nervous system disorders. *Fed Proc* 40:864.

Kirksey A, Keaton K, Abernathy RP, Greger JL. 1978. Vitamin B_6 nutritional status of a group of female adolescents. *Am J Clin Nutr* 31:946–954.

Kretsch MJ, Sauberlich HE, Newbrun E. 1991. Electroencephalographic changes and periodontal status during short-term vitamin B-6 depletion of young, non-pregnant women. *Am J Clin Nutr* 53:1266–1274.

Kretsch MJ, Sauberlich HE, Skala JH, Johnson HL. 1995. Vitamin B-6 requirement and status assessment: Young women fed a depletion diet followed by a plant- or animal-protein diet with graded amounts of vitamin B-6. *Am J Clin Nutr* 61:1091–1101.

Leklem JE. 1990. Vitamin B-6: A status report. *J Nutr* 120:1503–1507.

Leklem JE. 1991. Vitamin B_6. In: Machlin LJ ed. *Handbook of Vitamins*, 2nd edition. New York: Marcel Dekker. Pp. 341–392.

Leklem JE. 1994. Vitamin B_6. In: Shils ME, Olson JA, Shike M, eds. *Modern Nutrition in Health and Disease*. Philadelphia: Lea & Febiger. Pp. 383–394.

Leklem JE, Shultz TD. 1983. Increased plasma pyridoxal 5'-phosphate and vitamin B-6 in male adolescents after 4500-meter run. *Am J Clin Nutr* 38:541–548.

Lindberg AS, Leklem JE, Miller LT. 1983. The effect of wheat bran on the bioavailability of vitamin B-6 in young men. *J Nutr* 113:2578–2586.

Linkswiler HM. 1978. Vitamin B_6 requirements of men. In: *Human Vitamin B_6 Requirements: Proceedings of a Workshop*. Washington, DC: National Academy of Sciences. Pp. 279–290.

Lui A, Lumeng L, Aronoff GR, Li T-K. 1985. Relationship between body store of vitamin B_6 and plasma pyridoxal-P clearance: Metabolic balance studies in humans. *J Lab Clin Med* 106:491–497.

Lumeng L, Li TK. 1974. Vitamin B_6 metabolism in chronic alcohol abuse. Pyridoxal phosphate levels in plasma and the effects of acetaldehyde on pyridoxal phosphate synthesis and degradation in human erythrocytes. *J Clin Invest* 53:693–704.

Lumeng L, Cleary RE, Li TK. 1974. Effect of oral contraceptives on the plasma concentration of pyridoxal phosphate. *Am J Clin Nutr* 27:326–333.

Lumeng L, Cleary RE, Wagner R, Pao-Lo Y, Li TK. 1976. Adequacy of vitamin B_6 supplementation during pregnancy: A prospective study. *Am J Clin Nutr* 29:1376–1383.

Lumeng L, Ryan MP, Li TK. 1978. Validation of the diagnostic value of plasma pyridoxal 5'-phosphate measurements in vitamin B_6 nutrition of the rat. *J Nutr* 108:545–553.

Manore MM, Leklem JE. 1988. Effect of carbohydrate and vitamin B_6 on fuel substrates during exercise in women. *Med Sci Sports Exerc* 20:233–241.

Manore MN, Leklem JE, Walter MC. 1987. Vitamin B_6 metabolism as affected by exercise in trained and untrained women fed diets differing in carbohydrate and vitamin B_6 content. *Am J Clin Nutr* 46:995–1004.

Marcus RG. 1975. Suppression of lactation with high doses of pyridoxine. *S Afr Med J* 49:2155–2156.

Merrill AH Jr, Henderson JM, Wang E, McDonald BW, Millikan WJ. 1984. Metabolism of vitamin B-6 by human liver. *J Nutr* 114:1664–1674.

Meydani SN, Ribaya-Mercado JD, Russell RM, Sahyoun N, Morrow FD, Gershoff SN. 1991. Vitamin B-6 deficiency impairs interleukin 2 production and lymphocyte proliferation in elderly adults. *Am J Clin Nutr* 53:1275–1280.

Miller LT, Linkswiler H. 1967. Effect of protein intake on the development of abnormal tryptophan metabolism by men during vitamin B$_6$ depletion. *J Nutr* 93:53–59.

Miller LT, Johnson A, Benson EM, Woodring MJ. 1975. Effect of oral contraceptives and pyridoxine on the metabolism of vitamin B$_6$ and on plasma tryptophan and α-amino nitrogen. *Am J Clin Nutr* 28:846–853.

Miller LT, Leklem JE, Shultz TD. 1985. The effect of dietary protein on the metabolism of vitamin B$_6$ in humans. *J Nutr* 115:1663–1672.

Mitwalli A, Blair G, Oreopoulos DG. 1984. Safety of intermediate doses of pyridoxine. *Can Med Assoc J* 131:14.

Moss AJ, Levy AS, Kim I, Park YK. 1989. *Use of Vitamin and Mineral Supplements in the United States: Current Users, Types of Products, and Nutrients*. Advance Data, Vital and Health Statistics of the National Center for Health Statistics, No. 174. Hyattsville, MD: National Center for Health Statistics.

Mueller JF, Vilter RW. 1950. Pyridoxine deficiency in human beings induced by desoxypyridoxine. *J Clin Invest* 29:193–201.

Nakano H, Gregory JF 3rd. 1997. Pyridoxine and pyridoxine-5'-beta-D-glucoside exert different effects on tissue B-6 vitamins but similar effects of beta-glucosidase activity in rats. *J Nutr* 125:2751–2762.

Pannemans DL, van den Berg H, Westerterp KR. 1994. The influence of protein intake on vitamin B$_6$ metabolism differs in young and elderly humans. *J Nutr* 124:1207–1214.

Parry GJ, Bredesen DE. 1985. Sensory neuropathy with low-dose pyridoxine. *Neurology* 35:1466–1468.

Phillips WE, Mills JH, Charbonneau SM, Tryphonas L, Hatina GV, Zawidzka Z, Bryce FR, Munro IC. 1978. Subacute toxicity of pyridoxine hydrochloride in the beagle dog. *Toxicol Appl Pharmacol* 44:323–333.

Philpot J, Muntoni F, Skellett S, Dubowitz V. 1995. Congenital symmetrical weakness of the upper limbs resembling brachial plexus palsy: A possible sequel of drug toxicity in the first trimester of pregnancy. *Neuromuscul Disord* 5:67–69.

Raica N Jr, Sauberlich HE. 1964. Blood cell transaminase activity in human vitamin B$_6$ deficiency. *Am J Clin Nutr* 15:67–72.

Ribaya-Mercado JD, Russell RM, Sahyoun N, Morrow FD, Gershoff SN. 1991. Vitamin B-6 requirements of elderly men and women. *J Nutr* 121:1062–1074.

Riggs KM, Spiro A, Tucker K, Rush D. 1996. Relations of vitamin B-12, vitamin B-6, folate, and homocysteine to cognitive performance in the Normative Aging Study. *Am J Clin Nutr* 63:306–314.

Rimm EB, Willett WC, Hu FB, Sampson L, Colditz GA, Manson JE, Hennekens C, Stampfer MJ. 1998. Folate and vitamin B$_6$ from diet and supplements in relation to risk of coronary heart disease among women. *J Am Med Assoc* 279:359–364.

Robinson K, Mayer EL, Miller DP, Green R, van Lente F, Gupta A, Kottke-Marchant K, Savon SR, Selhub J, Nissen SE, Kutner M, Topol EJ, Jacobsen DW. 1995. Hyperhomocysteinemia and low pyridoxal phosphate. Common and independent reversible risk factors for coronary artery disease. *Circulation* 92:2825–2830.

Rose CS, Gyorgy P, Butler M, Andres R, Norris AH, Shock NW, Tobin J, Brin M, Spiegel H. 1976. Age differences in vitamin B$_6$ status of 617 men. *Am J Clin Nutr* 29:847–853.

Rose DP. 1978. Oral Contraceptives and Vitamin B_6. In: *Human Vitamin B_6 Requirements: Proceedings of a Workshop*. Washington, DC: National Academy Press. Pp. 193–201.

Sauberlich HE. 1964. Human requirements for vitamin B_6. *Vitam Horm* 22:807–823.

Sauberlich HE, Canham JE, Baker EM, Raica N Jr, Herman YF. 1972. Biochemical assessment of the nutritional status of vitamin B_6 in the human. *Am J Clin Nutr* 25:629–642.

Scaglione D, Vecchione A. 1982. Pyridoxine for the suppression of lactation—a clinical trial on 1592 cases. *Acta Vitaminol Enzymol* 4:207–214.

Schaeppi U, Krinke G. 1985. Differential vulnerability of three rapidly conducting somatosensory pathways in the dog with vitamin B_6 neuropathy. *Agents Actions* 16:567–579.

Schaumburg HH, Berger A. 1988. Pyridoxine neurotoxicity. In: *Clinical and Physiological Applications of Vitamin B_6*. New York: Alan R. Liss. Pp. 403–414.

Schaumburg H, Kaplan J, Windebank A, Vick N, Rasmus S, Pleasure D, Brown MJ. 1983. Sensory neuropathy from pyridoxine abuse. *N Engl J Med* 309:445–448.

Schumacher MF, Williams MA, Lyman RL. 1965. Effect of high intakes of thiamine, riboflavin and pyridoxine on reproduction in rats and vitamin requirements of the offspring. *J Nutr* 86:343–349.

Schuster K, Bailey LB, Mahan CS. 1981. Vitamin B_6 status of low-income adolescent and adult pregnant women and the condition of their infants at birth. *Am J Clin Nutr* 34:1731–1735.

Selhub J, Jacques PF, Wilson PWF, Rush D, Rosenberg IH. 1993. Vitamin status and intake as primary determinants of homocysteinemia in an elderly population. *J Am Med Assoc* 270:2693–2698.

Selhub J, Jacques PF, Bostom AG, D'Agostino RB, Wilson PW, Belanger AJ, O'Leary DH, Wolf PA, Schaefer EJ, Rosenberg IH. 1995. Association between plasma homocysteine concentrations and extracranial carotid-artery stenosis. *N Engl J Med* 332:286–291.

Shane B. 1978. Vitamin B_6 and blood. In: *Human Vitamin B_6 Requirements: Proceedings of a Workshop*. Washington, DC: National Academy Press. Pp. 111–128.

Shane B, Contractor SF. 1975. Assessment of vitamin B_6 status. Studies on pregnant women and oral contraceptive agent users. *Am J Clin Nutr* 28:739–747.

Shane B, Contractor SF. 1980. Vitamin B_6 status and metabolism in pregnancy. In: Tryfiates GP, ed. *Vitamin B_6 Metabolism and Role in Growth*. Westport, CT: Food & Nutrition Press. Pp. 137–171.

Sharma SK, Dakshinamurti K. 1992. Seizure activity in pyridoxine-deficient adult rats. *Epilepsia* 33:235–247.

Sharma SK, Bolster B, Dakshinamurti K. 1994. Picrotoxin and pentylene tetrazole induced seizure activity in pyridoxine-deficient rats. *J Neurol Sci* 121:1–9.

Shultz TD, Leklem JE. 1981. Urinary 4-pyridoxic acid, urinary vitamin B_6, and plasma pyridoxal phosphate as measures of vitamin B_6 status and dietary intake in adults. In: Leklem JE, Reynolds RD, eds. *Methods in Vitamin B_6 Nutrition*. New York: Plenum Press. Pp. 389–392.

Snell EE. 1958. Some aspects of the metabolism of vitamin B_6. In: *Fourth International Congress of Biochemistry-Vitamin Metabolism*, Vol. 11. New York: Pergamon. Pp. 250–265.

Snyderman SE, Holt LE, Carretero R, Jacobs K. 1953. Pyridoxine deficiency in the human infant. *Am J Clin Nutr* 1:200.

Stephens MC, Havlicek V, Dakshinamurti K. 1971. Pyridoxine deficiency and development of the central nervous system in the rat. *J Neurochem* 18:2407–2416.

Subbarao K, Kakkar VV. 1979. Thrombin induced surface changes of human platelets. *Biochem Biophys Res Commun* 88:470–476.

Tarr JB, Tamura T, Stokstad EL. 1981. Availability of vitamin B$_6$ and pantothenate in an average American diet in man. *Am J Clin Nutr* 34:1328–1337.

Tolis G, Laliberte R, Guyda H, Naftolin F. 1977. Ineffectiveness of pyridoxine (B$_6$) to alter secretion of growth hormone and prolactin and absence of therapeutic effects on galactorrhea-amenorrhea syndromes. *J Clin Endocrinol Metab* 44:1197–1199.

Ubbink JB, Vermaak WJ, Delport R, van der Merwe A, Becker PJ, Potgieter H. 1995. Effective homocysteine metabolism may protect South African blacks against coronary heart disease. *Am J Clin Nutr* 62:802–808.

Ubbink JB, van der Merwe A, Delport R, Allen RH, Stabler SP, Riezler R, Vermaak WJ. 1996. The effect of a subnormal vitamin B$_6$ status on homocysteine metabolism. *J Clin Invest* 98:177–184.

Unna K, Antopol W. 1940. Toxicity of vitamin B$_6$. *Proc Soc Exp Biol Med* 43:116–118.

van der Beek EJ, van Dokkum W, Wedel M, Schrijver J, van den Berg H. 1994. Thiamin, riboflavin and vitamin B$_6$: Impact of restricted intake on physical performance in man. *J Am Coll Nutr* 13:629–640.

Weinstein BB, Wohl Z, Mitchell GJ, Sustendal GF. 1944. Oral administration of pyridoxine hydrochloride in the treatment of nausea and vomiting of pregnancy. *Am J Obstet Gynecol* 47:389–394.

Weir MR, Keniston RC, Enriquez JI Sr, McNamee GA. 1991. Depression of vitamin B$_6$ levels due to dopamine. *Vet Hum Toxicol* 33:118–121.

West KD, Kirksey A. 1976. Influence of vitamin B$_6$ intake on the content of the vitamin in human milk. *Am J Clin Nutr* 29:961–969.

Willis RS, Winn WW, Morris AT, Newsome AA, Massey WE. 1942. Clinical observations in treatment of nausea and vomiting in pregnancy with vitamin B1 and B6. A preliminary report. *Am J Obstet Gynecol* 44:265–271.

Wozenski JR, Leklem JE, Miller LT. 1980. The metabolism of small doses of vitamin B$_6$ in men. *J Nutr* 110:275–285.

Yess N, Price JM, Brown RR, Swan PB, Linkswiler H. 1964. Vitamin B$_6$ depletion in man: Urinary excretion of tryptophan metabolites. *J Nutr* 84:229–236.

8

Folate

SUMMARY

Folate functions as a coenzyme in single-carbon transfers in the metabolism of nucleic and amino acids. The primary indicator used to estimate the Recommended Dietary Allowance (RDA) for folate is erythrocyte folate in conjunction with plasma homocysteine and folate concentrations. The RDA for both men and women is 400 µg/day of dietary folate equivalents (DFEs). DFEs adjust for the nearly 50 percent lower bioavailability of food folate compared with that of folic acid: 1 µg of dietary folate equivalent = 0.6 µg of folic acid from fortified food or as a supplement taken with meals = 1 µg of food folate = 0.5 µg of a supplement taken on an empty stomach. To reduce the risk of neural tube defects for women capable of becoming pregnant, the recommendation is to take 400 µg of folic acid daily from fortified foods, supplements, or both in addition to consuming food folate from a varied diet. The evidence available on the role of folate in reducing the risk of vascular disease, cancer, and psychiatric and mental disorders is not sufficiently conclusive to use risk reduction of these conditions as a basis for setting the Estimated Average Requirement (EAR) and the RDA.

In the U.S. adult population from 1988 to 1994, which was before cereal grains were fortified with folate, the reported median intake of folate from food was approximately 250 µg/day, but this value underestimates current intake. The ninety-fifth percentile of intake from food and supplements was close to 900 µg/day overall and nearly 1,700 µg/day for pregnant women. After the fortification of cereal grains with folate—which became mandatory for enriched grains in the United States as of January 1, 1998, and is now authorized in Canada—average intake of folate is expected to increase by about 80 to 100 µg/day for women and by more for men. The Tolerable Upper Intake Level (UL) for adults is set at 1,000 µg/day of folate from fortified food or as a supplement, exclusive of food folate.

BACKGROUND INFORMATION

Folate is a generic term for this water-soluble B-complex vitamin, which functions in single-carbon transfer reactions and exists in many chemical forms (Wagner, 1996). Folic acid (pteroylmonoglutamic acid), which is the most oxidized and stable form of folate, occurs rarely in food but is the form used in vitamin supplements and in fortified food products. Folic acid consists of a *p*-aminobenzoic acid molecule linked at one end to a pteridine ring and at the other end to one glutamic acid molecule. Most naturally occurring folates, called *food folate* in this report, are pteroylpolyglutamates, which contain one to six additional glutamate molecules joined in a peptide linkage to the γ-carboxyl of glutamate.

Function

The folate coenzymes are involved in numerous reactions that involve (1) deoxyribonucleic acid (DNA) synthesis, which depends on a folate coenzyme for pyrimidine nucleotide biosynthesis (methylation of deoxyuridylic acid to thymidylic acid) and thus is required for normal cell division; (2) purine synthesis (formation of glycinamide ribonucleotide and 5-amino-4-imidazole carboxamide ribonucleotide); (3) generation of formate into the formate pool (and utilization of formate); and (4) amino acid interconversions, including the catabolism of histidine to glutamic acid, interconversion of serine and glycine, and conversion of homocysteine to methionine. Folate-mediated transfer of single-carbon units from serine provides a major source of substrate in single-carbon metabolism. The conversion of homocysteine to methionine serves as a major source of methionine for the synthesis of *S*-adenosyl-methionine, an important in vivo methylating agent (Wagner, 1996).

Physiology of Absorption, Metabolism, and Excretion

Absorption, Transport, and Storage

Food folates (polyglutamate derivatives) are hydrolyzed to monoglutamate forms in the gut before absorption across the intestinal mucosa. This cleavage is accomplished by a γ-glutamylhydrolase, more commonly called folate conjugase. The monoglutamate form of folate is actively transported across the proximal small intestine by a saturable pH-dependent process. When pharmacological doses of the monoglutamate form of folate are consumed, it is also absorbed by a nonsaturable mechanism involving passive diffusion.

Monoglutamates, mainly 5-methyl-tetrahydrofolate, are present in the portal circulation. Much of this folate can be taken up by the liver, where it is metabolized to polyglutamate derivatives and retained or released into the blood or bile. Approximately two-thirds of the folate in plasma is protein bound. A variable proportion of plasma folate is bound to low-affinity protein binders, primarily albumin, which accounts for about 50 percent of bound folate. Low levels of high-affinity folate binders are also present in plasma.

Cellular transport of folate is mediated by a number of different folate transport systems, which can be characterized as either membrane carriers or folate-binding protein-mediated systems. These transport systems are not saturated by folate under physiological conditions, and folate influx into tissues would be expected after any elevation in plasma folate after supplementation.

Folate concentrations in liver of 4.5 to 10 μg/g were reported after liver biopsies (Whitehead, 1973). Because the adult male liver weighs approximately 1,400 g, the total quantity of folate in the liver would be approximately 6 to 14 mg. If the liver is assumed to contain 50 percent of the body stores of folate, the estimated total body folate store would be 12 to 28 mg. Using the same assumption, Hoppner and Lampi (1980) determined average liver folate concentrations to be approximately 8 μg/g (range 3.6 to 14.8 μg/g) after autopsy; the liver folate content would be approximately 11 mg and total body folate 22 mg.

Metabolism and Excretion

Before being stored in tissue or used as a coenzyme, folate monoglutamate is converted to the polyglutamate form by the enzyme folylpolyglutamate synthetase. When released from tissues into circulation, folate polyglutamates are reconverted to the monoglutamate form by γ-glutamylhydrolase. Folates must be reduced enzymatically and resynthesized to the polyglutamate form to function in single-carbon transfer reactions.

The metabolic interrelationship between folate and vitamin B_{12} may explain why a single deficiency of either vitamin leads to the same hematological changes. Both folate and vitamin B_{12} are required for the formation of 5,10-methylenetetrahydrofolate and involved in thymidylate synthesis by way of a vitamin B_{12}-containing enzyme. The formation of 5,10-methylene tetrahydrofolate depends on the regeneration of the parent compound (tetrahydrofolate) in the homocysteine-to-methionine conversion. This reaction involves the removal of a methyl group from methyl folate and the delivery of this group to homocysteine for the synthesis of methionine. Folate is involved as a substrate (5-methyl-tetrahydrofolate) and vitamin B_{12} as a coenzyme. The 5,10-methylenetetrahydrofolate delivers its methyl group to deoxyuridylate to convert it to thymidylate for incorporation into DNA. In either a folate or vitamin B_{12} deficiency, the megaloblastic changes occurring in the bone marrow and other replicating cells result from lack of adequate 5,10-methylenetetrahydrofolate.

The major route of whole-body folate turnover appears to be via catabolism to cleavage products. The initial step in folate catabolism involves the cleavage of intracellular folylpolyglutmate at the C9-N10 bond, and the resulting *p*-aminobenzoylpolyglutamates are hydrolyzed to the monoglutamate, which is N-acetylated before excretion.

Folate freely enters the glomerulus and is reabsorbed in the proximal renal tubule. The net effect is that most of the secreted folate is reabsorbed. The bulk of the excretion products in humans are folate cleavage products. Intact urinary folate represents only a very small percentage of dietary folate. Biliary excretion of folate has been estimated to be as high as 100 μg/day (Herbert and Das, 1993; Whitehead, 1986); however, much of this is reabsorbed by the small intestine (Weir et al., 1985). Fecal folate losses occur, but it is difficult to distinguish actual losses from losses of folate synthesized by the intestinal microflora (Krumdieck et al., 1978).

Clinical Effects of Inadequate Intake

Inadequate folate intake first leads to a decrease in serum folate concentration, then to a decrease in erythrocyte folate concentration, a rise in homocysteine concentration, and megaloblastic changes in the bone marrow and other tissues with rapidly dividing cells.

Within weeks of the development of early morphological abnormalities in the marrow, subtle changes appear in the peripheral blood (Eichner et al., 1971) when hypersegmentation of the neutrophils becomes apparent. The peripheral blood picture is variable before the development of a clearly increased mean cell volume or anemia (Lindenbaum et al., 1988). In some deficient individuals, macrocytes and macroovalocytes are seen on blood smears, but in others the erythrocytes may show only minimal anisocytosis or no abnormalities. When folate supply to the bone marrow becomes rate limiting for erythropoiesis, macrocytic cells are produced. However, because of the 120-day lifespan of normal erythrocytes, macrocytosis is not evident in the early stages of folate-deficient megaloblastosis.

As folate depletion progresses further, the mean cell volume increases above normal. Neutrophil hypersegmentation (defined as more than 5 percent five-lobed or any six-lobed cells per 100 granulocytes) is typically present in the peripheral blood at this stage of macrocytosis and the neutrophil lobe average is elevated.

Macrocytic anemia then develops, as first evidenced by a depression of the erythrocyte count. Eventually, all three measures of anemia (hematocrit, hemoglobin concentration, and erythrocyte concentration) are depressed. At this point, macroovalocytes and macrocytes are usually detectable in the peripheral blood, and hypersegmentation is more impressive (Lindenbaum et al., 1988).

Because the onset of anemia is usually gradual, compensating cardiopulmonary and biochemical mechanisms provide adaptive adjustments to the diminished oxygen-carrying capacity of the blood until anemia is moderate to severe. Symptoms of weakness, fatigue, difficulty concentrating, irritability, headache, palpitations, and shortness of breath therefore typically appear at an advanced stage of anemia. They may be seen at milder degrees of anemia in some patients, especially the elderly (Lindenbaum et al., 1988). Atrophic glossitis may also occur (Savage et al., 1994).

SELECTION OF INDICATORS FOR ESTIMATING THE REQUIREMENT FOR FOLATE

The primary indicator selected to determine folate adequacy is erythrocyte folate, which reflects tissue folate stores, as described in detail below. For some life stage or gender groups, this is used in conjunction with plasma homocysteine (which reflects the extent of the conversion of homocysteine to methionine) and plasma or serum folate. Other indicators are discussed briefly below; risk reduction of chronic disease or developmental abnormalities is covered in detail in a later section.

Erythrocyte Folate

Because folate is taken up only by the developing erythrocyte in the bone marrow and not by the circulating mature erythrocyte during its 120-day lifespan, erythrocyte folate concentration is an indicator of long-term status. Erythrocyte folate concentration was shown to be related to tissue stores by its correlation, although weak, with liver folate concentration determined by biopsy in the same individual in a study of 45 subjects (Wu et al., 1975).

Erythrocyte folate concentration does not reflect recent or transient changes in dietary folate intake. A value of 305 nmol/L (140 ng/mL) of folate was chosen as the cutoff point for adequate folate status on the basis of the following experiments: On a diet containing only 5 µg/day of folate, the appearance of hypersegmented neutrophils in the peripheral blood of one subject coincided with the approximate time when the erythrocyte folate concentration decreased to less than 305 nmol/L (140 ng/mL) (Herbert, 1962a). On a diet containing less than 20 µg/day of folate, the appearance of hypersegmented neutrophils in two subjects preceded the reduction in erythrocyte folate to concentrations below 305 nmol/L (140 ng/mL) by about 2 weeks (Eichner et al., 1971). In a group of 40 patients with megaloblastic anemia caused by folate deficiency, 100 percent had erythrocyte folate values less than 305 nmol/L (140 ng/mL); values were the lowest in the most anemic subjects and the highest mean lobe counts occurred in the subjects with the lowest erythrocyte folate concentrations (Hoffbrand et al., 1966). All 238 pregnant women with erythrocyte folate concentrations below 327 nmol/L (150 ng/mL) were found to have megaloblastic marrow (Varadi et al., 1966). Eight subjects with erythrocyte folate of less than 305 nmol/L (140 ng/mL) had eight- to ninefold greater incorporation of uracil into DNA than did 14 control subjects and had a threefold increase in frequency of cellular micronuclei (a measure of DNA and chromosome damage); folate supplementation reduced the abnormalities (Blount et al., 1997).

Plasma Homocysteine

In this report, plasma homocysteine concentration refers to total homocysteine concentration. Plasma homocysteine concentration increases when inadequate quantities of folate are available to donate the methyl group that is required to convert homocysteine to methionine. Controlled metabolic and epidemiological studies provide evidence that plasma homocysteine rises with reductions in blood folate indices. Different cutoff values have been used by various investigators to define elevated homocysteine concentrations. The cutoff value for plasma homocysteine cited most often is greater than 16 μmol/L, but 14 μmol/L (Selhub et al., 1993) and 12 μmol/L (Rasmussen et al., 1996) have also been used. Ubbink and coworkers (1995a) used a prediction model to define a reference range as 4.9 to 11.7 μmol/L. Other investigators have proposed age- and gender-specific reference intervals (Rasmussen et al., 1996).

Many investigators have reported that plasma homocysteine is significantly elevated in individuals who have been diagnosed as folate deficient on the basis of established serum folate, plasma folate, or erythrocyte folate norms (Allen et al., 1993; Chadefaux et al., 1994; Curtis et al., 1994; Kang et al., 1987; Savage et al., 1994; Stabler et al., 1988; Ubbink et al., 1993).

The evidence supporting the use of homocysteine as an ancillary indicator of folate status is summarized as follows:

- In 10 young men, folate depletion led to a rise in plasma homocysteine and a decrease in plasma folate (Jacob et al., 1994).
- In young women, a folate intake equivalent to 320 μg/day of dietary folate equivalents was associated with elevated plasma homocysteine (greater than 14 μmol/L); at this level of intake plasma homocysteine concentrations were inversely associated with erythrocyte and serum folate concentrations (O'Keefe et al., 1995).
- In a cross-sectional analysis involving elderly individuals, plasma homocysteine exhibited a strong inverse association with plasma folate after age, gender, and intakes of other vitamins were controlled for (Selhub et al., 1993); homocysteine values appeared to plateau at folate intakes greater than approximately 350 to 400 μg/day. A meta-analysis by Boushey and colleagues (1995) supports the existence of a plateau when adequate folate is consumed.

Thus, in studies of different types, a similar inverse relationship between folate intake and plasma homocysteine values is seen for pre- and postmenopausal women, adult men, and the elderly.

Ward and colleagues (1997) supplemented each of 30 male subjects with 100, 200, or 400 μg of folate. The men were consuming a regular diet that averaged 281 μg/day of folate. Plasma homocysteine, serum folate, and erythrocyte folate were assessed before, during, and 10 weeks after intervention. Results, expressed as tertiles of baseline plasma homocysteine, showed significant homocysteine lowering in the top (mean 11 µmol/L) and middle (mean 9 µmol/L) homocysteine tertiles but not in the bottom tertile (mean 7 µmol/L). All baseline homocysteine values were within the normal range; the highest was 12.3 µmol/L. Of the three folate doses, 200 µg appeared to be as effective as 400 µg whereas 100 µg was less effective at lowering homocysteine. These data suggest that there is a concentration of plasma homocysteine below which folate has no further lowering effect.

Maternal hyperhomocysteinemia has been implicated as a risk factor for complications during pregnancy (Burke et al., 1992; Goddijn-Wessel et al., 1996; Rajkovic et al., 1997; Steegers-Theunissen et al., 1992, 1994; Wouters et al., 1993), but the relationship between folate intake and the complications has not been established.

Although plasma homocysteine is a sensitive indicator of folate status, it is not a highly specific one: it can be influenced by vitamin B_{12} status (Stabler et al., 1996), vitamin B_6 status (Ubbink et al., 1995a), age (Selhub et al., 1993), gender (Selhub et al., 1993), race (Ubbink et al., 1995b), some genetic abnormalities (e.g., methyltetrahydrofolate reductase deficiency) (Jacques et al., 1996; Malinow et al., 1997), and renal insufficiency (Hultberg et al., 1993). Thus, plasma homocysteine alone is not an acceptable indicator on which to base the folate requirement.

Knowledge of the relationships of folate, homocysteine, and risk of vascular disease was judged too weak to use as the basis for deriving the Estimated Average Requirement (EAR) for folate. This topic is described in more detail in "Reducing Risk of Developmental Disorders and Chronic Degenerative Disease."

Serum Folate

A serum folate concentration of less than 7 nmol/L (3 ng/mL) indicates negative folate balance at the time the blood sample was drawn (Herbert, 1987). In all the experimental studies of human volunteers subjected to folate deprivation, a decrease in the serum folate concentration, usually occurring within 1 to 3 weeks, was the first event (Eichner and Hillman, 1971; Eichner et al., 1971; Halsted et al., 1973; Herbert 1962a; Sauberlich et al., 1987). This initial period of folate deprivation is followed by weeks or months when the serum folate concentration is low but there is no other evidence of deficiency. The circulating folate concentration may also be depressed in situations in which there is no detectable alteration in total body folate, such as acute alcohol ingestion (Eichner and Hillman, 1973).

In population surveys it is generally assumed that measuring serum folate alone does not differentiate between what may be a transitory reduction in folate intake or chronic folate deficiency accompanied by depleted folate stores and functional changes. Serum or plasma folate is, however, considered a sensitive indicator of dietary folate intake, as illustrated by the report of Jacques and colleagues (1993) in which plasma folate doubled across quartiles of folate intake assessed in a study of 140 people. In a controlled metabolic study, repeated measures over time in the same individual do reflect changes in status. Serum folate concentration may be a worthwhile diagnostic test if used and interpreted correctly in conjunction with other folate status indices (Lindenbaum et al., 1988).

Urinary Folate

Data from a metabolic study in which graded doses of folate were fed showed that urinary folate is not a sensitive indicator of folate status (Sauberlich et al., 1987). In that study, approximately 1 to 2 percent of dietary folate was excreted intact in the urine; excretion continued even in the face of advanced folate depletion. Other reports indicate that daily folate excretion on a normal diet ranges from 5 to 40 µg/day (Cooperman et al., 1970; Retief, 1969; Tamura and Stokstad, 1973).

The major route of whole-body folate turnover is by catabolism and cleavage of the C9-N10 bond producing pteridines and *p*-aminobenzoylglutamate (pABG) (Krumdieck et al. 1978; Saleh et al., 1982). Before excretion from the body, most pABG is N-acetylated to acetamidobenzoylglutamate (apABG). It is not known whether folate coenzymes are catabolized and excreted or whether they are recycled after metabolic utilization. In a study designed to estimate the folate requirements of pregnant and nonpregnant women, McPartlin and coworkers (1993) quantified the urinary excretion of pABG and apABG as a measure of daily folate utilization. This approach does not take into account endogenous fecal folate loss, which may be substantial (Krumdieck et al., 1978); thus, quantitation of urinary catabolites alone may result in an underestimation of the requirement.

Indicators of Hematological Status

The appearance of hypersegmented neutrophils, macrocytosis,

and other abnormal hematological findings occurs late in the development of deficiency (see "Clinical Effects of Inadequate Intake"). Thus, hematological findings were not used to derive the EAR.

Risk of Neural Tube Defects and of Chronic Degenerative Diseases

The role of folate in the prevention of neural tube defects (NTDs) was very carefully considered, but not in the context of setting an EAR. Although the evidence is strong that the risk of having a fetus with an NTD decreases with increasing intake of folate during the periconceptional period (about 1 month before to 1 month after conception), this type of risk reduction was judged inappropriate for use as an indicator for setting the EAR for folate for women of childbearing age. There are several reasons for this. The population at risk is all women capable of becoming pregnant, but only those women who become pregnant would benefit from an intervention aimed at reducing NTD risk. The risk of NTD in the U.S. population is about 1 per 1,000 pregnancies, but the critical period for prevention—the periconceptional period—is very short. The definition of EAR, which indicates that half of the individuals in the population have intakes sufficient to meet a particular criterion, does not accommodate NTD prevention as an appropriate criterion. Because of the importance of this topic, it is covered separately in the later section "Reducing Risk of Developmental Disorders and Chronic Degenerative Disease."

The possible use of criteria involving reduction of risk of vascular disease, certain types of cancer, and psychiatric and mental disorders was also carefully considered. The evidence was not judged sufficient to use prevention of any chronic disease or condition as a criterion for setting the EAR; this evidence is also presented in the section "Reducing Risk of Developmental Disorders and Chronic Degenerative Disease."

METHODOLOGICAL ISSUES

Measurement of Blood Folate Values

Substantial variation within and across methods was evident from the results of an international comparative study of the analysis of serum and whole-blood folate (Gunter et al., 1996). Results for whole-blood pools were more variable than for serum pools. The authors concluded that folate concentrations measured in one laboratory cannot be compared reliably with those measured in another laboratory without considering interlaboratory differences and that comparing data for different study populations measured by different methods is difficult.

The Bio-Rad Quantaphase Radioassay was used for the first 4 years of the Third National Health and Nutrition Examination Survey (NHANES III) (1988–1991). In 1991 it was determined that the Bio-Rad radioassay gave results that were 30 percent too high when external, purified pteroylglutamic acid (PGA) standard solutions were measured. The Bio-Rad assay was then recalibrated by using calibrator solutions of PGA concentrations of 2.3, 5.7, 11, 22.6, and 45 nmol/L (1.0, 2.5, 5.0, 10.0, and 20.0 ng/mL). The net effect of this recalibration was the expected 30 percent reduction in the measured folate concentrations of a sample. An analysis by another expert panel (LSRO/FASEB, 1994) provides further information. The NHANES III laboratory conducted a 19-day comparison study of NHANES III serum and erythrocyte specimens using the original and recalibrated Bio-Rad kits and confirmed the 30 percent reduction. Through the use of a regression equation developed from the comparison study, the correction was applied to the NHANES III data generated with the original assay (LSRO/FASEB, 1994).

The NHANES III data (Appendix K) have been corrected for this method problem associated with inappropriate calibration. Data from NHANES III are believed to "provide as accurate and precise an estimation of serum and RBC [red blood cell] folate levels in the United States population as is possible until a definitive method has been developed and [this should be considered] as a stand-alone data set, without applying cutoffs established using other laboratory methods" (E.W. Gunter, Division of Environmental Health Laboratory Sciences, National Center for Environmental Health, Centers for Disease Control and Prevention, personal communication, 1997).

Earlier, after NHANES II, similar issues were addressed by a Life Sciences Research Office expert panel (LSRO/FASEB, 1984). Such an effort is even more warranted related to NHANES III because this survey (unlike NHANES II) had been designed to provide an assessment of folate status of the entire U.S. population.

Measurement and Reporting of Food Folate

It is recognized that food folate composition data contained in currently used databases provide inaccurate estimations of folate intake of the U.S. population. Because of the limitations of traditional analytical methods used in generating the food composition data for folate, the database values underestimate actual folate content. Problems with the traditional methods include incomplete release of folate from the food matrix and possibly incomplete hydrolysis of polyglutamyl folates before quantitation. For example, buffer solutions widely used for sample homogenization in food analysis have been shown to yield incomplete recovery relative to a more effective extraction buffer (Gregory et al., 1990). As much as a twofold greater folate concentration is obtained when an improved extraction procedure rather than an older procedure is used in the analysis of foods such as green peas and liver (Tamura et al., 1997).

The use of a trienzyme approach (amylase and protease treatments in addition to folate conjugase) enhances folate yield in food analysis (Martin et al., 1990; Tamura et al., 1997). Pfeiffer and colleagues (1997b) confirmed the effectiveness of the trienzyme approach for the analysis of cereal-grain foods with or without folate fortification. The extent of differences among approaches varies from food to food, and there is no current means of predicting actual folate content from the existing database values. Analytical methods used to obtain food folate data for databases have used extraction procedures (Gregory et al., 1990) and enzyme digestion treatments that are not optimal for the specific food, resulting in a significant underestimation of food folate (DeSouza and Eitenmiller, 1990; Martin et al., 1990; Pfeiffer et al., 1997b; Tamura et al., 1997).

Many studies of population groups have used food composition databases and measures of food intake to estimate folate intake. The mean estimates in these studies are based on data largely or entirely from the U.S. Department of Agriculture nutrient database. For the analytical reasons indicated above, it is likely that all these estimates of dietary folate intake are underestimates of actual intake. Therefore, conclusions regarding the EAR for folate should not be based on estimates of folate intake from current food composition databases.

FACTORS AFFECTING THE FOLATE REQUIREMENT

Factors considered when estimating the folate requirement include the bioavailability of folic acid and food folate, nutrient-nutrient interactions, interactions with other food components, smoking, folate-drug interactions, and genetic variations.

Bioavailability

As explained below, the bioavailability of folate ranges from about 100 percent for folic acid supplements taken on an empty stomach to about 50 percent for food folate.

Bioavailability of Folic Acid

When consumed under fasting conditions, supplements of folic acid are nearly 100 percent bioavailable (Gregory, 1997). Daly and coworkers (1997) reported incremental increases in erythrocyte folate in response to graded doses of folic acid, which provides evidence for the high bioavailability of supplemental folate. Additional work may be necessary to improve the precision of the estimate of bioavailability (Pfeiffer et al., 1997a).

No published information was found regarding the effect of food on the bioavailability of folate supplements. Pfeiffer and coworkers (1997a) recently examined the bioavailability of C^{13}-labeled folic acid (administered in apple juice) when given with or without a serving of food; they found a slight (about 15 percent) but insignificant reduction when folic acid was consumed with a portion of food. From these experimental data the bioavailability of folic acid consumed with food is estimated to be 85 percent. Studies have not been conducted to define the bioavailability of folic acid consumed with entire meals. It is assumed that the bioavailability would be somewhat lower than that observed with folic acid alone or with a small portion of food.

Bioavailability of Folate Added to Foods

The recently approved U.S. fortification of breads and grains with folate has raised interest in the bioavailability of folate provided in the form of folic acid. On the basis of erythrocyte folate response over a 3-month study, it was concluded that the folate in a supplement and in fortified bread and breakfast cereal consumed in the context of normal diet was equally bioavailable (Cuskelly et al., 1996). Pfeiffer and colleagues (1997a) evaluated the bioavailability of folate from cereal-grain foods fortified experimentally with C^{13}-labeled folic acid. In a series of single-dose trials with human subjects, there was a slight but insignificant difference between the control (water with folic acid) and any of the tested foods (white and whole-wheat bread, pasta, and rice). This finding indicates high bioavailability of the folate in the form of added folic acid.

Overall, the very different studies of Cuskelly et al. (1996) and Pfeiffer et al. (1997a) complement each other and strongly indicate that folate added to cereal-grain foods is highly available and efficacious. These two studies contradict previous reports of low (30 to 60 percent) bioavailability of folate in experimentally fortified cereal-grain foods in South Africa (summarized by Colman [1982]). In the South African studies of folate-deficient pregnant women, the response criterion used to estimate bioavailability was either 2-hour changes in serum folate or changes in erythrocyte folate over time. The quantity of folate consumed in the fortified foods was not directly measured in these studies. If the amount was overestimated, that would explain the lower reported bioavailability (33 to 60 percent) compared with the recent estimates (85 to 100 percent) by Pfeiffer et al. (1997a) and Cuskelly et al. (1996). The experimental fortification of these South African foods in the 1970s may have little relevance to the current fortification process in the United States and Canada.

The value used in this report—85 percent bioavailability of folic acid consumed with a meal—is probably an underestimate, the effect of which may be an underestimation of the folate requirement.

Bioavailability of Food Folate

Perhaps the best data on which to base an estimate of the bioavailability of food folate are provided by Sauberlich and coworkers (1987). On the basis of changes in blood folate values, the authors concluded that the bioavailability of food folate was no more than 50 percent that of folic acid. Although this study was not designed as a quantitative study of food folate bioavailability, the results provide strong evidence in that regard. Similarly, the data of Cuskelly and colleagues (1996) suggest that food folate is less bioavailable than the synthetic form, as evidenced by a smaller increase in erythrocyte folate in the group that received an increased level of folate from food rather than from the synthetic form. The percentage bioavailability of folate could not be determined from this study because food consumption was not controlled.

A stable isotope investigation of the relative bioavailability of monoglutamyl and polyglutamyl folates consumed in water (control) or added to lima beans or tomatoes found that the relative bioavailability of deuterated polyglutamyl folates was equivalent to that of the monoglutamyl tracer (Wei et al., 1996). However, the bioavailability of polyglutamyl folate added to orange juice was approximately 33 percent lower ($p < 0.05$) than that of the monoglutamyl folate tracer. The authors concluded that naturally occurring polyglutamyl folates in orange juice are approximately 67 percent available—slightly more available than the food folate bioavailability estimate of Sauberlich. Related issues have been discussed in several reviews on this subject (Gregory, 1989, 1995, 1997).

Bioavailability Estimates and Assumptions

Many controlled studies on folate requirements have used a defined diet (food folate) supplemented with folic acid. Because folic acid taken with food is 85 percent bioavailable but food folate is only about 50 percent bioavailable, folic acid taken with food is 85/50 (i.e., 1.7) times more available. Thus, if a mixture of folic acid plus food folate has been fed, dietary folate equivalents (DFEs) are calculated as follows to determine the Estimated Average Requirement (EAR):

µg of DFEs provided = µg of food folate + (1.7 × µg of folic acid)

Expressed differently, to be comparable with food folate, only half as much folic acid is needed if taken on an empty stomach, or

1 µg of DFEs = 1 µg of food folate = 0.5 µg of folic acid taken on an empty stomach = 0.6 µg of folic acid with meals.

When food folate was the sole source of folate in studies used to determine requirements, no corrections were applied to convert to DFEs. Adjustments made for DFEs are indicated, if applicable, where folic acid was a source of folate. Adjustments cannot be made for epidemiological studies if data are lacking on the folate sources. If future research indicates that food folate is more than 50 percent bioavailable, this could lower the estimated requirements that appear later in the chapter.

Nutrient-Nutrient Interactions

No reports were found that demonstrate that the intake of other nutrients increases or decreases the requirement for folate. However, coexisting iron or vitamin B_{12} deficiency may interfere with the diagnosis of folate deficiency. In contrast to folate deficiency, iron deficiency leads to a decrease in mean cell volume. In the combined deficiency, interpretation of hematological changes may be unclear (Herbert, 1962a). A vitamin B_{12} deficiency results in the same hematological changes that occur with folate deficiency because the vitamin B_{12} deficiency results in a secondary folate deficiency (Selhub and Rosenberg, 1996).

Interactions with Other Food Components

Fiber

Experimental data do not support the hypothesis that dietary fiber per se reduces folate bioavailability (Bailey, 1988; Gregory, 1989). Human studies (Russell et al., 1976) confirmed the negative findings of both rat and chick bioassays regarding the identification of an inhibitory action of various dietary fiber sources. Certain forms of fiber (e.g., wheat bran) may decrease the bioavailability of certain forms of folate under some conditions (Bailey et al., 1988; Keagy et al., 1988), but many forms of fiber appear to have no adverse effects (Gregory, 1997).

Experimental evidence in rats indicates that synthesis of folate by intestinal bacteria influences folate status (Keagy and Oace, 1989; Krause et al., 1996). Rong and colleagues (1991) reported that bacterially synthesized folate in the rat large intestine is incorporated into host tissue polyglutamates. The applicability of these data to humans is unknown. Suggestive evidence of a positive association between dietary fiber intake and folate status in humans was reported by Houghton and coworkers (1997). Zimmerman (1990) provided evidence that the monoglutamate form can be transported into the mucosa of the human colon by facilitated diffusion, allowing for the possibility of subsequent absorption of folate synthesized in the large intestine.

Alcohol

Data from surveys of chronic alcoholics suggest that inadequate intake is a major cause of the folate deficiency that has often been observed in chronic alcohol users (Eichner and Hillman, 1971; Herbert et al., 1963). Ethanol intake may aggravate folate deficiency by impairing intestinal folate absorption and hepatobiliary metabolism (Halsted et al., 1967, 1971, 1973) and by increasing renal folate excretion (McMartin et al., 1986; Russell et al., 1983).

Cigarette Smoking

Although blood folate concentrations have been reported to be lower in smokers than in nonsmokers (Nakazawa et al., 1983; Ortega et al., 1994; Piyathilake et al., 1994; Senti and Pilch, 1985; Subar et al., 1990; Witter et al., 1982), data suggest that low intake (Subar et al., 1990) rather than an increased requirement may account for the poorer folate status of smokers.

Folate-Drug Interactions

The effects of drug use on folate status reviewed in this section are limited to effects seen in drugs used in chronic drug therapy of nonneoplastic diseases that affect a large percentage of the population and to oral contraceptive drugs. No information is available on the effects of these drugs on homocysteine values.

Nonsteroidal Anti-inflammatory Drugs

When taken in very large therapeutic doses (e.g., 3,900 mg/day), nonsteroidal anti-inflammatory drugs, including aspirin, ibuprofen, and acetaminophen, may exert antifolate activity (Baggott et al., 1992; Eichner et al., 1979; Lawrence et al., 1984; Willard et al., 1992). However, routine use of low doses of these drugs has not been reported to impair folate status.

Anticonvulsants

Numerous studies have cited evidence of impaired folate status associated with chronic use of the anticonvulsants diphenylhydantoin (phenytoin and Dilantin®) and phenobarbital (Collins et al., 1988; Klipstein, 1964; Malpas et al., 1966; Reynolds et al., 1966). Diphenylhydantoin is known to inhibit the intestinal absorption of folate (Elsborg, 1974; Young and Ghadirian, 1989). Few studies, however, have controlled for potential differences in dietary folate intake between groups of anticonvulsant users and nonusers (Collins et al., 1988). Thus, definitive conclusions cannot be drawn relative to adverse effects of these drugs on folate status.

Methotrexate

Methotrexate is a folate antagonist that has been used frequently and successfully in the treatment of nonneoplastic diseases such as rheumatoid arthritis, psoriasis, asthma, primary biliary cirrhosis, and inflammatory bowel disease (Morgan and Baggott, 1995). Methotrexate has been especially effective in the treatment of rheumatoid arthritis (Felson et al., 1990), with efficacy established in numerous trials (Morgan et al., 1994). Patients with rheumatoid arthritis are frequently reported to be folate deficient, and folate stores are decreased in patients with rheumatoid arthritis who take methotrexate (Morgan et al., 1987, 1994; Omer and Mowat, 1968). Some of the side effects of methotrexate administration, such as gastrointestinal intolerance, mimic severe folate deficiency (Jackson, 1984). When patients are also given high-folate diets or supplemental folate, there is a significant reduction in toxic side effects with no reduction in drug efficacy. It has been recommended that patients undergoing chronic methotrexate therapy for rheumatoid arthritis increase folate consumption (Morgan et al., 1994) or consider folate supplements (1 mg/day) (Morgan et al., 1997).

Other Drugs with Antifolate Activity

The following diseases have been treated with drugs having antifolate activity: malaria with pyrimethamine, bacterial infections with trimethoprim, hypertension with triamterene, *Pneumocystis carinii* infections with trimetrexate (Morgan and Baggott, 1995), and chronic ulcerative colitis with sulfasalazine (Mason, 1995).

Oral Contraceptives

A number of early studies of oral contraceptive agents containing high levels of estrogens suggested an adverse effect on folate status (Grace et al., 1982; Shojania et al., 1968, 1971; Smith et al., 1975). However, oral contraceptive use has not been reported to influence folate status in large-scale population surveys (LSRO/FASEB, 1984) or in metabolic studies in which dietary intake was controlled (Rhode et al., 1983).

Genetic Variations

Folic acid and its derivatives are involved in numerous biochemical reactions that are catalyzed by many different enzymes. As expected, folate metabolism is under genetic control, and genetic heterogeneity exists. To estimate the relative contribution of genetic and environmental factors in determining folate status, erythrocyte folate was measured in monozygotic and dizygotic twins (Mitchell et al., 1997); however, dietary intake was not assessed. The data were best described by a model in which 46 percent of the variance is attributable to additive genetic effects, 16 percent to age and sex, and 38 percent to random environmental effects including errors in measurement. A similar study was done for plasma homocysteine, and the estimated heritability was between 72 percent and 84 percent (Reed et al., 1991). In studies of twins, however, the influence of genetic factors may be overestimated, especially if environmental similarities are greater in monozygotic than in dizygotic twins.

A significant genetic heterogeneity in folate metabolism is related to the activity of 5,10-methylenetetrahydrofolate reductase (MTHFR). Severe MTHFR deficiency is rare; a reduced activity associated with a thermolabile form of the enzyme is much more common. A C667T polymorphism in the gene coding MTHFR has been linked with thermolability and reduced enzymatic activity (Frosst et al., 1995). Estimates of the frequency of homozygosity for the MTHFR T[677] allele in white populations vary from 2 to 16 percent (van der Put et al., 1995). Individuals homozygous for the MTHFR T[677] allele have significantly elevated plasma homocysteine (Frosst et al., 1995) and a tendency to have low plasma and erythrocyte folate concentrations (Ma et al., 1996; Molloy et al., 1997; Schmitz et al., 1996). In one study (Jacques et al., 1996) elevated fasting homocysteine was observed in individuals homozygous for the MTHFR T[677] allele who had plasma folate values below 15.4 nmol/L (7.07 ng/mL) but not in those with plasma folate values above this level. Because 5-methyl-tetrahydrofolate is a required substrate in the remethylation of homocysteine to methionine, reduced enzyme activity of the T[677] polymorphism increases dependence on an adequate folate supply. More detailed coverage of this genetic variation is provided in Appendix L.

FINDINGS BY LIFE STAGE AND GENDER GROUP

Infants Ages 0 through 12 Months

An Adequate Intake (AI) is used as the goal for folate intake by infants.

Method Used to Set the Adequate Intake

The AI reflects the observed mean folate intake of infants consuming exclusively human milk. Hematological and growth rate changes that have been measured in controlled studies of infants are not considered to be specifically attributable to the adequacy of dietary folate intake.

Serum and erythrocyte folate values of newborn infants are significantly higher than maternal blood concentrations, possibly reflecting an active transport process in utero (Ek, 1980; Landon and Oxley, 1971). These high blood folate values decline during the first 6 months in concert with the decline in the rate of cell division (Landon and Oxley, 1971).

The AI is the quantity of dietary folate that maintains blood folate concentrations comparable with those of the infant exclusively fed human milk. When human milk is consumed exclusively, the infant's serum or plasma folate concentration has been reported to range from 35 to over 60 nmol/L (16 to 30 ng/mL) whereas erythrocyte values averaged from 650 to over 930 nmol/L (300 to 430 ng/mL) (Ek and Magnus, 1979; Smith et al., 1985; Tamura et al., 1980). These values reported in infants are much higher than adult values (Smith et al., 1985), which makes the use of adult norms inappropriate for infants. Additionally there are no reports of full-term infants who are exclusively and freely fed human milk manifesting any signs of folate deficiency.

The folate concentration of human milk remains relatively constant regardless of maternal dietary folate intake unless there is a severe maternal deficiency (Metz, 1970). The reported concentration of folate in human milk varies with the methods used, and these have changed substantially over the past decade. However, recent data from the laboratories of Picciano and colleagues (Lim et al., 1997) are consistent with the data of Brown and colleagues (1986) and O'Connor and colleagues (1991), all of whom reported average human milk folate concentrations to be 85 µg/L. The human milk folate concentration used to estimate AIs for infants thus is 85 µg/L.

Ages 0 through 6 Months. The AI for infants 0 through 6 months of age, derived by using the average volume of milk of 0.78 L/day (see Chapter 2) for this age group and the average folate concentration in human milk after 1 month of lactation (85 µg/L), is 66 µg/day, which is rounded to 65 µg. This equals approximately 9.4 µg/kg of reference body weight. Because this is food folate, the amount is the same in dietary folate equivalents (DFEs).

Ages 7 through 12 Months. If the reference body weight ratio method described in Chapter 2 to extrapolate from the AI for folate for infants ages 0 through 6 months is used, the AI for folate for the older infants would be 80 µg/day after rounding. The second method (see Chapter 2), extrapolating from the Estimated Average Requirement (EAR) for adults and adjusting for the expected variance to estimate a recommended intake, gives a comparable AI of approximately 80 µg/day.

The five studies summarized in Table 8-1 illustrate the data from controlled studies that measured folate intake and assessed the infants' status. They include studies in which infants were fed either human milk or formula. Asfour and colleagues (1977) concluded that although the observed serum and erythrocyte concentrations in three groups of infants fed formula were borderline, the folate values were sufficient to maintain growth, hematopoiesis, and clinical well-being. However, the criteria of growth, hematopoiesis, and clinical well-being are too nonspecific for evaluating folate status. Therefore, these data suggest that none of the folate levels (3.6, 4.3, or 5.0 µg/kg) maintained folate adequacy in all the infants tested based on serum or erythrocyte folate concentrations. Ek and Magnus (1982) provided data that infant formula containing folate at 78 µg/L supported blood folate concentrations comparable with those of infants fed human milk. Smith and coworkers (1983) reported that serum folate values of infants fed human milk were approximately 45 nmol/L (20 ng/mL) at age 6 weeks and 65 nmol/ L (30 ng/mL) at age 12 weeks, whereas erythrocyte folate concentrations were approximately 1,000 nmol/L (460 ng/mL) at age 6 weeks and 940 nmol/L (430 ng/mL) at age 12 weeks. Smith and coworkers (1985) reported that throughout the first 6 months, serum folate concentrations were significantly higher in infants fed formula than in those fed human milk; erythrocyte folate concentrations of approximately 2,200 nmol/L (1,000 ng/mL) at age 4 months clearly show that 158 µg/L of formula is in excess of what is needed. Salmenpera and colleagues (1986) reported that infants fed exclusively human milk all maintained adequate plasma folate concentrations with values twofold to more than threefold higher than maternal concentrations throughout the study.

Folate AI Summary, Ages 0 through 12 Months

Data from the research studies included in Table 8-1 supports the AI of 65 µg/day of folate for young infants and of 80 µg/day for older infants.

AI for Infants
0–6 months   65 µg/day of dietary folate equivalents   ≈9.4 µg/kg
7–12 months   80 µg/day of dietary folate equivalents   ≈8.8 µg/kg

The extent to which the AIs for folate could be lowered and still meet the physiological needs for infants fed human milk is unknown.

Special Considerations

No data were found to support the need to adjust dietary intake of folate on the basis of the type of infant formula compared with human milk to achieve the same folate status other than that inherent in DFE equivalency.

Children Ages 1 through 8 Years

Method Used to Estimate the Average Requirement

No data were found on which to base an EAR for children. In the absence of additional information, EARs and RDAs for these ages have been estimated by using the method described in Chapter 2, which extrapolates from adult values. The resulting EARs are 120 and 160 µg/day of DFEs for children ages 1 through 3 and 4 through 8 years, respectively.

Folate EAR and RDA Summary, Ages 1 through 8 Years

EAR for Children
1–3 years   **120 µg/day of dietary folate equivalents**
4–8 years   **160 µg/day of dietary folate equivalents**

The RDA for folate is set by assuming a coefficient of variation (CV) of 10 percent (see Chapter 1) because information is not available on the standard deviation of the requirement for folate; the RDA is defined as equal to the EAR plus twice the CV to cover the needs of 97 to 98 percent of the individuals in the group (therefore, for folate the RDA is 120 percent of the EAR).

RDA for Children
1–3 years   **150 µg/day of dietary folate equivalents**
4–8 years   **200 µg/day of dietary folate equivalents**

TABLE 8-1　Folate Intake and Status of Infants by Study

Reference	Age of Infants	Number of Infants and Feeding Type	Folate Intake	Dietary Folate Equivalents
Asfour et al., 1977	2–11 mo	4 formula fed	3.6 µg/kg of body weight[a]	6.1 DFEs/kg
		4 formula fed	4.3 µg/kg[a]	7.3 DFEs/kg
		5 formula fed	5.0 µg/kg[a]	8.5 DFEs/kg
Ek and Magnus, 1982	0–12 mo	33 formula fed	39 µg/L[c]	66 DFEs/L
		31 formula fed	78 µg/L[c]	133 DFEs/L
Smith et al., 1983	6 wk	14 breastfed	45 µg/L	45 DFEs/L
	12 wk	14 breastfed	50 µg/L[d]	50 DFEs/L
Smith et al., 1985	1st 6 mo	14 breastfed	85 µg/L	85 DFEs/L
	3 and 6 wk	31 formula fed	162 µg/L	275 DFEs/L
	3 and 6 wk	22 formula fed	158 µg/L (plus iron)	269 DFEs/L
	12 mo	14 breastfed	85 µg/L	85 DFEs/L
	12 mo	31 formula fed	162 µg/L	275 DFEs/L
	12 mo	22 formula fed	158 µg/L (plus iron)	269 DFEs/L
Salmenpera et al., 1986	0–12 mo	200[f] exclusively breastfed	NA[g]	–

[a] Mean values.
[b] Mean values at age 4 mo.
[c] Volume consumed not reported.
[d] Analyzed using older methods that may have underestimated the folate content.
[e] Values were estimated from figures.

Serum Folate nmol/L (ng/mL)	Erythrocyte Folate nmol/L (ng/mL)	Comments
8.5 ± 3.5 (3.9 ± 1.6)[b]	353 ± 148 (162 ± 68)[b]	2 of 4 in deficient range.
11.1 ± 3.7 (5.1 ± 1.7)[b]	538 ± 170 (247 ± 78)[b]	1 of 4 had marginal erythrocyte folate.
10.7 ± 4.8 (4.9 ± 2.2)[b]	568 ± 244 (261 ± 112)[b]	2 of 5 had marginal erythrocyte and serum folate.
< 7 (3) (at 2 and 3 mo)	< 220 (100) (at 2 and 3 mo)	Values at other ages
> 41 (19) (at 2 and 3 mo)	> 435 (200) (at 2 and 3 mo)	were higher.
45 (20)[e]	1,000 (460)[e]	Within normal range.
65 (30)[e]	940 (430)[e]	Within normal range.
54–65 (25–30)[e]	1,090 down to 915 (500 down to 420)[e]	Within normal range.
> 130 (60)[e]	2,200 (1,000)[e]	Above usual normal range.
> 130 (60)[e]	2,200 (1,000)[e]	Above usual normal range.
> 35 (15)[e]	870 (400)[e]	Within normal range.
> 22 (10)[e]	760 (350)[e]	Within normal range.
> 22 (10)[e]	760 (350)[e]	Within normal range.
≥ 11 (5)[h]	NA	All infants had adequate plasma folate concentrations after 2 mo of age (> 7 nmol/ L [3 µg/L]).

[f] The number of infants in the study decreased from an initial 200 to 7 infants at the end.

[g] NA = not available.

[h] Lowest individual concentration reported for the study period.

Children and Adolescents Ages 9 through 18 Years

Method Used to Estimate the Average Requirement

As for younger children, EARs and RDAs for these ages have been extrapolated from adult values by using the method described in Chapter 2. Although body size varies because of gender in these age groups, no conclusive data indicating a difference in requirements for adults were determined, thus no difference based on gender is proposed for these age groups.

Folate EAR and RDA Summary, Ages 9 through 18 Years

EAR for Boys
9–13 years   250 µg/day of dietary folate equivalents
14–18 years   330 µg/day of dietary folate equivalents

EAR for Girls
9–13 years   250 µg/day of dietary folate equivalents
14–18 years   330 µg/day of dietary folate equivalents

The RDA for folate is set by assuming a coefficient of variation (CV) of 10 percent (see Chapter 1) because information is not available on the standard deviation of the requirement for folate; the RDA is defined as equal to the EAR plus twice the CV to cover the needs of 97 to 98 percent of the individuals in the group (therefore, for folate the RDA is 120 percent of the EAR).

RDA for Boys
9–13 years   300 µg/day of dietary folate equivalents
14–18 years   400 µg/day of dietary folate equivalents

RDA for Girls
9–13 years   300 µg/day of dietary folate equivalents
14–18 years   400 µg/day of dietary folate equivalents

Adults Ages 19 through 50 Years

Evidence Considered in Estimating the Average Requirement

No single indicator was judged a sufficient basis for deriving an EAR for adults. That is, it was not deemed appropriate to base the EAR on an examination limited to studies that provided data only on erythrocyte folate, plasma homocysteine, or any other single laboratory value. The main approach to determining the EAR for adults uses a combination of erythrocyte folate, plasma homocysteine, and plasma or serum folate. The focus used was on the adequacy of specific quantities of folate consumed under controlled metabolic conditions to maintain normal blood concentrations of these indicators. Cutoff points for the normal range were based on the occurrence of documented biochemical abnormalities.

The types of studies considered were primarily those in which maintenance or restoration of folate status was evaluated in controlled metabolic conditions. In these studies folate was provided either as food or as food plus folic acid. Intakes related to these status indicators were computed by calculating DFEs, which gives higher intakes when folic acid is used as part of the protocol than what the authors describe when reporting their work (see "Bioavailability").

In addition to data on maintenance or restoration of folate status, several other types of experimental data were critiqued and compared. These included kinetic estimates of body pool size and daily turnover (Gailani et al., 1970; Herbert, 1962b, 1968; Krumdieck et al., 1978; Russell et al., 1983; Stites et al., 1997; Von der Porten et al., 1992), quantitation of urinary folate catabolites as an index of folate turnover (McPartlin et al., 1993), and repletion of severe clinical folate deficiency (Hansen and Weinfeld 1962; Herbert, 1962a, 1968; Marshall and Jandl, 1960; Zalusky and Herbert, 1961). Analyses of relationships of dietary folate intake and biochemical indices of folate status from the Third National Health and Nutrition Examination Survey are in progress and were thus unavailable for use in this report.

Metabolic Studies

Two principal studies of healthy human subjects were critiqued and compared; the amounts of folate ranged from 100 to 489 µg/day of DFEs (O'Keefe et al., 1995; Sauberlich et al., 1987). Two additional studies (Jacob et al., 1994; Milne et al., 1983) were also considered but were given less weight because of the study design. These four studies are summarized in Table 8-2.

100 to 150 µg/day of DFEs. Sauberlich and colleagues (1987) conducted a controlled depletion-repletion metabolic study (28 days of depletion followed by 64 days of graded repletion phases, each phase lasting 21 days) with nonpregnant women. Plasma and eryth-

TABLE 8-2 Key Controlled Metabolic Studies Providing Evidence Used to Derive the Estimated Average Requirement (EAR)[a]

Reference	Type of Controlled Metabolic Study	Number and Age of Subjects	Baseline Folate Intake (µg)	Duration of Study
Milne et al., 1983	Maintenance	40 men, 19–54 y	NA[c]	2–8 mo
Sauberlich et al., 1987	Depletion-repletion	3 women, 21–40 y	400[d]	28 d depletion 21 d repletion
		2 women	400[d]	28 d depletion 21 d repletion
Jacob et al., 1994	Depletion-repletion	10 men, 33–46 y	440[e]	30 d depletion 15 d repletion
O'Keefe et al., 1995	Maintenance	5 women, 21–27 y	NA	70 d
		6 women, 21–27 y	NA	70 d

[a] The EAR is the intake that meets the estimated nutrient needs of 50% of the individuals in a group.

[b] To compute dietary folate equivalents, use the formula µg food folate + (1.7 × µg folic acid).

rocyte folate concentrations continued to fall in response to repletion with 100 µg of food folate (equal to 100 µg DFEs) for 21 days. This continued depletion led to the conclusion that 100 µg of dietary folate is below the average requirement.

Jacob and coworkers (1994) conducted a controlled depletion-repletion metabolic study (30 days depletion at 25 µg/day of folate

Folate Intake During Repletion or Maintenance			
Food Folate (µg)	Folic Acid (µg)	Dietary Folate Equivalents[b] (µg)	Results
200	0	200	Serum and erythrocyte folate decreased significantly over time, but not below normal cutoff values.
0 100	0 0	0 100	Plasma and erythrocyte folate decreased throughout.
0 200	0 0	0 200	Plasma folate stabilized. Erythrocyte folate decreased throughout.
25 25	0 74	25 151	Plasma homocysteine did not normalize. Plasma folate (and erythrocyte folate) did not return to predepletion values.
30	170	319	Homocysteine rose above 16 µmol/L in 2 of 5; erythrocyte and serum folate values were low in 3 of 5 (<362 nmol/L [166 ng/mL] and <7 nmol/L [3 ng/mL], respectively).
30	270	489	Erythrocyte folate and plasma homocysteine were maintained in all.

[c] NA = not applicable.
[d] Analyzed value.
[e] Calculated value for 9-d baseline diet.

[25 µg DFEs] followed by 15 days repletion at 151 µg/day of DFEs) with adult males. Although 150 µg of DFEs was insufficient to decrease the elevated plasma homocysteine concentration below 16 µmol in 4 of the 10 subjects or to return plasma folate to predepletion concentrations, the repletion period was too short to allow appropriate evaluation of the primary response variables. Thus, no conclusion about the adequacy of 150 µg/day of DFEs can be reached from this study.

Approximately 200 µg/day of DFEs. Sauberlich and colleagues (1987) evaluated the repletion response of two subjects and reported that erythrocyte folate continued to fall in response to 200 µg of food folate (200 µg DFEs) for 21 days. Data are not sufficient for estimating the erythrocyte folate response to a longer repletion phase.

Milne and coworkers (1983) used serum and erythrocyte folate to evaluate maintenance of folate status in 40 men consuming 200 µg/day of food folate (200 µg DFEs) for periods of 2 to 8 months. Both serum and erythrocyte folate decreased significantly over time regardless of initial status but not below the cutoff values of 7 nmol/L (3 ng/mL) and 305 nmol/L (140 ng/ml), respectively. This study was designed primarily for a different purpose, however, and had several limitations for the estimation of average requirements: the diet was changed during the study, subjects were included for different periods of time, and some of the subjects resumed their normal diet (for 10 days to 2 months) during the study. Thus, the findings from this study were judged equivocal.

Approximately 320 µg/day of DFEs. O'Keefe and colleagues (1995) conducted a controlled metabolic study in which five women were fed a diet that provided 319 µg/day of DFEs (30 µg from food sources and 170 µg from folic acid). Three of the five had erythrocyte folate concentrations less than 305 nmol/L (140 ng/mL) and serum folate concentrations less than 7 nmol/L (3 ng/mL). Two of the subjects had elevated homocysteine concentrations (greater than 16 µmol/L) and a third subject had a homocysteine concentration greater than 14 µmol/L. (These data were obtained directly from the investigators of the published study.) These findings suggest that approximately half would have had normal erythrocyte folate and plasma homocysteine concentrations if 320 µg/day of DFEs had been consumed.

Approximately 500 µg/day of DFEs. O'Keefe and coworkers (1995) fed subjects 270 µg as folic acid with 30 µg of food folate, corresponding to 489 µg of DFEs. This level of intake maintained normal plasma homocysteine, erythrocyte folate, and serum folate values with no significant increase or decrease throughout the 70-day maintenance study. Therefore, 489 µg/day of DFEs could be considered to be above the average requirement.

Summary. Of the controlled metabolic studies reviewed above, greatest weight was given to the study by O'Keefe for five reasons: (1) it was designed as a maintenance study for the purpose of estimating the folate requirement; (2) although it included only five subjects, this sample size exceeds that in the Sauberlich study, which was also rigorously controlled; (3) it evaluated the metabolic response of homocysteine in addition to erythrocyte and serum folate; (4) the diet was fed for 70 days in contrast to very short repletion phases in other metabolic studies (i.e., 15 days [Jacob et al., 1994], and 21 days [Sauberlich et al., 1987]); and (5) it provided folate largely in the form of folic acid, thus minimizing the possibility that folate intake was underestimated. Moreover, considering the evidence that problems with methods have led to underestimates of the folate content of food, it is likely that the subjects in the Sauberlich et al. (1987) and Milne et al. (1983) studies received more folate than reported.

Other Evidence Considered

Epidemiological data support an Estimated Average Requirement (EAR) of approximately 320 µg/day of DFEs. A primary example is the study by Selhub and colleagues (1993). In this study the prevalence of a homocysteine value greater than 14 µmol/L was significantly greater among individuals in the lowest four deciles of folate intake (less than 280 µg/day) as determined from a food frequency questionnaire. Reported intakes in this study were obtained prior to folate fortification and do not include supplements, but they include synthetic folic acid from ready-to-eat or cooked cereals (which frequently contained added folate) and thus would be higher if expressed in DFEs.

The amount of folate utilized daily has been estimated by measuring the catabolic products excreted in the urine and then expressing the sum as *folate equivalents* by multiplying the value by two (because the molecular weight of folate is approximately two times that of catabolites) (McPartlin et al., 1993). This approach may underestimate folate requirements because folate coenzymes may be recycled and not catabolized when utilized and because measurement of urinary catabolites does not account for endogenous folate lost from the body as a mixture of catabolites and intact folates in the feces (Caudill et al., 1998).

Results of other studies were considered (Table 8-3). Several (Gailani et al., 1970; Herbert, 1962a, b; Zalusky and Herbert, 1961) were found less useful than the previously cited metabolic studies

TABLE 8-3 Additional Studies of the Folate Status of Adults

Reference	Type of Study	Type of Dietary Assessment	Age of Subjects (y)	Number of Subjects
Zalusky and Herbert, 1961	Depletion-repletion	Folate-free synthetic diet	60	1 male
Herbert, 1962a	Depletion	Folate-free diet	35	1 male
Herbert, 1962b	Depletion-repletion	Defined folate-deficient diet	NA[c]	1 female
				1 female
				1 female
Krumdieck et al., 1978	Kinetic	Not reported	36	1 female
Von der Porten et al., 1992	Kinetic	Self-selected diets	22–31	6 males
Stites et al., 1997	Kinetic	Self-selected, folate-adequate diets	20–30	4 males

[a] IM = intramuscular.

[b] DFEs = dietary folate equivalents. To compute DFEs, use the formula μg folic acid × 2 for IM injections or μg food folate + (1.7 × μg folic acid) for a combination of food folate and folic acid.

[c] Not available.

for estimating the folate requirement because the diets were deficient in more than one nutrient. Ancillary information is provided by the studies using stable isotope methods to estimate in vivo folate pool size and the rate of daily utilization. With use of the estimate of the total body pool folate of 20 mg as extrapolated from liver folate measurements (Hoppner and Lampi, 1980; Whitehead, 1973), and

Dietary Folate Intake (µg/d)	Other Folate Source	Comments
None	50 µg of folic acid IM[a] (100 µg of DFEs[b])	Subject was folate deficient and scorbutic at the beginning of the study; folate injection produced a reticulocyte response.
5	None	Signs and symptoms of deficiency coincided with a fall in the serum folate level to < 7 nmol/L (3 ng/mL) and a decrease in erythrocyte folate concentration to < 305 nmol/L (140 ng/mL). Diet was also deficient in potassium.
5	25 µg of folic acid p.o.[d] (48 µg of DFEs)	Serum folate activity fell below normal levels in the subject supplemented with 25 µg/d. Test period was only 42 d. Subjects were on a low-calorie diet.
5	50 µg of folic acid p.o. (90 µg of DFEs)	
5	100 µg of folic acid p.o. (170 µg of DFEs)	
None	320 µg of labeled folic acid	Turnover rate estimated to be ≈1% of total body folate pool per day. Subject was Hodgkin's disease patient in remission.
200[e]	1.6 mg/d of labeled folic acid	Turnover rate was 4.5% of the total body folate pool per day.
443[f]	100 µg of labeled folic acid + 100 µg of unlabeled folic acid	Turnover rate was estimated to be ≈1%.

[d] p.o. = by mouth.
[e] Typical intake assessed by diet records.
[f] Average value.

the assumption of a 1 percent daily turnover rate of folate (Krumdieck et al., 1978; Stites et al., 1997; Von der Porten et al., 1992), the daily quantity of folate utilized is calculated to be approximately 200 µg. When 200 µg/day is corrected for the 50 percent bioavailability of food folate, the DFE is 400 µg/day.

Folate EAR and RDA Summary, Ages 19 through 50 Years

With greatest weight given to the metabolic maintenance study by O'Keefe along with data considered from the other studies reviewed above, it was concluded that the data support an EAR of approximately 320 µg/day of DFEs for the age group 19 through 50 years. A special recommendation is made for women capable of becoming pregnant (see "Recommendations for Neural Tube Defects Risk Reduction").

EAR for Men
19–30 years   320 µg/day of dietary folate equivalents
31–50 years   320 µg/day of dietary folate equivalents

EAR for Women
19–30 years   320 µg/day of dietary folate equivalents
31–50 years   320 µg/day of dietary folate equivalents

The RDA for folate is set by assuming a coefficient of variation (CV) of 10 percent (see Chapter 1) because information is not available on the standard deviation of the requirement for folate; the RDA is defined as equal to the EAR plus twice the CV to cover the needs of 97 to 98 percent of the individuals in the group (therefore, for folate the RDA is 120 percent of the EAR).

RDA for Men
19–30 years   400 µg/day of dietary folate equivalents
31–50 years   400 µg/day of dietary folate equivalents

RDA for Women
19–30 years   400 µg/day of dietary folate equivalents
31–50 years   400 µg/day of dietary folate equivalents

Adults Ages 51 Years and Older

The aging process has not been associated with a reduction in the ability to utilize folate (Bailey et al., 1984). Folate status as measured by serum folate or erythrocyte folate has not been shown to decline as a function of age (Rosenberg, 1992; Selhub et al., 1993). In contrast, numerous reports indicate that homocysteine concentration increases as a function of age (Selhub et al., 1993). It has been postulated (Selhub et al., 1993) that this increase may result from an age-related decline in cystathionine β-synthase and possibly other enzymes involved in homocysteine metabolism (Gartler et al., 1981).

Evidence Considered in Estimating the Average Requirement

The EAR for men and women ages 51 years and older is based on evaluation of three types of studies: metabolic (Jacob et al., 1998), observational folate status assessment of population subgroups (Bates et al., 1980; Garry et al., 1982, 1984; Jägerstad, 1977; Jägerstad and Westesson, 1979; Koehler et al., 1996; Ortega et al., 1993; Rosenberg, 1992; Sahyoun et al., 1988), and epidemiological (Selhub et al., 1993).

Jacob and colleagues (1998) conducted a depletion-repletion metabolic study in eight post-menopausal women aged 49 to 63 years. A folate depletion diet (56 μg/day [56 μg/day of DFEs]) was fed for 35 days, followed by three repletion periods in which graded amounts of folate were added to the diet. After being converted to DFEs, the three repletion amounts were 150, 450, and 850 μg/day for 28, 13, and 8 days, respectively. Plasma homocysteine concentrations remained elevated (greater than 12 μmol/L) in five of the eight women in response to either 150 or 450 μg/day of DFEs. Plasma folate remained low (less than 7 nmol/L [3 ng/mL]) in five of the eight subjects in response to 150 μg/day of DFEs but returned to normal in all subjects in response to 450 μg/day. The short repletion periods limit conclusions regarding the adequacy of these intake levels. From the plasma folate changes, which do respond quickly, it appears that 450 μg/day was adequate for all subjects and 150 μg/day was inadequate for a large percentage of the group. Extrapolating from these data, approximately 300 μg/day would result in normal folate status in approximately 50 percent of the group and would therefore be consistent with an EAR of 320 μg/day.

The observational folate status assessment studies that provide data on both folate intake and biochemical measures of folate status (Table 8-4) provide evidence that tends to support an EAR for older adults that is equivalent to that for younger adults: 320 μg/day of DFEs. Data from Selhub and colleagues (1993) (Figure 8-1) show that the mean homocysteine concentration begins to stabilize when folate intakes are approximately 300 μg/day. Figure 8-2 presents data showing the relationship of plasma homocysteine to plasma folate concentrations (Lewis et al., 1992).

TABLE 8-4 Observational Status Assessment Studies
Considered in Setting the Estimated Average Requirement
(EAR) for Folate in the Elderly

Reference	Number and Age of Subjects	Folate Intake Assessment
Studies suggesting an EAR greater than 150–200 µg of dietary folate equivalents[a]		
Jägerstad, 1977; Jägerstad and Westesson, 1979	37 Swedish men and women, 67 y	Microbiological analysis
Bates et al., 1980	21 elderly men and women	Dietary record
Ortega et al., 1993	72 men and women, 65–89 y	5-d food records
Studies suggesting an EAR of 250–300 µg of dietary folate equivalents		
Selhub et al., 1993; Tucker et al., 1996	1,000 men and women, 67–80 y	Food frequency questionnaire
Koehler et al., 1996	44 men and women, 68–96 y (nonsupplement users)	Food frequency questionnaire
Other studies		
Garry et al., 1982, 1984	304 Caucasian men and women, ≥ 60 y	3-d diet records, prospective
Sahyoun et al., 1988; Sahyoun, 1992; Rosenberg, 1992	686 free-living adults, ≥ 60 y	3-d food records

NOTE: In these studies, it is impossible to calculate dietary folate equivalents because intake of foods fortified with folic acid was not specified. Moreover, on the basis of data from Tamura et al. (1997) and Martin et al. (1990), it is believed that folate intakes are underestimated.

Results

Median intake of folate was 150 µg/d for males and 125 µg/d for females. Erythrocyte folate values ranged from approximately 175 to 350 nmol/L (80 to 160 ng/mL)

Mean intake was 135 µg of folate/d; 40% had an erythrocyte folate value < 305 nmol/L (140 ng/mL).

Intake averaged 214 µg/d of folate. Mean erythrocyte folate was 250 nmol/L (115 ng/mL); 85% of the values were < 327 nmol/L (150 ng/mL).

Plasma hcy[b] plateaued in normal range (< 14 µmol/L) at folate intakes of 350–400 µg/d and serum folate of 15 nmol/L (7 ng/mL). See Figure 8-1.

Mean erythrocyte folate was ≈1,035 nmol/L (475 ng/mL) and plasma hcy was 11.2 µmol/L for those not taking supplements (average intake ≈300 µg/d).

Values for supplement users were not distinguished from those for nonusers. For nonusers, 75% had folate intakes < 250 µg/d. Overall, < 3% had erythrocyte folate of < 305 nmol/L (140 ng/mL).

Median folate intakes of nonsupplement users were 254 µg for men and 216 µg for women. Median plasma folate was ≈19 nmol/L (9 ng/mL) for both.

[a] Dietary folate equivalents: 1µg food folate = 0.6 µg of folic acid from fortified food or as a supplement consumed with food = 0.5 µg of a supplement taken on an empty stomach.

[b] hcy = total homocysteine.

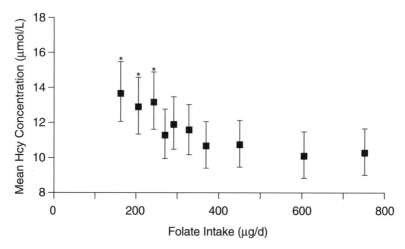

FIGURE 8-1 Mean plasma homocysteine (Hcy) concentrations (and 95% confidence intervals) by deciles of intake of folate. Means are adjusted for age, gender, and other vitamin intakes. Asterisk indicates significantly different from mean in the highest decile ($p < 0.01$). Reprinted with permission, from Selhub et al. (1993). Copyright 1993 by the American Medical Association.

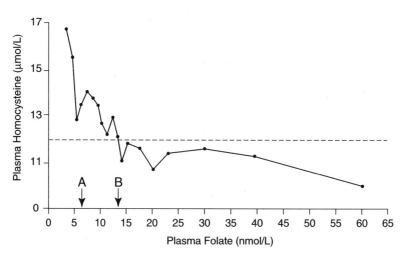

FIGURE 8-2 Relationship of plasma homocysteine concentrations to plasma folate concentrations in 209 adult males. *A* indicates lower limit of normal plasma folate as used by the Second National Health and Nutrition Examination Survey (6.8 nmol/L). *B* indicates lower limit of normal plasma folate as used by the World Health Organization (13.6 nmol/L). Homocysteine concentrations above the dotted line (12 µmol/L) are considered elevated. Reprinted with permission, from Lewis et al. (1992). Copyright 1992 by the New York Academy of Sciences.

Folate EAR and RDA Summary, Ages 51 Years and Older

Data from metabolic folate status assessment and epidemiological studies support an EAR for adults ages 51 years and older of 320 µg/day of DFEs. The EAR for this age group is expected to be the same as that for younger age groups because the aging process does not appear to impair folate absorption or utilization nor do studies separate those over age 70 from those 51 to 70 years.

EAR for Adults
51–70 years	**320 µg/day of dietary folate equivalents**
> 70 years	**320 µg/day of dietary folate equivalents**

The RDA for folate is set by assuming a coefficient of variation (CV) of 10 percent (see Chapter 1) because information is not available on the standard deviation of the requirement for folate; the RDA is defined as equal to the EAR plus twice the CV to cover the needs of 97 to 98 percent of the individuals in the group (therefore, for folate the RDA is 120 percent of the EAR).

RDA for Adults
51–70 years	**400 µg/day of dietary folate equivalents**
> 70 years	**400 µg/day of dietary folate equivalents**

Pregnancy

Folate requirements increase substantially during pregnancy because of the marked acceleration in single-carbon transfer reactions, including those required for nucleotide synthesis and thus cell division. During pregnancy, cells multiply in association with uterine enlargement, placental development, expansion of maternal erythrocyte number, and fetal growth (Cunningham et al., 1989). Additionally, folate is actively transferred to the fetus as indicated by elevated folate concentrations in cord blood relative to that of maternal blood. When folate intake is inadequate, maternal serum and erythrocyte folate concentrations decrease and megaloblastic marrow changes may occur (Picciano, 1996). If inadequate intake continues, megaloblastic anemia may develop. This section does not address the reduction of risk of neural tube defects because the neural tube is formed before most women know that they are pregnant (see "Neural Tube Defects").

Evidence Considered in Estimating the Average Requirement

For pregnant women the maintenance of erythrocyte folate, which reflects tissue stores, was selected as the primary indicator of adequacy. When this indicator was not measured, serum folate was evaluated with the recognition that hemodilution contributes to a normal reduction in serum folate concentration during gestation. Homocysteine concentrations have not been shown to reflect folate status during pregnancy, possibly because of hormonal changes, hemodilution, or other unknown factors associated with pregnancy (Andersson et al., 1992; Bonnette et al., 1998).

Population-Based Studies. A number of population-based studies confirm that folic acid consumed in conjunction with diet prevents folate deficiency in pregnant women as assessed by maintenance of normal folate concentration in erythrocytes, serum, or both. The folate has been provided either by supplements (Chanarin et al., 1968; Dawson, 1966; Hansen and Rybo, 1967; Lowenstein et al., 1966; Qvist et al., 1986; Willoughby, 1967; Willoughby and Jewel, 1966) or fortified food (Colman et al., 1975) (see Table 8-5).

Willoughby and Jewel (1966, 1968) conducted a series of studies involving approximately 3,500 pregnant women beginning at 12 weeks of gestation who were assigned to different levels of folate supplementation (0, 100, 350, or 450 µg/day). Their dietary folate was estimated to be less than 100 µg/day. A supplementation level of 100 µg/day in conjunction with the low-folate diet was insufficient to prevent deficient (less than 7 nmol/L [3 ng/mL]) serum concentrations in 33 percent of the group (Willoughby and Jewel, 1966) or to prevent megaloblastic anemia in 5 percent of the group (Willoughby, 1967). In contrast, 300 µg/day of supplemental folate was sufficient to maintain a mean serum folate concentration that was comparable with the mean in healthy nonpregnant control subjects (Willoughby and Jewel, 1966) and to prevent megaloblastic anemia (Willoughby, 1967). These data agree with those of Dawson (1966), who found that taking 150 µg/day of folate supplements (beginning at 28 weeks) in addition to diet resulted in low serum folate concentrations (less than 7 nmol/L [3 ng/mL]) in 30 percent of the group at delivery. Also confirming these findings, Hansen and Rybo (1967) reported that 100 µg of folic acid plus diet was not sufficient to prevent serum folate reduction (defined as less than 4 nmol/L [2 ng/mL]) in 15 percent of the group whereas a folate supplement of 500 µg/day resulted in a mean serum folate concentration of 13 nmol/L(6 ng/mL) at 36 to 38 weeks of gestation.

Lowenstein and colleagues (1966) compared serum and erythrocyte folate and bone marrow morphology of women taking 500 µg/day of supplemental folate with those of women taking a placebo. In the folate-supplemented women, mean serum and erythrocyte folate levels were approximately 21 and 870 nmol/L (10 and 400 ng/mL), respectively, at 36 and 38 weeks of gestation and postpartum. Bone marrow aspirates at 38 weeks were essentially normal. In contrast, a large percentage of the placebo-treated subjects had serum and erythrocyte folate concentrations that were less than normal.

Chanarin and colleagues (1968) compared erythrocyte folate concentrations in 103 pregnant women supplemented with 100 µg of folate from 25 weeks of gestation until delivery with those of 103 unsupplemented pregnant control subjects. Dietary intake was analyzed in 111, 24-hour duplicate diets and reported to be 676 µg/day. Supplementation of the usual diet with 100 µg/day resulted in maintenance of erythrocyte folate concentration throughout pregnancy whereas a significant reduction in erythrocyte folate was observed in the unsupplemented subjects.

Colman et al. (1975) evaluated the efficacy of folate-fortified maize to maintain erythrocyte folate concentrations in 70 pregnant women. Erythrocyte folate response was compared between women receiving maize fortified to provide 300, 500, or 1,000 µg of folic acid and a control group. (Additional groups consumed folic acid in tablet form to assess the relative bioavailability of the fortified maize.) Maize containing 300 µg of folic acid in addition to dietary folate, the lowest level tested, was effective in preventing the progression of folate depletion in the eighth month of pregnancy.

Controlled Metabolic Study. Caudill and colleagues (1997) conducted a metabolic study in which either of two levels of folate was consumed for 12 weeks by pregnant women during the second trimester (14 weeks to 25 weeks of gestation). Their folate status was compared with that of nonpregnant control subjects. Folate was provided as a combination of dietary folate (120 µg/day) and folic acid (either 330 or 730 µg/day) consumed with the diet. After correcting for bioavailability, the intakes of the groups were approximately 60 µg/day of DFEs and more than 1,300 µg/day of DFEs, respectively. Folate status was normal (serum folate greater than 7 nmol/L [3 ng/mL] and erythrocyte folate values greater than 305 nmol/L [140 ng/mL]) in all subjects consuming the diet with 680 µg/day of DFEs and was not different from that of the nonpregnant control subjects with the same folate intake.

TABLE 8-5 Supplementation Studies in Pregnancy

Reference	Number of Subjects	Folate Diet	Folate Supplement
Dawson, 1966	20	Not reported	150
Lowenstein et al., 1966	311	82–92	0 500
Willoughby and Jewell, 1966	350	< 50	0 100 300 450
Hansen and Rybo, 1967	95	Not reported	50 100 200 500
Willoughby and Jewell, 1968	48	Not reported	330
Chanarin et al., 1968	103 103	676	0 100
Colman et al., 1975	122	Not reported	0 300 500 1,000

The data provided by the only diet-controlled metabolic study that has been conducted in pregnant women (Caudill et al., 1997) agree with the findings from the population studies and confirm that a combination of approximately 300 µg of folate from supplements, fortified food, or both plus dietary folate (assumed to be approximately 100 µg/day before folate fortification) has been shown to be sufficient to maintain normal folate status during pregnancy. When expressed as DFEs, the consistent finding across the numerous population studies and the controlled metabolic study is that 600 µg/day of DFEs is adequate to maintain normal folate status.

Total in Dietary Folate Equivalents	Results
300 plus diet	Serum folate low in 40%
Diet 1,000 plus diet	Increase 40–60% abnormal erythrocyte folate normal level compared with 10–20% in supplemented group
≤ 100 200 plus diet 600 plus diet 900 plus diet	Prevented deficiency in 72%, 84%, and 94%, respectively, comparable with nonpregnancy control
100 plus diet 200 plus diet 400 plus diet 1,000 plus diet	Decrease in serum folate in 15%; normal level serum folate
660 plus diet	Prevented deficiency in supplemented groups
Diet 200 plus diet	Maintained normal levels erythrocyte folate
Diet 510 plus diet 850 plus diet 1,700 plus diet	Folate depletion No apparent folate depletion

Three of the four studies provided data that 100 to 150 µg/day of supplemental folate plus a low-folate diet was inadequate to maintain normal serum and hematological indices, which were the only outcomes measured in all of the subjects. The accuracy of the dietary estimates could not be ascertained, but they were lower than the one analyzed intake estimate (676 µg/day) reported by Chanarin and coworkers (1968).

Other Evidence Considered. McPartlin and colleagues (1993) quantitated the urinary excretion of the major folate catabolites in six pregnant women and six nonpregnant control subjects. These investigators converted the quantity of urinary catabolites to urinary folate equivalents and estimated that the recommended folate intake for second-trimester pregnant women would be 660 µg/day.

Folate EAR and RDA Summary, Pregnancy

From these data, low dietary folate intake plus 100 µg of supplemental folate (equivalent to approximately 200 µg/day of DFEs) is inadequate to maintain normal folate status in a significant percentage of population groups assessed. The EAR therefore was derived by adding this quantity in DFEs (200 µg/day) to the EAR for non-pregnant women (320 µg/day) to provide an EAR of 520 µg/day of DFEs.

EAR for Pregnancy
14–18 years	**520 µg/day of dietary folate equivalents**
19–30 years	**520 µg/day of dietary folate equivalents**
31–50 years	**520 µg/day of dietary folate equivalents**

The RDA for folate is set by assuming a coefficient of variation (CV) of 10 percent (see Chapter 1) because information is not available on the standard deviation of the requirement for folate; the RDA is defined as equal to the EAR plus twice the CV to cover the needs of 97 to 98 percent of the individuals in the group (therefore, for folate the RDA is 120 percent of the EAR). Data from the controlled metabolic study support an RDA of 600 µg/day of DFEs based on maintenance of normal erythrocyte folate concentrations and agree with the findings from the series of population studies that 600 µg/day of DFEs is adequate to maintain normal folate status in groups of pregnant women.

RDA for Pregnancy
14–18 years	**600 µg/day of dietary folate equivalents**
19–30 years	**600 µg/day of dietary folate equivalents**
31–50 years	**600 µg/day of dietary folate equivalents**

Lactation

Method Used to Estimate the Average Requirement

The EAR for the lactating woman is estimated as the folate intake necessary to replace the folate secreted daily in human milk plus the amount required by the nonlactating woman to maintain folate status. The average daily amount of folate secreted in human milk is estimated to be 85 µg/L, as described in the previous section "Infants Ages 0 through 12 Months" (see "Human Milk"). The dietary intake needed to provide this amount must account for the estimated 50 percent bioavailability of food folate (see "Bioavailability").

Other Evidence Considered

There are no metabolic studies in which lactating women consumed controlled amounts of dietary folate. It is unclear whether the reduction in maternal folate concentration observed in lactating women (Keizer et al., 1995; Qvist et al., 1986; Smith et al., 1983) is related to the discontinuation of use of prenatal folate supplements, loss of maternal body folate stores, or other factors. For example, in a recent study of lactating adolescents (Keizer et al., 1995), both breastfeeding mothers and mothers of formula-fed infants showed a decline in erythrocyte folate between 4 and 12 weeks postpartum, suggesting that the postpartum decline in folate status may not be related to lactation. The decrease was prevented by supplemental folate (300 µg/day).

In a recent study in which folate status was compared in supplemented and nonsupplemented lactating women (Mackey et al., 1997), dietary folate intake was estimated to be 400 µg/day. In the unsupplemented lactating women, plasma homocysteine concentrations increased significantly but remained well within the normal range (6 to 7 µmol/L); this increase, therefore, does not appear to be of nutritional significance.

Folate EAR and RDA Summary, Lactation

The calculation used to obtain the extra amount of folate needed to cover lactation is

0.78 L (milk volume) × 85 µg/L (folate concentration) ×
2 (bioavailability correction factor) = 133 µg/day.

When this quantity is added to the EAR for the nonlactating nonpregnant woman (320 µg/day), the result is rounded down, giving an EAR of 450 µg/day of DFEs. Women who are only partially breastfeeding would need less.

EAR for Lactation

14–18 years	450 µg/day of dietary folate equivalents
19–30 years	450 µg/day of dietary folate equivalents
31–50 years	450 µg/day of dietary folate equivalents

The RDA for folate is set by assuming a coefficient of variation (CV) of 10 percent (see Chapter 1) because information is not available on the standard deviation of the requirement for folate; the RDA is defined as equal to the EAR plus twice the CV to cover the needs of 97 to 98 percent of the individuals in the group (therefore, for folate the RDA is 120 percent of the EAR).

RDA for Lactation

14–18 years	500 µg/day of dietary folate equivalents
19–30 years	500 µg/day of dietary folate equivalents
31–50 years	500 µg/day of dietary folate equivalents

Special Considerations

Intakes higher than the RDA may be needed by women who are pregnant with more than one fetus, mothers nursing more than one infant, individuals with chronic heavy intake of alcohol, and individuals on chronic anticonvulsant or methotrexate therapy. Folate from supplements or fortified foods in addition to dietary folate is recommended for women capable of becoming pregnant.

REDUCING RISK OF DEVELOPMENTAL DISORDERS AND CHRONIC DEGENERATIVE DISEASE

Neural Tube Defects

Neural tube defects (NTDs) constitute an important public health problem in terms of mortality, morbidity, social cost, and human suffering. Many studies have been conducted regarding the association between folate intake and the occurrence of NTDs. The aim of this section is to review the evidence linking folate with the etiology and occurrence of NTDs in humans, estimate the risk of NTDs according to various levels of folate intake in the U.S. and Canadian populations, and develop a folate recommendation for women capable of becoming pregnant. Survey data indicate that fewer than half of U.S. females aged 15 to 44 years are at any appreciable risk of conceiving (Abma et al., 1997). Approximately 22 percent of them are permanently sterile (most often as a result of a specific operation done for that purpose); 20 percent are using highly effective contraceptives, usually long-term in nature; 5 percent are pregnant or immediately postpartum at any particular point in time; and 11 percent have never had sexual intercourse.

Classification, Anatomy, and Embryology

NTDs are the most common major congenital malformations of the central nervous system. They arise as a result of a disturbance of the embryonic process of neurulation and are midline defects that affect neural tissues, their coverings anywhere along the neuraxis, or both. They are heterogeneous malformations, and the terms used to define them here (see Box 8-1) are based on clinical descriptions and the presumed embryological defect (Lindseth, 1996; Volpe, 1995). The terminology in the literature may vary.

NTDs are not to be confused with spina bifida occulta, a common radiographic finding that does not involve neural elements, or encephalocele, a protrusion of meninges and brain tissue outside the cranium, most frequently in the occipital region.

BOX 8-1 Forms of Neural Tube Defects

- *Anencephaly:* a fatal form characterized by partial absence of brain tissue, presumably caused by failure of closure of the anterior neuropore.

- *Meningomyelocele:* a midline defect of the spinal cord in which the neural tissue is dysplastic and the overlying meninges form a cystic expansion, presumably because of failure of closure of the neural tube at this site. This defect is most often in the lumbosacral region, usually results in peripheral neurological deficit, and may be called spina bifida aperta or cystica. Myelomeningocele often is associated with the Arnold Chiari malformation and hydrocephalus.

- *Meningocele:* a less severe result of the embryological defect that causes meningomyelocele, involving only the meninges.

- *Craniorachischisis:* a fatal form in which the entire neuraxis—from the brain to distal spinal cord—is dysplastic and lacks covering by dura, muscle, or skin. This is presumably due to total failure of neurulation.

In the less-severe forms of NTD a child may otherwise be normal and, with appropriate surgical and medical care, can lead a productive life, including parenthood. Less than 20 percent of NTDs show associations with malformations in nonneural tissues, chromosomal defects, or specific genetic syndromes (Khoury et al., 1982). Techniques have been established for prenatal screening for NTDs by measuring maternal serum and amniotic fluid α-fetoprotein and by fetal ultrasound (Hobbins, 1991).

Neurulation is the first organogenetic process to be initiated and completed. It begins in the human at approximately 21 days postfertilization and is complete by 28 days. Thus, neurulation is ongoing at the time that a woman may first recognize her pregnancy by a missed menstrual period. Closure of the neural tube begins separately and consecutively in at least three sites: the cervical-hindbrain boundary, the forebrain-midbrain boundary, and the rostral extremity of the forebrain. Closure spreads to the intervening regions with completion of neural tube formation at neuropores in the forebrain (anterior neuropore), the hindbrain, and the lumbosacral region (posterior neuropore). NTDs appear to arise from failure or inadequacy of this closure process. Different forms of NTD could arise at different times in neurulation, possibly from distinct mechanisms. Although many specific molecules are involved in the successful completion of neurulation, none have been implicated in the mechanisms underlying the common human NTDs.

Prevalence of NTDs

United States and Canada. National birth-defect registry data are not available, but a decrease in the prevalence of NTDs at birth has been observed during the past 30 years. This is not entirely explained by increased widespread prenatal screening and diagnostic techniques (De Wals et al., 1999; Yen et al., 1992). Although the comparison of results of studies using different methods for case identification should be made cautiously, regional variation in the risk of NTDs is likely (Table 8-6). It is not known whether the especially low rate observed in Hawaii is caused by genetic or by environmental factors (Cragan et al., 1995). The populations studied included women who had taken vitamin supplements at the time of conception, but the frequency of folate supplementation was estimated only in California (Velie and Shaw, 1996).

Other Countries. The incidence of the common forms of NTD varies worldwide from less than 1 to approximately 9 per 1,000 total births,

TABLE 8-6 Total Prevalence Rates of Neural Tube Defect in Selected Areas of North America, from Birth Defect Registry Data, 1985–1994

Area	Prevalence Rate per 1,000	95% Confidence Interval
Arkansas	1.03	0.85–1.24
Atlanta	0.99	0.78–1.23
California	0.94	0.87–1.01
Iowa	0.90	0.78–1.07
Hawaii	0.72	0.59–0.87
Québec, Canada	1.41	0.95–2.01

SOURCES: Cragan et al. (1995), De Wals et al. (1999), and Velie and Shaw (1996).

with the highest incidence reported in Great Britain and Ireland (Copp and Bernfield, 1994). Other populations with high incidence include northern Chinese and Australian Aborigines (Bower et al., 1984; Moore et al., 1997). Although Sikhs have a high NTD incidence, the defects are often thoracic and associated with minimal deficit, suggesting a distinct etiology (Baird, 1983). The decrease in NTDs among Irish immigrants to the United States could be explained by genetic dilution through interethnic marriages. However, some studies of migrant populations in which NTD incidence decreases with changes in locale suggest a nutritional etiology (Borman et al., 1986; Carter, 1974).

Etiology of NTDs

The causes of these abnormalities have been the subject of intensive research over many decades. Differences in the pathogenesis and the epidemiology of different categories of NTD have led to the idea that NTDs are highly heterogeneous in etiology (Dolk et al., 1991). Substantial familial aggregation indicates that anencephaly, myelomeningocele, and craniorachischisis are related pathogenetically and genetically. Evidence from epidemiological studies of NTDs indicates that heredity is a major contributor. Indeed, the recurrence risk in a sibling birth is 3 to 5 percent (Laurence, 1990). For the most cases the inheritance is believed to be polygenic, potentially involving multiple genes. Such polygenic traits are influenced by environmental factors, thus the etiology appears to be multifactorial (Laurence, 1990). Recently, attention has turned to assessing the genetic basis of NTD and to evaluating the role of vitamins, specifically folate, in the prevention of NTDs. Coverage of other risk factors for NTDs is beyond the scope of this report.

Genetic Evidence. An assessment of heritability for common forms of NTDs has been put at 60 percent (Emery, 1986). Data from demographic, family, and mouse model studies have prompted a search for candidate genes that predispose individuals to an NTD. A defect in enzymes involved in homocysteine metabolism is suggested by altered folate, vitamin B_{12}, homocysteine, and methylmalonate values in mothers of infants with NTDs (Mills et al., 1995; Steegers-Theunissen et al., 1994); the prevention of some human NTDs by folate administration; and the prevention of NTDs in some rodent models by methionine (Essien, 1992; Vanaerts et al., 1994). These enzymes are 5,10-methylenetetrahydrofolate reductase (MTHFR), cystathionine β-synthase, and methionine synthase (Figure 8-3). Interestingly, families with homocystinuria caused by severe mutations in genes for each of these enzymes do not exhibit NTDs

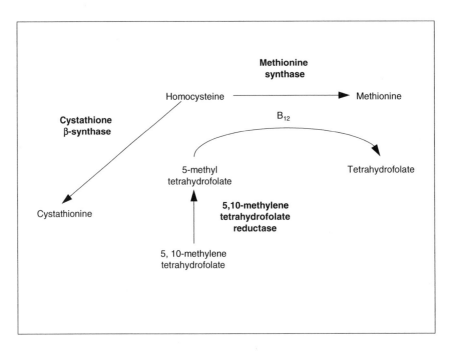

FIGURE 8-3 Major pathways depicting involvement of vitamin B_{12} and folate in homocysteine metabolism.

(Haworth et al., 1993; Kang et al., 1991b). Moreover, deletion of cystathionine β-synthase in the mouse yields a phenocopy of human homocystinuria but not an NTD (Watanabe et al., 1995). Subsequent work on other genetic markers of risk has not provided conclusive results (van der Put et al., 1997a, b).

Two studies have shown a statistically significant association between mothers of children with NTDs and a common variation within the gene for MTHFR (van der Put et al., 1995; Whitehead et al., 1995). This gene codes for a thermolabile variant of the enzyme that shows about 50 percent enzymatic activity and is associated with elevated serum homocysteine concentrations (Kang et al., 1991a). This association would account for approximately 15 percent of NTD cases (van der Put et al., 1995). The polymorphism was recently associated with low erythrocyte folate values (Molloy et al., 1997), which suggests that these values by themselves could account for the increased NTD risk. However, similar studies using linkage assessments are needed in suitable (genetically homogeneous) NTD populations with adequate numbers and types of controls.

No correlation has been found between two common mutations in cystathionine β-synthase and NTD prevalence in an Irish population (Ramsbottom et al., 1997). The gene for methionine synthase has only recently been cloned in mammals (Chen et al., 1997; Li et al., 1996). There are no reports of an association between mutations of this gene and NTDs.

The genetic mouse models of NTD suggest a variety of other candidate genes for human NTDs (Baldwin et al., 1992; Tassabehji et al., 1993). However, no reports assess whether any of the causative genes for mouse models show linkage with the common forms of human NTD. Because the likely heterogeneity of human NTDs may make it impossible to demonstrate linkage for any one candidate gene, candidate genes will need to be assessed in individuals with NTDs by sequence analysis.

A summary of evidence from animal studies on the etiology of NTD appears in Appendix M. Animal models of NTD have been examined and manipulated to elucidate the mechanisms of abnormal neurulation and to test etiologic hypotheses suggested by human epidemiological data.

Nutrition Evidence. Studies of migrant populations suggest a nutritional etiology for NTD (Borman et al., 1986; Carter, 1974). Lower socioeconomic class also correlates with NTD incidence (Elwood and Elwood, 1980; Laurence, 1990). Differences in diet and in supplement use could contribute to the inverse relationship of socioeconomic status with incidence of NTD, but this has not been analyzed in recent years.

Nutritional markers, particularly maternal serum vitamin B_{12} and serum and erythrocyte folate, as well as their metabolic indicators of adequacy, have been assessed in relation to the risk of NTD. The results have been inconsistent, some showing no association with NTD prevalence (Wald, 1994). Others have demonstrated low or low normal levels of both vitamin B_{12} and erythrocyte folate and suggested that both vitamins represent independent risk factors for NTD (Kirke et al., 1993). Methylmalonic acid is elevated in maternal serum of midterm NTD pregnancies (Adams et al., 1995). Some of the women who gave birth to infants with NTDs had elevated homocysteine values (Mills et al., 1995; Steegers-Theunissen et al., 1994). These studies support the proposition that NTD is associated with altered status of vitamin B_{12}, folate, or both during pregnancy.

Teratology Studies. Drugs identified as causes of NTD in humans include folate antagonists (specifically aminopterin, previously used as an antitumor agent) (Thiersch, 1952); carbamazepine (Rosa, 1991) and valproate (commonly used antiepileptic drugs) (Blaw and Woody, 1983; Gomez, 1981; Stanley and Chambers, 1982); and retinoids, including isotretinoin (used to treat acne) (Dai et al., 1989; Hill, 1984) and etretinate (used to treat psoriasis) (Happle et al., 1984). Clomiphene (an oocyte maturation agent) is also suspected as a teratological cause of human NTDs (Wilson, 1973). These agents account for less than 0.1 percent of all NTDs. In general, the induced malformations are not restricted to NTD, and the precise mechanisms of these teratological effects are not clear. Indeed, as with other teratogens, the pharmacological and teratological mechanisms may differ because the embryo, especially at neurulation stages or earlier, is a qualitatively different organism from other developmental stages and the adult.

Risk of NTD According to Maternal Intake of Folate

The possibility that folate might be involved in NTD was first raised by Hibbard (1964). This was followed by observational studies of the effect of both dietary folate and folate supplements on NTD (Table 8-7), nonrandomized intervention studies of folate supplementation (Table 8-8), and randomized prevention studies, most of which were conducted with women who had prior NTD pregnancies (Table 8-8). The best evidence comes from the four randomized prevention trials (Czeizel and Dudas, 1992; Kirke et al., 1992;

Laurence et al., 1981; Wald et al., 1991), but the observational evidence (Table 8-7) strongly supports the intervention studies and provides the only evidence concerning dietary folate.

NTD Risk Associated with Different Levels of Dietary Folate. Data from two observational studies (Shaw et al., 1995c; Werler et al., 1993) indicate a statistically significant decreasing risk of NTD with increasing dietary folate in unsupplemented women. In both studies, the median dietary folate in the control group was approximately 300 µg/day. In the hospital-based case-control study carried out from 1988 to 1991 (Werler et al., 1993) (Tables 8-7 and 8-9), mothers were interviewed by telephone within 6 months after delivery. The interview included detailed questions about use of vitamin supplements and a semiquantitative food frequency questionnaire. In the population-based study from 1989 to 1991 (Shaw et al., 1995c) (Tables 8-7 and 8-9), folate supplement use and dietary folate intake during the periconceptional period were retrospectively assessed by using a face-to-face interview and a semiquantitative food frequency questionnaire with mothers of children with NTDs and randomly selected controls. Interviews were completed an average of 5 months after delivery. The proportion of women reporting no use of a folate supplement before conception or in the first trimester was 39 percent (207/526) for cases and 29 percent (149/523) for controls (Velie and Shaw, 1996). From these data, the average risk of NTD in the fraction of the population taking no supplement can be estimated to be 1.3 per 1,000.

In a case-control study in Australia, a negative association was found between NTD occurrence and free and total folate intake in early pregnancy (Bower and Stanley, 1989). However, the published results contain only the combined data on supplement use and dietary intake.

The results from the studies of Werler et al. (1993) and Shaw et al. (1995c) can be used to draw a tentative risk curve. Point estimates in the two studies are remarkably concordant (Figure 8-4). There is a quasilinear decreasing NTD risk for dietary folate values between 100 and 400 µg/day but no further decrease is observed for higher intake values. A possible explanation for the risk observed at higher intakes could be overreporting of consumption of folate-rich foods by some women. Also, imprecision in risk estimates because of the small sample numbers cannot be excluded.

NTD Risk Associated with Periconceptional Folate Supplement Use. The only randomized trial on the effect of periconceptional vitamin sup-

TABLE 8-7 Observational Studies of Folate and Risk of Neural Tube Defect

Study	Design	Subjects
Mulinare et al., 1988 (as reported in CDC, 1992)	Case/control in metropolitan Atlanta	NTD[a] case infants and normal control infants Pregnant women without a prior NTD-affected pregnancy
Bower and Stanley, 1989 (as reported in CDC, 1992)	Case/control in Western Australia	Spina bifida case infants and normal control infants Pregnant women without a prior NTD-affected pregnancy
Mills et al., 1989 (as reported in CDC, 1992)	Case/control in California and Illinois	NTD case infants and normal control infants Pregnant women without a prior NTD-affected pregnancy
Milunsky et al., 1989 (as reported in CDC, 1992)	Prospective cohort in New England	NTD case infants and normal control infants Pregnant women without a prior NTD-affected pregnancy

Exposure	Results	Comments
Multivitamin supplement containing 0–0.8 mg of folic acid at least 1 mo before conception through the 1st trimester	24 NTD cases in infants from women supplemented and 157 cases in infants from women unsupplemented 405 normal cases in infants from supplemented mothers and 1,075 normal cases in infants from unsupplemented women controls Odds ratio = 0.40, $p < 0.05$	60% reduction in risk
Dietary folate and multivitamin supplement at least 1 mo before conception through the 1st trimester	77 NTD cases and 154 control mothers in study. The highest folate quartile was compared with the lowest. An increasing protective effect was observed from the lowest to the highest quartile. Odds ratio = 0.25, $p < 0.05$	75% reduction in risk
Multivitamin plus folate supplement containing up to 0.8 mg of folic acid plus diet at least 1 mo before conception through the 1st trimester	89 NTD cases in infants from supplemented women and 214 cases in infants from unsupplemented women 90 normal infants from supplemented women and 196 normal infants from unsupplemented women controls Odds ratio = 0.91, not statistically significant	No protective effect
Multivitamin plus folate supplement containing 0.1–1.0 mg of folic acid plus diet at least 1 mo before conception through the 1st trimester	10 NTD pregnancies among 10,713 women who took multivitamin plus folate 39 NTD pregnancies among 11,944 women who took multivitamins without folate Relative risk = 0.28, $p < 0.05$	72% reduction in risk

continued

TABLE 8-7 Continued

Study	Design	Subjects
Werler et al., 1993	Case/control in Boston, Philadelphia, and Toronto	NTD cases and controls with other major malformations Mothers of cases and controls
Shaw et al., 1995c	Case/control in California	NTD cases and normal control infants

[a] NTD = neural tube defect.

TABLE 8-8 Controlled Trials Relating Folate Supplementation and Risk of Neural Tube Defect in the Periconceptual Period

Study	Design	Subjects
Randomized controlled trials—previous NTD pregnancy		
Laurence et al., 1981	Randomized controlled trial in Wales	Pregnant women with prior NTD[a]-affected pregnancy; supplemented mothers took 4 mg of folic acid daily Unsupplemented mothers took a placebo
Wald et al., 1991	Randomized controlled multicenter trial in United Kingdom and Hungary	Pregnant women with prior NTD-affected pregnancy Supplemented mothers took 4 mg of folic acid daily Unsupplemented mothers took a placebo

Exposure	Results	Comments
Daily use of multivitamins, mostly 0.4 mg of folic acid, from 28 d before through 28 d after last menstrual period	34 supplemented and 250 unsupplemented NTD case women 339 supplemented and 1,253 unsupplemented women controls Adjusted odds ratio = 0.6 (95% CI = 0.4–0.8)	40% reduction in risk
Any use of folate-containing vitamins in the 3 mo before conception	88 supplemented and 207 unsupplemented NTD case women 98 supplemented and 149 unsupplemented women controls Odds ratio = 0.65 (95% confidence interval = 0.45–0.94)	35% reduction in risk

Exposure	Results	Comments
4 mg of folic acid or placebo daily at least 1 mo before conception through the 1st trimester	2 NTD pregnancies in 60 supplemented women 4 NTD pregnancies in 51 placebo-treated women Relative risk = 0.40, not statistically significant	60% reduction in risk
4 mg of folic acid or placebo daily at least 1 mo before conception through the 1st trimester	6 NTD pregnancies in 593 supplemented women 21 NTD pregnancies in 602 unsupplemented women Relative risk = 0.28, $p < 0.05$	72% reduction in risk

continued

TABLE 8-8 Continued

Study	Design	Subjects
Kirke et al., 1992	Randomized controlled multicenter trial in Ireland	Pregnant women with prior NTD-affected pregnancy Supplemented women took 0.36 mg of folic acid with or without multivitamins daily Unsupplemented women took multivitamins daily excluding folic acid

Nonrandomized controlled trials—previous NTD pregnancy

Smithells et al., 1983	Nonrandomized controlled multicenter trial in UK	Pregnant women with prior NTD-affected pregnancy Supplemented mothers took 0.36 mg of folic acid plus multivitamins daily Unsupplemented mothers took nothing
Vergel et al., 1990	Nonrandomized controlled trial in Cuba	Pregnant women with prior NTD-affected pregnancy Supplemented mothers took 5 mg of folic acid daily Unsupplemented mothers took nothing

Randomized controlled trial—all women planning pregnancy

Czeizel and Dudas, 1992	Randomized controlled trial in Hungary	Women planning a pregnancy Supplemented women took 0.8 mg of folic acid plus multivitamins daily Unsupplemented women took a trace-element supplement

[a] NTD = neural tube defect.
SOURCE: Adapted from CDC (1992).

Exposure	Results	Comments
Supplements taken for at least 2 mo before conception and until the date of the third missed menstrual period	0 NTD in 172 infants/fetuses of supplemented women 1 NTD in 89 infants/fetuses of unsupplemented women Indeterminate protective effect, not statistically significant	Trial was prematurely terminated
0.36 mg of folic acid plus multivitamins or no use from 1 mo before conception through the 1st trimester	3 NTD pregnancies in 454 supplemented women 24 NTD pregnancies in 519 unsupplemented women Relative risk = 0.14, $p < 0.05$	86% reduction in risk
5 mg of folic acid or no use from 1 mo before conception through the 1st trimester	0 NTD pregnancies in 81 supplemented women 4 NTD pregnancies in 114 untreated women Indeterminant protective effect, not statistically significant	Complete protective effect
Supplements taken for at least 1 mo before conception and until the date of the second missed period	0 NTD pregnancies in 2,104 supplemented women 6 NTD pregnancies in 2,052 unsupplemented women Relative risk = 0.0, $p = 0.029$	Complete protective effect

TABLE 8-9 Relative Risk of Neural Tube Defect Based on Reported Folate Intake During the Periconception Period

Intake Category (μg/d)	n (cases/controls)	Average Value (μg/d)	Relative Risk, 95% Confidence Interval	Adjusted Relative Risk, 95% Confidence Interval
Werler et al., 1993				
Quintiles: Did not use supplements (intake from foods)				
31–196	58/262	114	Reference (1.0)	1.52
197–252	62/260	225	1.0 (0.7–1.5)	1.52 (1.07–2.28)
253–310	46/258	282	0.7 (0.4–1.1)	1.07 (0.61–1.67)
311–391	38/237	351	0.6 (0.3–0.9)	0.91 (0.46–1.37)
392–2,195	46/236	1,294	0.6 (0.4–1.1)	0.91 (0.61–1.67)
Quintiles: Did use supplements (intake from supplements only)				
0	214/1,236		Reference (1.0)	
< 400	3/50		0.5 (0.2–1.5)	
400	8/185		0.3 (0.1–0.6)	
500–900	2/15		0.9 (0.2–4.2)	
≥ 1,000	3/52		0.4 (0.1–1.3)	
Shaw et al., 1995b				
Quartiles: Did not use supplements (intake from foods)				
10–227	140/116	119	Reference (1.0)	1.24
228–312	117/115	270	0.89 (0.62–1.3)	1.10 (0.77–1.61)
313–428	98/115	371	0.69 (0.47–1.0)	0.86 (0.58–1.24)
429–1,660	105/115	1,045	0.69 (0.47–1.0)	0.86 (0.58–1.24)
Quartiles: Did use supplements (intake from supplements only)				
0 (and 1st quartile of dietary intake)	55/30		Reference (1.0)	
228–399	75/54		0.76 (0.41–1.40)	
400–999	89/74		0.66 (0.37–1.20)	
≥ 1,000	33/46		0.39 (0.20–0.77)	

NOTE: Dietary data were obtained by using semiquantitative food frequency questionnaires. The data do not allow the computation of dietary folate equivalents in toto because intake of folic acid from fortified foods is not available. Both studies were conducted in the United States prior to mandatory folate fortification.

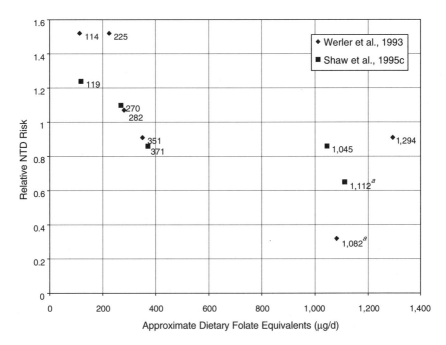

FIGURE 8-4 Risk of neural tube defect (NTD) according to folate intake based on two retrospective studies. Intake values appear next to each point, in micrograms. Midpoint values for each reference intake category have been used for defining folate intake, and relative NTD risks have been linearly adjusted to a baseline absolute risk of 1.29 per 1,000 for a folate intake of 312 µg/d. Values that include folate supplements (indicated by [a]) are estimated in dietary folate equivalents (1 dietary folate equivalent = 1 µg food folate = 0.6 µg folate from fortified food or as a supplement taken with food = 0.5 µg supplemented folate when fasting). SOURCE: Data from Shaw et al. (1995c) and Werler et al. (1993).

plementation (800 µg/day of folic acid) on the risk of a first occurrence of NTD was prematurely terminated after 4,753 women had been enrolled (Czeizel and Dudas, 1992) (Table 8-8). No case of NTD was observed in the group taking multivitamins containing 800 µg of folate daily compared with six cases in the group receiving a trace element supplement ($p = 0.029$). The effect of supplemental folate alone was not assessed. Although no protective effect was observed in one case-control study (Mills et al., 1989), a significant reduction in risk associated with supplementation was seen in one cohort (Milunsky et al., 1989) and four other case-control studies

(Bower and Stanley, 1992b; Mulinare et al., 1988; Shaw et al., 1995c; Werler et al., 1993) (Table 8-7). The optimal timing for supplementation seems to be during the 4 weeks before and after conception (Mulinare et al., 1988).

In the United States the risk reduction achieved with a daily supplement of 400 μg of folate, the most usual dose in multivitamins, was 70 percent (relative risk 0.3; 95 percent confidence interval [CI], 0.2 to 0.6) in New England (Werler et al., 1993) and 35 percent (relative risk 0.65; 95 percent CI, 0.45 to 0.94) in California (Shaw et al., 1995c) in unselected populations with average daily dietary folate intake of about 300 μg. The intake values assume a high level of compliance with supplementation, and the points represent values adjusted for bioavailability because they are given in dietary folate equivalents. It is not clear whether supplements at doses lower than 400 μg/day of folic acid provide the same level of protection as 400 μg/day or whether higher doses are associated with increased risk reduction.

The Medical Research Council Trial (Wald et al., 1991) (Table 8-8), which addressed reduction of the recurrence of NTD, used a factorial design to investigate folate in a dose of 4.0 mg/day and a mixture of other vitamins (A, thiamin, riboflavin, B_6, C, D, and nicotinamide). This study found a 71 percent decreased NTD incidence in offspring of women taking the folate supplement relative to those on no supplements but no reduction with the other vitamins. In a nonrandomized trial on the risk of NTD recurrence conducted in the United Kingdom, a daily supplement of 360 μg in addition to normal diet was apparently protective (Smithells et al., 1981).

Because NTDs represent a heterogeneous group of congenital malformations both etiologically and pathogenetically, it is probable that there are cases not preventable even by large doses of folate, as was the case in the Medical Research Council Vitamin Study (Wald et al., 1991). More studies are needed to evaluate whether fortification of foods is similarly associated with reduced risk or with a valid proxy for NTD risk.

NTD Risk According to Maternal Folate Status. The erythrocyte folate concentration is a marker of long-term folate status. Studies looking for an association of erythrocyte folate with NTD risk based on estimating erythrocyte folate levels in blood samples taken early in pregnancy are preferred because maternal folate status is likely to change during pregnancy and postpartum. In four studies with blood specimens taken during pregnancy, erythrocyte folate values were higher in women with a normal pregnancy than in women carrying a fetus with an NTD (Kirke et al., 1993; Laurence et al., 1981; Smithells et al., 1976) or with a fetus having another type of malformation (Bunduki et al., 1995). This difference was not found in one study with only eight NTD cases (Economides et al., 1992).

In a case-control study in three maternity hospitals in Dublin, Ireland, from 1986 to 1990, erythrocyte folate values were measured in frozen samples taken at a median gestational age of 15 weeks (Daly et al., 1995; Kirke et al., 1993). The percentage of women using folate supplements was 5 percent. A negative apparently non-linear association was observed between NTD risk and erythrocyte folate concentration (Table 8-10). It is not known whether the risk would continue to decrease as erythrocyte folate values increased to higher than 1,241 nmol/L (570 ng/mL), which was the mean erythrocyte concentration of the controls who had concentrations in the highest category in Table 8-10. However, the population studied had a relatively high incidence of NTD, around 2 per 1,000 births. Extrapolation of results should be made with great care because the NTD risk in the U.S. population could be lower at every level of erythrocyte folate.

Determinants of Erythrocyte Folate. In a recent study in women aged 22 to 35 years in the Minneapolis-St. Paul area, folate supplements and folate-fortified cereals were found to be independent predic-

TABLE 8-10 Distribution of Cases and Controls and Risks of Neural Tube Defect (NTD) by Erythrocyte Folate Concentration

Erythrocyte Folate nmol/L (ng/mL)[a]	N (%) of Cases	N (%) of Controls	Risk of NTD per 1,000 Births[b]	95% Confidence Interval
0–339 (0–149)	11 (13.1)	10 (3.8)	6.6	3.3–11.7
340–452 (150–199)	13 (15.5)	24 (9.0)	3.2	1.7–5.5
453–679 (200–299)	29 (34.5)	75 (28.2)	2.3	1.6–3.3
680–903 (300–399)	20 (23.8)	77 (29.0)	1.6	1.0–2.4
≥ 906 (400)	11 (13.1)	80 (30.0)	0.8	0.4–1.5
Total	84 (100.0)	266 (100.0)	1.9	1.5–2.3

[a] 1 ng/mL = 2.27 nmol/L, as reported in the original study. This conversion factor differs from that used in the rest of this report.

[b] Absolute NTD risk has been extrapolated from the odds ratio computed in a case-control study.

SOURCE: Adapted from Daly et al. (1995).

tors of erythrocyte folate levels (Brown et al., 1997). An apparently nonlinear correlation was observed between folate intake from various sources and erythrocyte folate. These results are concordant with those of a controlled experiment in women who were randomly assigned to receive 0, 100, 200, or 400 µg/day of supplemental folate (Daly et al., 1997). In this randomized placebo trial, Daly and coworkers estimated the quantity of additional folate associated with an erythrocyte folate concentration of greater than 870 nmol/L (400 ng/mL), which is the amount previously shown to be associated with a significant reduction in NTD risk. The initial erythrocyte folate concentrations of the women were in the normal range (327 to 870 nmol/L [150 to 400 ng/mL], median 707 nmol/L [325 ng/mL]). The median incremental changes in erythrocyte folate concentration in the 100-, 200-, and 400-µg/day groups were + 146 nmol/L (67 ng/mL), + 283 nmol/L (130 ng/mL), and + 435 nmol/L (200 ng/mL), respectively.

The relative effectiveness of different interventions in increasing erythrocyte folate concentrations was evaluated in a 3-month randomized trial in 62 healthy women aged 17 to 40 years in Northern Ireland (Cuskelly et al., 1996). Erythrocyte folate concentrations improved significantly only in the groups taking folate supplements or food fortified with folate; there was no increase in the group provided extra food folate or dietary advice. Because food intake was not controlled, further studies are needed to evaluate more precisely the relative efficacy of different supplementation regimens in reducing NTD risk.

Mechanism. The mechanism by which folate could reduce NTD risk is not known. Increasing folate intake and thus the concentrations of folate derivatives in tissues might overcome a metabolic deficiency in the production of proteins or in DNA synthesis at the time of neural tube closure (Mills et al., 1995). Another hypothesis is that folate does not prevent the occurrence of NTD but selectively increases the abortion rate of affected fetuses (Hook and Czeizel, 1997). Certainly, more research is needed to understand the effect of folate on embryonic and fetal development.

Recommendations for NTD Risk Reduction

To summarize the data, a reduced risk of NTD has been observed for women who took a folate supplement of 360 to 800 µg/day in addition to a dietary folate intake of 200 to 300 µg/day. Folate intake is positively associated with erythrocyte folate concentration

(Bower and Stanley, 1989; Brown et al., 1997; Cuskelly et al., 1996; Daly et al., 1997), and NTD risk is inversely associated with both folate intake (Bower and Stanley, 1989; Shaw et al., 1995c; Werler et al., 1993) and erythrocyte folate concentration (Daly et al., 1995).

Although it is recognized that there are still uncertainties about the relationships among folate intake, erythrocyte folate, and NTD risk and the extent to which there are differences in the absorption of folate from food compared with supplements, the evidence is still judged sufficient to support a recommendation to reduce the risk of NTD. The recommendation made here for women capable of becoming pregnant is for intake that exceeds the Recommended Dietary Allowance (RDA) for folate. In particular, it is recommended that women capable of becoming pregnant consume 400 µg of folate daily from supplements, fortified foods, or both in addition to consuming food folate from a varied diet. At this time the evidence for a protective effect from folate supplements is much stronger than that for food folate. It is certainly conceivable that, if taken in adequate quantity, food folate will be shown to be as effective as folic acid, but it remains to be demonstrated. When more data are available, this recommendation will be revised.

An even larger dose of folate has been recommended to prevent recurrence in women with a previous NTD-affected pregnancy (CDC, 1991). However, some NTDs are not prevented by increasing folate intake.

To date there is no conclusive evidence to support any population screening for genetic markers of NTD risk. In the event that the correlation between the 5,10-MTHFR T^{677} allele and NTD is confirmed, screening women for the gene that codes for the thermolabile variant would identify only about 15 percent of those at risk for NTD. Thus, recommending consumption of 400 µg folate daily from supplementation or fortified foods for all women capable of becoming pregnant would be a more effective prevention measure than screening for the variant (Mills and Conley, 1996).

Other Congenital Anomalies

Folate may also prevent the occurrence of other types of congenital anomalies. In one randomized trial (Czeizel, 1993) and several case-control studies (Botto et al., 1996; Czeizel et al., 1996; Hayes et al., 1996; Munger et al., 1997; Shaw et al., 1995a, b; Tolarova and Harris, 1995), a reduction in the frequency of orofacial clefts and cardiovascular malformations was observed in women taking vitamin supplements and folate-fortified food. The results, however, were not always consistent across the studies, and negative findings have also been reported (Bower and Stanley, 1992a; Hayes et al., 1996; Scanlon et al., 1998). Because multivitamins were used in all these studies, it is difficult to disentangle the effect of folate from that of other constituents. Also, the presence of unmeasured confounding cannot be excluded.

Vascular Disease

The link between homocysteine and the risk of vascular disease was derived from the study of homocystinuria. Classical homocystinuria is a rare autosomal recessive disease caused by a deficiency of cystathionine β-synthase and characterized by excessively elevated plasma homocysteine (Mudd et al., 1985). Clinical manifestations include mental retardation, skeletal abnormalities, lens dislocation, and a marked tendency to develop premature and severe atherosclerosis with thromboembolic events. In 1976 a study first showed a significant difference in homocysteine plasma concentration between patients with vascular disease and normal control subjects (Wilcken and Wilcken, 1976).

Since then, many observational and experimental studies have been published on the risk of vascular disease associated with elevated homocysteine levels. In 1995 Boushey and coworkers first published a meta-analysis; this work has recently been updated and includes a total of 20 studies (Beresford and Boushey, 1997). The relative increase in risk of coronary heart disease (CHD) as estimated by the combined odds ratio was 1.6 (95 percent CI, 1.5 to 1.7) for men and 1.5 (95 percent CI, 1.3 to 1.7) for women for each increment of 5 mmol/L in total plasma or serum homocysteine. The combined odds ratio for data from both men and women was 1.8 (95 percent CI, 1.6 to 2.0) for cerebrovascular disease and 2.0 (95 percent CI, 1.5 to 2.6) for peripheral vascular disease. There is evidence that hyperhomocysteinemia is a risk factor for CHD independent of other known risk factors such as smoking, cholesterol, body mass index, age, high blood pressure, and diabetes (Beresford and Boushey, 1997; Graham et al., 1997; Mayer et al., 1996; Verhoef et al., 1996). Similarly, Nygård and colleagues (1997) reported that plasma homocysteine values were a strong predictor of mortality in patients with angiographically confirmed coronary artery disease.

The mechanism by which elevated homocysteine might increase the risk of developing vascular disease is unclear. Several hypotheses have been proposed, and the subject was reviewed by Mayer et al. (1996). Homocysteine can exert a direct toxic effect on endothelial cells and promote the growth of smooth muscle cells, leading to atherosclerotic lesions (Tsai et al., 1994). It can also increase adhesiveness of platelets and affect several factors involved in the clotting cascade (Harpel et al., 1996). Folate is required in the form of methyltetrahydrofolate as a substrate for methionine synthase. Therefore, the remethylation of homocysteine depends on adequate quantities of folate. Homocysteine levels can be markedly elevated in folate deficiency (Kang et al., 1987; Stabler et al., 1988). Negative correlations between serum and plasma folate and homocysteine have been seen in studies of normal subjects (Bates et al., 1997; Jacobsen et al., 1994; Pancharuniti et al., 1994; Selhub et al., 1993; Ubbink et al., 1993). Beresford and Boushey (1997) reviewed 14 intervention studies, 2 metabolic studies, and 1 observational study that included folate supplementation to reduce homocysteine levels. Results demonstrated an effect of various doses of supplemental folate in reducing homocysteine levels. Apparently, the effect of a given dose of folate was greater at higher pretreatment homocysteine values (Landgren et al., 1995; Ubbink et al., 1995b). The latter observation could, however, be partially explained by a regression to the mean of higher-than-usual values in some individuals.

As seen in Figure 8-5, the inverse association between mean dietary folate and mean homocysteine concentration is not linear but seems to reach a plateau at total folate intake levels greater than 300 mg/day. A review of seven studies indicates that homocysteine concentrations are also inversely correlated with plasma folate concentrations, and there seems to be a serum folate concentration around 9 nmol/L (4 ng/mL) above which homocysteine values do not decrease significantly (Beresford and Boushey, 1997).

In two case-control studies, concentrations of plasma and serum folate were significantly lower in patients with early-onset vascular disease than in control subjects (Pancharuniti et al., 1994; Verhoef et al., 1996). Such an association was not found or was only observed in a subset of patients in six other studies (Bergmark et al., 1993; Brattstrom et al., 1984, 1990; Dalery et al., 1995; Giles et al., 1995; Molgaard et al., 1992). In a retrospective cohort study of participants in the Nutrition Canada Survey, a statistically significant association between serum folate concentration and risk of fatal CHD was found, with a rate ratio of 1.69 (95 percent CI: 1.10 to 2.61) for individuals in the lowest serum folate category compared with those in the highest category (Morrison et al., 1996). For participants in the U.S. Physicians' Health Study, the reported inverse association of plasma folate concentrations with risk of myocardial

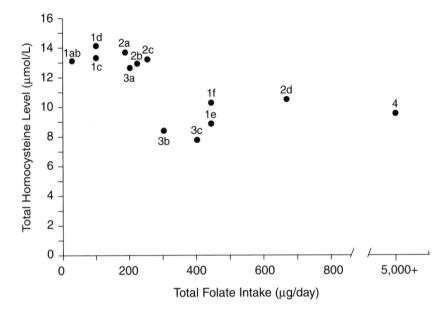

FIGURE 8-5 Mean intakes of folate and plasma levels of homocysteine (Hcy). Data points based on groups from Jacob et al. (1994) (1a–1f), mean levels of Hcy at three levels of dietary folate; Selhub et al. (1993) (2a–2d), mean levels of Hcy from lowest deciles and highest decile of folate intake; O'Keefe et al. (1995) (3a–3c), mean values of Hcy at three levels of dietary folate; Brattstrom et al. (1988) (4), mean Hcy level after added supplements of 5,000 μg of folate, pretreatment folate intake unknown. Reprinted with permission, from Beresford and Boushey (1997). Copyright 1997 by Humana Press.

infarction was not statistically significant (Chasan-Taber et al., 1996). For participants in the Multiple Risk Factor Intervention Trial, no association was observed between homocysteine concentration and the risk of heart disease (Evans et al., 1997).

A recent prospective observational study examined the effect of self-selection for intake of folate and vitamin B_6 on the incidence of myocardial infarction and CHD (Rimm et al., 1998). After controlling for other risk factors for CHD and adjusting vitamin intake on the basis of energy intake, about a twofold reduction in CHD was found for individuals in the quintile with the highest folate and vitamin B_6 intakes compared with those with the lowest intakes. When intakes of each of the vitamins were considered separately, the multivariate analyses suggested a reduction of about 30 percent in disease incidence between the highest and lowest quintiles of intake for each of the vitamins. For folate the data are compatible with the Framingham study (Selhub et al., 1993), in which the lowest deciles of folate intake were associated with higher circulating homocysteine. In the Rimm et al. (1998) study, although multivariate analysis indicated a trend in risk reduction across the quintiles of intake, the major reduction appeared to occur between the first and second quintiles of intake (median intakes 158 and 217 mg of folate). In the subgroup analysis there appeared to be no risk reduction beyond the second quintile of folate intake (217 mg) in nondrinkers, but in alcohol consumers risk reduction increased over the quintiles of intake. Although these data are consistent with the hypothesis that self-selection for increased folate reduces vascular disease risk, other variables associated with lifestyle differences of individuals who consume higher vitamin intakes may also have influenced CHD risk. Some of these variables were not or could not have been considered in the analysis. Several ongoing randomized trials are addressing whether supplements per se will decrease risk of CHD.

Individuals homozygous for the 5,10-MTHFR T^{677} allele tend to have high homocysteine concentrations as a result of reduced enzymatic activity (Frosst et al., 1995). A few investigations found the risk of vascular disease to be increased in persons homozygous for the T^{677} allele (deFranchis et al., 1996; Gallagher et al., 1996; Kluijtmans et al., 1996) but the association has not been found in most studies (Ma et al., 1996; Schmitz et al., 1996; Schwartz et al., 1997; Verhoef et al., 1997a; Wilcken et al., 1996). In a meta-analysis the combined odds ratio of CHD associated with homozygosity for T^{677} allele was 0.98 (95 percent CI, 0.83 to 1.17) (Verhoef et al., 1997b).

The inverse relationship between folate intake and homocysteine concentration is well established. However, there are conflicting data on the association among indicators of folate status or metabolism, homocysteine concentration, and risk of vascular disease. Whether increasing intake of folate could reduce the risk of vascular disease remains to be demonstrated. Folate may reduce the risk of cardiovascular disease through other mechanisms. For example, the data from the study by Verhaar and colleagues (1998) support a direct effect of folate catabolites in restoring or preserving the endothelium function and integrity by affecting cellular oxidative metabolism. More evidence concerning a causal relationship between folate status and vascular disease will be provided by data from prospective controlled intervention trials that are currently under way. At present it is premature to consider vascular disease risk reduction as an indicator for setting the Estimated Average Requirement (EAR) and Recommended Dietary Allowance (RDA) for folate.

Cancer

Experimental data indicate that changes in folate status may influence the process of neoplastic changes in certain epithelial tissue: a negative change in folate status may stimulate carcinogenesis. It is unclear if supraphysiological doses obtained from supplements afford any protection.

Dysplasia and Metaplasia

Dysplastic and metaplastic changes have been reported to reverse in response to high-dose supplemental folate. Butterworth and co-workers (1982) conducted a prospective, controlled clinical intervention trial giving supplements of 10 mg/day of folate to 47 women with dysplastic changes in the epithelium of the uterine cervix. They observed a significant attenuation of dysplasia, but the alteration in cytology may have been an attenuation of dysplasia or simply a reduction in megaloblastic cellular changes. A subsequent intervention trial by the same research group (Butterworth et al., 1992b) was unable to reproduce the results. However, the subjects in this second intervention study initially had the lowest grade of dysplasia, which has a greater than 60 percent spontaneous rate of reversion to normal. Heimburger and colleagues (1988) observed a significant reduction in metaplastic change in bronchial epithelial tissue in response to 10 mg of folate with 500 µg of vitamin B_{12} given daily for 4 months to 36 subjects compared with changes in 37 subjects given a placebo. These findings may be questioned because of spontaneous variation in bronchial cytology, small sample size, short duration of trial, and the very high doses of folate and vitamin B_{12} used.

It has been hypothesized that poor folate status by itself is not carcinogenic but may enhance an underlying predisposition to cancer (Heimburger et al., 1987; Mason and Levesque, 1996). Support for this theory includes data from a case-control intervention trial of patients with cervical dysplasia who also were at significantly higher risk for cervical cancer because of cervical infection with human papilloma virus-16 (HPV-16) (Butterworth et al., 1992a). Subjects with the HPV-16 infection had a fivefold greater risk of having dysplasia if they also had diminished erythrocyte folate values (660 nmol/L [303 ng/mL]) (Butterworth et al., 1992a). On the basis of these data and other data from study of the colorectum (Lashner, 1993), Mason and Levesque (1996) suggested that even a minor decrease in folate status may promote carcinogenesis.

Potential mechanisms for folate-related enhancement of carcinogenesis include the induction of DNA hypomethylation (Kim et al., 1997), increased chromosomal fragility or diminished DNA repair (Kim et al., 1997), secondary choline deficiency, diminution in natural killer cell surveillance, misincorporation of uridylate for thymidylate in DNA synthesis, and facilitation of tumorigenic virus metabolism (Mason and Levesque, 1996).

Cervical Neoplasia

Although several studies suggest that increased consumption of folate reduces the relative risk of cervical neoplasia (Brock et al., 1988; Potischman et al., 1991; Verreault et al., 1989; Ziegler et al., 1990, 1991), statistical significance was not attained in these studies after adjustments were made for confounding variables. These studies had several limitations: folate intake was assessed with a food frequency instrument that had not been validated for folate intake (Mason and Levesque, 1996); because subjects were not stratified for HPV infections as was done by Butterworth and colleagues (1992a), the inverse association between folate intake and cervical neoplasia in high-risk subjects was not examined; and the subjects had either carcinoma in situ or invasive cancer—advanced stages of neoplasia that may be unresponsive to folate (Heimburger et al., 1987; Mason and Levesque, 1996). Therefore, the effect of folate status on carcinogenesis in the cervix remains uncertain.

Colorectal Cancer

Data supporting the modulation of carcinogenesis by folate status are the strongest for the colorectum. Patients with chronic ulcerative colitis are at increased risk for colonic cancer and also coexisting folate deficiency. Sulfasalazine, a drug taken chronically by these patients, inhibits folate absorption (Halsted et al., 1981) and metabolism (Selhub et al., 1978). Lashner and coworkers (1989) observed that the rate of colonic neoplasia was 62 percent lower in folate-supplemented patients with chronic ulcerative colitis than in unsupplemented patients and that sulfasalazine therapy was associated with an increase in the risk of dysplasia. These observations were not statistically significant but pointed to an important area of investigation. Lashner (1993) subsequently compared prospectively the erythrocyte folate concentrations in patients with neoplastic changes in the colorectum with those for disease-matched control patients without neoplasia. The mean erythrocyte concentration was significantly lower in the individuals with neoplasia (988 nmol/L [454 ng/mL]) than in the control patients (1,132 nmol/L [520 ng/mL]) but was still well within the normal range, which is in line with observations of erythrocyte folate concentrations and dysplasia in the uterine cervix (Butterworth et al., 1992a). Meenan and colleagues (1996) described the lack of association between erythrocyte folate levels and colonic biopsy specimens in healthy individuals, indicating the potential difficulty in predicting localized folate deficiency. In a subsequent report (Meenan et al., 1997), epithelial cell folate depletion occurred in neoplastic but not adjacent normal colonic mucosa.

In general, epidemiological studies support an inverse relationship between folate status and the rate of colorectal neoplasia (Mason and Levesque, 1996). Two large, well-controlled prospective studies support the inverse association between folate intake and incidence of colorectal adenomatous polyps (Giovannucci et al., 1993) and colorectal cancers (Giovannucci et al., 1995). In these two studies, moderate-to-high alcohol intake greatly increased the neoplastic risk of a low-folate diet. There was a significant 35 percent lower risk of adenoma in those in the highest quintile of folate intake (approximately 800 µg/day) relative to those in the lowest quintile (approximately 200 µg/day, relative risk approximately 0.65). The adverse effect of high alcohol intake coupled with a low-folate diet was confirmed by Glynn and colleagues (1996), who observed a significant fourfold increase in risk of colorectal cancer. Physicians' Health Study participants with the MTHFR polymorphism had reduced risk of colon cancer, but low folate intake or high alcohol consumption appeared to negate some of the protective effect (Ma et al., 1997) (see Appendix L for further discussion of MTHFR polymorphism).

More evidence for or against a causal relationship between folate status and colorectal cancer will be provided by data from prospective controlled intervention trials that are currently under way.

Lung, Esophageal, and Stomach Cancer

As reviewed by Mason and Levesque (1996), data are not sufficient for making conclusions regarding the possible role of folate in reducing the risk of cancer of the lung, esophagus, or stomach.

Psychiatric and Mental Disorders

The suggestion that folate deficiency might produce psychiatric disturbances was made more than 30 years ago (Herbert, 1962a). Since then the issue has been examined by three approaches: assessment of the incidence of psychiatric disturbances in patients presenting with a medical condition related to folate deficiency (e.g., megaloblastic anemia), assessment of the incidence of folate deficiency in patients presenting with a psychiatric condition (of any etiology), and evaluation of the efficacy of folate treatment in the resolution of psychiatric disorders. In general, the database linking folate to altered mental function is not large but appears sufficient to suggest the likelihood of a causative association. However, it is still unclear whether reduced folate intake is the cause or an effect of the mental disorders.

The most unambiguous observation suggesting this link is derived from studying patients with megaloblastic anemia. Shorvon and coworkers (1980) reported that among such patients having a clear folate deficiency (plasma folate 3.4 ± 1.4 [standard deviation] nmol/L [1.5 ± 0.6 ng/mL]) in the absence of vitamin B_{12} deficiency, the prevalence of an affective (mood) disturbance was 56 percent. Other studies of nonpsychiatric patients are consistent with this observation, showing changes in mood and in mental function (Goodwin et al., 1983; Herbert, 1962a; Reynolds et al., 1973).

Most studies that attempt to link folate deficiency and mental disorder are in psychiatric patients. The studies involved measurements of serum, plasma, or erythrocyte folate concentrations in patients on long-term drug therapy, some of whom were drug free when examined. No patients with a psychiatric diagnosis appear to have been assessed at first admission before drug therapy was instituted. Coppen and Abou-Saleh (1982), for example, measured serum folate concentrations in unipolar and bipolar depressed patients: mean plasma folate concentrations were significantly lower than those in a group of control subjects (13 vs. 15 nmol/L [6 vs. 7 ng/mL]). They further observed that in the psychiatric subjects, morbidity was significantly higher in individuals with plasma folate concentrations below 9 nmol/L (4 ng/mL) than in those with values at or above 18 nmol/L (8 ng/mL). In subjects with depression, the prevalence of folate deficiency (plasma folate less than 5.7 nmol/L [2.5 ng/mL]) was found to be 15 to 17 percent, a value substantially higher than the 2 percent found in control subjects (Abou-Saleh and Coppen, 1989); erythrocyte folate was also measured and found to correlate highly with plasma folate concentrations.

In a study of erythrocyte folate concentrations, Carney and colleagues (1990) observed that among patients admitted to a psychiatric unit with endogenous depression, 20 percent had erythrocyte folate concentrations below 327 nmol/L (150 ng/mL), a prevalence markedly higher than that observed in euthymic, manic, schizophrenic, or alcoholic patients. A recent study involving plasma folate determinations suggests that the prevalence of folate deficiency may not be this high (Fava et al., 1997). Nevertheless, patients with low plasma folate levels responded less well to standard antidepressant (fluoxetine) therapy than did those with normal folate values. In these studies, there appears to be no uniform definition of folate deficiency (as indexed via the plasma or erythrocyte folate determination); moreover, folate assays (and absolute folate values) differed among laboratories (and within studies, e.g., Coppen et al. [1986]), making any blood deficiency threshold difficult to standardize (Young and Ghadirian, 1989).

Two double-blind studies (Coppen et al., 1986; Godfrey et al., 1990) evaluated the efficacy of folate supplementation in the recovery from psychiatric illness, but the use of nutrients for treatment is not relevant to this report and will not be discussed here.

Although the connection between folate and mental function has been most strongly made for depression and affective state, intake of the vitamin has also been linked (though less convincingly at present) to other psychiatric conditions and to deficits in learning and memory, particularly in the elderly (Joyal et al., 1993; Riggs et al., 1996; Wahlin et al., 1996).

The mechanism by which folate modifies brain functions has been sought for more than two decades and is generally hypothesized to be related to its role in single-carbon metabolism (Alpert and Fava, 1997). In particular, methylene tetrahydrofolate is the methyl donor in methionine synthesis from homocysteine and is postulated to be important in maintaining adequate methionine pools for S-adenosylmethionine (SAM) biosynthesis (Bottiglieri et al., 1994). SAM is the cofactor in key methylation reactions in catecholamine synthesis and metabolism in brain (Turner, 1977); catecholamines are transmitters known to be important in maintaining affective state, and exogenous SAM has been shown by some to elevate mood (Bell et al., 1988). Folate has also been linked to the maintenance of adequate brain levels of tetrahydropterin (Hamon et al., 1986), a key cofactor in the hydroxylation reactions leading to the synthesis of transmitters such as serotonin and the catecholamines (Turner, 1977). Methylation reactions involving folate may be important in maintaining neuronal and glial membrane lipids (Hirata and Axelrod,

1980), which could have effects on more general brain functions as reflected in changes in mood, irritability, and sleep.

Although available information may suggest that a link exists between folate deficiency and abnormal mental function, more than three decades of research have not produced a definitive connection. There is a clear need to evaluate folate supplementation more completely at multiple doses, under double-blind conditions, and in individuals with mental disease as well those having nonpsychiatric illnesses in order to make this connection more convincing (Joyal et al., 1993). Furthermore, coexistent conditions in subjects with low folate status rather than the folate deficiency may account for observed deficits in mental function and affective state. These conditions include chronic disease, drug history, alcohol use, age, education, and family history and must be more carefully considered in future studies (Young and Ghadirian, 1989).

Summary of Evidence Concerning the Risk of Developmental Disorders and Chronic Degenerative Disease

Reducing the Risk of NTD

Uncertainties still exist about the relationships among folate intake, erythrocyte folate, and NTD risk and about the extent to which the effect of food folate should be distinguished from the effect of folate from supplements or fortified foods, but the evidence is judged sufficient to support a specific recommendation to reduce the risk of NTD.

Reducing the Risk of Cardiovascular Disease, Cancer, and Psychiatric and Mental Disorders

The evidence that folate may reduce the risk of cardiovascular disease, certain types of cancer, and psychiatric and mental disorders is provocative and promising. However, it is not yet sufficiently substantiated and is somewhat conflicting. It is premature to consider reduction of any of these risks as a basis for setting an EAR or Adequate Intake (AI).

INTAKE OF FOLATE

Food Sources

Data obtained from the Continuing Survey of Food Intakes by

Individuals (CSFII) indicates that the greatest contribution to folate intake of the U.S. adult population in 1992–1994 came from fortified ready-to-eat cereals and a category called "other vegetables" (see footnote *d* in Table 8-11 for the list of vegetables in this category). Although many of the vegetables in the "other vegetables" category have lower folate content than dark green vegetables such as spinach, some of them (e.g., green beans and vegetable soup) are so commonly eaten that their contribution to total folate intake is relatively high.

As of January 1, 1998, all enriched cereal grains (e.g., enriched bread, pasta, flour, breakfast cereal, and rice) are required to be fortified with folate at 1.4 mg/kg of grain (DHHS, 1996). During the period when data were collected in CSFII, with few exceptions the only grain products that were fortified with folate were ready-to-eat cereals (most kinds) and cooked cereals. Because enriched cereal grains are widely consumed in the United States, they are now an even more important contributor of folate than is indicated in Table 8-11. In Canada the fortification of flour and cornmeal with folate is proposed at a level of 1.5 mg/kg and fortification of

TABLE 8-11   Food Groups Providing Folate in the Diets of U.S. Men and Women Aged 19 Years and Older, CSFII, 1995[a]

	Contribution to Total Folate Intake[b] (%)		Foods Within the Group that Provide at Least 80 µg of Folate[c] per Serving	
Food Group	Men	Women	80–160 µg	> 160 µg
Food groups providing at least 5% each of total folate intake				
Ready-to-eat cereals	16.1	18.6	Moderately fortified	Highly fortified
Other vegetables[d]	11.5	12.4	Green beans, green peas, lettuce, cabbage, and vegetable soup	—
Bread and bread products	8.1	7.6	—	—
Citrus fruits and juices	6.3	7.5	Orange juice	—

continued

TABLE 8-11 Continued

Food Group	Contribution to Total Folate Intake[b] (%)		Foods Within the Group that Provide at Least 80 µg of Folate[c] per Serving	
	Men	Women	80–160 µg	> 160 µg
Mixed foods[e]	6.0	4.3	NA[f]	NA
Legumes	5.6	4.9	Chickpeas; pink, pinto, red kidney, mung, and fava beans; and black-eye peas	Cowpeas and lentils
Mixed foods, main ingredient is grain	5.6	4.4	NA	NA
Folate from other foods				
Pasta, rice, and cooked cereals	3.2	3.2	—	Fortified oatmeal
Dark green vegetables	2.7	4.3	Spinach and turnip greens	—
Organ meats	0.5	0.6	Kidney	Liver

NOTE: The fortification in the United States of all enriched cereal grains with folate that began in 1998 and the recommended use of mg of dietary folate equivalents in place of mg of folate, which takes into account bioavailability of the various sources, would cause major changes in the relative contributions of each food group to total folate intake following mandatory fortification of enriched cereals and grains.

[a] CSFII = Continuing Survey of Food Intakes by Individuals.

[b] Contribution to total intake reflects both the concentration of the nutrient in the food and the amount of the food consumed. It refers to the percentage contribution to the American diet for both men and women, based on 1995 CSFII data.

[c] 80 µg represents 20% of the Reference Daily Intake (400 µg) of folate—a value set by the Food and Drug Administration. Values do not represent dietary folate equivalents; expressed as dietary folate equivalents, values for ready-to-eat cereals or other food fortified with folate would be higher.

[d] Includes artichoke, asparagus, green beans, fresh lima beans, beets, Brussels sprouts, cabbage, cauliflower, corn, cucumber, eggplant, kohlrabi, lettuce, mushrooms, onions, okra, green peas, peppers, rutabaga, snowpeas, squash, turnips, vegetable salads, vegetable combinations, and vegetable soups.

[e] Includes sandwiches and other foods with meat, poultry, or fish as main ingredient.

[f] NA = not applicable. Mixed foods were not considered for this table.

SOURCE: Unpublished data from the Food Surveys Research Group, Agricultural Research Service, U.S. Department of Agriculture, 1997.

alimentary paste is proposed at a level of at least 2.0 mg/kg (Health Canada, 1997). It is estimated that folate fortification will increase the folate intake of most U.S. women by 80 µg/day (136 µg of dietary folate equivalents [DFEs]) or more. This amount would be provided by one cup of pasta plus one slice of bread. Depending on what cereal grains are chosen and how much is consumed, five servings daily might add 220 µg/day or more of folate from fortified foods (nearly 400 µg of DFEs) to the diet (see Chapter 13).

Dietary Intake

According to the U.S. Department of Agriculture's CSFII (Appendix G), the mean dietary folate intake by young women in the United States in 1994 through 1995 was approximately 200 µg/day. Intake data from the Third National Health and Nutrition Examination Survey (NHANES III) (Appendix H) gathered from 1988 to 1994 indicate a mean dietary intake of approximately 220 µg for young women and a total intake (including supplements) that was only slightly higher (250 µg). These values substantially underestimate actual current intake, partly because of the problems with analysis of the folate content of food (DeSouza and Eitenmiller, 1990; Pfeiffer et al., 1997b; Tamura et al., 1997), partly because of underreporting of intake (LSRO/FASEB, 1995), and partly because of the change in fortification discussed above. Thus, it is not possible to use these data to accurately assess the adequacy of current folate intake by Americans.

Survey data from the early 1990s from two Canadian provinces found similar or lower mean dietary intakes of folate for young women (approximately 200 µg/day in Québec and 160 µg/day in Nova Scotia) (Appendix I).

The Boston Nutritional Status Survey (Appendix F) conducted from 1981 to 1984 estimated that this relatively advantaged group of people over age 60 who were not taking supplements had median folate intakes of 254 µg/day for men and 208 µg/day for women.

Intake from Supplements

Results of a nationwide telephone survey conducted during January and February 1997 indicated that 43 percent of women of childbearing age reported taking some form of vitamin supplement containing folate. Thirty-two percent reported taking a folate supplement daily and 12 percent reported taking a supplement less frequently (CDC, 1998).

Information from the Boston Nutritional Status Survey on folate supplement use by a free-living elderly population from 1981 to 1984 is given in Appendix F. Both the fiftieth percentile and seventy-fifth percentiles of folate intake from supplements were 400 µg for supplement users. Largely because of supplement use, the median folate intake by pregnant women in NHANES III in 1988 to 1994 was nearly 1,000 µg/day (Appendix H). Supplements containing 1,000 µg or more of folate are available only by prescription in the United States and Canada. Smaller doses, usually 400 (g, are available over the counter.

TOLERABLE UPPER INTAKE LEVELS

Hazard Identification

This section reviews the potential hazards associated with high intake of folate as one of the primary steps in developing a Tolerable Upper Intake Level (UL). In reviewing potential hazards, careful consideration was given to the metabolic interrelationships between folate and vitamin B_{12}, which include shared participation of the two vitamins in an enzymatic reaction; identical hematological complications resulting from deficiency of either nutrient; amelioration by folate administration of the hematological complications caused by either folate or vitamin B_{12} deficiency; and in vitamin B_{12} deficiency, the occurrence of neurological complications that do not respond to folate administration.

Adverse Effects

No adverse effects have been associated with the consumption of the amounts of folate normally found in fortified foods (Butterworth and Tamura, 1989). Therefore, this review is limited to evidence concerning intake of supplemented folate. The experimental data in animal studies and in vitro tissue and cell culture studies were considered briefly to determine whether they supported the limited human data.

Neurological Effects. The risk of neurological effects described in this section applies to individuals with vitamin B_{12} deficiency. Vitamin B_{12} deficiency is often undiagnosed but may affect a substantial percentage of the population, especially older adults (see Chapter 9). Three types of evidence suggest that excess supplemental folate intake may precipitate or exacerbate the neurological damage of vitamin B_{12} deficiency. First, numerous human case reports show onset or progression of neurological complications in vitamin B_{12}-deficient individuals receiving supplemental folate (Table 8-12). Second, studies in monkeys (Agamanolis et al., 1976) and fruit bats (van der Westhuyzen and Metz, 1983; van der Westhuyzen et al., 1982) show that vitamin B_{12}-deficient animals receiving supplemental folate develop signs of neuropathology earlier than do controls. The monkey studies used dietary methods to induce vitamin B_{12} deficiency whereas the fruit bat studies used a well-described method involving nitrous oxide (Metz and van der Westhuyzen, 1987). Third, a metabolic interaction between folate and vitamin B_{12} is well documented (Chanarin et al., 1989). Although the association between folate treatment and neurological damage observed in human case reports does not provide proof of causality, the hazard associated with excess supplemental folate cannot be ruled out. The hazard remains plausible given the findings from animal studies and the demonstrated biochemical interaction of the two nutrients. The resulting neurological damage may be serious, irreversible, and crippling.

For many years, it has been recognized that excessive intake of folate supplements may obscure or mask and potentially delay the diagnosis of vitamin B_{12} deficiency. Delayed diagnosis can result in an increased risk of progressive, unrecognized neurological damage.

Evidence from animal as well as in vitro tissue and cell culture studies (Baxter et al., 1973; Hommes and Obbens, 1972; Kehl et al.,

TABLE 8-12 Dose and Duration of Oral Folate Administration and the Occurrence of Neurological Manifestations in Patients with Pernicious Anemia

Study	Number of Subjects	Dose (mg/d)	Duration	Occurrence of Neurological Manifestations[a]
Crosby, 1960	1	0.35	2 y	1 of 1
Ellison, 1960	1	0.33–1	3 mo	1 of 1
Allen et al., 1990	3	0.4–1	3–18 mo	3 of 3
Baldwin and Dalessio, 1961	1	0.5	16 mo	1 of 1
Ross et al., 1948	4	1.25	9–23 mo	1 of 4
Chodos and Ross, 1951	4	1.25[b]	3.5–26 mo	3 of 4
Victor and Lear, 1956	2	1.5–2.55	10–39 mo	2 of 2
Conley and Krevans, 1951	1	4.5	3 y	1 of 1

continued

TABLE 8-12 Continued

Study	Number of Subjects	Dose (mg/d)	Duration	Occurrence of Neurological Manifestations[a]
Schwartz et al., 1950	48	5	48 mo	32 of 48
Ross et al., 1948	2	5	20–23 mo	1 of 2
Conley and Krevans, 1951	2	5–8	2–2.5 y	2 of 2
Will et al., 1959	36	5–10	1–10 y	16 of 36
Bethell and Sturgis, 1948	15	5–20	12 mo	4 of 15
Chodos and Ross, 1951	11	5–30	3–25 mo	7 of 11
Israels and Wilkinson, 1949	20	5–40	35 mo	16 of 20
Wagley, 1948	10	5–600	12 mo	8 of 10
Ellison, 1960	1	5.4–6.4	2 y	1 of 1
Victor and Lear, 1956	1	6.68	2.5 y	1 of 1
Berk et al., 1948	12	10	> 17 mo	3 of 12
Best, 1959	1	10	26 mo	1 of 1
Spies and Stone, 1947	1	10	22 d	1 of 1
Ross et al., 1948	6	10–15	≤ 12 mo	4 of 6
Hall and Watkins, 1947	14	10–15	2–5 mo	3 of 14
Heinle et al., 1947	16	10–40	≤ 12 mo	2 of 16
Jacobson et al., 1948	1	10–65	5 mo	1 of 1
Heinle and Welch, 1947	1	10–100	4 mo	1 of 1
Spies et al., 1948	38	≥10	24 mo	28 of 38
Ross et al., 1948	7	15	28–43 mo[c]	3 of 7
Chodos and Ross, 1951	1	15	10.5 mo[c]	1 of 1
Fowler and Hendricks, 1949	2	15–20	4–5 mo	2 of 2
Vilter et al., 1947	21	50–500	10–40 d	4 of 4

NOTE: All studies except Allen et al. (1990) were conducted before folate was added to any foods as a fortificant. In most of the case reports for which hematological status was reported, some degree of hematological improvement occurred. Studies are presented in increasing order by dose. When different doses were reported within a study, there is more than one entry for that study. Case reports that covered hematological rather than neurological effects were excluded, namely, Alperin (1966), Heinle and Welch (1947), Herbert (1963), Reisner and Weiner (1952), Ritz et al. (1951), Sheehy et al. (1961), and Thirkettle et al. (1964). The exception was the study by Allen et al. (1990) in which the subjects were vitamin B_{12} deficient but did not have pernicious anemia.

[a] Refers to neurological relapses or progression of preexisting neurological manifestations while on folate therapy.

[b] In two patients, the neurological progression was characterized as minimal or slight. Neurological progression was also observed when the dose was increased to 15 mg/d in these patients.

[c] The initial dosage of 1.25 mg/d was increased to 15 mg/d after variable durations of treatment. Neurological progression occurred only at 15 mg/d in these patients.

1984; Loots et al., 1982; Olney et al., 1981; Spector, 1972; Weller et al., 1994) suggests that folate in the form of folic acid is neurotoxic and epileptogenic in animals; however, there is no clear evidence of folate-induced neurotoxicity in humans. Concerns have been raised about the possibility of decreased effectiveness of treatment if individuals treated with anticonvulsant drugs take high doses of folate. However, the UL does not apply to drug-drug interactions or to high doses taken under medical supervision (see "Anticonvulsants" and "Methotrexate").

General Toxicity. In one nonblinded uncontrolled trial, oral doses of 15 mg/day of folate for 1 month were associated with mental changes, sleep disturbances, and gastrointestinal effects (Hunter et al., 1970). However, studies using comparable or higher doses, longer durations, or both failed to confirm these findings (Gibberd et al., 1970; Hellstrom, 1971; Richens, 1971; Sheehy, 1973; Suarez et al., 1947).

Reproductive and Developmental Effects. Many studies have evaluated the periconceptional use of supplemental folate (in doses of approximately 0.4 to 5.0 mg) to prevent neural tube defects (Table 8-13). No adverse effects have been demonstrated, but the studies were not specifically designed to assess adverse effects. No reports were found of adverse effects attributable to folate in long-term folate supplement users or in infants born each year to mothers who take supplements, but this has not been investigated systematically. Because it is possible that subtle effects might have been missed, investigations designed to detect adverse effects are needed.

Carcinogenicity. In a large epidemiological study, positive associations were found between supplemental folate intake and the incidence of cancer of the oropharynx and hypopharynx and of total cancer (Selby et al., 1989). However, the authors of this study suggest that these associations might have been related to unmeasured confounding variables such as alcohol and smoking. Additionally, other studies suggest that folate might be anticarcinogenic (see "Cancer") (Campbell, 1996).

Hypersensitivity. Individual cases of hypersensitivity reactions to oral and parenteral folate administration were reported (Gotz and Lauper, 1980; Mathur, 1966; Mitchell et al., 1949; Sesin and Kirschenbaum, 1979; Sparling and Abela, 1985). Such hypersensitivity is rare, but reactions have occurred at supplemental folate doses as low as 1 mg/day (Sesin and Kirschenbaum, 1979).

Intestinal Zinc Absorption. Although there has been some controversy regarding whether supplemental folate intake adversely affects intestinal zinc absorption (Butterworth and Tamura, 1989), a comprehensive review of the literature reveals that folate supplementation has either no effect on zinc nutriture or an extremely subtle one (Arnaud et al., 1992; Butterworth et al., 1988; Hambidge et al., 1993; Keating et al., 1987; Milne et al., 1984; Tamura, 1995; Tamura et al., 1992). In a study of prenatal folate supplementation, Mukherjee et al. (1984) noted a significant association between the occurrence of fetomaternal complications and the combination of low maternal plasma zinc and high maternal plasma folate concentrations. However, this study may have failed to control for potential confounding factors. Furthermore, these findings are not supported by Tamura and colleagues (1992), who found high serum folate concentrations to be associated with favorable pregnancy outcomes including higher birth weight and Apgar scores of newborns, reduced prevalence of fetal growth retardation, and lower incidence of maternal infection close to the time of delivery.

Summary

The weight of the limited but suggestive evidence that excessive folate intake may precipitate or exacerbate neuropathy in vitamin B_{12}-deficient individuals justifies the selection of this endpoint as the critical endpoint for the development of a UL for folate.

Dose-Response Assessment

Adults

Data Selection. To evaluate a dose-response relationship and derive a UL for folate, case reports were used that involved oral administration of folate in patients with vitamin B_{12} deficiency who showed development or progression of neurological complications. Because a number of apparently healthy individuals are vitamin B_{12} deficient (see Chapter 9), these individuals are considered part of the general population in setting a UL.

Identification of a No-Observed-Adverse-Effect Level (NOAEL) and a Lowest-Observed-Adverse-Effect Level (LOAEL). The literature was re-

TABLE 8-13 Assessing Adverse Reproductive Effects from Studies Involving Supplemental Folate

Reference	Subjects	Duration of Study	Study Design
Laurence et al., 1981	95 women	≥ 9 wk	Clinical trial: randomized, controlled, double-blinded
Smithells et al., 1981	550 women	110 d (mean duration)	Clinical trial: controlled
Mukherjee et al., 1984	450 pregnant women	≥ 9 mo	Prospective cohort study
Vergel et al., 1990	81 women	≥ 3 mo	Clinical trial: controlled
Wald et al., 1991	910 women	A few months[d]	Clinical trial: randomized, double-blinded, controlled
Czeizel and Dudas, 1992	4,753 women (< 35 y)	3 mo	Clinical trial: randomized, controlled
Holmes-Siedle et al., 1992	100 women	Periconceptional period; 7–10 y follow-up	Observational study
Kirke et al., 1992	354 pregnant women	5 mo	Clinical trial: randomized, controlled
Czeizel et al., 1994	5,502 women	3 mo	Randomized, controlled trial

[a] NR = not reported. Study was not designed to assess adverse effects.

[b] Plasma folate was measured at different times in pregnancy, but compliance with prenatal vitamin use was not recorded.

[c] There was no control of confounding variables making it difficult to interpret the results.

[d] The average duration of exposure is not indicated in the publication but was likely a few months.

viewed to find cases in which vitamin B_{12}-deficient patients who were receiving oral doses of folate experienced progression of neurological disorders. Data were not available on which to set a NOAEL. A LOAEL of 5 mg of folate is based on the data presented in Table 8-12:

• at doses of folate of 5 mg/day and above, there were more than 100 reported cases of neurological progression;
• at doses of less than 5 mg/day of folate (0.33 to 2.5 mg/day), there are only eight well-documented cases;

Folate Dose (mg/d)	Adverse Effects Observed	Methods for Assessing Associations and Adverse Effects
4	None	NR[a]
1	None	NR
0.4–1[b]	Pregnancy complications, fetal distress[c]	Statistical association between 12 indices of nutrient status and 7 poorly defined categories of complications
5	None	NR
4	None	Medical exams performed[b]
0.8	None	NR
1	Frequency of developmental anomalies not greater than expected[e]	NR
0.36	None	NR
0.8	13.4% fetal death rate in supplemented group compared with 11.5% fetal death rate on controls[f]	Documentation for all pregnancy outcomes was collected. Statistical evaluation based on two-tailed chi-square test.

[e] The frequency of developmental anomalies was not greater than expected but parental reports of worries, fearfulness, and fussiness in the children were greater than expected.

[f] This may be a chance finding resulting from multiple comparisons. It has been reported that prenatal multivitamin supplementation (which includes folic acid) can reduce preterm deliveries, causing an apparent increase in recognized abortions as the duration of all pregnancies increases (Scholl et al., 1997).

• in most cases throughout the dose range, folate supplementation maintained the patients in hematological remission over a considerable time span; and

• the background intake of folate from food was not specified, but all except for three cases (those reported by Allen and coworkers [1990]) occurred before the fortification of breakfast cereal with added folate.

Uncertainty Assessment. An uncertainty factor (UF) of 5 was selected. Compared with the UFs used to date for other nutrients for which there was also a lack of controlled, dose-response data, a UF of 5 is large. The selection of a relatively large UF is based primarily on the severity of the neurological complications observed but also on the use of a LOAEL rather than a NOAEL to derive the UL. The UF is not larger than 5 on the basis of the uncontrolled observation that millions of people have been exposed to self-treatment with about one-tenth of the LOAEL (i.e., 400 µg in vitamin pills) without reported harm.

Derivation of a UL. The LOAEL of 5 mg/day of folate was divided by a UF of 5 to obtain the UL for adults of 1 mg/day or 1,000 µg/day of folate from supplements for fortified food. A UL of 1,000 µg/day is set for all adults rather than just for the elderly because of (1) the devastating and irreversible nature of the neurological consequences, (2) data suggesting that pernicious anemia may develop at a younger age in some racial or ethnic groups (Carmel and Johnson, 1978), and (3) uncertainty about the occurrence of vitamin B_{12} deficiency in younger age groups. In general, the prevalence of vitamin B_{12} deficiency in females in the childbearing years is very low and the consumption of supplemental folate at or above the UL in this subgroup is unlikely to produce adverse effects.

Folate UL Summary, Adults

UL for Adults
19 years and older   1,000 µg/day of folate from fortified
                   food or supplements

Other Life Stage Groups

There are no data on other life stage groups that can be used to identify a NOAEL or LOAEL and derive a UL. For infants the UL was judged not determinable because of a lack of data on adverse effects in this age group and concern about the infant's ability to handle excess amounts. To prevent high levels of intake, the only source of intake for infants should be from food. No data were found to suggest that other life stage groups have increased susceptibility to adverse effects of high supplemental folate intake. Therefore, the UL of 1,000 µg/day is also set for adult pregnant and lactating women. The UL of 1,000 µg/day for adults was adjusted for children and adolescents on the basis of relative body weight as described in Chapter 3. Values have been rounded down.

ULs for Infants
  0–12 months   Not possible to establish for supplemental folate

ULs for Children
  1–3 years   300 µg/day of folate from fortified foods or supplements
  4–8 years   400 µg/day of folate from fortified foods or supplements
  9–13 years   600 µg/day of folate from fortified foods or supplements
  14–18 years   800 µg/day of folate from fortified foods or supplements

ULs for Pregnancy
  14–18 years   800 µg/day of folate from fortified foods or supplements
  19 years and older   1,000 µg/day of folate from fortified foods or supplements

ULs for Lactation
  14–18 years   800 µg/day of folate from fortified foods or supplements
  19 years and older   1,000 µg/day of folate from fortified foods or supplements

Special Considerations

Individuals who are at risk of vitamin B_{12} deficiency (e.g., those who eat no animal foods [vegans] and other individuals identified in Table 9-4) may be at increased risk of the precipitation of neurological disorders if they consume excess folate.

Intake Assessment

It is not possible to use data from the Third National Health and Nutrition Examination Survey (NHANES III) or the Continuing Survey of Food Intakes by Individuals to determine the population's exposure to folic acid. Currently, survey data do not distinguish between food folate and folic acid added as a fortificant or taken as a supplement. Based on data from NHANES III and excluding pregnant women (for whom folate supplements are often prescribed), the highest reported total folate intake from food and supplements at the ninety-fifth percentile, 983 μg/day, was found in females aged 30 through 50 years. This intake was obtained from food (which probably included fortified ready-to-eat cereals, a few of which contain as much as 400 μg of folic acid per serving) and supplements. For the same group of women, the reported intake at the ninety-fifth percentile from food alone (which also probably included fortified ready-to-eat cereal) was 438 μg/day. In Canada, the contribution of ready-to-eat cereals is expected to be lower because the maximum amount of folic acid that can be added to breakfast cereal is 60 μg/100 g (Health Canada, 1996).

It would be possible to exceed the UL of 1,000 μg/day of folic acid through the ingestion of fortified foods, supplements, or both, as indicated by the information on the folate content of foods in Table 8-14.

Risk Characterization

The intake of folate is currently higher than indicated by NHANES III because enriched cereal grains in the U.S. food supply, to which no folate was added previously, are now fortified with folate at 140 μg/100 g of cereal grain. Using data from the 1987–1988 U.S. Department of Agriculture's Nationwide Food Consump-

TABLE 8-14 Folate Content of Selected Fortified Cereal-Grain Products[a] and Commonly Used Folate-Containing Supplements

Food Product	Serving Size	Folic Acid Content per Serving (μg)
Ready-to-eat cereal		
Highly fortified	30 g	400
Moderately fortified	Varies[b]	100
Noodles, pasta, rice (prepared)	140 g (1 cup)	60
Bread	25 g (1 slice)	20
Over-the-counter supplements	1 unit	400

[a] Other products containing grains (such as prepared macaroni and cheese, crackers, cookies, donuts, and hot cereal) may also be fortified with folate. See DHHS, 1993b.

[b] Serving size ranges from 15 g for puffed cereals to 55 g for dense cereals (e.g., biscuit types); the volume of a serving would be approximately 1 cup to 1/2 cup, respectively.

SOURCE: Data adapted from DHHS (1996).

tion Survey, the U.S. Food and Drug Administration (FDA) estimated that the 95th percentile of folate intakes for males aged 11 to 18 years would be 950 µg of total folate at this level of fortification; this value assumes that these young males would also take supplements containing 400 µg of folate (DHHS, 1993a). Excluding pregnant women, for whom estimates were not provided, the 95th percentile for total folate for all other groups would be lower, and folic acid intake would be lower still. Using a different method of analysis, the FDA estimated that those who follow the guidance of the Food Guide Pyramid and consume cereal grains at the upper end of the recommended range would obtain an additional 440 µg of folate under the new U.S. fortification regulations (DHHS, 1993a). (This estimate assumes 8 servings [16 slices] of bread at 40 µg of folic acid per serving and approximately two 1-cup servings of noodles or pasta at 60 µg of folic acid per serving.) Those who eat other fortified foods (such as cookies, crackers, and donuts) instead of bread might ingest a comparable amount of folic acid. By either method of analysis and with the assumption of regular use of an over-the-counter supplement that contains folate (ordinarily 400 µg per dose), it is unlikely that intake of folate added to foods or as supplements would regularly exceed 1,000 µg for any of the life stage or gender groups.

RESEARCH RECOMMENDATIONS FOR FOLATE

High-Priority Recommendations

Priority should be given to four topics of research related to folate:

• Determination of the mechanisms and magnitude of relationships of folate intake with risk reduction for the occurrence of neural tube defects (NTDs) and vascular disease and the influence of related factors, including genetic polymorphism, on these relationships. Targeted intervention programs need a clearer understanding of the mechanisms by which adequate folate intake ensures normal embryogenesis and may reduce vascular disease risk.

• Estimation of folate requirements in high-risk groups for which data are limited and for which public health problems may result from deficiencies. These groups include children, adolescents, women of reproductive age (including pregnant women by trimester and lactating women), and the elderly. These studies should identify and use new folate status indicators that are linked to metabolic function and traditional indices of folate status.

• Development of more precise and reproducible methods of analysis for the estimation of both blood and food folate and for the estimation of folate bioavailability. Improved methods would allow for comparison of status indicators among laboratories, revision of the food folate databases, and improved estimation of how dietary requirements are influenced by the food matrix and the source of folate (food or synthetic).

• Identification and quantitation of adverse effects of high intakes. Further investigation is needed on the effect of increasing folate intake from supplements and fortified foods on the onset and progression of vitamin B_{12} deficiency.

Other Research Areas

Other areas of recommended research are as follows:

• Determination of the mechanisms by which maternal folate sufficiency reduces the occurrence of NTDs in the infant, including the establishment of which genes are responsible for the heritability and folate-responsiveness of NTD.

• Determination of the effect of folate fortification on folate intake and occurrence of NTD and vascular disease.

• Determination of whether folate status affects the risk of birth defects other than NTDs and of chronic diseases other than vascular disease (e.g., cancer).

REFERENCES

Abma J, Chandra A, Mosher W, Peterson L, Piccinino L. 1997. Fertility, Family Planning, and Women's Health: New Data from the 1995 National Survey of Family Growth. National Center for Health Statistics. *Vital Health Stat Series* 23, Number 19.

Abou-Saleh MT, Coppen A. 1989. Serum and red blood cell folate in depression. *Acta Psychiatr Scand* 80:78–82.

Adams MJ Jr, Khoury MJ, Scanlon KS, Stevenson RE, Knight GJ, Haddow JE, Sylvester GC, Cheek JE, Henry JP, Stabler SP. 1995. Elevated midtrimester serum methylmalonic acid levels as a risk factor for neural tube defects. *Teratology* 51:311–317.

Agamanolis DP, Chester EM, Victor M, Kark JA, Hines JD, Harris JW. 1976. Neuropathology of experimental vitamin B_{12} deficiency. *Neurology* 26:905–914.

Allen RH, Stabler SP, Savage DG, Lindenbaum J. 1990. Diagnosis of cobalamin deficiency. I. Usefulness of serum methylmalonic acid and total homocysteine concentrations. *Am J Hematol* 34:90–98.

Allen RH, Stabler SP, Lindenbaum J. 1993. Serum betaine, N,N-dimethylglycine and N-methylglycine levels in patients with cobalamin and folate deficiency and related inborn errors of metabolism. *Metabolism* 42:1448–1460.

Alperin JB. 1966. Response to varied doses of folic acid and vitamin B_{12} in megaloblastic anemia. *Clin Res* 14:52.

Alpert JE, Fava M. 1997. Nutrition and depression: The role of folate. *Nutr Rev* 55:145–149.

Andersson A, Hultberg B, Brattstrom L, Isaksson A. 1992. Decreased serum homocysteine in pregnancy. *Eur J Clin Chem Clin Biochem* 30:377–379.

Arnaud J, Favier A, Herrmann MA, Pilorget JJ. 1992. Effect of folic acid and folinic acid on zinc absorption. *Ann Nutr Metab* 36:157–161.

Asfour R, Wahbeh N, Waslien CI, Guindi S, Darby WJ. 1977. Folacin requirement of children. 3. Normal infants. *Am J Clin Nutr* 30:1098–1105.

Baggott JE, Morgan SL, Ha TS, Vaughn WH, Hine RJ. 1992. Inhibition of folate-dependent enzymes by non-steroidal anti-inflammatory drugs. *Biochem J* 282:197–202.

Bailey LB. 1988. Factors that affect folate bioavailability. *Food Technol* 42:206–212, 238.

Bailey LB, Cerda JJ, Bloch BS, Busby MJ, Vargas L, Chandler CJ, Halsted CH. 1984. Effect of age on poly- and monoglutamyl folacin absorption in human subjects. *J Nutr* 114:1770–1776.

Bailey LB, Barton LE, Hillier SE, Cerda JJ. 1988. Bioavailability of mono and polyglutamyl folate in human subjects. *Nutr Rep Int* 38:509–518.

Baird PA. 1983. Neural tube defects in the Sikhs. *Am J Med Genet* 16:49–56.

Baldwin CT, Hoth CF, Amos JA, da Silva EO, Milunsky A. 1992. An exonic mutation in the HuP2 paired domain gene causes Waardenburg's syndrome. *Nature* 355:637–638.

Baldwin JN, Dalessio DJ. 1961. Folic acid therapy and spinal-cord degeneration in pernicious anemia. *N Engl J Med* 264:1339–1342.

Bates CJ, Fleming M, Paul AA, Black AE, Mandal AR. 1980. Folate status and its relation to vitamin C in healthy elderly men and women. *Age Ageing* 9:241–248.

Bates CJ, Mansoor MA, van der Pols J, Prentice A, Cole TJ, Finch S. 1997. Plasma total homocysteine in a representative sample of 972 British men and women aged 65 and over. *Eur J Clin Nutr* 51:691–697.

Baxter MG, Millar AA, Webster RA. 1973. Some studies on the convulsant action of folic acid. *Br J Pharmacol* 48:350–351.

Bell KM, Plon L, Bunney WE, Potkin SG. 1988. S-Adenosylmethionine treatment of depression: A controlled clinical trial. *Am J Psychiatry* 145:1110–1114.

Beresford SA, Boushey CJ. 1997. Homocysteine, folic acid, and cardiovascular disease risk. In: Bendich A, Deckelbaum RJ, eds. *Preventive Nutrition: The Comprehensive Guide for Health Professionals*. Totowa, NJ: Humana Press.

Bergmark C, Mansoor MA, Swedenborg J, de Faire U, Svardal AM, Ueland PM. 1993. Hyperhomocysteinemia in patients operated for lower extremity ischeamia below the age of 50—effect of smoking and extent of disease. *Eur J Vasc Surg* 7:391–396.

Berk L, Bauer JL, Castle WB. 1948. A report of 12 patients treated with synthetic pteroylglutamic acid with comments on the pertinent literature. *S Afr Med J* 22:604–611.

Best CN. 1959. Subacute combined degeneration of spinal cord after extensive resection of ileum in Crohn's disease: Report of a case. *Br Med J* 2:862–864.

Bethell FH, Sturgis CC. 1948. The relations of therapy in pernicious anemia to changes in the nervous system. Early and late results in a series of cases observed for periods of not less than ten years, and early results of treatment with folic acid. *Blood* 3:57–67.

Blaw ME, Woody RC. 1983. Valproic acid embryopathy? *Neurology* 33:255.

Blount BC, Mack MM, Wehr CM, MacGregor JT, Hiatt RA, Wang G, Wickramasinghe SN, Everson RB, Ames BN. 1997. Folate deficiency causes uracil misincorporation into human DNA and chromosome breakage: Implications for cancer and neuronal damage. *Proc Natl Acad Sci USA* 94:3290–3295.

Bonnette RE, Caudill MA, Bailey LB. 1998. Plasma homocysteine response to controlled folate intake in pregnant women. *Obstet Gynecol* 92:167-170.

Borman GB, Smith AH, Howard JK. 1986. Risk factors in the prevalence of anencephalus and spina bifida in New Zealand. *Teratology* 33:221–230.

Bottiglieri T, Hyland K, Reynolds EH. 1994. The clinical potential of ademetionine (*S*-adenosylmethionine) in neurological disorders. *Drugs* 48:137–152.

Botto LD, Khoury MJ, Mulinare J, Erickson JD. 1996. Periconceptional multivitamin use and the occurrence of conotruncal heart defects: Results from a population-based, case-control study. *Pediatrics* 98:911–917.

Boushey CJ, Beresford SA, Omenn GS, Motulsky AG. 1995. A quantitative assessment of plasma homocysteine as a risk factor for vascular disease. Probable benefits of increasing folic acid intakes. *J Am Med Assoc* 274:1049–1057.

Bower C, Stanley FJ. 1989. Dietary folate as a risk factor for neural-tube defects: Evidence from a case-control study in Western Australia. *Med J Aust* 150:613–619.

Bower C, Stanley FJ. 1992a. Dietary folate and nonneural midline birth defects: No evidence of an association from a case-control study in Western Australia. *Am J Med Genet* 44:647–650.

Bower C, Stanley FJ. 1992b. Periconceptional vitamin supplementation and neural tube defects; evidence from a case-control study in Western Australia and a review of recent publications. *J Epidemiol Community Health* 46:157–161.

Bower C, Hobbs M, Carney A, Simpson D. 1984. Neural tube defects in Western Australia 1966–81 and a review of Australian data 1942–81. *J Epidemiol Community Health* 38:208–213.

Brattstrom LE, Herbebo JE, Hultberg BL. 1984. Moderate homocysteinemia—a possible risk factor for arteriosclerotic cerebrovascular disease. *Stroke* 15:1012–1016.

Brattstrom LE, Israelsson B, Jeppsson JO, Hultberg BL. 1988. Folic acid—an innocuous means to reduce plasma homocysteine. *Scand J Clin Lab Invest* 48:215–221.

Brattstrom LE, Israelsson B, Norrving B, Bergkvist D, Thorne J, Hultberg B, Hamfelt A. 1990. Impaired homocysteine metabolism in early-onset cerebral and peripheral occlusive arterial disease. Effects of pyridoxine and folic acid treatment. *Atherosclerosis* 81:51–60.

Brock KE, Berry G, Mock PA, MacLennan R, Truswell AS, Brinton LA. 1988. Nutrients in diet and plasma and risk of in situ cervical cancer. *J Natl Cancer Inst* 80:580–585.

Brown CM, Smith AM, Picciano MF. 1986. Forms of human milk folacin and variation patterns. *J Pediatr Gastroenterol Nutr* 5:278–282.

Brown JE, Jacobs DR, Hartman TJ, Barosso GM, Stang JS, Gross MD, Zeuske MA. 1997. Predictors of red cell folate level in women attempting pregnancy. *J Am Med Assoc* 277:548–552.

Bunduki V, Dommergues M, Zittoun J, Marquet J, Muller F, Dumez Y. 1995. Maternal-fetal folate status and neural tube defects: A case control study. *Biol Neonate* 67:154–159.

Burke G, Robinson K, Refsum H, Stuart B, Drumm J, Graham I. 1992. Intrauterine growth retardation, perinatal death, and maternal homocysteine levels. *N Engl J Med* 326:69–70.

Butterworth CE, Tamura T. 1989. Folic acid safety and toxicity: A brief review. *Am J Clin Nutr* 50:353–358.

Butterworth CE Jr, Hatch K, Gore H, Meuller H, Krumdieck C. 1982. Improvement in cervical dysplasia associated with folic acid therapy in users of oral contraceptives. *Am J Clin Nutr* 35:73–82.

Butterworth CE Jr, Hatch K, Cole P, Sauberlich HE, Tamura T, Cornwell PE, Soong S-J. 1988. Zinc concentration in plasma and erythrocytes of subjects receiving folic acid supplementation. *Am J Clin Nutr* 47:484–486.

Butterworth CE Jr, Hatch K, Macaluso M, Cole P, Sauberlich HE, Soong S-J, Borst M, Baker V. 1992a. Folate deficiency and cervical dysplasia. *J Am Med Assoc* 267:528–533.

Butterworth CE Jr, Hatch K, Soong S-J, Cole P, Tamura T, Sauberlich HE, Borst M, Macaluso M, Baker V. 1992b. Oral folic acid supplementation for cervical dysplasia: A clinical intervention trial. *Am J Obstet Gynecol* 166:803–809.

Campbell NR. 1996. How safe are folic acid supplements? *Arch Intern Med* 156:1638–1644.

Carmel R, Johnson CS. 1978. Racial patterns in pernicious anemia: Early age at onset and increased frequency of intrinsic-factor antibody in black women. *N Engl J Med* 298:647–650.

Carney MW, Chary TK, Laundy M, Bottiglieri T, Chanarin I, Reynolds EH, Toone B. 1990. Red cell folate concentrations in psychiatric patients. *J Affect Disord* 19:207–213.

Carter CO. 1974. Clues to the aetiology of neural tube malformations. *Dev Med Child Neurol* 16:3–15.

Caudill MA, Cruz AC, Gregory JF 3rd, Hutson AD, Bailey LB. 1997. Folate status response to controlled folate intake in pregnant human subjects. *J Nutr* 127:2363–2370.

Caudill MA, Gregory JF, Hutson AD, Bailey LB. 1998. Folate catabolism in pregnant and nonpregnant women with controlled folate intakes. *J Nutr* 128:204–208.

CDC (Centers for Disease Control and Prevention). 1991. Use of folic acid for prevention of spina bifida and other neural tube defects—1983–1991. *Morb Mortal Wkly Rep* 40:513–516.

CDC (Centers for Disease Control and Prevention). 1992. Recommendations for the use of folic acid to reduce the number of cases of spina bifida and other neural tube defects. *Morb Mortal Wkly Rep* 41:1–7.

CDC (Centers for Disease Control and Prevention). 1998. Use of folic acid-containing supplements among women of childbearing age—United States, 1997. *Morb Mortal Wkly Rep* 47:131–134.

Chadefaux B, Cooper BA, Gilfix BM, Lue-Shing H, Carson W, Gavsie A, Rosenblatt DS. 1994. Homocysteine: Relationship to serum cobalamin, serum folate, erythrocyte folate, and lobation of neutrophils. *Clin Invest Med* 17:540–550.

Chanarin I, Rothman D, Ward A, Perry J. 1968. Folate status and requirement in pregnancy. *Br Med J* 2:390–394.

Chanarin I, Deacon R, Lumb M, Perry J. 1989. Cobalamin-folate interrelations. *Blood Rev* 3:211–215.

Chasan-Taber L, Selhub J, Rosenberg IH, Malinow MR, Terry M, Tishler PV, Willett W, Hennekens CH, Stampfer MJ. 1996. A prospective study of folate and vitamin B_6 and risk of myocardial infarction in U.S. physicians. *J Am Coll Nutr* 15:136–143.

Chen LH, Liu ML, Hwang HY, Chen LS, Korenberg J, Shane B. 1997. Human methionine synthase, cDNA cloning, gene localization, and expression. *J Biol Chem* 272:3628–3634.

Chodos RB, Ross JF. 1951. The effects of combined folic acid and liver extract therapy. *Blood* 6:1213–1233.

Collins CS, Bailey LB, Hillier S, Cerda JJ, Wilder BJ. 1988. Red blood cell uptake of supplemental folate in patients on anticonvulsant drug therapy. *Am J Clin Nutr* 48:1445–1450.

Colman N. 1982. Addition of folic acid to staple foods as a selective nutrition intervention strategy. *Nutr Rev* 40:225–233.

Colman N, Larsen JV, Barker M, Barker EA, Green R, Metz J. 1975. Prevention of folate deficiency by food fortification. 3. Effect in pregnant subjects of varying amounts of added folic acid. *Am J Clin Nutr* 28:465–470.

Conley CL, Krevans JR. 1951. Development of neurologic manifestations of pernicious anemia during multivitamin therapy. *N Engl J Med* 245:529–531.

Cooperman JM, Pesci-Bourel A, Luhby AL. 1970. Urinary excretion of folic acid activity in man. *Clin Chem* 16:375–381.

Copp AJ, Bernfield M. 1994. Etiology and pathogenesis of human neural tube defects: Insights from mouse models. *Curr Opin Pediatr* 6:624–631.

Coppen A, Abou-Saleh MT. 1982. Plasma folate and affective morbidity during long-term lithium therapy. *Br J Psychiatr* 141:87–89.

Coppen A, Chaudhry S, Swade C. 1986. Folic acid enhances lithium prophylaxis. *J Affect Disord* 10:9–13.

Cragan JD, Roberts HE, Edmonds LD, Khoury MJ, Kirby RS, Shaw GM, Velie EM, Merz RD, Forrester MB, Williamson RA. 1995. Surveillance for anencephaly and spina bifida and the impact of prenatal diagnosis—United States, 1985–1994. *CDC Surveill Summ* 44:1–13.

Crosby WH. 1960. The danger of folic acid in multivitamin preparations. *Mil Med* 125:233–235.

Cunningham FG, MacDonald PC, Grant NF. 1989. *Williams Obstetrics.* Norwalk, Conn.: Appleton & Lange.

Curtis D, Sparrow R, Brennan L, Van der Weyden MB. 1994. Elevated serum homocysteine as a predictor for vitamin B_{12} or folate deficiency. *Eur J Haematol* 52:227–232.

Cuskelly GJ, McNulty H, Scott JM. 1996. Effect of increasing dietary folate on red-cell folate: Implications for prevention of neural tube defects. *Lancet* 347:657–659.

Czeizel A. 1993. Prevention of congenital abnormalties by periconceptional multivitamin supplementation. *Br Med J* 306:1645–1648.

Czeizel AE, Dudas I. 1992. Prevention of the first occurrence of neural-tube defects by periconceptional vitamin supplementation. *N Engl J Med* 327:1832–1835.

Czeizel AE, Dudas I, Metneki J. 1994. Pregnancy outcomes in a randomized controlled trial of periconceptional multivitamin supplementation. Final report. *Arch Gynecol Obstet* 255:131–139.

Czeizel AE, Toth M, Rockenbauer M. 1996. Population-based case contol study of folic acid supplementation during pregnancy. *Teratology* 53:345–351.

Dai WS, Hsu M-A, Itri LM. 1989. Safety of pregnancy after discontinuation of iso-tretinoin. *Arch Dermatol* 125:362–365.

Dalery K, Lussier-Cacan S, Selhub J, Davignon J, Latour Y, Genest J. 1995. Homo-cysteine and coronary artery disease in French Canadian subjects: Relation with vitamins B_{12}, B_6, pyridoxal phosphate, and folate. *Am J Cardiol* 75:1107–1111.

Daly LE, Kirke PN, Molloy A, Weir DG, Scott JM. 1995. Folate levels and neural tube defects. Implications for prevention. *J Am Med Assoc* 274:1698–1702.

Daly S, Mills JL, Molloy AM, Conley M, Lee YJ, Kirke PN, Weir DG, Scott JM. 1997. Minimum effective dose of folic acid for food fortification to prevent neural tube defects. *Lancet* 350:1666–1669.

Dawson DW. 1966. Microdoses of folic acid in pregnancy. *J Obstet Gynaecol Br Commonw* 73:44–48.

DeSouza S, Eitenmiller R. 1990. Effects of different enzyme treatments on extrac-tion of total folate from various foods prior to microbiological assay and radioassay. *J Micronutr Anal* 7:37–57.

deFranchis R, Mancini FP, D'Angelo A, Sebastio G, Fermo I, DeStefano V, Mar-gaglione M, Mazzola G, DiMinno G, Andria G. 1996. Elevated total plasma homocysteine and 677C→T mutation of the 5,10-methylenetetrahydrofolate reductase gene in thrombotic vascular disease. *Am J Hum Genet* 59:262–264.

De Wals P, Trochet C, Pinsonneault L. 1999. Prevalence of neural tube defects in the province of Quebec, 1992. *Can J Public Health* 90:237–239.

DHHS (U.S. Department of Health and Human Services). 1993a. Food and Drug Administration. Folic acid; proposed rules. *Fed Regist* 21:53293–53294.

DHHS (U.S. Department of Health and Human Services). 1993b. Food and Drug Administration. Food standards: Amendment of the standards of identity for enriched grain products to require the addition of folic acid. *Fed Regist* 58:53305.

DHHS (U.S. Department of Health and Human Services). 1996. Food and Drug Administration. Food standards: Amendment of the standards of identity for enriched grain products to require addition of folic acid. *Fed Regist* 61:8781–8807.

Dolk H, De Wals P, Gillerot Y, Lechat MF, Ayme S, Cornel M, Cuschieri A, Garne E, Goujard J, Laurence KM. 1991. Heterogeniety of neural tube defects in Europe: The significance of site of defect and presence of other major anom-alies in relation to geographic differences in prevalence. *Teratology* 44:547–559.

Economides DL, Ferguson J, Mackenzie IZ, Darley J, Ware II, Holmes-Siedle M. 1992. Folate and vitamin B_{12} concentrations in maternal and fetal blood, and amniotic fluid in second trimester pregnancies complicated by neural tube defects. *Br J Obstet Gynaecol* 99:23–25.

Eichner ER, Hillman RS. 1971. The evolution of anemia in alcoholic patients. *Am J Med* 50:218–232.

Eichner ER, Hillman RS. 1973. Effect of alcohol on serum folate level. *J Clin Invest* 52:584–591.

Eichner ER, Pierce HI, Hillman RS. 1971. Folate balance in dietary-induced mega-loblastic anemia. *N Engl J Med* 284:933–938.

Eichner ER, Loewenstein JE, McDonald CR, Dickson VL. 1979. Effect of common drugs on serum level and binding of folate in man. In: Kisliuk RL, Brown GM, eds. *Chemistry and Biology of Pteridines*. New York: Elsevier North Holland. Pp. 537–542.

Ek J. 1980. Plasma and red cell folate values in newborn infants and their mothers in relation to gestational age. *J Pediatr* 97:288–292.

Ek J, Magnus EM. 1979. Plasma and red blood cell folate in breastfed infants. *Acta Paediatr Scand* 68:239–243.

Ek J, Magnus E. 1982. Plasma and red cell folate values and folate requirements in formula-fed term infants. *J Pediatr* 100:738–744.

Ellison ABC. 1960. Pernicious anemia masked by multivitamins containing folic acid. *J Am Med Assoc* 173:240–243.

Elsborg L. 1974. Inhibition of intestinal absorption of folate by phenytion. *Acta Haematol* 52:24–28.

Elwood JM, Elwood JH. 1980. *Epidemiology of Anencephalus and Spina Bifida.* Oxford: Oxford University Press.

Emery AE. 1986. *Methodology in Medical Genetics: An Introduction to Statistical Methods.* 2nd ed. Edinburgh: Churchill Livingstone.

Essien FB. 1992. Maternal methionine supplementation promotes the remediation of axial defects in *Axd* mouse neural tube mutants. *Teratology* 45:205–212.

Evans RW, Shaten BJ, Hempel JD, Cutler JA, Kuller LH. 1997. Homocyst(e)ine and risk of cardiovascular disease in the Multiple Risk Factor Intervention Trial. *Arterioscler Thromb Vasc Biol* 17:1947–1953.

Fava M, Borus JS, Alpert JE, Nierenberg AA, Rosenbaum JF, Bottiglieri T. 1997. Folate, vitamin B_{12}, and homocysteine in major depressive disorder. *Am J Psychiatry* 154:426–428.

Felson DT, Anderson JJ, Meenan RF. 1990. The comparative efficacy and toxicity of second-line drugs in rheumatoid arthritis. Results of two metaanalyses. *Arthritis Rheum* 33:1449–1461.

Fowler WM, Hendricks AB. 1949. Folic acid and the neurologic manifestations of pernicious anemia. *Am Pract* 3:609–613.

Frosst P, Blom HJ, Milos R, Goyette P, Sheppard CA, Matthews RG, Boers GJ, den Heijer M, Kluijtmans LA, van den Heuvel LP, Rozen R. 1995. A candidate genetic risk factor for vascular disease: A common mutation in methylenetetrahydrofolate reductase. *Nat Genet* 10:111–113.

Gailani SD, Carey RW, Holland JF, O'Malley JA. 1970. Studies of folate deficiency in patients with neoplastic diseases. *Cancer Res* 30:327–333.

Gallagher PM, Meleady R, Shields DC, Tan KS, McMaster D, Rozen R, Evans A, Graham IM, Whitehead AS. 1996. Homocysteine and risk of premature coronary heart disease. Evidence for a common gene mutation. *Circulation* 94:2154–2158.

Garry PJ, Goodwin JS, Hunt WC, Hooper EM, Leonard AG. 1982. Nutritional status in a healthy elderly population: Dietary and supplemental intakes. *Am J Clin Nutr* 36:319–331.

Garry PJ, Goodwin JS, Hunt WC. 1984. Folate and vitamin B_{12} status in a healthy elderly population. *J Am Geriatr Soc* 32:719–726.

Gartler SM, Hornug SK, Motulsky AG. 1981. Effect of chronologic age on induction of cystathionine synthase, uroporphyrinogen I synthase, and glucose 6-phosphate dehydrogenase activities in lymphocytes. *Proc Natl Acad Sci USA* 78:1916–1919.

Gibberd FB, Nicholls A, Dunne JF, Chaput de Saintonge DM. 1970. Toxicity of folic acid. *Lancet* 1:360–361.

Giles WH, Kittner SJ, Anda RF, Croft JB, Casper ML. 1995. Serum folate and risk for ischemic stroke. First National Health and Nutrition Examination Survey epidemiology follow-up study. *Stroke* 26:1166–1170.

Giovannucci E, Stampfer MJ, Colditz GA, Rimm EB, Trichopolous D, Rosner BA, Speizer FE, Willett WC. 1993. Folate, methionine, and alcohol intake and risk of colorectal adenoma. *J Natl Cancer Inst* 85:875–884.

Giovannucci E, Rimm EB, Ascherio A, Stampfer MJ, Colditz GA, Willett WC. 1995. Alcohol, low-methionine–low folate diets and risk of colon cancer in men. *J Natl Cancer Inst* 87:265–273.

Glynn SA, Albanes D, Pietinen P, Brown CC, Rautalahti M, Tangrea JA, Gunter EW. 1996. Colorectal cancer and folate status: A nested case control study among male smokers. *Cancer Epidemiol Biomarkers Prev* 5:487–494.

Goddijn-Wessel TA, Wouters MG, van de Molen EF, Spuijbroek MD, Steegers-Theunissen RP, Blom HJ, Boers GH, Eskes TK. 1996. Hyperhomocysteinemia: A risk factor for placental abruption or infarction. *Eur J Obstet Gynecol Reprod Biol* 66:23–29.

Godfrey PS, Toone BK, Carney MW, Flynn TG, Bottiglieri T, Laundy M, Chanarin I, Reynolds EH. 1990. Enhancement of recovery from psychiatric illness by methylfolate. *Lancet* 336:392–395.

Gomez MR. 1981. Possible teratogenicity of valproic acid. J Pediatr 98:508–509.

Goodwin JS, Goodwin JM, Garry PJ. 1983. Association between nutritional status and cognitive functioning in a healthy elderly population. *J Am Med Assoc* 249:2917–2921.

Gotz VP, Lauper RD. 1980. Folic acid hypersensitivity or tartrazine allergy? *Am J Hosp Pharm* 37:1470–1474.

Grace E, Emans SJ, Drum DE. 1982. Hematologic abnormalities in adolescents who take oral contraceptive pills. *J Pediatr* 101:771–774.

Graham IM, Daly LE, Refsum HM, Robinson K, Brattstrom LE, Ueland PM, Palma-Reis RJ, Boers GH, Sheahan RG, Israelsson B, Uiterwaal CS, Meleady R, McMaster D, Verhoef P, Witteman J, Rubba P, Bellet H, Wautrecht JC, de Valk HW, Sales Luis AC, Parrot-Rouland FM, Tan KS, Higgins I, Garcon D, Andria G. 1997. Plasma homocysteine as a risk factor for vascular disease. The European Concerted Action Project. *J Am Med Assoc* 277:1775–1781.

Gregory JF 3rd. 1989. Chemical and nutritional aspects of folate research: Analytical procedures, methods of folate synthesis, stability and bioavailability of dietary folates. *Adv Food Nutr Res* 33:1–101.

Gregory JF 3rd. 1995. The bioavailability of folate. In: Bailey LB, ed. *Folate in Health and Disease.* New York: Marcel Dekker. Pp. 195–235.

Gregory JF 3rd. 1997. Bioavailability of folate. *Eur J Clin Nutr* 51: S54–S59.

Gregory JF 3rd, Engelhardt R, Bhandari SD, Sartain DB, Gustafson SK. 1990. Adequacy of extraction techniques for determination of folate in foods and other biological materials. *J Food Comp Anal* 3:134–144.

Gunter EW, Bowman BA, Caudill SP, Twite DB, Adams MJ, Sampson EJ. 1996. Results of an international round robin for serum and whole-blood folate. *Clin Chem* 42:1689–1694.

Hall BE, Watkins CH. 1947. Experience with pteroylglutamic (synthetic folic acid) in the treatment of pernicious anemia. *J Lab Clin Med* 32:622–634.

Halsted CH, Griggs RC, Harris JW. 1967. The effect of alcoholism on the absorption of folic acid (H^3-PGA) evaluated by plasma levels and urine excretion. *J Lab Clin Med* 69:116–131.

Halsted CH, Robles EA, Mezey E. 1971. Decreased jejunal uptake of labeled folic acid (H^3-PGA) in alcoholic patients: Roles of alcohol and malnutrition. *N Engl J Med* 285:701–706.

Halsted CH, Robles EA, Mezey E. 1973. Intestinal malabsorption in folate-deficient alcoholics. *Gastroenterology* 64:526–532.

Halsted CH, Gandhi G, Tamura T. 1981. Sulfasalazine inhibits the absorption of folates in ulcerative colitis. *N Engl J Med* 305:1513–1517.

Hambidge M, Hackshaw A, Wald N. 1993. Neural tube defects and serum zinc. *Br J Obstet Gynecol* 100:746–749.

Hamon CG, Blair JA, Barford PA. 1986. The effect of tetrahydrofolate on tetrahydrobiopterin metabolism. *J Ment Defic Res* 30:179–183.

Hansen H, Rybo G. 1967. Folic acid dosage in profylactic treatment during pregnancy. *Acta Obstet Gynecol Scand* 46:107–112.

Hansen HA, Weinfeld A. 1962. Metabolic effects and diagnostic value of small doses of folic acid and B_{12} in megaloblastic anemias. *Acta Med Scand* 172:427–443.

Happle R, Traupe H, Bounameaux Y, Fisch T. 1984. Teratogenic effects of etretinate in humans. *Dtsch Med Wochenschr* 109:1476–1480.

Harpel PC, Zhang X, Borth W. 1996. Homocysteine and hemostasis: Pathogenic mechanisms predisposing to thrombosis. *J Nutr* 126:12855–12895.

Haworth JC, Dilling LA, Surtees RA, Seargeant LE, Lue-Shing H, Cooper BA, Rosenblatt DS. 1993. Symptomatic and asymptomatic methylenetetrahydrofolate reductase deficiency in two adult brothers. *Am J Med Genet* 45:572–576.

Hayes C, Werler MM, Willett WC, Mitchell AA. 1996. Case-control study of periconceptional folic acid supplementation and oral clefts. *Am J Epidemiol* 143:1229–1234.

Health Canada. 1996. *Departmental Consolidation of the Food and Drugs Act and the Food and Drug Regulations with Amendments to December 19, 1996.* Ottawa: Canada Communications Group.

Health Canada. 1997. Regulations amending the Food and Drug Regulations (1066). *Canada Gazette, Part I,* November 29. Pp. 3702-3705.

Heimburger DC, Krumdieck CL, Alexander CB, Birch R, Dill SR, Bailey WC. 1987. Localized folic acid deficiency and bronchial metaplasia in smokers: Hypothesis and preliminary report. *Nutr Int* 3:54–60.

Heimburger DC, Alexander CB, Birch R, Butterworth CE, Bailey WC, Krumdieck CL. 1988. Improvement in bronchial squamous metaplasia in smokers treated with folate and vitamin B_{12}. Report of a preliminary randomized, double-blind intervention trial. *J Am Med Assoc* 259:1525–1530.

Heinle RW, Welch AD. 1947. Folic acid in pernicious anemia: Failure to prevent neurologic relapse. *J Am Med Assoc* 133:739–741.

Heinle RW, Dingle JT, Weisberger AS. 1947. Folic acid in the maintenance of pernicious anemia. *J Lab Clin Med* 32:970–981.

Hellstrom L. 1971. Lack of toxicity of folic acid given in pharmacological doses to healthy volunteers. *Lancet* 1:59–61.

Herbert V. 1962a. Experimental nutritional folate deficiency in man. *Trans Assoc Am Physicians* 75:307–320.

Herbert V. 1962b. Minimal daily adult folate requirement. *Arch Intern Med* 110:649–652.

Herbert V. 1963. Current concepts in therapy: Megaloblastic anemia. *N Engl J Med* 268:201–203, 368–371.

Herbert V. 1968. Nutritional requirements for vitamin B_{12} and folic acid. *Am J Clin Nutr* 21:743–752.

Herbert V. 1987. Making sense of laboratory tests of folate status: Folate requirements to sustain normality. *Am J Hem* 26:199–207.

Herbert V, Das KC. 1993. Folic acid and vitamin B_{12}. In: Shils ME, Olson JA, Shike M, eds. *Modern Nutrition in Health and Disease, 8th ed.* Philadelphia: Lea & Febiger. Pp. 402–425.

Herbert V, Zalusky R, Davidson CS. 1963. Correlates of folate deficiency with alcoholism and associated macrocytosis, anemia, and liver disease. *Ann Intern Med* 58:977–988.

Hibbard BM. 1964. The role of folic acid in pregnancy. *J Obstet Gynaecol Br Commonw* 71:529–542.

Hill RM. 1984. Isotretinoin teratogenicity. *Lancet* 1:1465.

Hirata F, Axelrod J. 1980. Phospholipid methylation and biological signal transmission. *Science* 209:1082–1090.

Hobbins JC. 1991. Diagnosis and management of neural tube defects today. *N Engl J Med* 324:690–691.

Hoffbrand AV, Newcombe BF, Mollin DL. 1966. Method of assay of red cell folate activity and the value of the assay as a test for folate deficiency. *J Clin Pathol* 19:17–28.

Holmes-Siedle M, Lindenbaum RH, Galliard A. 1992. Recurrence of neural tube defect in a group of at risk women: A 10 year study of Pregnavite Forte F. *J Med Genet* 29:134–135.

Hommes OR, Obbens EA. 1972. The epileptogenic action of Na-folate in the rat. *J Neurol Sci* 16:271–281.

Hook EB, Czeizel AE. 1997. Can terathanasia explain the protective effect of folic-acid supplementation on birth defects? *Lancet* 350:513–515.

Hoppner K, Lampi B. 1980. Folate levels in human liver from autopsies in Canada. *Am J Clin Nutr* 33:862–864.

Houghton LA, Green TJ, Donovan UM, Gibson RS, Stephen AM, O'Connor DL. 1997. Association between dietary fiber intake and the folate status of a group of female adolescents. *Am J Clin Nutr* 66:1414–1421.

Hultberg B, Andersson A, Sterner G. 1993. Plasma homocysteine in renal failure. *Clin Nephrol* 40:230–235.

Hunter R, Barnes J, Oakeley HF, Matthews DM. 1970. Toxicity of folic acid given in pharmacological doses to healthy volunteers. *Lancet* 1:61–3.

Israels MC, Wilkinson JF. 1949. Risk of neurological complications in pernicious anemia treated with folic acid. *Br Med J* 2:1072–1075.

Jackson RC. 1984. Biological effects of folic acid antagonists with antineoplastic activity. *Pharmacol Ther* 25:61–82.

Jacob RA, Wu M-M, Henning SM, Swendseid ME. 1994. Homocysteine increases as folate decreases in plasma of healthy men during short-term dietary folate and methyl group restriction. *J Nutr* 124:1072–1080.

Jacob RA, Gretz DM, Taylor PC, James SJ, Pogribny IP, Miller BJ, Henning SM, Swendseid ME. 1998. Moderate folate depletion increases plasma homocysteine and decreases lymphocyte DNA methylation in postmenopausal women. *J Nutr* 128:1204–1212.

Jacobsen DW, Gatautis VJ, Green R, Robinson K, Savon SR, Secic M, Ji J, Otto JM, Taylor LM. 1994. Rapid HPLC determination of total homocysteine and other thiols in serum and plasma: Sex differences and correlation with cobalamin and folate concentrations in healthy subjects. *Clin Chem* 40:873–881.

Jacobson SD, Berman L, Axelrod AR, Vonder Heide EC. 1948. Folic acid therapy: Its effect as observed in two patients with pernicious anemia and neurologic symptoms. *J Am Med Assoc* 137:825–827.

Jacques PF, Sulsky SI, Sadowski JA, Phillips JC, Rush D, Willett WC. 1993. Comparison of micronutrient intake measured by a dietary questionnaire and biochemical indicators of micronutrient status. *Am J Clin Nutr* 57:182–189.

Jacques PF, Bostom AG, Williams RR, Ellison RC, Eckfeldt JH, Rosenberg IH, Selhub J, Rozen R. 1996. Relation between folate status, a common mutation in methylenetetrahydrofolate reductase, and plasma homocysteine concentrations. *Circulation* 93:7–9.

Jägerstad M. 1977. Folate intake and blood folate in elderly subjects, a study using the double sampling portion technique. *Nutr Metab* 21:29–31.

Jägerstad M, Westesson A-K. 1979. Folate. *Scand J Gastroenterol* 14:196–202.

Joyal CC, Lalonde R, Vikis-Freibergs V, Botez MI. 1993. Are age-related behavioral disorders improved by folate administration? *Exp Aging Res* 19:367–376.

Kang SS, Wong PW, Norusis M. 1987. Homocysteinemia due to folate deficiency. *Metabolism* 36:458–462.

Kang SS, Wong PW, Bock HG, Horwitz A, Grix A. 1991a. Intermediate hyperhomocysteinemia resulting from compound heterozygosity of methylenetetrahydrofolate reductase mutations. *Am J Hum Genet* 48:546–551.

Kang SS, Wong PW, Susmano A, Sora J, Norusis M, Ruggie N. 1991b. Thermolabile methylenetetrahydrofolate reductase: An inherited risk factor for coronary artery disease. *Am J Hum Genet* 48:536–545.

Keagy PM, Oace SM. 1989. Rat bioassay of wheat bran folate and effects of intestinal bacteria. *J Nutr* 119:1932–1939.

Keagy PM, Shane B, Oace SM. 1988. Folate bioavailability in humans: Effects of wheat bran and beans. *Am J Clin Nutr* 47:80–88.

Keating JN, Wada L, Stokstad EL, King JC. 1987. Folic acid: Effect of zinc absorption in humans and in the rat. *Am J Clin Nutr* 46:835–839.

Kehl SJ, McLennan H, Collingridge GL. 1984. Effects of folic and kainic acids on synaptic responses of hippocampal neurones. *Neuroscience* 11:111–124.

Keizer SE, Gibson RS, O'Connor DL. 1995. Postpartum folic acid supplementation of adolescents: Impact on maternal folate and zinc status and milk composition. *Am J Clin Nutr* 62:377–384.

Khoury MJ, Erickson JD, James LM. 1982. Etiologic heterogeneity of neural tube defects: Clues from epidemiology. *Am J Epidemiol* 115:538–548.

Kim Y-I, Pogribny IP, Basnakian AG, Miller JW, Selhub J, James SJ, Mason JB. 1997. Folate deficiency in rats induces DNA strand breaks and hypomethylation with the p53 tumor suppressor gene. *Am J Clin Nutr* 65:46–52.

Kirke PN, Daly LE, Elwood JH. 1992. A randomized trial of low-dose folic acid to prevent neural tube defects. *Arch Dis Child* 67:1442–1446.

Kirke PN, Molloy AM, Daly LE, Burke H, Weir DG, Scott JM. 1993. Maternal plasma folate and vitamin B_{12} are independent risk factors for neural tube defects. *Q J Med* 86:703–708.

Klipstein FA. 1964. Subnormal serum folate and macrocytosis associated with anticonvulsant drug therapy. *Blood* 23:68–86.

Kluijtmans LA, van den Heuvel LP, Boers GH, Frosst P, Stevens EM, van Oost BA, den Heijer M, Trijbels FJ, Rozen R, Blom HJ. 1996. Molecular genetic analysis in mild hyperhomocysteinemia: A common mutation in the methylenetetrahydrofolate reductase gene is a genetic risk factor for cardiovascular disease. *Am J Hum Genet* 58:35–41.

Koehler KM, Romero LJ, Stauber PM, Pareo-Tubbeh SL, Liang HC, Baumgartner RN, Garry PJ, Allen RH, Stabler SP. 1996. Vitamin supplementation and other variables affecting serum homocysteine and methylmalonic acid concentrations in elderly men and women. *J Am Coll Nutr* 15:364–376.

Krause LJ, Forsberg CW, O'Connor DL. 1996. Feeding human milk to rats increases bifidobacterium in the cecum and colon which correlates with enhanced folate status. *J Nutr* 126:1505–1511.

Krumdieck CL, Fukushima K, Fukushima T, Shiota T, Butterworth CE Jr. 1978. A long-term study of the excretion of folate and pterins in a human subject after ingestion of ^{14}C folic acid, with observations on the effect of diphenylhydantoin administration. *Am J Clin Nutr* 31:88–93.

Landgren F, Israelsson B, Lindgren A, Hultberg B, Andersson A, Brattstrom L. 1995. Plasma homocysteine in acute myocardial infarction: Homocysteine-lowering effect of folic acid. *J Intern Med* 237:381–388.

Landon MJ, Oxley A. 1971. Relation between maternal and infant blood folate activities. *Arch Dis Child* 46:810–814.

Lashner BA. 1993. Red blood cell folate is associated with the development of dysplasia and cancer in ulcerative colitis. *J Cancer Res Clin Oncol* 119:549–554.

Lashner BA, Heidenreich PA, Su GL, Kane SV, Hanauer SB. 1989. The effect of folate supplementation on the incidence of dysplasia and cancer in chronic ulcerative colitis. A case-control study. *Gastroenterology* 97:255–259.

Laurence KM. 1990. The genetics and prevention of neural tube defects and "uncomplicated" hydrocephalus. In: Emery AE, Rimoin DL, eds. *Principles and Practice of Medical Genetics*, 2nd ed., Vol. 1. Edinburgh: Churchill Livingstone. Pp. 323–346.

Laurence KM, James N, Miller MH, Tennant GB, Campbell H. 1981. Double-blind randomized controlled trial of folate treatment before conception to prevent recurrence of neural tube defects. *Br Med J* 282:1509–1511.

Lawrence VA, Loewenstein JE, Eichner ER. 1984. Aspirin and folate binding: In vivo and in vitro studies of serum binding and urinary excretion of endogenous folate. *J Lab Clin Med* 103:944–948.

Lewis CA, Pancharuniti N, Sauberlich HE. 1992. Plasma folate adequacy as determined by homocysteine level. *Ann NY Acad Sci* 669:360–362.

Li YN, Gulati S, Baker PJ, Brody LC, Banerjee R, Kruger WD. 1996. Cloning, mapping and RNA analysis of the human methionine synthase gene. *Hum Mol Genet* 5:1851–1858.

Lim HS, Mackey AD, Tamura T, Picciano MF. 1997. Measurable folates in human milk are increased by treatment with α-amylase and protease. *FASEB J* 11:A395.

Lindenbaum J, Healton EB, Savage DG, Brust JC, Garrett TJ, Podell ER, Marcell PD, Stabler SP, Allen RH. 1988. Neuropsychiatric disorders caused by cobalamin deficiency in the absence of anemia or macrocytosis. *N Engl J Med* 318:1720–1728.

Lindseth RE. 1996. Myelomeningocele. In: Morrissy RT, Weinstein SL, eds. *Lovell and Winter's Pediatric Orthopaedics*, 4th ed., Vol. 1. Philadelphia: Lippincott-Raven. Pp. 503–505.

Loots JM, Kramer S, Brennan MJW. 1982. The effect of folates on the reflex activity in the isolated hemisected frog spinal cord. *J Neural Transm* 54:239–249.

Lowenstein L, Cantlie G, Ramos O, Brunton L. 1966. The incidence and prevention of folate deficiency in a pregnant clinic population. *Can Med Assoc J* 95:797–806.

LSRO/FASEB (Life Sciences Research Office/Federation of American Societies for Experimental Biology). 1984. *Assessment of the Folate Nutritional Status of the U.S. Population Based on Data Collected in the Second National Health and Nutrition Examination Survey, 1976–1980*. Senti FR, Pilch SM, eds. Bethesda, MD: LSRO/FASEB.

LSRO/FASEB (Life Sciences Research Office/Federation of American Societies for Experimental Biology). 1994. *Assessment of the Folate Methodology Used in the Third National Health and Nutrition Examination Survey (NHANES III, 1988–1994)*. Raiten DJ, Fisher KD, eds. Bethesda, MD: LSRO/FASEB.

LSRO/FASEB (Life Sciences Research Office/Federation of American Societies for Experimental Biology). 1995. *Third Report on Nutrition Monitoring in the United States*. Washington DC: U.S. Government Printing Office.

Ma J, Stampfer MJ, Hennekens CH, Frosst P, Selhub J, Horsford J, Malinow MR, Willett WC, Rozen R. 1996. Methylenetetrahydrofolate reductase polymorphism, plasma folate, homocysteine, and risk of myocardial infarction in U.S. physicians. *Circulation* 94:2410–2416.

Ma J, Stampfer MJ, Giovannucci E, Artigas C, Hunter DJ, Fuchs C, Willett WC, Selhub J, Hennekens CH, Rozen R. 1997. Methylenetetrahydrofolate reductase polymorphism, dietary interactions, and risk of colorectal cancer. *Cancer Res* 57:1098–1102.

Mackey AD, Lim HS, Picciano MF, Smiciklas-Wright H. 1997. Biochemical indices of folate adequacy diminish in women during lactation. *FASEB J* 11:A179.

Malinow MR, Nieto FJ, Kruger WD, Duell PB, Hess DL, Gluckmann RA, Block PC, Holzgang CR, Anderson PH, Seltzer D, Upson B, Lin QR. 1997. The effects of folic acid supplementation on plasma total homocysteine are modulated by multivitamin use and methylenetetrahydrofolate reductase genotypes. *Arterioscler Thromb Vasc Biol* 17:1157–1162.

Malpas JS, Spray GH, Witts LJ. 1966. Serum folic acid and vitamin B_{12} levels in anticonvulsant therapy. *Br Med J* 1:955–957.

Marshall RA, Jandl JH. 1960. Response to "physiologic" doses of folic acid on megaloblastic anemia. *AMA Arch Intern Med* 105:352–360.

Martin JI, Landen WO, Soliman A-G, Eitenmiller RR. 1990. Application of a tri-enzyme extraction for total folate determination in foods. *J Assoc Offic Anal Chem* 73:805–808.

Mason JB. 1995. Folate status: Effects on carcinogenesis. In: Bailey LB, ed. *Folate in Health and Disease*. New York: Marcel Dekker. Pp. 361–378.

Mason JB, Levesque T. 1996. Folate: Effects on carcinogenesis and the potential for cancer chemoprevention. *Oncology* 10:1727–1736.

Mathur BP. 1966. Sensitivity of folic acid: A case report. *Indian J Med Sci* 20:133–134.

Mayer EL, Jacobsen DW, Robinson K. 1996. Homocysteine and coronary atherosclerosis. *J Am Coll Cardiol* 27:517–527.

McMartin KE, Collins TD, Shiao CQ, Vidrine L, Redetzki HM. 1986. Study of dose-dependence and urinary folate excretion produced by ethanol in humans and rats. *Alcohol Clin Exp Res* 10:419–424.

McPartlin J, Halligan A, Scott JM, Darling M, Weir DG. 1993. Accelerated folate breakdown in pregnancy. *Lancet* 341:148–149.

Meenan J, O'Hallinan E, Lynch S, Molloy A, McPartlan J, Scott J, Weir DG. 1996. Folate status of gastrointestinal epithelial cells is not predicted by serum and red cell folate values in replete subjects. *Gut* 38:410–413.

Meenan J, O'Hallinan E, Scott J, Weir DG. 1997. Epithelial cell folate depletion occurs in neoplastic but not adjacent normal colon mucosa. *Gastroenterology* 112:1163–1168.

Metz J. 1970. Folate deficiency conditioned by lactation. *Am J Clin Nutr* 23:843–847.

Metz J, van der Westhuyzen J. 1987. The fruit bat as an experimental model of the neuropathy of cobalamin deficiency. *Comp Biochem Physiol* 88A:171–177.

Mills JL, Conley MR. 1996. Folic acid to prevent neural tube defects: Scientific advances and public health issues. *Curr Opin Obstet Gynecol* 8:394–397.

Mills JL, Rhoads GG, Simpson JL, Cunningham GC, Conley MR, Lassman MR, Walden ME, Depp DR, Hoffman HJ. 1989. The absence of a relation between the periconceptional use of vitamins and neural tube defects. National Institute of Child Health and Human Development Neural Tube Defects Study Group. *N Engl J Med* 321:430–435.

Mills JL, McPartlin JM, Kirke PN, Lee YJ, Conley MR, Weir DG, Scott JM. 1995. Homocysteine metabolism in pregnancies complicated by neural-tube defects. *Lancet* 345:149–151.

Milne DB, Johnson LK, Mahalko JR, Sandstead HH. 1983. Folate status of adult males living in a metabolic unit: Possible relationships with iron nutriture. *Am J Clin Nutr* 37:768–773.

Milne DB, Canfield WK, Mahalko JR, Sandstead HH. 1984. Effect of oral folic acid supplements on zinc, copper, and iron absorption and excretion. *Am J Clin Nutr* 39:535–539.

Milunsky A, Jick H, Jick SS, Bruell CL, MacLaughlin DS, Rothman KJ, Willett W. 1989. Multivitamin/folic acid supplementation in early pregnancy reduces the prevalence of neural tube defects. *J Am Med Assoc* 262:2847–2852.

Mitchell DC, Vilter RW, Vilter CF. 1949. Hypersensivity to folic acid. *Ann Intern Med* 31:1102–1105.

Mitchell LE, Duffy DL, Duffy P, Bellingham G, Martin NG. 1997. Genetic effects on variation in red-blood-cell folate in adults: Implications for the familial aggregation of neural tube defects. *Am J Hum Genet* 60:433–438.

Molgaard J, Malinow MR, Lassvik C, Holm A-C, Upson B, Olsson AG. 1992. Hyperhomocyst(e)inaemia: An independent risk factor for intermittent claudication. *J Intern Med* 231:273–279.

Molloy AM, Daly S, Mills JL, Kirke PN, Whitehead AS, Ramsbottom D, Conley MR, Weir DG, Scott JM. 1997. Thermolabile variant of 5,10-methylenetetrahydrofolate reductase associated with low red-cell folates: Implications for folate intake recommendations. *Lancet* 349:1591–1593.

Moore CA, Li S, Li Z, Hong SX, Gu HQ, Berry RJ, Mulinare J, Erickson JD. 1997. Elevated rates of severe neural tube defects in a high-prevalence area in northern China. *Am J Med Genet* 73:113–118.

Morgan SL, Baggott JE. 1995. Folate antagonists in nonneoplastic disease: Proposed mechanisms of efficacy and toxicity. In: Bailey LB, ed. *Folate in Health and Disease*. New York: Marcel Dekker. Pp. 405–433.

Morgan SL, Baggott JE, Altz-Smith M. 1987. Folate status of rheumatoid arthritis patients receiving long term, low-dose methotrexate therapy. *Arthritis Rheum* 30:1348–1356.

Morgan SL, Baggott JE, Vaughn WH, Austin JS, Veitch TA, Lee JY, Koopman WJ, Krumdieck CL, Alarcon GS. 1994. Supplementation with folic acid during methotrexate therapy for rheumatoid arthritis. A double-blind, placebo-controlled trial. *Ann Intern Med* 121:833–841.

Morgan SL, Baggott JE, Alarcon GS. 1997. Methotrexate in rheumatoid arthritis. Folate supplementation should always be given. *Bio Drugs* 8:164–175.

Morrison HI, Schaubel D, Desmeules M, Wigle DT. 1996. Serum folate and risk of fatal coronary heart disease. *J Am Med Assoc* 275:1893–1896.

Mudd SH, Skovby F, Levy HL, Pettigrew KD, Wilcken B, Pyeritz RE, Andria G, Boers GH, Bromberg IL, Cerone R, Fowler B, Gröbe H, Schmidt H, Schweitzer L. 1985. The natural history of homocystinuria due to cystathionine β-synthase deficiency. *Am J Hum Genet* 37:1–31.

Mukherjee MD, Sandstead HH, Ratnaparkhi MV, Johnson LK, Milne DB, Stelling HP. 1984. Maternal zinc, iron, folic acid and protein nutriture and outcome of human pregnancy. *Am J Clin Nutr* 40:496–507.

Mulinare J, Cordero JF, Erickson JD, Berry RJ. 1988. Periconceptional use of multivitamins and the occurrence of neural tube defects. *J Am Med Assoc* 260:3141–3145.

Munger R, Romitti P, West N, Murray J, Hanson J. 1997. Maternal intake of folate, vitamin B_{12}, and zinc and risk of orofacial cleft birth defects. *Am J Epidemiol* 145:S30.

Nakazawa Y, Chiba K, Imatoh N, Kotoroii T, Sakamoto T, Ishizaki T. 1983. Serum folic acid levels and antipyrine clearance rates in smokers and nonsmokers. *Drug Alcohol Depend* 11:201–207.

Nygård O, Nordrehaug JE, Fefsum H, Ueland PM, Farstad M, Vollset SE. 1997. Plasma homocysteine levels and mortality in patients with coronary artery disease. *N Engl J Med* 327:230–236.

O'Connor DL, Tamura T, Picciano MF. 1991. Pteroylpolyglutamates in human milk. *Am J Clin Nutr* 53:930–934.

O'Keefe CA, Bailey LB, Thomas EA, Hofler SA, Davis BA, Cerda JJ, Gregory JF 3rd. 1995. Controlled dietary folate affects folate status in nonpregnant women. *J Nutr* 125:2717–2725.

Olney JW, Fuller TA, de Gubareff T, Labruyere J. 1981. Intrastriatal folic acid mimics the distant but not local brain damaging properties of kainic acid. *Neurosci Lett* 25:207–210.

Omer A, Mowat AG. 1968. Nature of anaemia in rhematoid arthritis. 9. Folate metabolism in patients with rheumatoid arthritis. Ann Rheum Dis 27:414–424.

Ortega RM, Redondo R, Andres P, Eguileor I. 1993. Nutritional assessment of folate and cyanocobalamin status in a Spanish elderly group. *Int J Vitam Nutr Res* 63:17–21.

Ortega RM, Lopez-Sobaler AM, Gonzalez-Gross MM, Redondo RM, Marzana I, Zamora MJ, Andres P. 1994. Influence of smoking on folate intake and blood folate concentrations in a group of elderly Spanish men. *J Am Coll Nutr* 13:68–72.

Pancharuniti N, Lewis CA, Sauberlich HE, Perkins LL, Go RC, Alvarez JO, Macaluso M, Acton RT, Copeland RB, Cousins AL, Gore TB, Cornwell PE, Roseman JE. 1994. Plasma homocyst(e)ine, folate, and vitamin B_{12} concentrations and risk for early-onset coronary artery disease. *Am J Clin Nutr* 59:940–948.

Pfeiffer CM, Rogers LM, Bailey LB, Gregory JF 3rd. 1997a. Absorption of folate from fortified cereal-grain products and of supplemental folate consumed with or without food determined by using a dual-label stable-isotope protocol. *Am J Clin Nutr* 66:1388–1397.

Pfeiffer CM, Rogers LM, Gregory JF 3rd. 1997b. Determination of folate in cereal-grain food products using trienzyme extraction and combined affinity and reversed-phase liquid chromatography. *J Agric Food Chem* 45:407–413.

Picciano MF. 1996. Pregnancy and lactation. In: Ziegler EE, Filer LJ Jr., eds. *Present Knowledge in Nutrition*. Washington, DC: ILSI Press. Pp. 384–395.

Piyathilake CJ, Macaluso M, Hine RJ, Richards EW, Krumdieck CL. 1994. Local and systemic effects of cigarette smoking on folate and vitamin B-12. *Am J Clin Nutr* 60:559–566.

Potischman N, Brinton LA, Laiming VA, Reeves WC, Brenes MM, Herroro R, Tenorio F, de Britton RC, Gaitan E. 1991. A case-control study of serum folate levels and invasive cervical cancer. *Cancer Res* 51:4785–4789.

Qvist I, Abdulla M, Jägerstad M, Svensson S. 1986. Iron, zinc and folate status during pregnancy and two months after delivery. *Acta Obstet Gynecol Scand* 65:15–22.

Rajkovic A, Catalano PM, Malinow MR. 1997. Elevated homocyst(e)ine levels with preeclampsia. *Obstet Gynecol* 90:168–171.

Ramsbottom D, Scott JM, Molloy A, Weir DG, Kirke PN, Mills JL, Gallagher PM, Whitehead AS. 1997. Are common mutations of cystathionine beta-synthase involved in the aetiology of neural tube defects? *Clin Genet* 51:39–42.

Rasmussen K, Moller J, Lyngbak M, Pedersen A-M, Dybkjaer L. 1996. Age- and gender-specific reference intervals for total homocysteine and methylmalonic acid in plasma before and after vitamin supplementation. *Clin Chem* 42:630–636.

Reed T, Malinow MR, Christian JC, Upson B. 1991. Estimates of heritability of plasma homocyst(e)ine levels in aging adult male twins. *Clin Genet* 39:425–428.

Reisner EH Jr, Weiner L. 1952. Studies on mutual effect of suboptimal oral doses of vitamin B_{12} and folic acid in pernicious anemia. *N Engl J Med* 247:15–17.

Retief FP. 1969. Urinary folate excretion after ingestion of pteroylmonoglutamic acid and food folate. *Am J Clin Nutr* 22:352-355.

Reynolds EH, Chanarin I, Milner G, Matthews DM. 1966. Anticonvulsant therapy, folic acid and vitamin B_{12} metabolism and mental symptoms. *Epilepsia* 7:261–270.

Reynolds EH, Rothfeld P, Pincus JH. 1973. Neurological disease associated with folate deficiency. *Br Med J* 2:398–400.

Rhode BM, Cooper BA, Farmer FA. 1983. Effect of orange juice, folic acid, and oral contraceptives on serum folate in women taking a folate-restricted diet. *J Am Coll Nutr* 2:221–230.

Richens A. 1971. Toxicity of folic acid. *Lancet* 1:912.

Riggs KM, Spiro A, Tucker K, Rush D. 1996. Relations of vitamin B-12, vitamin B-6, folate, and homocysteine to cognitive performance in the Normative Aging Study. *Am J Clin Nutr* 63:306–314.

Rimm EB, Willett WC, Hu FB, Sampson L, Colditz GA, Manson JE, Hennekens C, Stampfer MJ. 1998. Folate and vitamin B_6 from diet and supplements in relation to risk of coronary heart disease among women. *J Am Med Assoc* 279:359–364.

Ritz ND, Meyer LM, Brahin C, Sawitsky A. 1951. Further observations on the oral treatment of pernicious anemia with subminimal doses of folic acid and vitamin B_{12}. *Acta Hematol* 5:334–338.

Rong N, Selhub J, Goldin BR, Rosenberg IH. 1991. Bacterially synthesized folate in rat large intestine is incorporated into host tissue folyl polyglutamates. *J Nutr* 121:1955–1959.

Rosa FW. 1991. Spina bifida in infants of women treated with carbamazepine during pregnancy. *N Engl J Med* 324:674–677.

Rosenberg IH. 1992. Folate. In: Hartz SC, Russell RM, Rosenberg IH, eds. *Nutrition in the Elderly*. The Boston Nutritional Status Survey. London: Smith-Gordon. Pp. 135–139.

Ross JF, Belding H, Paegel BL. 1948. The development and progression of sub-acute combined degeneration of the spinal cord in patients with pernicious anemia treated with synthetic pteroylglutamic (folic) acid. *Blood* 3:68–90.

Russell RM, Ismail-Beigi F, Reinhold JG. 1976. Folate content of Iranian breads and the effect of their fiber content on the intestinal absorption of folic acid. *Am J Clin Nutr* 29:799–802.

Russell RM, Rosenberg IH, Wilson PD, Iber FL, Oaks EB, Giovetti AC, Otradovec CL, Karwoski PA, Press AW. 1983. Increased urinary excretion and prolonged turnover time of folic acid during ethanol ingestion. *Am J Clin Nutr* 38:64–70.

Sahyoun N. 1992. Nutrient intake by the NSS elderly population. In: Hartz SC, Russell RM, Rosenberg IH, eds. *Nutrition in the Elderly*. The Boston Nutritional Status Survey. London: Smith-Gordon. Pp. 31–44.

Sahyoun NR, Otradovec CL, Hartz SC, Jacob RA, Peters H, Russell RM, McGandy RB. 1988. Dietary intakes and biochemical indicators of nutritional status in an elderly, institutionalized population. *Am J Clin Nutr* 47:524–533.

Saleh AM, Pheasant AE, Blair JA, Allan RN, Walters J. 1982. Folate metabolism in man: The effect of malignant disease. *Br J Cancer* 46:346–353.

Salmenpera L, Perheentupa J, Siimes MA. 1986. Folate nutrition is optimal in exclusively breast-fed infants but inadequate in some of their mothers and in formula-fed infants. *J Pediatr Gastroenterol Nutr* 5:283–289.

Sauberlich HE, Kretsch MJ, Skala JH, Johnson HL, Taylor PC. 1987. Folate requirement and metabolism in nonpregnant women. *Am J Clin Nutr* 46:1016–1028.

Savage DG, Lindenbaum J, Stabler SP, Allen RH. 1994. Sensitivity of serum methylmalonic acid and total homocysteine determinations for diagnosing cobalamin and folate deficiencies. *Am J Med* 96:239–246.

Scanlon KS, Ferencz C, Loffredo CA, Wilson PD, Correa-Villaseñor A, Khoury MJ, Willett WC. 1998. Preconceptional folate intake and malformations of the cardiac outflow tract. *Epidemiology* 9:95–98.

Schmitz C, Lindpaintner K, Verhoef P, Gaziano JM, Buring J. 1996. Genetic polymorphism of methylenetetrahydrofolate reductase and myocardial infarction. A case-control study. *Circulation* 94:1812–1814.

Scholl TO, Hediger ML, Bendich A, Schall JI, Smith WK, Krueger PM. 1997. Use of multivitamin/mineral prenatal supplements: Influence on the outcome of pregnancy. *Am J Epidemiol* 146:134–141.

Schwartz SM, Siscovick DS, Malinow MR, Rosendaal FR, Beverly RK, Hess DL, Psaty BM, Longstreth WT, Koepsell TD, Raghunathan TE, Reitsma PH. 1997. Myocardial infarction in young women in relation to plasma total homocysteine, folate, and a common variant in the methylenetetrahydrofolate reductase gene. *Circulation* 96:412–417.

Schwartz SO, Kaplan SR, Armstrong BE. 1950. The long-term evaluation of folic acid in the treatment of pernicious anemia. *J Lab Clin Med* 35:894–898.

Selby JV, Friedman GD, Fireman BH. 1989. Screening prescription drugs for possible carcinogenecity: Eleven to fifteen years of follow-up. *Cancer Res* 49:5736–5747.

Selhub J, Rosenberg IH. 1996. Folic acid. In: Ziegler EE, Filer LJ Jr., eds. *Present Knowledge in Nutrition*. Washington, DC: ILSI Press. Pp. 206–219.

Selhub J, Dhar G, Rosenberg IH. 1978. Inhibition of folate enzymes by sulfasalazine. *J Clin Invest* 61:221–224.

Selhub J, Jacques PF, Wilson PWF, Rush D, Rosenberg IH. 1993. Vitamin status and intake as primary determinants of homocysteinemia in an elderly population. *J Am Med Assoc* 270:2693–2698.

Senti FR, Pilch SM. 1985. Analysis of folate data from the Second National Health and Nutrition Examination Survey (NHANES II). *J Nutr* 115:1398–1402.

Sesin GP, Kirschenbaum H. 1979. Folic acid hypersensitivity and fever: A case report. *Am J Hosp Pharm* 36:1565–1567.

Shaw GM, Lammer EJ, Wasserman CR, O'Malley CD, Tolarova MM. 1995a. Risks of orofacial clefts in children born to women using multivitamins containing folic acid periconceptionally. *Lancet* 346:393–396.

Shaw GM, O'Malley CD, Wasserman CR, Tolarova MM, Lammer EJ. 1995b. Maternal periconceptional use of multivitamins and reduced risk for conotruncal heart defects and limb deficiencies among offspring. *Am J Med Genet* 59:536–545.

Shaw GM, Schaffer D, Velie EM, Morland K, Harris JA. 1995c. Periconceptional vitamin use, dietary folate, and the occurrence of neural tube defects. *Epidemiology* 6:219–226.

Sheehy TW. 1973. Folic acid: Lack of toxicity. *Lancet* 1:37.

Sheehy TW, Rubini ME, Perez-Santiago E, Santini R Jr, Haddock J. 1961. The effect of "minute" and "titrated" amounts of folic acid on the megaloblastic anemia of tropical sprue. *Blood* 18:623–636.

Shojania AM, Hornady G, Barnes PH. 1968. Oral contraceptives and serum-folate level. *Lancet* 1:1376–1377.

Shojania AM, Hornady GJ, Barnes PH. 1971. The effect of oral contraceptives on folate metabolism. *Am J Obstet Gynecol* 111:782–791.

Shorvon SD, Carney MW, Chanarin I, Reynolds EH. 1980. The neuropsychiatry of megaloblastic anaemia. *Br Med J* 281:1036–1038.

Smith AM, Picciano MF, Deering RH. 1983. Folate supplementation during lactation: Maternal folate status, human milk folate content, and their relationship to infant folate status. *J Pediatr Gastroenterol Nutr* 2:622–628.

Smith AM, Picciano MF, Deering RH. 1985. Folate intake and blood concentrations of term infants. *Am J Clin Nutr* 41:590–598.

Smith JL, Goldsmith GA, Lawrence JD. 1975. Effect of oral contraceptive steroids on vitamin and lipid levels in serum. *Am J Clin Nutr* 28:371–376.

Smithells RW, Sheppard S, Schorah CJ. 1976. Vitamin deficiencies and neural tube defects. *Arch Dis Child* 51:944–950.

Smithells RW, Sheppard S, Schorah CJ, Seller MJ, Nevin NC, Harris R, Read AP, Fielding DW. 1981. Apparent prevention of neural tube defects by periconceptional vitamin supplementation. *Arch Dis Child* 56:911–918.

Smithells RW, Nevin NC, Seller MJ, Sheppard S, Harris R, Read AP, Fielding DW, Walker S, Schorah CJ, Wild J. 1983. Further experience of vitamin supplementation for prevention of neural tube defect recurrences. *Lancet* 1:1027–1031.

Sparling R, Abela M. 1985. Hypersensitivity to folic acid therapy. *Clin Lab Haematol* 7:184–185.

Spector RG. 1972. Influence of folic acid on exitable tissues. *Nature* 240:247–249.

Spies TD, Stone RE. 1947. Liver extract, folic acid, and thymine in pernicious anemia and subacute combined degeneration. *Lancet* 1:174–176.

Spies TD, Stone RE, Lopez GG, Milanes F, Aramburu T, Toca RL. 1948. The association between gastric achlorhydria and subacute combined degeneration of the spinal cord. *Postgrad Med* 4:89–95.

Stabler SP, Marcell PD, Podell ER, Allen RH, Savage DG, Lindenbaum J. 1988. Elevation of total homocysteine in the serum of patients with cobalamin or folate deficiency detected by capillary gas chromatography-mass spectrometry. *J Clin Invest* 81:466–474.

Stabler SP, Lindenbaum J, Allen RH. 1996. The use of homocysteine and other metabolites in the specific diagnosis of vitamin B-12 deficiency. *J Nutr* 126:1266S–1272S.

Stanley OH, Chambers TL. 1982. Sodium valproate and neural tube defects. *Lancet* 2:1282–1283.

Steegers-Theunissen RP, Boers GH, Blom HJ, Trijbels FJ, Eskes TK. 1992. Hyperhomocysteinaemia and recurrent spontaneous abortion or abruptio placentae. *Lancet* 339:1122–1123.

Steegers-Theunissen RP, Boers GH, Trijbels FJ, Finkelstein JD, Blom HJ, Thomas CM, Borm GF, Wouters MG, Eskes TK. 1994. Maternal hyperhomocysteinemia: A risk factor for neural tube defects? *Metabolism* 43:1475–1480.

Stites TE, Bailey LB, Scott KC, Toth JP, Fisher WP, Gregory JF 3rd. 1997. Kinetic modeling of folate metabolism through use of chronic administration of deuterium-labeled folic acid in men. *Am J Clin Nutr* 65:53–60.

Suarez RM, Spies TD, Suarez RM Jr. 1947. The use of folic acid in sprue. *Ann Intern Med* 26:643–677.

Subar AF, Harlan LC, Mattson ME. 1990. Food and nutrient intake differences between smokers and non-smokers in the U.S. *Am J Publ Health* 80:1323–1329.

Tamura T. 1995. Nutrient interaction of folate and zinc. In: Bailey LB, ed. *Folate in Health and Disease.* New York: Marcel Dekker. Pp. 287–312.

Tamura T, Stokstad EL. 1973. The availability of food folate in man. *Br J Haematol* 25:513–532.

Tamura T, Yoshimura Y, Arakawa T. 1980. Human milk folate and folate status in lactating mothers and their infants. *Am J Clin Nutr* 33:193–197.

Tamura T, Goldenberg RL, Freeberg LE, Cliver SP, Cutter GR, Hoffman HJ. 1992. Maternal serum folate and zinc concentrations and their relationships to pregnancy outcome. *Am J Clin Nutr* 56:365–370.

Tamura T, Mizuno Y, Johnston KE, Jacob RA. 1997. Food folate assay with protease, α-amylase, and folate conjugase treatments. *J Agric Food Chem* 45:135–139.

Tassabehji M, Read AP, Newton VE, Patton M, Gross P, Harris R, Strachan T. 1993. Mutations in the *PAX3* gene causing Waardenburg syndrome type I and type 2. *Nat Genet* 3:26–30.

Thiersch JB. 1952. Therapeutic abortions with folic acid antagonists, 4-amino pteroylglutamic acid administration by the oral route. *Am J Obstet Gynecol* 63:1298–1304.

Thirkettle JL, Gough KR, Read AE. 1964. Diagnostic value of small oral doses of folic acid in megaloblastic anemia. *Br Med J* 1:1286–1289.

Tolarova M, Harris J. 1995. Reduced recurrence of orofacial clefts after periconceptional supplementation with high-dose folic acid and multivitamins. *Teratology* 51:71–78.

Tsai JC, Perrella MA, Yoshizumi M, Hsieh CM, Haber E, Schlegel R, Lee ME. 1994. Promotion of vascular smooth muscle cell growth by homocysteine: A link to atherosclerosis. *Proc Natl Acad Sci USA* 91:6369–6373.

Tucker KL, Selhub J, Wilson PW, Rosenberg IH. 1996. Dietary intake pattern relates to plasma folate and homocysteine concentrations in the Framingham Heart Study. *J Nutr* 126:3025–3031.

Turner AJ. 1977. Commentary: The roles of folate and pteridine derivatives in neurotransmitter metabolism. *Biochem Pharmacol* 26:1009–1014.

Ubbink JB, Vermaak WJ, van der Merwe A, Becker PJ. 1993. Vitamin B-12, vitamin B-6, and folate nutritional status in men with hyperhomocysteinemia. *Am J Clin Nutr* 57:47–53.

Ubbink JB, Becker PJ, Vermaak WJ, Delport R. 1995a. Results of B-vitamin supplementation study used in a prediction model to define a reference range for plasma homocysteine. *Clin Chem* 41:1033–1037.

Ubbink JB, Vermaak WJ, Delport R, van der Merwe A, Becker PJ, Potgieter H. 1995b. Effective homocysteine metabolism may protect South African blacks against coronary heart disease. *Am J Clin Nutr* 62:802–808.

Vanaerts LA, Blom HJ, Deabreu RA, Trijbels FJ, Eskes TK, Copius Peereboom-Stegeman JH, Noordhoek J. 1994. Prevention of neural tube defects by and toxicity of L-homocysteine in cultured postimplantation rat embryos. *Teratology* 50:348–360.

van der Put NM, Steegers-Theunissen RP, Frosst P, Trijbels FJ, Eskes TK, van den Heuvel LP, Mariman EC, den Heyer M, Rozen R, Blom HJ. 1995. Mutated methylenetetrahydrofolate reductase as a risk factor for spina bifida. *Lancet* 346:1070–1071.

van der Put NM, Thomas CM, Eskes TK, Trijbels FJ, Steegers-Theunissen RP, Mariman EC, De Graaf-Hess A, Smeitink JA, Blom HJ. 1997a. Altered folate and vitamin B_{12} metabolism in families with spina bifida offspring. *Q J Med* 90:505–510.

van der Put NM, van der Molen EF, Kluijtmans LA, Heil SG, Trijbels JM, Eskes TK, Van Oppenraaij-Emmerzaal D, Banerjee R, Blom HJ. 1997b. Sequence analysis of the coding region of human methionine synthase: Relevance to hyperhomocysteinaemia in neural-tube defects and vascular disease. *Q J Med* 90:511–517.

van der Westhuyzen J, Metz J. 1983. Tissue S-adenosylmethionine levels in fruit bats with N_2O-induced neuropathy. *Br J Nutr* 50:325–330.

van der Westhuyzen J, Fernandes-Costa F, Metz J. 1982. Cobalamin inactivation by nitrous oxide produces severe neurological impairment in fruit bats: Protection by methionine and aggravation by folates. *Life Sci* 31:2001–2010.

Varadi S, Abbott D, Elwis A. 1966. Correlation of peripheral white cell and bone marrow changes with folate levels in pregnancy and their clinical significance. *J Clin Pathol* 19:33–36.

Velie EM, Shaw GM. 1996. Impact of prenatal diagnosis and elective termination on prevalence and risk estimates of neural tube defects in California, 1989–1991. *Am J Epidemiol* 144:473–479.

Vergel RG, Sanchez LR, Heredero BL, Rodriguez PL, Martinez AJ. 1990. Primary prevention of neural tube defects with folic acid supplementation: Cuban experience. *Prenat Diagn* 10:149–152.

Verhaar MC, Wever RM, Kastelein JJ, van Dam T, Koomans HA, Rabelink TJ. 1998. 5-Methyltetrahydrofolate, the active form of folic acid, restores endothelial function in familial hypercholesterolemia. *Circulation* 97:237–241.

Verhoef P, Stampfer MJ, Buring JE, Gaziano JM, Allen RH, Stabler SP, Reynolds RD, Kok FJ, Hennekens CH, Willett WC. 1996. Homocysteine metabolism and risk of myocardial infarction: Relation with vitamins B_6, B_{12}, and folate. *Am J Epidemiol* 143:845–859.

Verhoef P, Kok FJ, Kluijtmans LA, Blom HJ, Refsum H, Ueland PM, Kruyssen DA. 1997a. The 677C→T mutation in the methylenetetrahydrofolate reductase gene: Associations with plasma total homocysteine levels and risk of coronary atherosclerotic disease. *Atherosclerosis* 132:105–113.

Verhoef P, Rimm EB, Hunter DJ, Chen J, Willett WC, Kelsey K, Stampfer MJ. 1997b. Methylenetetrahydrofolate reductase polymorphism and risk of coronary heart disease: Results from health professionals study and meta-analysis. *Am J Epid* 145:307.

Verreault R, Chu J, Mandelson M, Shy K. 1989. A case-control study of diet and invasive cervical cancer. *Int J Cancer* 43:1050–1054.

Victor M, Lear AA. 1956. Subacute combined degeneration of the spinal cord. Current concepts of the disease process. Value of serum vitamin B_{12} determinations in clarifying some of the common clinical problems. *Am J Med* 20:896–911.

Vilter CF, Vilter RW, Spies TD. 1947. The treatment of pernicious and related anemias with synthetic folic acid. 1. Observations on the maintenance of a normal hematologic status and on the occurrence of combined system disease at the end of one year. *J Lab Clin Med* 32:262–273.

Volpe JJ. 1995. *Neurology of the Newborn*, 3rd ed. Philadelphia: WB Saunders.

Von der Porten AE, Gregory JF 3rd, Toth JP, Cerda JJ, Curry SH, Bailey LB. 1992. In vivo folate kinetics during chronic supplementation of human subjects with deuterium-labeled folic acid. *J Nutr* 122:1293–1299.

Wagley PF. 1948. Neurologic disturbances with folic acid therapy. *N Engl J Med* 238:11–15.

Wagner C. 1996. Symposium on the subcellular compartmentation of folate metabolism. *J Nutr* 126:1228S–1234S.

Wahlin A, Hill RD, Winblad B, Backman L. 1996. Effects of serum vitamin B_{12} and folate status on episodic memory performance in very old age: A population-based study. *Psychol Aging* 11:487–496.

Wald NJ. 1994. Folic acid and neural tube defects: The current evidence and implications for prevention. *Ciba Found Symp* 181:192–208.

Wald N, Sneddon J, Densem J, Frost C, Stone R. 1991. Prevention of neural tube defects: Results of the Medical Research Council vitamin study. *Lancet* 338:131–137.

Watanabe M, Osada J, Aratani Y, Kluckman K, Reddick R, Malinow MR, Maeda N. 1995. Mice deficient in cystathionine beta-synthase: Animal models for mild and severe homocyst(e)inemia. *Proc Natl Acad Sci USA* 92:1585–1589.

Ward M, McNulty H, McPartlin J, Strain JJ, Weir DG, Scott JM. 1997. Plasma homocysteine, a risk factor for cardiovascular disease, is lowered by physiological doses of folic acid. *Q J Med* 90:519–524.

Wei M-M, Bailey LB, Toth JP, Gregory JF 3rd. 1996. Bioavailability for humans of deuterium-labeled monoglutamyl and polyglutamyl folates is affected by selected foods. *J Nutr* 126:3100–3108.

Weir DG, McGing PG, Scott JM. 1985. Commentary: Folate metabolism, the enterohepatic circulation and alcohol. *Biochem Pharmacol* 34:1–7.

Weller M, Marini AM, Martin B, Paul SM. 1994. The reduced unsubstituted pteroate moiety is required for folate toxicity of cultured cerebellar granule neurons. *J Pharmacol Exp Ther* 269:393–401.

Werler MM, Shapiro S, Mitchell AA. 1993. Periconceptional folic acid exposure and risk of occurrent neural tube defects. *J Am Med Assoc* 269:1257–1261.

Whitehead AS, Gallagher P, Mills JL, Kirke PN, Burke H, Molloy AM, Weir DG, Shields DC, Scott JM. 1995. A genetic defect in 5, 10 methylenetetrahydrofolate reductase in neural tube defects. *Q J Med* 88:763–766.

Whitehead VM. 1973. Polygammaglutamyl metabolites of folic acid in human liver. *Lancet* 1:743–745.

Whitehead VM. 1986. Pharmacokinetics and physiological disposition of folate and its derivatives. In: Blakely RL, Whitehead VM, eds. *Folates and Pterins, Vol. 3*. New York: John Wiley & Sons. Pp. 177–205.

Wilcken DE, Wilcken B. 1976. The pathogenesis of coronary artery disease. A possible role for methionine metabolism. *J Clin Invest* 57:1079–1082.

Wilcken DE, Wang XL, Sim AS, McCredie RM. 1996. Distribution in healthy and coronary populations of the methylenetetrahydrofolate reductase (MTHFR) *C677T* mutation. *Arterioscler Thromb Vasc Biol* 16:878–882.

Will JJ, Mueller JF, Brodine C, Kiely CE, Friedman B, Hawkins VR, Dutra J, Vilter RN. 1959. Folic acid and vitamin B_{12} in pernicious anemia. Studies on patients treated with these substances over a ten-year period. *J Lab Clin Med* 53:22–38.

Willard JE, Lange RA, Hillis LD. 1992. The use of aspirin in ischemic heart disease. *N Engl J Med* 327:175–181.

Willoughby ML. 1967. An investigation of folic acid requirements in pregnancy. II. *Br J Haematol* 13:503–509.

Willoughby ML, Jewell FJ. 1966. Investigation of folic acid requirements in pregnancy. *Br Med J* 2:1568–1571.

Willoughby ML, Jewell FG. 1968. Folate status throughout pregnancy and in postpartum period. *Br Med J* 4:356–360.

Wilson JG. 1973. *Environment and Birth Defects*. New York: Academic Press

Witter FR, Blake DA, Baumgardner R, Mellits ED, Niebyl JR. 1982. Folate, carotene, and smoking. *Am J Obstet Gynecol* 144:857.

Wouters MG, Boers GH, Blom HJ, Trijbels FJ, Thomas CM, Borm GF, Steegers-Theunissen RP, Eskes TK. 1993. Hyperhomocysteinemia: A risk factor in women with unexplained recurrent early pregnancy loss. *Fertil Steril* 60:820–825.

Wu A, Chanarin I, Slavin G. Levi AJ. 1975. Folate deficiency in the alcoholic—its relationship to clinical and haematological abnormalities, liver disease and folate stores. *Br J Haematol* 29:469–478.

Yen IH, Khoury MJ, Erickson JD, James LM, Waters GD, Berry RJ. 1992. The changing epidemiology of neural tube defects: United States, 1968–1989. *Am J Dis Child* 146:857–861.

Young SN, Ghadirian AM. 1989. Folic acid and psychopathology. *Prog Neuropsychopharmacol Biol Psychiatry* 13:841–863.

Zalusky R, Herbert V. 1961. Megaloblastic anemia in scurvy with response to 50 micrograms of folic acid daily. *N Engl J Med* 265:1033–1038.

Ziegler RG, Brinton LA, Hammon RF, Lehman HF, Levine RS, Mallin K, Norman SA, Rosenthal JF, Trumble AC, Hoover RN. 1990. Diet and risk of invasive cervical cancer among white women in the United States. *Am J Epidemiol* 132:432–445.

Ziegler RG, Jones CJ, Brinton LA, Norman SA, Mallin K, Levine RS, Lehman HF, Hammon RF, Trumble AC, Rosenthal JF. 1991. Diet and risk of in situ cervical cancer among white women in the United States. *Cancer Causes Control* 2:17–29.

Zimmerman J. 1990. Folic acid transport in organ-cultured mucosa of human intestine. Evidence for distinct carriers. *Gastroenterology* 99:964–972.

9

Vitamin B$_{12}$

SUMMARY

Vitamin B$_{12}$ (cobalamin) functions as a coenzyme for a critical methyl transfer reaction that converts homocysteine to methionine and for a separate reaction that converts L-methylmalonyl-coenzyme A (CoA) to succinyl-CoA. The Recommended Dietary Allowance (RDA) for vitamin B$_{12}$ is based on the amount needed for the maintenance of hematological status and normal serum vitamin B$_{12}$ values. An assumed absorption of 50 percent is included in the recommended intake. The RDA for adults is 2.4 µg/day of vitamin B$_{12}$. Because 10 to 30 percent of older people may be unable to absorb naturally occurring vitamin B$_{12}$, it is advisable for those older than 50 years to meet their RDA mainly by consuming foods fortified with vitamin B$_{12}$ or a vitamin B$_{12}$-containing supplement. Individuals with vitamin B$_{12}$ deficiency caused by a lack of intrinsic factor require medical treatment. The median intake of vitamin B$_{12}$ from food in the United States was estimated to be approximately 5 µg/day for men and 3.5 µg/day for women. The ninety-fifth percentile of vitamin B$_{12}$ intake from both food and supplements was approximately 27 µg/day. In one Canadian province the mean dietary intake was estimated to be approximately 7 µg/day for men and 4 µg/day for women. There is not sufficient scientific evidence to set a Tolerable Upper Intake Level (UL) for vitamin B$_{12}$ at this time.

BACKGROUND INFORMATION

Cobalamin is the general term used to describe a group of cobalt-containing compounds (corrinoids) that have a particular structure that contains the sugar ribose, phosphate, and a base (5, 6-dimethyl benzimidazole) attached to the corrin ring. Vitamin B$_{12}$ can be converted to either of the two cobalamin coenzymes that are active in human metabolism: methylcobalamin and 5-deoxyadenosylcobalamin. Although the preferred scientific use of the term *vitamin B$_{12}$* is usually restricted to cyanocobalamin, in this report, B$_{12}$ will refer to all potentially biologically active cobalamins.

In the United States, cyanocobalamin is the only commercially available B$_{12}$ preparation used in supplements and pharmaceuticals. It is also the principal form used in Canada (B. A. Cooper, Department of Hematology, Stanford University, personal communication, 1997). Another form, hydroxocobalamin, has been used in some studies of B$_{12}$. Compared with hydroxocobalamin, cyanocobalamin binds to serum proteins less well and is excreted more rapidly (Tudhope et al., 1967).

Function

B$_{12}$ is a cofactor for two enzymes: methionine synthase and L-methylmalonyl-CoA mutase. Methionine synthase requires methylcobalamin as a cofactor for the methyl transfer from methyltetrahydrofolate to homocysteine to form methionine and tetrahydrofolate. L-Methymalonyl-CoA mutase requires adenosylcobalamin to convert L-methymalonyl-CoA to succinyl-CoA in an isomerization reaction. In B$_{12}$ deficiency, folate may accumulate in the serum as a result of slowing of the B$_{12}$-dependent methyltransferase. An adequate supply of B$_{12}$ is essential for normal blood formation and neurological function.

Physiology of Absorption, Metabolism, Storage, and Excretion

Small amounts of B$_{12}$ are absorbed via an active process that requires an intact stomach, intrinsic factor (a glycoprotein that the parietal cells of the stomach secrete after being stimulated by food), pancreatic sufficiency, and a normally functioning terminal ileum. In the stomach, food-bound B$_{12}$ is dissociated from proteins in the presence of acid and pepsin. The released B$_{12}$ then binds to R proteins (haptocorrins) secreted by the salivary glands and the gastric mucosa. In the small intestine, pancreatic proteases partially degrade the R proteins, releasing B_{12} to bind with intrinsic factor. The resulting complex of intrinsic factor and B_{12} attaches to specific receptors in the ileal mucosa; after internalization of the complex, B_{12} enters the enterocyte. Approximately 3 to 4 hours later, B_{12} enters the circulation. All circulating B_{12} is bound to the plasma binding proteins—transcobalamin I, II, or III (TCI, TCII, or TCIII). Although TCI binds approximately 80 percent of the B_{12} carried in the blood, TCII is the form that delivers B_{12} to the tissues through specific receptors for TCII (Hall and Finkler, 1966; Seetharam and Alpers, 1982). The liver takes up approximately 50 percent of the B_{12} and the remainder is transported to other tissues.

If there is a lack of intrinsic factor (as is the case in the condition called pernicious anemia), malabsorption of B_{12} results; if this is untreated, potentially irreversible neurological damage and life-threatening anemia develop.

The average B_{12} content of liver tissue is approximately 1.0 $\mu g/g$ of tissue in healthy adults (Kato et al., 1959; Stahlberg et al., 1967). Estimates of the average total-body B_{12} pool in adults range from 0.6 (Adams et al., 1972) to 3.9 mg (Grasbeck et al., 1958), but most estimates are between 2 and 3 mg (Adams, 1962; Adams et al., 1970; Heinrich, 1964; Reizenstein et al., 1966). The highest estimate found for an individual's total body B_{12} store was 11.1 mg (Grasbeck et al., 1958). Excretion of B_{12} is proportional to stores (see "Excretion").

Absorption

Studies to measure the actual absorption of B_{12} involve whole-body counting of radiolabeled B_{12}, counting of radiolabeled B_{12} in the stool, or both. No data are available on whether B_{12} absorption varies with B_{12} status, but fractional absorption decreases as the oral dose is increased (Chanarin, 1979). Total absorption increases with increasing intake. Adams and colleagues (1971) measured fractional absorption of radiolabeled cyanocobalamin and reported that nearly 50 percent was retained at a 1-μg dose, 20 percent at a 5-μg dose, and just over 5 percent at a 25-μg dose. The second of two doses of B_{12} given 4 to 6 hours apart is absorbed as well as the first (Heyssel et al., 1966). When large doses of crystalline B_{12} are ingested, up to approximately 1 percent of the dose may be absorbed by mass action even in the absence of intrinsic factor (Berlin et al., 1968; Doscherholmen and Hagen, 1957).

Absorption from Food. The approximate percentage absorption of B$_{12}$ from a few foods is presented in Table 9-1. These values apply to normal, healthy adults. No studies were found on the absorption of B$_{12}$ from dairy foods or from red meat other than mutton and liver. The absorption efficiency of B$_{12}$ from liver reportedly was low because of its high B$_{12}$ content. Although evidence indicates that a B$_{12}$ content of 1.5 to 2.5 µg/meal saturates ileal receptors and thus limits further absorption (Scott, 1997), absorption of as much as 7 µg in one subject (18 percent) was reported from a serving of liver paste that contained 38 µg of B$_{12}$ (average absorption was 4.1 µg or 11 percent) (Heyssel et al., 1966).

Assumptions Used in this Report. Because of the lack of data on dairy foods and most forms of red meat and fish, a conservative adjustment for the bioavailability of naturally occurring B$_{12}$ is used in this report. In particular, it is assumed that 50 percent of dietary B$_{12}$ is absorbed by healthy adults with normal gastric function. A smaller fractional absorption would apply, however, if a person consumed a large portion of foods rich in B$_{12}$. Different levels of absorption are assumed under various conditions, as shown in Table 9-2. Crystalline B$_{12}$ appears in the diet only in foods that have been fortified with B$_{12}$, such as breakfast cereals and liquid meal replacements.

Enterohepatic Circulation

B$_{12}$ is continually secreted in the bile. In healthy individuals most of this B$_{12}$ is reabsorbed and available for metabolic functions. El Kholty et al. (1991) demonstrated that the secretion of B$_{12}$ into the bile averaged 1.0 ± 0.44 nmol/day (1.4 µg/day) in eight cholecystectomized patients, and this represented 55 percent of total corrinoids. If approximately 50 percent of this B$_{12}$ is assumed to be

TABLE 9-1 Percentage Absorption of Vitamin B$_{12}$ from Foods by Healthy Adults

Reference	Food	Absorption (%)
Heyssel et al., 1966	Mutton	65
Heyssel et al., 1966	Liver	11
Doscherholmen et al., 1975	Eggs	24–36
Doscherholmen et al., 1978	Chicken	60
Doscherholmen et al., 1981	Trout	25–47

TABLE 9-2 Assumed Vitamin B_{12} Absorption under Different Conditions

Form of Vitamin B_{12}	Normal Gastric Function (%)	Pernicious Anemia[a] (%)
Naturally occurring[b]	50	0
Crystalline, low dose (< 5 µg)[b]	60	0
Crystalline, high dose (≥ 500 µg) with water[c]	1	1
Crystalline, high dose with food[c]	0.5	≤ 0.5

[a] A disorder in which lack of intrinsic factor severely limits the absorption of vitamin B_{12}.
[b] Heyssel et al. (1966).
[c] Berlin et al. (1968).

reabsorbed, the average loss of biliary B_{12} in the stool would be 0.5 nmol/day (0.7 µg/day). Research with baboons (Green et al., 1982) suggests that the form of B_{12} present in bile may be absorbed more readily than is cyanocobalamin, but the absorption of both forms was enhanced by intrinsic factor. Both Green and colleagues (1982) and Teo and coworkers (1980) reported data suggesting that bile enhances B_{12} absorption. However, in the absence of intrinsic factor, essentially all the B_{12} from the bile is excreted in the stool rather than recirculated. Thus, B_{12} deficiency develops more rapidly in individuals who have no intrinsic factor or who malabsorb B_{12} for other reasons than it does in those who become complete vegetarians and thus ingest no B_{12}.

Excretion

If the circulating B_{12} exceeds the B_{12} binding capacity of the blood, the excess is excreted in the urine. This typically occurs only after injection of B_{12}. The highest losses of B_{12} ordinarily occur through the feces. Sources of fecal B_{12} include unabsorbed B_{12} from food or bile, desquamated cells, gastric and intestinal secretions, and B_{12} synthesized by bacteria in the colon. Other losses occur through the skin and metabolic reactions. Fecal (Reizenstein, 1959) and urinary losses (Adams, 1970; Heinrich, 1964; Mollin and Ross, 1952) decrease when B_{12} stores decrease. Various studies have indicated losses of 0.1 to 0.2 percent of the B_{12} pool per day (Amin et al., 1980; Boddy and Adams, 1972; Bozian et al., 1963; Heinrich, 1964; Heyssel et al., 1966; Reizenstein et al., 1966) regardless of the size of the store, with the 0.2 percent value generally applicable to those with pernicious anemia.

Clinical Effects of Inadequate Intake

Hematological Effects of Deficiency

The major cause of clinically observable B$_{12}$ deficiency is pernicious anemia (see "Pernicious Anemia"). The hematological effects of B$_{12}$ deficiency are indistinguishable from those of folate deficiency (see Chapter 8). These include pallor of the skin associated with a gradual onset of the common symptoms of anemia, such as diminished energy and exercise tolerance, fatigue, shortness of breath, and palpitations. As in folate deficiency, the underlying mechanism of anemia is an interference with normal deoxyribonucleic acid (DNA) synthesis. This results in megaloblastic change, which causes production of larger-than-normal erythrocytes (macrocytosis). This leads first to an increase in the erythrocyte distribution width index and ultimately to an elevated mean cell volume. Oval macrocytes and other abnormally shaped erythrocytes are present in the blood. Typically, as with folate deficiency, the appearance of hypersegmentation of polymorphonuclear leukocytes precedes the development of macrocytosis. However, the sensitivity of this finding has recently been questioned (Carmel et al., 1996). By the time anemia has become established, there is usually also some degree of neutropenia and thrombocytopenia because the megaloblastic process affects all rapidly dividing bone marrow elements. The hematological complications are completely reversed by treatment with B$_{12}$.

Neurological Effects of Deficiency

Neurological complications are present in 75 to 90 percent of individuals with clinically observable B$_{12}$ deficiency and may, in about 25 percent of cases, be the only clinical manifestation of B$_{12}$ deficiency. Evidence is mounting that the occurrence of neurological complications of B$_{12}$ deficiency is inversely correlated with the degree of anemia; patients who are less anemic show more prominent neurological complications and vice versa (Healton et al., 1991; Savage et al., 1994a). Neurological manifestations include sensory disturbances in the extremities (tingling and numbness), which are worse in the lower limbs. Vibratory and position sense are particularly affected. Motor disturbances, including abnormalities of gait, also occur. Cognitive changes may occur, ranging from loss of concentration to memory loss, disorientation, and frank dementia, with or without mood changes. In addition, visual disturbances, insomnia, impotency, and impaired bowel and bladder control may develop. The progression of neurological manifestations is variable but generally gradual. Whether neurological complications are reversible after treatment depends on their duration. The neurological complications of B_{12} deficiency occur at a later stage of depletion than do the indicators considered below and were, therefore, not used for estimating the requirement for B_{12}. Moreover, neurological complications are not currently amenable to easy quantitation nor are they specific to B_{12} deficiency.

Gastrointestinal Effects of Deficiency

B_{12} deficiency is also frequently associated with various gastrointestinal complaints, including sore tongue, appetite loss, flatulence, and constipation. Some of these complaints may be related to the underlying gastric disorder in pernicious anemia.

SELECTION OF INDICATORS FOR ESTIMATING THE REQUIREMENT FOR VITAMIN B_{12}

Search of the literature revealed numerous indicators that could be considered as the basis for deriving an Estimated Average Requirement (EAR) for vitamin B_{12} for adults. These include but are not limited to hematological values such as erythrocyte count, hemoglobin concentration or hematocrit, and mean cell volume (MCV), blood values such as plasma B_{12}, and the metabolite methylmalonic acid (MMA).

Indicators of Hematological Response

Measurements used to indicate a hematological response that could be considered as indicative of B_{12} sufficiency have consisted of either a minimal but significant increase in hemoglobin, hematocrit, and erythrocyte count; a decrease in MCV; or an optimal rise in reticulocyte number.

In the earliest studies, MCV was a calculated value that was derived from relatively imprecise erythrocyte counts. Although MCV is now directly measured and precise, the response time of this measurement to changes in dietary intake is slow because of the 120-day longevity of erythrocytes. Consequently, the MCV is of limited usefulness. The erythrocyte count, hemoglobin, and hematocrit values are all robust measurements of response. Again, however, the response time is slow before an improvement in B_{12} status leads to a return to normal values. Partial responses are of limited value because they do not predict the ultimate completeness or maintenance of response.

The reticulocyte count is a useful measure of hematological response because an increase is apparent within 48 hours of B$_{12}$ administration and reaches a peak at 5 to 8 days.

Serum or Plasma Vitamin B$_{12}$

The concentration of B$_{12}$ in the serum or plasma reflects both the B$_{12}$ intake and stores. The lower limit is considered to be approximately 120 to 180 pmol/L (170 to 250 pg/mL) for adults but varies with the method used and the laboratory conducting the analysis. As deficiency develops, serum values may be maintained at the expense of B$_{12}$ in the tissues. Thus, a serum B$_{12}$ value above the cutoff point does not necessarily indicate adequate B$_{12}$ status (see the section "Vitamin B$_{12}$ Deficiency") but a low value may represent a long-term abnormality (Beck, 1991) or prolonged low intake.

Methylmalonic Acid

The range that represents expected variability (2 standard deviations) for serum MMA is 73 to 271 nmol/L (Pennypacker et al., 1992). The concentration of MMA in the serum rises when the supply of B$_{12}$ is low. Elevation of MMA may also be caused by renal failure or intravascular volume depletion (Stabler et al., 1988), but Lindenbaum and coworkers (1994) reported that moderate renal dysfunction in the absence of renal failure does not affect MMA values as strongly as does inadequate B$_{12}$ status. MMA values tend to rise in the elderly (Joosten et al., 1996); in most cases this appears to reflect inadequate B$_{12}$ intake or absorption. Lindenbaum and coworkers (1988) reported that elevated serum MMA concentrations are present in many patients with neuropsychiatric disorders caused by B$_{12}$ deficiency. Pennypacker and colleagues (1992) found that intramuscular injections of B$_{12}$ reduced the elevated MMA values in their elderly subjects. The reduction of elevated MMA values with B$_{12}$ therapy has also been reported in other studies (Joosten et al., 1993; Naurath et al., 1995; Norman and Morrison, 1993). Increased activity of anaerobic flora in the intestinal tract may increase serum MMA values; treatment with antibiotics decreases the serum MMA concentration in this situation (Lindenbaum et al., 1990). Because the presence of elevated concentrations of MMA in serum represents a metabolic change that is highly specific to B$_{12}$ deficiency, the serum MMA concentration is a preferred indicator of B_{12} status. However, data were not sufficient to use MMA as the criterion on which to base the EAR in this report. Serum MMA values from older studies may not be comparable with those obtained recently because of improvements of methods over time (Beck, 1991; Green and Kinsella, 1995). More importantly, no studies were found that examined directly the relationship of B_{12} intake and MMA concentrations.

Homocysteine

Serum total homocysteine concentration is commonly elevated in elderly persons whose folate status is normal but who have a clinical response to treatment with B_{12} (Stabler et al., 1996). Because a lack of folate, vitamin B_6, or both also results in an elevated serum and plasma homocysteine concentration, this indicator has poor specificity and does not provide a useful basis for deriving an EAR.

Formiminoglutamic Acid, Propionate, and Methylcitrate

Although most patients with untreated B_{12} deficiency excrete an increased amount of formiminoglutamic acid (FIGLU) in the urine after an oral loading dose of histidine, FIGLU excretion is also almost invariably increased in folate deficiency as well. The test, therefore, lacks specificity for the diagnosis of either vitamin deficiency. Concentrations of propionate, the metabolic precursor of methylmalonate, also rise with B_{12} deficiency. Propionate may be converted to 2-methylcitrate, serum and cerebrospinal fluid concentrations of which also rise in B_{12} deficiency (Allen et al., 1993). However, the measurement of either propionate or methyl citrate offers no advantages over serum MMA for the detection of B_{12} deficiency.

Holotranscobalamin II

Among the three plasma B_{12} binding proteins, transcobalamin II (TCII) is responsible for receptor-mediated uptake of B_{12} into cells. However, only a small fraction of the plasma B_{12} (10 to 20 percent) is present as the TCII-B_{12} complex. This fraction, termed *holoTCII*, may provide a good indication of B_{12} status, and methods have been described to measure this fraction (Herzlich and Herbert, 1988; Vu et al., 1993). These methods are currently considered to be insufficiently robust for routine clinical use.

METHODOLOGICAL ISSUES

Vitamin B$_{12}$ Content

The two primary microbial organisms used to determine the vitamin B$_{12}$ content of serum, urine, and stool are *Euglena gracilis* and *Lactobacillus leichmannii*. Although either organism will yield essentially similar results, *L. leichmannii* is the preferred method for reasons of convenience (Chanarin, 1969). Microbiological assays have been largely supplanted by radioligand binding assays. Until 1978 radioligand binding assays frequently gave higher results; the binding protein for B$_{12}$ used in these assays would also bind analogues of B$_{12}$ (Beck, 1991; Russell, 1992). Since 1978 the use of purified intrinsic factor as the binder in commercial radioisotope dilution assay kits has resulted in serum concentrations of B$_{12}$ comparable with those obtained from microbiological assays. More recently, nonisotopic serum B$_{12}$ assays have been introduced, which has resulted in cutoff levels for B$_{12}$ deficiency again rising. Care must be taken in comparing studies because much variation has been noted across laboratories, and different cutoff points have been used to identify deficiency (Beck, 1991; Green and Kinsella, 1995; Miller et al., 1991; Rauma et al., 1995; WHO, 1970; Winawer et al., 1967).

The serum B$_{12}$ value may be misleading as an indicator because it includes all the B$_{12}$ regardless of the protein to which it is bound. Transcobalamin II (TCII) is the key transport protein, and it has been proposed that only the TCII-bound fraction of the serum B$_{12}$ (holoTCII) is important in relation to B$_{12}$ nutritional and metabolic status (Herzlich and Herbert, 1988; Vu et al., 1993). However, at this time, there is no reliable method to determine holoTCII.

Retention

Studies of the retention of parenterally administered B$_{12}$ indicate that percentage retention depends on the dose and the route of administration (intramuscular [IM] or intravenous). The expected percentage retention of IM cyanocobalamin is shown in Table 9-3. These values, which vary from 15 to 100 percent, are useful when IM doses of B$_{12}$ are used to estimate the B$_{12}$ requirement.

TABLE 9-3   Change in Percentage Retention of Vitamin B_{12} with Increasing Intramuscular Dose

Vitamin B_{12} Dose (µg)	Retention (%)
3	100
10	97
25	95
40	93
1,000	15

SOURCE: Chanarin (1969).

DIAGNOSIS

Vitamin B_{12} Deficiency

Early detection of vitamin B_{12} deficiency depends on biochemical measurements. Lindenbaum and colleagues (1990) reported that metabolites that arise from B_{12} insufficiency are more sensitive indicators of B_{12} deficiency than is the serum B_{12} value. This was found in patients with pernicious anemia or previous gastrectomy who experienced early hematological relapse: serum methylmalonic acid (MMA), total homocysteine, or both were elevated in 95 percent of the instances of relapse whereas the serum B_{12} value was low (less than 150 pmol/L [200 pg/mL]) in 69 percent. Similarly, serum B_{12} was found to be an insensitive indicator in a review of records of patients with clinically significant B_{12} deficiency. Five deficient individuals had neurological disorders that were responsive to B_{12} and had elevated serum MMA and homocysteine values even though their serum B_{12} values were greater than 150 pmol/L (200 pg/mL) and anemia was absent or mild. In a recent series of 173 patients, 5.2 percent of those with recognized B_{12} deficiency had serum B_{12} values in the normal range. Similar findings were reported elsewhere (e.g., Carmel, 1988; Pennypacker et al., 1992; Stabler et al., 1996). At present, the techniques developed to measure serum MMA and homocysteine (capillary gas chromatography and mass spectrometry) are costly and may be beyond the scope of routine laboratories. Conditions that may warrant assessment of B_{12} status because they may result in B_{12} deficiency are summarized in Table 9-4.

TABLE 9-4 Conditions That May Result in Vitamin B$_{12}$ Deficiency

Cause	Pathogenesis
Dietary deficiency	Insufficient B$_{12}$ intake, as seen in complete vegetarians
Pernicious anemia	Lack of intrinsic factor
Gastrectomy	Lack of intrinsic factor
Atrophic gastritis	Inability to digest protein-bound B$_{12}$ and bacterial uptake and/or conversion
Bacterial overgrowth of the small intestine	Bacterial uptake and/or conversion of B$_{12}$
Infection with *Diphyllobothrium latum*	Uptake of B$_{12}$ by the parasite
Terminal ileal disease or resection	Inability to absorb B$_{12}$
Pancreatic insufficiency	Inability to digest protein-bound B$_{12}$

Pernicious Anemia

Pernicious anemia is the end stage of an autoimmune disorder in which parietal cell autoantibodies against H$^+$K$^+$-adenosine triphosphatase cause loss of gastric parietal cells. The loss of parietal cells reduces and then completely prevents production of intrinsic factor. In addition, blocking autoantibodies can bind to the B$_{12}$ binding site for intrinsic factor and prevent the formation of the B$_{12}$-intrinsic factor complex. Deficiency of intrinsic factor gradually results in B$_{12}$ deficiency (see "Clinical Effects of Inadequate Intake").

The prevalence of undiagnosed, untreated pernicious anemia was recently estimated to be approximately 2 percent in a nonrandom sample of free-living elderly aged 60 years or older in Southern California (Carmel, 1996). Rates were higher for white and black women than for Latin American or Asian women and for all men. These estimates are consistent with the 2.9 percent prevalence of intrinsic factor antibody in individuals older than 60 years (Krasinski et al., 1986). Earlier studies reported a higher prevalence of anti–intrinsic factor antibody in blacks with pernicious anemia than in whites with pernicious anemia (Carmel, 1992) and an earlier onset of pernicious anemia in blacks (Carmel et al., 1987; Houston et al., 1985) and Hispanics (Carmel et al., 1987). Approximately 20 percent of relatives of patients with pernicious anemia also have pernicious anemia (Toh et al., 1997). Pernicious anemia carries an excess risk of gastric carcinoma (1 to 3 percent) and of gastric carcinoid tumors (Hsing et al., 1993).

A flow sheet for the diagnosis of pernicious anemia appears in Figure 9-1. Autoantibodies to gastric parietal cells should be measured along with intrinsic factor. The demonstration of circulating intrinsic factor autoantibodies is almost diagnostic of type A gastritis and pernicious anemia (Toh et al., 1997).

FACTORS AFFECTING THE VITAMIN B_{12} REQUIREMENT

Aging

Plasma vitamin B_{12} tends to decrease and serum methylmalonic acid (MMA) concentration tends to increase with age. These changes may represent a decline in B_{12} status. Factors that may contribute to these changes include a decrease in gastric acidity, the presence of atrophic gastritis and of bacterial overgrowth accompanied by food-bound B_{12} malabsorption, severity of atrophic gastritis, compromised functional and structural integrity of the B_{12} binding proteins, and a lack of liver B_{12} stores (van Asselt et al., 1996). Percentage absorption of crystalline B_{12} does not appear to decrease with age (McEvoy et al., 1982). In a study of 38 healthy subjects each 76 years old taken from a larger cohort study (Nilsson-Ehle et al., 1986), cyanocobalamin absorption was found to be comparable with that reported in eight other studies of healthy younger people.

Studies of absorption in the elderly have yielded somewhat contradictory results. van Asselt and coworkers (1996) found no significant difference in cobalamin absorption (either free or protein bound) between subjects younger than 64 years (median 57) and those 65 years and older (median 75 years). These investigators could not explain the high prevalence of low cobalamin values in the elderly by either the aging process or the occurrence of mild-to-moderate atrophic gastritis. In contrast Krasinski and coworkers (1986) demonstrated that although a small proportion of the elderly with atrophic gastritis have a low serum concentration of B_{12} (less than 88 pmol/L [120 pg/mL]), those with lowest serum B_{12} values tend to have severe atrophic gastritis. Scarlett and colleagues (1992) reported a reduction in dietary B_{12} absorption with age that was associated with elevated serum gastrin, which indicates reduced gastric acidity.

Prevalence of Atrophic Gastritis

Large differences in the prevalence of atrophic gastritis in the elderly, ranging from approximately 10 to 30 percent, have been

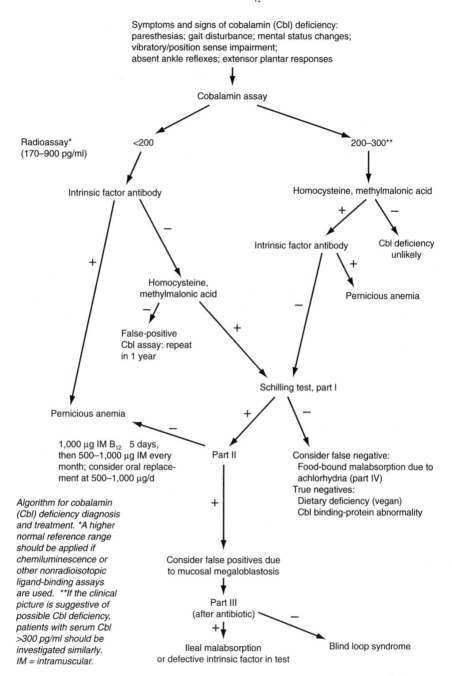

FIGURE 9-1 The diagnosis of pernicious anemia. Reprinted with permission from Green and Kinsella (1995). Copyright 1995 by Lippincott-Raven Publishers.

reported in Australia (Andrews et al., 1967), Missouri (Hurwitz et al., 1997), Scandinavia (Johnsen et al., 1991), and Boston (Krasinski et al., 1986). In the general elderly population, many cases of atrophic gastritis may remain undiagnosed.

Food-Bound B_{12} Malabsorption

Testing of individuals who have low serum B_{12} values but who do not have pernicious anemia reveals a substantial proportion with malabsorption of protein-bound B_{12} (Carmel et al., 1987, 1988; Jones et al., 1987). More importantly, Carmel and coworkers (1988) found that 60 percent of those with neurological, cerebral, or psychological abnormalities malabsorbed food-bound B_{12}. Food-bound malabsorption is found in persons with certain gastric dysfunctions (e.g., hypochlorhydria or achlorhydria with an intact stomach, postgastric surgery such as Billroth I or II, and postvagotomy with pyloroplasty) and in some persons with initially unexplained low serum B_{12} (Carmel et al., 1988; Doscherholmen et al., 1983). Suter and colleagues (1991) reported that subjects with atrophic gastritis absorb significantly less B_{12} than do healthy control subjects but that the difference disappears after antibiotic therapy.

Miller and colleagues (1992) studied the absorption of radiolabeled B_{12} in patients who had not had gastric surgery but who had low B_{12} values. All patients with elevated serum gastrin levels absorbed food-bound B_{12} poorly compared with 21 percent of all those with normal serum gastrin values. In this study normal values were specified as greater than 12 percent absorption of food-bound B_{12} and greater than 33 percent absorption of free B_{12} as measured by direct body radioactivity measurements. Control subjects with normal serum B_{12} values (median 173 pmol/L [234 pg/mL], range 125 to 284 pmol/L [170 to 385 pg/mL]) absorbed 12 to 39 percent of food-bound B_{12} and 54 to 97 percent of free B_{12} (median 75 percent). The median age of this group was 61 years (range 49 to 69 years). Available evidence does not indicate that aging or atrophic gastritis increases the amount of B_{12} that must actually be absorbed to meet the body's needs.

Smoking

The high cyanide intake that occurs with cigarette smoking may disturb the metabolism of B_{12}. In a study of healthy adults (Linnell et al., 1968), mean urinary B_{12} excretion was significantly higher in the 16 smokers than in the 16 nonsmokers (81.2 ± 8.7 [standard error] and 60.3 ± 7.9, respectively, $p < 0.02$), and urinary thiocyanate excretion (an index of the exogenous cyanide load) was inversely associated with serum B$_{12}$. Similarly, in a study of pregnant women, the distribution of values of serum B$_{12}$ was slightly lower for smokers than for nonsmokers. However, in a cross-sectional study, differences in B$_{12}$ concentrations of smokers and nonsmokers were not significant in multivariate analyses. The effect of smoking on the B$_{12}$ requirement thus appears to be negligible.

Gender

In a cross-sectional study of 77 young men and 82 young women (Fernandes-Costa et al., 1985), the women were found to have significantly higher serum B$_{12}$ values and unsaturated cobalamin binding capacity than did the men ($p < 0.001$ and 0.05, respectively). Subjects were excluded if they were taking vitamin supplements, oral contraceptive agents, or other medications other than patent analgesics. Mean serum B$_{12}$ values were 477 and 604 pmol/L (647 and 819 pg/mL) for men and women, respectively—well above the cutoff of adequacy. Other investigators have reported similar findings (Low-Beer et al., 1968; Metz et al., 1971). Studies that have found no difference in mean B$_{12}$ values were smaller and less well-controlled for other factors that could influence B$_{12}$ values (Rosner and Schreiber, 1972; Scott et al., 1974). Taken together, these studies do not provide sufficient evidence on which to quantitate a difference in B$_{12}$ requirements by gender.

Nutrient-Nutrient Interactions

Folate with B$_{12}$

Although adequate or high folate intake may mitigate the effects of a B$_{12}$ deficiency on normal blood formation, there is no evidence that folate intake or status changes the requirement for B$_{12}$.

Vitamin C with B$_{12}$

Low serum B$_{12}$ values reported in persons receiving megadoses of vitamin C are likely to be artifacts of the effect of ascorbate on the radioisotope assay for B$_{12}$ (Herbert et al., 1978)—and thus not a true nutrient-nutrient interaction.

Other Food Components

Although it is clear that protein-bound B_{12} is less well absorbed than crystalline B_{12}, the effect varies greatly with the specific protein and may be modified by gastric factors (see "Food-Bound B_{12} Malabsorption"). Data on absorption from different types of diets (e.g., high in dairy products or beef) are not sufficient to use as a basis for adjusting the estimated requirement for B_{12}.

No evidence was found that a high-fiber diet increases the amount of B_{12} that should be consumed. A single study (Doi et al., 1983) was found that examined the effect of dietary fiber (specifically, konjac mannan, or glucomannan) on the absorption of B_{12}. A 3.9-g dose of the fiber with a meal did not change the rate of B_{12} absorption in either normal subjects or those with diabetes mellitus.

Genetic Defects

Underutilization of B_{12} has been reported in individuals with genetic defects that involve deletions or defects of MMA-CoA mutase, transcobalamin II, or enzymes in the pathway of cobalamin adenosylation (Kano et al., 1985; Rosenberg and Fenton, 1989).

FINDINGS BY LIFE STAGE AND GENDER GROUP

Infants Ages 0 through 12 Months

Methods Used to Set the Adequate Intake

An Adequate Intake (AI) is set for the recommended intake for infants. The AI reflects the observed average vitamin B_{12} intake of infants fed principally with human milk.

Reported values for the concentration of the vitamin in human milk vary widely, partly because of differences in methods of analysis and partly because of differences in maternal B_{12} status and current intake. Despite high intraindividual diurnal variability within a group of lactating women, no consistent effect on B_{12} concentration of time of day, breast, or time within a feed has been demonstrated. Thus, casual samples of human milk can be used to represent concentrations for the group (Trugo and Sardinha, 1994). However, the wide intraindividual variability may lead to inaccuracies in reported mean values if the number of individuals sampled is small. Median values are substantially lower than average values (Casterline et al., 1997; Donangelo et al., 1989). Acceptable methods of analysis include *Euglena gracilis* after pretreatment with papain to release the vitamin from the R protein in milk and radioassays in which the vitamin is released by heating (Areekul et al., 1977; Trugo and Sardinha, 1994). Studies used for estimating the concentration of the vitamin in human milk are limited to those that used one of these two methods.

The single longitudinal study of the change in B$_{12}$ concentration in human milk over time (Trugo and Sardinha, 1994) suggests somewhat higher concentrations in colostrum than in mature milk ($\leq$ 21 days postpartum) but little change after the first month of lactation.

Ages 0 through 6 Months

The AI for infants ages 0 through 6 months is based on the B$_{12}$ intake of infants fed human milk. B$_{12}$ deficiency does not occur in infants fed milk from mothers with adequate B$_{12}$ status. In samples collected from nine well-nourished Brazilian mothers who were not taking supplements and whose infants were receiving human milk exclusively, the average concentration of the vitamin was 0.42 µg/L at 2 months; this decreased to an average of 0.34 µg/L at 3 months (Trugo and Sardinha, 1994). Milk collected at least 2 months postpartum from 13 unsupplemented American mothers who were vegetarians was lower in B$_{12}$ content, averaging 0.31 µg/L (Specker et al., 1990). The B$_{12}$ content of milk in a large group of low-income Brazilian mothers ($n = 83$) who had received prenatal supplements containing B$_{12}$ was much higher, averaging 0.91 µg/L after 1 month of lactation (Donangelo et al., 1989). Given that the average concentration at 2 months postpartum of well-nourished mothers whose infants received exclusively human milk was higher than those on vegetarian diets, the higher value of 0.42 µg/L is chosen in order to be sure adequate amounts are available. Using the average human milk volume of 0.78 L/day during the first 6 months and the higher average B$_{12}$ content of 0.42 µg/L, the AI for B$_{12}$ for the infant 0 through 6 months of age fed human milk would be 0.33 µg/day, rounded up to 0.4 µg.

Maintenance of Normal Methylmalonic Acid Concentrations. Data on methylmalonic acid (MMA) excretion is also available for infants. An infant may be born with low B$_{12}$ stores and may consume human milk that is low in B$_{12}$ if its mother is a vegan (a person who avoids all animal foods) or has untreated pernicious anemia. Such infants begin to show clinical signs of B$_{12}$ deficiency at about 4 months postpartum. In the study of 13 vegan mothers and their infants, Specker and colleagues (1990) found increased urinary MMA in 2- to 14-month old (mean 7.3) infants predominantly fed human milk when the B_{12} concentration in human milk was below 0.49 µg/L. Assuming an average volume of human milk consumption of 0.78 L/day during the first 6 months, the infant of a vegan mother would be receiving an average of 0.24 µg/day of B_{12} (0.31 µg/L × 0.78 L/ day).

In these infants, urinary MMA concentrations were strongly correlated with those of their mothers and inversely related to maternal plasma B_{12} concentrations, supporting the assumption that the elevations in infant urinary MMA were caused by poor maternal B_{12} status (Specker et al., 1988). Although these infants were probably born with depleted stores of the vitamin, the data suggest that a mean intake of 0.24 µg/day is inadequate to maintain B_{12} balance in infants.

Clinical signs of B_{12} deficiency are usually seen if the mother has been a strict vegetarian for at least 3 years. The B_{12} status of the infant is clearly abnormal by about 4 to 6 months of age. In case studies of infants born to strict vegetarians who were identified because of clinical signs of B_{12} deficiency, human milk concentrations have been reported to be 0.02 (Hoey et al., 1982), 0.037 (Gambon et al., 1986), 0.032 and 0.042 (Jadhav et al., 1962), 0.051 (Johnson and Roloff, 1982), and 0.085 (Kuhne et al., 1991) µg/L. If clinical signs appear within 9 months in infants consuming milk containing 0.085 µg/L of B_{12}, this intake (approximately 0.07 µg/day) cannot support B_{12} requirements of the infant during the first year. However, from these data it cannot be determined how far this estimate falls below the average requirement for these infants.

Rate of Depletion of Stores. The liver of a well-nourished newborn infant contains 18 to 22 pmol (25 to 30 µg) of B_{12} (Baker et al., 1962; Loria et al., 1977; Vaz Pinto et al., 1975). There are no data on liver B_{12} content at birth in full-term infants born to depleted mothers, only data for two infants who died prematurely (Baker et al., 1962); thus, the utilization of B_{12} by infants remains speculative.

Summary. The AI for infants ages 0 through 6 months is 0.33 µg/ day based on the average concentration of B_{12} in the milk of mothers with adequate B_{12} status. This value is rounded up to 0.4 µg/day. The adequacy of this intake is supported by evidence that it is above the intake level that has been associated with increased urinary MMA excretion.

Ages 7 through 12 Months

If the reference body weight ratio method described in Chapter 2 to extrapolate from the AI for B$_{12}$ for infants ages 0 through 6 months is used, the AI for B$_{12}$ for the older infants would be 0.5 µg/day after rounding up. This is a somewhat lower value than that obtained from the second method (see Chapter 2) by extrapolating down from the Estimated Average Requirement (EAR) for adults and adjusting for the expected variance to estimate a recommended intake, which results in an AI for B$_{12}$ of 0.6 µg/day.

In one study of three infants exclusively fed human milk who had clinically observable B$_{12}$ deficiency caused by low maternal consumption of animal products, one infant was treated parenterally with B$_{12}$ whereas two infants were treated with small oral B$_{12}$ doses (Jadhav et al., 1962). At 9 months of age, 0.1 µg/day of oral B$_{12}$ normalized bone marrow within 5 days in one of the two infants given oral doses and produced profound improvements in behavior by 18 days (after a total of 1.8 µg of B$_{12}$ had been given). The mother's milk contained 0.032 µg/L. In the second infant, who was 7 months old, 0.1 µg/day of B$_{12}$ caused abnormal pigmentation to disappear and "an adequate hematologic response." His mother's milk contained 0.042 µg/L. Although evidence of sustained recovery was not provided, it appears from these limited data that 0.1 µg/day may be adequate to improve clinical and hematological signs of deficiency in infants at this age. However, it is not known whether this level of intake is adequate to sustain normal plasma B$_{12}$ and MMA concentrations or hematological response.

In the study of infants of vegan mothers (Specker et al., 1990) the mean age of infants was 7.3 months (ranged 2 to 14 months). As discussed, a mean intake of 0.23 µg/day was not adequate to maintain B$_{12}$ balance in this group as determined by urinary MMA excretion.

B$_{12}$ AI Summary, Ages 0 through 12 Months

AI for Infants

0–6 months	0.4 µg/day of vitamin B$_{12}$	≈0.05 µg/kg
7–12 months	0.5 µg/day of vitamin B$_{12}$	≈0.05 µg/kg

Special Considerations

Infants of vegan mothers should be supplemented with B$_{12}$ at the AI from birth on the basis of evidence that their stores at birth are low and their mother's milk may supply very small amounts of the vitamin.

Children and Adolescents Ages 1 through 18 Years

Method Used to Estimate the Average Requirement

Only one study is available to provide data regarding B_{12} status and intake in young children. Plasma MMA was elevated in 11- to 22-month-old (mean 16.8 months) infants of Dutch vegan mothers. The sensitivity of plasma MMA to distinguish the group of infants born to macrobiotic mothers from those born to omnivorous mothers was 85 percent (Schneede et al., 1994). The average intake of B_{12} by these infants, who were exclusively fed human milk for a mean of 4.8 months and then at least partially fed human milk for 13.6 ± 6.6 (standard deviation) months and fed macrobiotic foods, was 0.3 ± 0.2 µg/day for the first 6 to 16 months compared with 2.9 ± 1.2 µg/day for well-nourished control infants (Dagnelie et al., 1991). These data suggest that an intake of 0.3 µg/day of B_{12} between 6 and 16 months of age was inadequate to prevent elevated plasma MMA concentrations of infants born to vegan mothers.

No other direct data were found on which to base an Estimated Average Requirement (EAR) for B_{12} for children or adolescents. In the absence of additional information, EARs and RDAs for children and adolescents have been estimated by using the method described in Chapter 2, which extrapolates down from adult values, and rounded up.

B_{12} EAR and RDA Summary, Ages 1 through 18 Years

EAR for Children	1–3 years	0.7 µg/day of vitamin B_{12}
	4–8 years	1.0 µg/day of vitamin B_{12}
EAR for Boys	9–13 years	1.5 µg/day of vitamin B_{12}
	14–18 years	2.0 µg/day of vitamin B_{12}
EAR for Girls	9–13 years	1.5 µg/day of vitamin B_{12}
	14–18 years	2.0 µg/day of vitamin B_{12}

The RDA for B_{12} is set by assuming a coefficient of variation (CV) of 10 percent (see Chapter 1) because information is not available on the standard deviation of the requirement for B_{12}; the RDA is defined as equal to the EAR plus twice the CV to cover the needs of

97 to 98 percent of the individuals in the group (therefore, for B$_{12}$ the RDA is 120 percent of the EAR).

RDA for Children	**1–3 years**	**0.9 µg/day of vitamin B$_{12}$**
	4–8 years	**1.2 µg/day of vitamin B$_{12}$**
RDA for Boys	**9–13 years**	**1.8 µg/day of vitamin B$_{12}$**
	14–18 years	**2.4 µg/day of vitamin B$_{12}$**
RDA for Girls	**9–13 years**	**1.8 µg/day of vitamin B$_{12}$**
	14–18 years	**2.4 µg/day of vitamin B$_{12}$**

Adults Ages 19 through 50 Years

Method Used to Estimate the Average Requirement

No single indicator was judged to be a sufficient basis for deriving an EAR for adults. It was not deemed appropriate to base the EAR on an examination limited to studies that provided data on mean cell volume (MCV) or serum B$_{12}$ or any other single laboratory value. Data on men and women were examined together because of small numbers. Three general approaches were considered to derive the EAR for adults: determination of the amount of B$_{12}$ needed to maintain adequate hematological status (as measured by stable hemoglobin value, normal MCV, and normal reticulocyte response) and serum B$_{12}$ values in persons with pernicious anemia or with known intakes that were very low in dietary B$_{12}$; use of daily B$_{12}$ turnover to estimate the amount of B$_{12}$ needed to maintain body stores at a specified level; and estimation of the dietary B$_{12}$ intake by healthy adults that corresponds to adequate serum values of B$_{12}$ and of MMA.

The first approach was chosen as the primary method for deriving an EAR because it is the only approach for which there are sufficient and reliable data for estimating need. A low serum B$_{12}$ value in persons with pernicious anemia was assumed to indicate incomplete response to treatment.

Primary Criterion: Maintenance of Hematological Status and Serum B$_{12}$ Values. The primary method used to derive the EAR for adults estimates the amount of B$_{12}$ needed for the maintenance of hematological status and serum B$_{12}$ values, primarily by using data derived from patients with pernicious anemia in remission. Data from studies of vegetarians were also examined to determine whether they

---

BOX 9-1   Assumptions Made in Estimating the Amount of Vitamin B_{12} Needed for Maintenance of Hematological Status and Serum Vitamin B_{12} Values

• Maintenance of hematological status requires a relatively stable hemoglobin value upon administration of B_{12} and a normal mean cell volume, not just a reticulocyte response.
• Normal serum B_{12} is $\geq$ 150 pmol/L (200 pg/mL).
• Because B_{12} is not absorbed from the bile, the estimated extra loss of B_{12} by a person with pernicious anemia in remission is 0.4 nmol/day (0.5 µg/day) based on data from Bozian et al. (1963), El Kholty et al. (1991), Heyssel et al. (1966), and Reizenstein (1959).
• The average fractional absorption of B_{12} from food by healthy individuals is approximately 50 percent (see "Absorption").

---

provided information on levels of B_{12} intake needed to maintain hematological status. In some cases, neurological manifestations may be the earliest clinical sign of low B_{12} values (Beck, 1991; Karnaze and Carmel, 1990; Lindenbaum et al., 1988; Martin et al., 1992). Assumptions that were integral to the application of this method are shown in Box 9-1.

In brief, this method involves estimating the amount of B_{12} required daily to maintain hematological and serum B_{12} status of individuals with pernicious anemia in remission; subtracting the amount of endogenous B_{12} lost from the bile in excess of that lost by a healthy individual; and, because the value is to be used for individuals with normal ability to absorb B_{12} from food, correcting for bioavailability. The result is shown in Box 9-2.

BOX 9-2   Steps Used to Estimate the Vitamin B_{12} Requirement by Using Data Obtained from Subjects with Pernicious Anemia

Step 1	Estimate the average intramuscular requirement for maintenance of person with pernicious anemia	1.5 µg/day
Step 2	Subtract estimate of extra losses due to lack of reabsorption of biliary B_{12}	– 0.5 µg/day
Subtotal	Estimate average requirement of normal person for absorbed B_{12}	1.0 µg/day
Step 3	Correct for bioavailability (50 percent)	÷ 0.5
Result	Average requirement of normal person for B_{12} from food: Estimated Average Requirement (EAR)	2.0 µg/day

The following studies provide the basis for the estimate used in Step 1. These studies do not provide ideal data on which to base an EAR, but they bracket the requirement by providing values that are obviously too low or too high to meet the needs of 50 percent of the individuals in an age group.

Studies of Patients with Pernicious Anemia. Darby and coworkers (1958) studied the effects of various intramuscular (IM) doses of B_{12} in 20 subjects with pernicious anemia who had not previously been treated or who were in relapse. The diagnosis of pernicious anemia had been based on the clinical history and on the findings of macrocytic anemia, megaloblastic hyperplasia of bone marrow, histamine-fast achlorhydria, and a negative radiological examination of the gastrointestinal tract. These diagnoses were not made based on results of the Schilling test, first published as a method in 1953 (Schilling, 1953). The extent of the disease differed among the subjects; 14 had neurological manifestations. Of the 18 subjects who received doses of 1 µg/day of B_{12} or less for 2 weeks, 5 or fewer responded satisfactorily according to the standards used for erythrocytes (Isaacs et al., 1938) and reticulocytes (Isaacs and Friedman, 1938). At B_{12} dosages of less than 0.5 µg/day, no patient met those standards. Dosages used for maintenance were increased to 1 to 4 µg/day for a period of months to years. MCVs greater than 100 were considered macrocytic. No reticulocyte counts or serum B_{12} values were reported. According to the authors' interpretation, the data indicated that subjects achieved and maintained maximum erythropoiesis as indicated in Table 9-5. Approximately half (4 of 7) did so at a B_{12} intake of 1.4 µg/day IM.

TABLE 9-5 Effectiveness of Intramuscular Vitamin B_{12} Doses for Maintenance of Maximum Erythropoiesis

Daily B_{12} Dose, Intramuscular (µg)	Number of Subjects Achieving Maximum Erythropoiesis ($n = 7$)	Cumulative Number Achieving Maximum Erythropoiesis ($n = 7$)
0.5	1	1
1.0	2	3
1.4	1	4
2.0	2	6
4.0	1	7

SOURCE: Darby et al. (1958).

Results of other studies of patients with pernicious anemia are presented in Table 9-6. The short-term study by Hansen and Weinfeld (1962) used relatively high B_{12} doses to restore normal status but did not assess maintenance requirement. The long-term studies by Bastrup-Madsen et al. (1983) and Lindenbaum et al. (1990) used different dosages and methods of reporting that make it impossible to draw precise conclusions. Nonetheless, the results indicate that 0.8 to 1.0 µg/day of B_{12} IM will maintain normal hematological, serum B_{12}, and serum metabolite status in nearly half of the individuals over time and that 1.7 µg will maintain it in all individuals. The study conducted by Best and colleagues (1956) was designed to determine the effective dosage of intrinsic factor concentrates, not to estimate the B_{12} requirement, but it suggests that 1.4 µg of B_{12} exceeds the requirement for absorbed B_{12} in most of the subjects tested. The often-cited study of Sullivan and Herbert (1965) was interpreted as providing evidence that 0.1 µg/day of B_{12} was not sufficient for treating pernicious anemia and maintaining adequate B_{12} status. Similarly, the 0.6 to 0.7 µg/day of B_{12} supplied IM in the study by Will and coworkers (1959) was also judged too low to maintain a normal serum B_{12} concentration.

The study by Darby and colleagues (1958), which indicates an average requirement in such patients of approximately 1.5 µg, is supported by the supplementary data from the other studies described in Table 9-6. These studies provide support for a physiological average requirement of 1.0 µg/day of B_{12} after adjustment for the extra loss of B_{12} by subjects with pernicious anemia (0.5 µg/ day) (Step 2 in Box 9-2). Adjusting for incomplete absorption of B_{12} from food of 50 percent (Step 3) converts this value to an EAR for B_{12} of 2.0 µg/day.

Studies of Individuals with Low B_{12} Intake. Studies of individuals with low B_{12} intake were examined to determine whether these reports (Table 9-7) supported the findings for subjects with pernicious anemia. Because B_{12} is not a component of plant foods, diets containing little or no animal food may lead to B_{12} deficiency. Deficiency develops slowly because of efficient reabsorption of biliary B_{12}. It is also possible but not certain that vegans consume some B_{12} from animal products that contaminate plant food or from bacterial action. Studies of vegetarians generally have not analyzed the B_{12} content of the food, and accurate data are not available for some of the foods (e.g., certain algae) consumed by vegetarians. Without actual analyses it is not clear what B_{12} content should be assumed for vegans.

The studies covered by Table 9-7 suggest that the B$_{12}$ requirement is higher than the amounts reported to be consumed by the subjects and more than that provided by the treatments that were described. In three studies (Baker and Mathan, 1981; Jathar et al., 1975; Winawer et al., 1967), all adults required more than 1 µg/day of B$_{12}$ by mouth. Two studies (Narayanan et al., 1991; Stewart et al., 1970) give evidence that 1.5 µg/day of dietary B$_{12}$ is not sufficient to maintain hematological status and serum B$_{12}$ in half of the subjects studied. The meager data provided by the studies of vegetarians indicate that the B$_{12}$ average requirement should probably be at least 1.5 µg/day, but a higher average requirement is not ruled out.

Supportive Data: Maintenance of B$_{12}$ Body Stores

Various studies have indicated losses of 0.1 to 0.2 percent/day of the B$_{12}$ pool (e.g., Amin et al., 1980; Boddy and Adams, 1972; Heyssel et al., 1966; Reizenstein et al., 1966) regardless of the size of the pool. A loss of 0.2 percent appears to be typical for individuals who do not reabsorb biliary B$_{12}$ because of pernicious anemia (Boddy and Adams, 1972). A person with a B$_{12}$ pool of 1,000 µg and a loss of 0.1 percent would excrete 1 µg of B$_{12}$ daily, and a person with a 3,000-µg pool would excrete 3 µg daily. If only 50 percent of dietary B$_{12}$ is absorbed, the amounts required daily to replenish the pools are 2 and 6 µg of B$_{12}$, respectively. The higher value would lead to less efficient use of B$_{12}$, but the larger store of B$_{12}$ would cover a longer period of inadequate B$_{12}$ intake or absorption.

With a 0.1 percent loss, the period of protection afforded by the B$_{12}$ pool can be estimated if the lowest pool size consistent with health is also known. If it is assumed that this value is 300 µg (derived from Bozian and coworkers [1963]), there is no absorption of B$_{12}$ from food or supplements, and the enterohepatic circulation is intact, then stores of 1 mg would be expected to meet the body's needs for 3 years, 2 mg for about 5 years, and 3 mg for about 6 years. A 1.5 percent loss would reduce these estimates to 2, 3.6, and 4 years (see Appendix N for the method used to obtain these values).

The extent of the supply of reserve B$_{12}$ may be an important consideration when persons approach the age of 50 and the risk increases for food-bound B$_{12}$ malabsorption secondary to atrophic gastritis (see "Factors Affecting the Vitamin B$_{12}$ Requirement" and section "Adults Ages 51 Years and Older"). Because the absorption of B$_{12}$ from fortified foods, oral supplements, or the bile does not

TABLE 9-6 Other Studies of Subjects with Pernicious Anemia Considered in Setting the Estimated Average Requirement for Vitamin B_{12} for Adults

Reference	Number of Subjects	Age Range (y)
Suggested IM[a] requirement > 2.0 µg		
Hansen and Weinfeld, 1962	14	
Suggested IM requirement of 1.0–2.0 µg		
Bastrup-Madsen et al., 1983	112	33–78
Lindenbaum et al., 1990	44	NA[c]
Other Studies		
Best et al., 1956	6	NA
Will et al., 1959	40	NA
Sullivan and Herbert, 1965	8	46–86

[a] IM = intramuscular.
[b] 1,000 µg × 0.15 retention/90 d.
[c] NA = not available.
[d] 1,000 µg × 0.15 retention/182 d.

Treatment	Results
2–5 µg of B$_{12}$ IM for 8–15 d.	Five persons who were given 3 µg/d of B$_{12}$ for 15 d had a reticulocyte response that was not followed by a further response to more B$_{12}$. This amount allowed restoration of status.
1 mg of slow-release B$_{12}$ IM every 2 or 3 mo for at least 8 y. The less-frequent dose was equivalent to 1.7 µg of B$_{12}$/d.[b]	Serum B$_{12}$ values were well above the cutoff of 180 pmol/L (250 pg/mL) early in the study and complete hematological remission occurred in all.
35 received 1 mg of B$_{12}$ IM every 5–6 mo, 6 received it every 3–4 mo, 3 received it every 2 mo. Smallest and most frequent dose was equivalent to 0.8–1.0 µg of B$_{12}$/d.[d]	From total group analyses, 14 subjects had mild hematological relapse on 42 occasions; 34 subjects had at least one abnormal serum B$_{12}$ or metabolite value on 146 occasions when there was no evidence of hematological relapse.
2.0 µg oral dose of B$_{12}$ Co^{60} given with intrinsic factor.	With 70% absorption, complete hematological response, and adequate plasma B$_{12}$ concentration, 1.4 µg of absorbed B$_{12}$ met the requirements of two-thirds of the subjects.
10 µg of B$_{12}$ given IM every 2 wk or 20 µg of B$_{12}$ given IM monthly for 10 y (equivalent average of 0.7 µg/d).	None of the subjects maintained serum B$_{12}$ concentration above the 180 pmol/L (250 pg/mL) lower limit of normal for the *Lactobacillus leichmannii* method.
0.1 µg/d of cyanocobalamin IM for 10 d; 0.1 µg/d of coenzyme B$_{12}$ IM for 10 d.	Posttreatment serum B$_{12}$ was 85 pmol/L (60 pg/mL) (range, 20–200 pmol/L [14–139 pg/mL]); 6 of 8 had reticulocyte response, but macrocytosis persisted in all and hypersegmentation did in many. In some, neurological abnormalities progressed until at least 1 µg of B$_{12}$ was given daily. All but one were later given higher doses of B$_{12}$.

TABLE 9-7   Studies of Individuals with Low Vitamin B_{12} Intake Considered in Setting the Estimated Average Requirement for B_{12} for Adults

Reference	Description	Dietary B_{12} Intake
Suggested dietary B_{12} average requirement > 1.5 µg/d		
Stewart et al., 1970	1 Hindu woman with megaloblastic anemia	0.5 µg/d (analyzed homogenate)
Narayanan et al., 1991	10 subjects with serum B_{12} values below the 2.5 percentile (< 120 pmol/L [162 pg/mL]) not caused by disease or vegetarianism	1.5 ± 0.4 (SD[c]) µg/d of B_{12} (range 0.6–1.9)
Suggested dietary B_{12} average requirement > 1.0 µg/d		
Winawer et al., 1967	1 64-y-old vegan with B_{12}-deficient megaloblastic anemia, gastritis on biopsy, and normal gastric acidity	Assumed to be negligible
Jathar et al., 1975	7 East Indian lactovegetarians	0.3–0.8 µg/d of B_{12} from milk, assuming that it was not boiled
Baker and Mathan, 1981	4 East Indians with B_{12} deficiency anemia secondary to diet	NA[d]

[a] p.o. = by mouth.
[b] Based on USDA data (URL http://www.nal.usda.gov/fnic/foodcomp/).

appear to be impaired, the combination of stores and absorbed crystalline B_{12} may cover needs for an extended period.

The estimates above for the period of protection afforded by body stores are consistent with the periods required to develop overt signs of B_{12} deficiency after a total gastrectomy; for example, megaloblastic anemia has been typically diagnosed 2 to 5 years after a total gastrectomy (Chanarin, 1990).

Treatment	Comments
1 µg/d of B$_{12}$ p.o.[a]	Serum B$_{12}$ rose to 121 pmol/L (164 pg/mL) (said to be normal) and hemoglobin stabilized at 10.7 g/100 mL
1 pint/d of fresh milk (≈1.5 µg of B$_{12}$)[b]	Serum B$_{12}$ maintained at 100 pmol/L (134 pg/mL)
Not specified	Seven fulfilled at least one criterion for tissue B$_{12}$ deficiency
1 µg/d of B$_{12}$ p.o.	Serum B$_{12}$ rose to 64 pmol/L (87 pg/mL), well below normal; gastritis may have decreased absorption of any B$_{12}$ inadvertently present in the food
None	Half had serum B$_{12}$ values < 74 pmol/L (100 pg/mL)
0.07–0.25 µg/d of B$_{12}$	Judged inadequate
0.3–0.65 µg/d of B$_{12}$	Hematological responses seen but serum B$_{12}$ ≤ 74 pmol/L (100 pg/mL) in all Interpretation complicated by transfusions and intramuscular injections

[c] SD = standard deviation.
[d] NA = not available.

Possible Ancillary Method: Maintenance of a Serum B$_{12}$ Concentration That Is Consistent with a Normal Circulating MMA Value

Several investigators have urged the use of the serum MMA concentration as the most sensitive indicator of B$_{12}$ status (Lindenbaum et al., 1990; Moelby et al., 1990; Savage et al., 1994b; Stabler et al., 1996). This indicator could not be used as the criterion for setting the EAR for B_{12} because of a lack of direct data. At least one study (Lindenbaum et al., 1994) relates serum B_{12} to circulating MMA values. None link MMA with B_{12} intake. Moreover, although MMA is a metabolite that accumulates abnormally when the B_{12} supply is low, studies have not yet convincingly demonstrated that elevated MMA caused by insufficient B_{12} intake has adverse health consequences. However, because MMA values hold promise as a criterion for estimating the B_{12} requirement in the future, an indirect approach was used to estimate a requirement for B_{12} as a means of confirming or refining the EAR value derived by using the primary approach. For example, because the serum B_{12} value of 150 pmol/L (200 pg/mL) appears to be the level at which half the population would have an elevated MMA value (Lindenbaum et al., 1994), one could select the dietary intake that would maintain this value in healthy individuals in that population.

In a study of 548 surviving members of the original Framingham Heart Study cohort, aged 67 to 96 years, and 117 healthy control subjects younger than 65 years, Lindenbaum and colleagues (1994) reported on serum B_{12}, MMA, and homocysteine values (Table 9-8). These investigators used a cutoff value equal to or greater than 260 pmol/L (350 pg/mL) of B_{12} as adequate; more than 15 percent of subjects below the cutoff value had elevated MMA concentrations whereas fewer than 10 percent of subjects above the cutoff did. Serum creatinine was elevated in 10 of those with both increased MMA and low B_{12} values, which would indicate confounding abnormal renal function. Slightly more than 40 percent of the 70 elderly

TABLE 9-8   Vitamin B_{12} Status: Occurrence of Low Serum Values for Two Age Groups

Serum Values of Subjects	Healthy Younger Control Subjects[a] (Number [%])	Elderly Subjects from Framingham Study[b] (Number [%])
Vitamin B_{12} < 148 pmol/L (201 pg/mL)	2 (1.7)	29 (5.3)
Vitamin B_{12} < 258 pmol/L (351 pg/mL)	21 (17.9)	222 (40.5)
Methylmalonic acid > 376 nmol/L	Not available	82 (15.0)

NOTE: 1 pmol/L of B_{12} = 1.36 pg/mL.
[a] Aged < 65 years; n = 117.
[b] Aged 67–96 years; n = 548.
SOURCE: Lindenbaum et al. (1994).

subjects with serum B$_{12}$ less than 150 pmol/L (200 pg/mL) had an elevated MMA concentration.

Studies of B$_{12}$ intake and serum B$_{12}$ concentration provide very limited information on the relationship of the two. In Finland, vegans consuming an uncooked ("living food") diet were estimated to consume a mean of 1.8 µg/day of B$_{12}$ (range 0 to 12.8 µg) (Rauma et al., 1995), but the accuracy of the dietary intake data is uncertain. The 16 vegans who ate seaweed (the main source of B$_{12}$ reported) had B$_{12}$ concentrations twice as high as those not eating seaweed (mean of 220 pmol/L [300 pg/mL] compared with 105 pmol/L [142 pg/mL]). On this diet 57 percent of the vegans had serum B$_{12}$ concentrations less than 200 pmol/L (270 pg/mL). A study by Draper and colleagues (1993) provided dietary data on vegans that were not sufficient for drawing conclusions about diet-B$_{12}$ relationships. Neither Garry and coworkers (1984) nor Sahyoun and colleagues (1988) separated data with regard to supplement use, so their data are not interpretable for setting EARs. A study of a macrobiotic population (Miller et al., 1991) revealed that more than half of the adults had low serum B$_{12}$ concentrations and nearly one-third were excreting high amounts of MMA, but dietary information from the study was not sufficient for drawing conclusions. Moreover, studies need to be conducted in younger persons in whom B$_{12}$ absorption is more likely to be normal.

B$_{12}$ EAR and RDA Summary, Ages 19 through 50 Years

On the basis of hematological evidence and serum B$_{12}$ values, the EAR for B$_{12}$ is estimated to be 2 µg/day for men and women ages 19 through 50 years. Sufficient data were not available to enable differences in requirements to be discerned for men and women in these age groups.

EAR for Men	19–30 years	2 µg/day of vitamin B$_{12}$
	31–50 years	2 µg/day of vitamin B$_{12}$
EAR for Women	19–30 years	2 µg/day of vitamin B$_{12}$
	31–50 years	2 µg/day of vitamin B$_{12}$

The RDA for B$_{12}$ is set by assuming a coefficient of variation (CV) of 10 percent (see Chapter 1) because information is not available on the standard deviation of the requirement for B$_{12}$; the RDA is defined as equal to the EAR plus twice the CV to cover the needs of

97 to 98 percent of the individuals in the group (therefore, for B_{12} the RDA is 120 percent of the EAR).

RDA for Men	**19–30 years**	**2.4 µg/day of vitamin B_{12}**
	31–50 years	**2.4 µg/day of vitamin B_{12}**
RDA for Women	**19–30 years**	**2.4 µg/day of vitamin B_{12}**
	31–50 years	**2.4 µg/day of vitamin B_{12}**

Adults Ages 51 Years and Older

Evidence Considered in Estimating the Average Requirement

Because 10 to 30 percent of people older than 50 years are estimated to have atrophic gastritis with low stomach acid secretion (Andrews et al., 1967; Hurwitz et al, 1997; Johnsen et al., 1991; Krasinski et al., 1986), they may have decreased bioavailability of B_{12} from food. Therefore, because of the high prevalence of this condition, 50 percent bioavailability of dietary B_{12} (see Box 9-2) cannot be assumed for this age group, and the EAR would be higher than 2.0 µg. Similarly, 2.4 µg of B_{12}, which is the RDA for younger adults, might not meet the needs of 97 percent of this large age group. There is not sufficient information on which to base a bioavailability correction factor for persons with atrophic gastritis who obtain their B_{12} from animal foods. However, because the bioavailability of crystalline B_{12} is not altered in people with atrophic gastritis, the same EAR and RDA would apply if the dietary sources of B_{12} were foods fortified with B_{12}, supplements, or a combination of both.

B_{12} EAR and RDA Summary, Ages 51 Years and Older

The EAR and RDA for B_{12} for adults ages 51 years and older are the same as for younger adults but with the recommendation that B_{12}-fortified foods (such as fortified ready-to-eat cereals) or B_{12}-containing supplements be used to meet much of the requirement.

EAR for Men	**51–70 years**	**2 µg/day of vitamin B_{12}***
	> 70 years	**2 µg/day of vitamin B_{12}***
EAR for Women	**51–70 years**	**2 µg/day of vitamin B_{12}***
	> 70 years	**2 µg/day of vitamin B_{12}***

The RDA for B$_{12}$ is set by assuming a coefficient of variation (CV) of 10 percent (see Chapter 1) because information is not available on the standard deviation of the requirement for B$_{12}$; the RDA is defined as equal to the EAR plus twice the CV to cover the needs of 97 to 98 percent of the individuals in the group (therefore, for B$_{12}$ the RDA is 120 percent of the EAR).

RDA for Men	51–70 years	2.4 μg/day of vitamin B$_{12}$*
	> 70 years	2.4 μg/day of vitamin B$_{12}$*
RDA for Women	51–70 years	2.4 μg/day of vitamin B$_{12}$*
	> 70 years	2.4 μg/day of vitamin B$_{12}$*

*It is advisable for most of this amount to be obtained by consuming foods fortified with B$_{12}$ or a B$_{12}$-containing supplement.

Pregnancy

Evidence Considered in Estimating the Average Requirement

Absorption and Utilization of B$_{12}$. There is some evidence that the absorption of B$_{12}$ may increase during pregnancy. An increase in the number of intrinsic factor–B$_{12}$ receptors was observed in pregnant mice and found to be regulated by placental lactogen (Robertson and Gallagher, 1983). A greater absorption of oral B$_{12}$ was reported from the single study of pregnant women (Hellegers et al., 1957), but the methods used do not permit quantification of the increase.

Serum total B$_{12}$ concentrations begin to decline early in the first trimester. In a longitudinal Dutch study of 23 subjects, serum B$_{12}$ fell significantly by the end of the first trimester, more than could be accounted for by hemodilution (Fernandes-Costa and Metz, 1982). There were further decreases through the sixth month to about half of nonpregnancy concentrations. Some of the later decrease was due to hemodilution. However, transcobalamin I and III increase during the second and third trimesters, and transcobalamin II increases sharply in the third trimester to about one-third more than in nonpregnant, nonlactating control subjects (Fernandes-Costa and Metz, 1982).

Transfer to the Fetus. The serum B$_{12}$ concentration of the newborn is twice that of the mother, decreasing to adult concentrations at about 6 to 7 months postpartum (Luhby et al., 1958). The placenta concentrates B_{12}, which is then transferred to the fetus down a concentration gradient. Fetal and maternal B_{12} serum concentrations are quite strongly correlated (Fréry et al., 1992). It appears that only newly absorbed B_{12} is readily transported across the placenta and that maternal liver stores are a less important source of the vitamin for the fetus (Luhby et al., 1958). This implies that current maternal intake and absorption of the vitamin during pregnancy have a more important influence on the B_{12} status of the infant than do maternal B_{12} stores. The importance of adequate maternal intake during pregnancy is supported by the appearance of B_{12} deficiency in infants at 4 to 6 months when their mothers have been strict vegetarians for only 3 years (Specker et al., 1990).

Fetal Accumulation. The human fetus accumulates an average of 0.07 to 0.14 nmol/day (0.1 to 0.2 µg/day) of B_{12}, a range based on three studies of the liver content of infants born to women who were adequate in B_{12} (Baker et al., 1962; Loria et al., 1977; Vaz Pinto et al., 1975) and an assumption that the liver contains half the total body B_{12} content. Placental B_{12} is negligible (0.01 nmol/L [14 ng/L]) (Muir and Landon, 1985). The low body content of B_{12} in the newborn implies that pregnancy is unlikely to deplete maternal stores.

B_{12} EAR and RDA Summary, Pregnancy

On the basis of a fetal deposition of 0.1 to 0.2 µg/day throughout pregnancy and evidence that maternal absorption of the vitamin becomes more efficient during pregnancy, the EAR is increased by 0.2 µg/day during pregnancy. No distinction is made for the age of the mother.

EAR for Pregnancy	**14–18 years**	**2.2 µg/day of vitamin B_{12}**
	19–30 years	**2.2 µg/day of vitamin B_{12}**
	31–50 years	**2.2 µg/day of vitamin B_{12}**

The RDA for B_{12} is set by assuming a coefficient of variation (CV) of 10 percent (see Chapter 1) because information is not available on the standard deviation of the requirement for B_{12}; the RDA is defined as equal to the EAR plus twice the CV to cover the needs of 97 to 98 percent of the individuals in the group (therefore, for B_{12} the RDA is 120 percent of the EAR).

RDA for Pregnancy   **14–18 years**   **2.6 µg/day of vitamin B$_{12}$**
                          19–30 years   **2.6 µg/day of vitamin B$_{12}$**
                          31–50 years   **2.6 µg/day of vitamin B$_{12}$**

Lactation

Evidence Considered in Estimating the Average Requirement

As described earlier, the average amount of B$_{12}$ secreted in the milk of mothers with adequate B$_{12}$ status is approximately 0.33 µg/day during the first 6 months of lactation. During the second 6 months, the average amount of B$_{12}$ secreted is slightly less: 0.25 µg/day.

The concentration of B$_{12}$ in milk is usually similar to that in maternal plasma. In some studies, human milk and maternal plasma concentrations are strongly correlated (Srikantia and Reddy, 1967) but in others they are not (Casterline et al., 1997; Donangelo et al., 1989). The correlation appears to be stronger when maternal B$_{12}$ status is marginal (Fréry et al., 1992).

Current maternal intake of the vitamin may have an important influence on secretion of the vitamin in milk. In several studies of infants with clinical signs of B$_{12}$ deficiency caused by low maternal intake or absorption of the vitamin, maternal plasma concentrations of the vitamin were found to be normal or low normal, suggesting that maternal B$_{12}$ stores are less important than current maternal intake (Hoey et al., 1982; Johnson and Roloff, 1982; Kuhne et al., 1991; Sklar, 1986). This is also indicated by the observation that the length of time that mothers had been strict vegetarians was not correlated with the urinary MMA concentrations of their infants (Specker et al., 1988).

Low B$_{12}$ concentrations in human milk occur commonly in two situations involving inadequate intake: when the mother is a strict vegetarian and in developing countries where the usual consumption of animal products is low. When the B$_{12}$ status of the mother is marginal, further maternal depletion may occur as reflected in decreasing concentrations of maternal plasma B$_{12}$ (Black et al., 1994; Shapiro et al., 1965).

B$_{12}$ EAR and RDA Summary, Lactation

To estimate the EAR for lactation, 0.33 µg/day of B$_{12}$ is added to the EAR of 2 µg/day for adolescent girls and adult women; the result is rounded up.

EAR for Lactation **14–18 years**   **2.4 µg/day of vitamin B$_{12}$**
                          19–30 years   **2.4 µg/day of vitamin B$_{12}$**
                          31–50 years   **2.4 µg/day of vitamin B$_{12}$**

The RDA for B$_{12}$ is set by assuming a coefficient of variation (CV) of 10 percent (see Chapter 1) because information is not available on the standard deviation of the requirement for B$_{12}$; the RDA is defined as equal to the EAR plus twice the CV to cover the needs of 97 to 98 percent of the individuals in the group (therefore, for B$_{12}$ the RDA is 120 percent of the EAR).

RDA for Lactation **14–18 years**   **2.8 µg/day of vitamin B$_{12}$**
                          19–30 years   **2.8 µg/day of vitamin B$_{12}$**
                          31–50 years   **2.8 µg/day of vitamin B$_{12}$**

Special Considerations

Persons with any malabsorption syndrome will likely require increased amounts of B$_{12}$. Patients with pernicious anemia or Crohn's disease involving the terminal ileum and patients who have had a gastrectomy, gastric bypass surgery, or ileal resection will require B$_{12}$ under a physician's direction. Persons who are positive for human immunodeficiency virus with chronic diarrhea may also require either increased oral or parenteral B$_{12}$.

Patients with atrophic gastritis, pancreatic insufficiency, or prolonged omeprazole treatment (Bellou et al., 1996; Gueant et al., 1990; Suter et al., 1991; Termanini et al., 1998) will have decreased bioavailability of food-bound B$_{12}$ and will require normal amounts of crystalline B$_{12}$ (either in foods fortified with B$_{12}$ or as a supplement).

INTAKE OF VITAMIN B$_{12}$

Food Sources

Ordinarily, humans obtain vitamin B$_{12}$ from animal foods. Unlike other B vitamins, B$_{12}$ is not a normal constituent of plant foods except for certain algae (Ford and Hutner, 1955). B$_{12}$ is not supplied by commonly eaten plant foods unless they have been exposed to bacterial action that has produced the vitamin; contaminated with soil, insects, or other substances that contain B$_{12}$; or fortified with B$_{12}$ (e.g., fortified ready-to-eat breakfast cereals and meal replacement formulas).

Data obtained from the 1995 Continuing Survey of Food Intakes by Individuals (CSFII) indicate that the greatest contribution to B$_{12}$ intake of the U.S. adult population comes from the category of mixed foods (including sandwiches) with meat, fish, or poultry as the main ingredient (Table 9-9). For women, the second category contributing the most B$_{12}$ is milk and milk drinks, whereas beef is the second category of B$_{12}$ for men. Fortified ready-to-eat cereals contribute a greater proportion of dietary B$_{12}$ for women than for men. The foods that are the richest sources of B$_{12}$—shellfish, organ meats such as liver, some game meat, and a few kinds of fish (see Table 9-9)—are not a regular part of many people's diets.

Analyses of CSFII 1994 to 1995 intake data for food fortified with B$_{12}$ for adults aged 51 through 70 years and older than 70 years were provided by the U.S. Department of Agriculture (A. Moshfegh, Agricultural Research Service, U.S. Department of Agriculture, personal communication, 1997). Because of the higher bioavailability of synthetic B$_{12}$ than of protein-bound B$_{12}$ for a substantial proportion of older adults, these results were examined to determine whether fortified foods contributed differently to the B$_{12}$ content of the diet for different age groups (Table 9-10). These cross-sectional data suggest that fortified foods provide a larger proportion of the B$_{12}$ consumed by older than by younger adults, especially men.

Few studies report cooking losses. However, Stewart and coworkers (1970) tested one sample and found that boiling milk for 10 minutes reduced its B$_{12}$ content by about 50 percent. Reconstituted evaporated milk contains only about 25 percent of the B$_{12}$ content of fluid whole milk (USDA, 1997). Such cooking losses may seriously limit B$_{12}$ intake by vegetarians. Boiling milk, for example, was described as a common cooking practice among Hindu women in the United Kingdom (Stewart et al., 1970). With a B$_{12}$ content of 0.4 mg/100 mL (0.9 mg/8 oz), fresh pasteurized fluid milk may be an important source of B$_{12}$ for vegetarians.

Dietary Intake

Because a generous intake of animal foods is common in the United States and Canada, median B$_{12}$ intake from food is well above the EAR. For example, in the United States the median daily intake from food by young adult men has been reported to be approximately 4 to 5 µg and by young adult women, 3 µg (Appendixes G and H). In one Canadian province, the mean dietary intake was reported as approximately 7 µg/day for men and 4 µg/day for women (Appendix I).

TABLE 9-9   Food Groups Providing Vitamin B_{12} in the Diets of U.S. Men or Women Aged 10 Years and Older, CSFII, 1995[a]

Food Group	Contribution to Total B_{12} Intake[b] (%)		Foods Within the Group that Provide at Least 1 µg of B_{12}[c] per Serving	
	Men	Women	1–2 µg	> 2 µg
Food groups providing at least 5% of total vitamin B_{12} intake				
Mixed foods[d]	18.5	16.4	NA[e]	NA
Beef	15.0	12.0	Beef	—
Milk and milk drinks	10.6	14.6	Plain and flavored yogurt[f]	—
Shellfish	9.4	4.9	Crayfish and scallops	Clams, oysters, mussels, crab, and lobster
Mixed foods, main ingredient is grain	7.1	5.7	NA	NA
Processed meats[g]	7.0	5.0	—	—
Organ meats	5.5	6.9	—	Liver, kidney, heart, brains, and tongue
Ready-to-eat cereals	4.7	8.2	Moderately fortified	Highly fortified
Finfish	3.4	5.7	Catfish, pike, whiting, perch, swordfish, carp, porgy, and flounder	Herring, sardines, trout, mackerel, salmon, and canned tuna
Vitamin B_{12} from other food groups				
Lamb, veal, game, and other carcass meat	0.8	0.8	Lamb and veal	Venison, rabbit, and squirrel
Soy-based supplements and meal replacements	0.7	0.2	Soy-based meat substitutes	—

[a] CSFII = Continuing Survey of Food Intakes by Individuals.

[b] Contribution to total intake reflects both the concentration of the nutrient in the food and the amount of the food consumed. It refers to the percentage contribution to the American diet for both men and women, based on 1995 CSFII data.

[c] 1 µg represents 20% of the Recommended Daily Intake (6.0 µg) of B_{12}—a value set by the Food and Drug Administration.

[d] Includes sandwiches and other foods with meat, poultry, or fish as the main ingredient.

[e] NA = not applicable. Mixed foods were not considered for this table.

[f] Whole, low fat, and nonfat.

[g] Includes frankfurters, sausages, lunch meats, and meat spreads.

SOURCE: Unpublished data from the Food Surveys Research Group, Agricultural Research Service, U.S. Department of Agriculture, 1997.

TABLE 9-10 Contribution of Fortified Foods to the Vitamin B$_{12}$ Intake of U.S. Men and Women by Age Group, CSFII, 1995[a]

| | Contribution of Food Group to Total B$_{12}$ Intake[b] (%) | | | | | |
| | Adults ≥ 19 Years | | Ages 51–70 Years | | Ages 70+ Years | |
Food Group	Men	Women	Men	Women	Men	Women
Ready-to-eat cereals	4.7	8.2	7.8	10.3	10.9	11.9
Soy-based supplements and meal replacements	0.7	0.5	0.9	0.5	1.2	0.3
Milk-based supplements and meal replacements	0.2	0.2	0.2	0.3	0.5	0.3
Total	5.6	8.9	8.9	12.1	12.6	12.5

[a] CSFII = Continuing Survey of Food Intake by Individuals.

[b] Refers to the percentage contribution to the American diet for both men and women, based on 2-day weighted 1995 CSFII data.

SOURCE: Unpublished data from the Food Surveys Research Group, Agricultural Research Service, U.S. Department of Agriculture, 1997.

Intake by the elderly continues to be high relative to the EAR and RDA (Appendix F); however, quantitative data are not available on the amount of B$_{12}$ provided by fortified foods. In a study of Boston elderly aged 60 to more than 90 years (Russell, 1992), median B$_{12}$ intake by males who were not taking supplements was 3.4 µg/day. The median plasma B$_{12}$ concentration for this unsupplemented group was 286 pmol/L (388 pg/mL). For females not taking supplements, the median B$_{12}$ intake was 2.6 µg/day and the median plasma B$_{12}$ concentration was 272 pmol/L (369 pg/mL). B$_{12}$ intake was correlated with serum levels, but the actual correspondence of intake with plasma values was not determined.

Quinn and Basu (1996) reported on the dietary B$_{12}$ intake estimated from 3-day (nonconsecutive) food records of 156 elderly males and females aged 65 to 77 years residing in Northern Alberta, Canada. Supplement users were excluded from the sample. The mean daily B$_{12}$ intake by males was 3.7 ± 0.3 (standard error of the mean) µg and by females was 4.3 ± 1.0 µg. Mean plasma B$_{12}$ was 286 ± 24 pmol/L (388 ± 33 pg/mL) for males and 335 ± 37 pmol/L (454 ± 50 pg/mL) for females, which is consistent with the difference in reported dietary intake. None of the males and 7 percent of the females had estimated intakes of less than 1.3 µg/day.

Intake from Supplements

Information from the Boston Nutritional Status Survey on supplement use of B_{12} by a free-living elderly population is given in Appendix F. For those taking supplements, the fiftieth percentile of supplemental B_{12} intake was 5.0 µg for men and 6.0 µg for women. Approximately 26 percent of all adults reported taking a B_{12}-containing supplement in 1986 (Moss et al., 1989).

TOLERABLE UPPER INTAKE LEVELS

Hazard Identification

Adverse Effects

No adverse effects have been associated with excess B_{12} intake from food or supplements in healthy individuals. There is very weak evidence from animal studies suggesting that B_{12} intake enhances the carcinogenesis of certain chemicals (Day et al., 1950; Georgadze, 1960; Kalnev et al., 1977; Ostryanina, 1971). These findings are contradicted by evidence that increased B_{12} intake inhibits tumor induction in the human liver, colon, and esophagus (Rogers, 1975). Some studies suggest a possible association between high-dose, parenterally administered B_{12} (0.5 to 5 mg) and acne formation (Berlin et al., 1969; Dugois et al., 1969; Dupre et al., 1979; Puissant et al., 1967; Sherertz, 1991). However, the acne lesions were primarily associated with hydroxocobalamin rather than cyanocobalamin, the form used in the United States and Canada. Furthermore, iodine particles in commercial B_{12} preparations may have been responsible for the acne. In conclusion, the evidence from these data was considered not sufficient for deriving a Tolerable Upper Intake Level (UL).

Studies involving periodic parenteral administration of B_{12} (1 to 5 mg) to patients with pernicious anemia provide supportive evidence for the lack of adverse effects at high doses (Boddy and Adams, 1968; Mangiarotti et al., 1986; Martin et al., 1992). Periodic doses of 1 mg are used in standard clinical practice to treat patients with pernicious anemia. As indicated earlier, when high doses are given orally (see "Absorption") only a small percentage of B_{12} can be absorbed from the gastrointestinal tract, which may explain the apparent low toxicity.

Special Considerations

B$_{12}$-deficient individuals who are at risk for Leber's optic atrophy should not be given cyanocobalamin to treat the B$_{12}$ deficiency. Leber's optic atrophy is a genetic disorder caused by chronic cyanide intoxication (present in tobacco smoke, alcohol, and some plants). Reduced serum B$_{12}$ concentrations have been associated with a reduced ability to detoxify the cyanide in exposed individuals (Foulds, 1968, 1969a, b, 1970; Wilson and Matthews, 1966). Cyanocobalamin may increase the risk of irreversible neurological damage (from the optic atrophy). Hydroxocobalamin is a cyanide antagonist and therefore not associated with adverse effects when given to these individuals.

Dose-Response Assessment

The data on adverse effects of B$_{12}$ intake were considered not sufficient for a dose-response assessment and derivation of a UL.

Intake Assessment

In 1986 approximately 26 percent of adults in the United States took a supplement containing B$_{12}$ (Moss et al., 1989). Although no UL can be set for B$_{12}$, an exposure assessment is provided here for possible future use. Based on data from the Third National Health and Nutrition Examination Survey (see Appendix H), the highest median intake of B$_{12}$ from diet and supplements for any life stage and gender group was for males aged 31 through 50 years: 17 µg/day. The highest reported intake at the ninety-fifth percentile was 37 µg/day for pregnant females aged 14 through 55 years.

Risk Characterization

On the basis of the review of data involving high-dose intakes of B$_{12}$, there appear to be essentially no risks of adverse effects to the general population even at the current ninety-fifth percentile of intake noted above. Furthermore, there appear to be no risks associated with intakes of supplemental B$_{12}$ that are more than two orders of magnitude higher than the ninety-fifth percentile of intake. Although there are extensive data showing no adverse effects associated with high intakes of supplemental B$_{12}$, the studies in which such intakes were reported were not designed to assess adverse effects.

RESEARCH RECOMMENDATIONS FOR VITAMIN B_{12}

High-Priority Recommendations

Priority should be given to three topics of research related to vitamin B_{12}:

- The prevalence of B_{12} deficiency as diagnosed by biochemical, neurological, or hematological abnormalities (e.g., methylmalonic acid and holotranscobalamin II).
- Improved, economical, and sensitive methods to detect B_{12} malabsorption and deficiency before adverse neurological and hematological changes occur.
- Effective methods to reduce the risk of suboptimal B_{12} status resulting from B_{12} malabsorption or vegetarian diets. For elderly persons with food-bound malabsorption, research is needed on the form and amount of B_{12} that can normalize and maintain B_{12} stores. For vegetarians, information is needed about the absorption of B_{12} from dairy products, algae, and fortified food products.

Other Research Areas

Two additional topics also merit attention:

- The feasibility and potential benefits and adverse effects of fortification of cereal grain foods with B_{12}, considering stability, identity of any degradation products, and bioavailability for normal individuals and those who malabsorb protein-bound B_{12}.
- The contribution of bacterial overgrowth to elevated serum methylmalonic acid.

REFERENCES

Adams JF. 1962. The measurement of the total assayable vitamin B_{12} in the body. In: Heinrich HC, ed. *Vitamin B_{12} und Intrinsic Faktor.* Stuttgart, Germany: Ferdinand Enke. Pp. 397–403.

Adams JF. 1970. Correlation of serum and urine vitamin B_{12}. *Br Med J* 1:138–139.

Adams JF, Tankel HI, MacEwan F. 1970. Estimation of the total body vitamin B_{12} in the live subject. *Clin Sci* 39:107–113.

Adams JF, Ross SK, Mervyn RL, Boddy K, King P. 1971. Absorption of cyanocobalamin, coenzyme B_{12}, methylcobalamin, and hydroxocobalamin at different dose levels. *Scand J Gastroenterol* 6:249–252.

Adams JF, Boddy K, Douglas AS. 1972. Interrelation of serum vitamin B_{12}, total body vitamin B_{12}, peripheral blood morphology and the nature of erythropoiesis. *Br J Haematol* 23:297–305.

Allen RH, Stabler SP, Lindenbaum J. 1993. Serum betaine, *N,N*-dimethylglycine and *N*-methylglycine levels in patients with cobalamin and folate deficiency and related inborn errors of metabolism. *Metabolism* 42:1448–1460.

Amin S, Spinks T, Ranicar A, Short MD, Hoffbrand AV. 1980. Long-term clearance of [57Co]cyanocobalamin in vegans and pernicious anaemia. *Clin Sci* 58:101–103.

Andrews GR, Haneman B, Arnold BJ, Booth JC, Taylor K. 1967. Atrophic gastritis in the aged. *Australas Ann Med* 16:230–235.

Areekul S, Oumarum K, Dougbarn J. 1977. Determination of vitamin B$_{12}$ and vitamin B$_{12}$-binding protein in human and cow's milk. *Mod Med Asia* 13:17–23.

Baker SJ, Mathan VI. 1981. Evidence regarding the minimal daily requirement of dietary vitamin B$_{12}$. *Am J Clin Nutr* 34:2423–2433.

Baker SJ, Jacob E, Rajan KT, Swaminathan SP. 1962. Vitamin B$_{12}$ deficiency in pregnancy and the puerperium. *Br Med J* 1:1658–1661.

Bastrup-Madsen P, Helleberg-Rasmussen I, Norregaard S, Halver B, Hansen T. 1983. Long term therapy of pernicious anaemia with the depot cobalamin preparation Betolvex®. *Scand J Haematol* 31:57–62.

Beck WS. 1991. Neuropsychiatric consequences of cobalamin deficiency. *Adv Intern Med* 36:33–56.

Bellou A, Aimone-Gastin I, De Korwin JD, Bronowicki JP, Moneret-Vautrin A, Nicolas JP, Bigard MA, Gueant JL. 1996. Cobalamin deficiency with megaloblastic anaemia in one patient under long-term omeprazole therapy. *J Intern Med* 240:161–164.

Berlin H, Berlin R, Brante G. 1968. Oral treatment of pernicious anemia with high doses of vitamin B$_{12}$ without intrinsic factor. *Acta Med Scand* 184:247–258.

Berlin H, Berlin R, Brante G. 1969. Treatment with high oral doses of vitamin B$_{12}$. Five years experience. *Lakartidningen* 66:153–158.

Best WR, White WF, Robbins KC, Landmann WA, Steelman SL. 1956. Studies on urinary excretion of vitamin B$_{12}$Co^{60} in pernicious anemia for determining effective dosage of intrinsic factor concentrates. *Blood* 11:338–351.

Black AK, Allen LH, Pelto GH, de Mata M, Chávez A. 1994. Iron, vitamin B-12 and folate status in Mexico: Associated factors in men and women and during pregnancy and lactation. *J Nutr* 124:1179–1188.

Boddy K, Adams JF. 1968. Excretion of cobalamins and coenzyme B$_{12}$ following massive parenteral doses. *Am J Clin Nutr* 21:657–664.

Boddy K, Adams JF. 1972. The long-term relationship between serum vitamin B$_{12}$ and total body vitamin B$_{12}$. *Am J Clin Nutr* 25:395–400.

Bozian RC, Ferguson JL, Heyssel RM, Meneely GR, Darby WJ. 1963. Evidence concerning the human requirement for vitamin B$_{12}$. Use of the whole body counter for determination of absorption of vitamin B$_{12}$. *Am J Clin Nutr* 12:117–129.

Carmel R. 1988. Pernicious anemia. The expected findings of very low serum cobalamin levels, anemia, and macrocytosis are often lacking. *Arch Intern Med* 148:1712–1714.

Carmel R. 1992. Reassessment of the relative prevalences of antibodies to gastric parietal cell and to intrinsic factor in patients with pernicious anaemia: Influence of patient age and race. *Clin Exp Immunol* 89:74–77.

Carmel R. 1996. Prevalence of undiagnosed pernicious anemia in the elderly. *Arch Intern Med* 156:1097–1100.

Carmel R, Sinow RM, Karnaze DS. 1987. Atypical cobalamin deficiency. Subtle biochemical evidence of deficiency is commonly demonstrable in patients without megaloblastic anemia and is often associated with protein-bound cobalamin malabsorption. *J Lab Clin Med* 109:454–463.

Carmel R, Sinow RM, Siegel ME, Samloff IM. 1988. Food cobalamin malabsorption occurs frequently in patients with unexplained low serum cobalamin levels. *Arch Intern Med* 148:1715–1719.

Carmel R, Green R, Jacobsen DW, Qian GD. 1996. Neutrophil nuclear segmentation in mild cobalamin deficiency: Relation to metabolic tests of cobalamin status and observations on ethnic differences in neutrophil segmentation. *Am J Clin Pathol* 106:57–63.

Casterline JE, Allen LH, Ruel MT. 1997. Vitamin B-12 deficiency is very prevalent in lactating Guatemalan women and their infants at three months postpartum. *J Nutr* 127:1966–1972.

Chanarin I. 1969. *The Megaloblastic Anaemias*, 1st ed. Oxford: Blackwell Scientific.

Chanarin I. 1979. *The Megaloblastic Anaemias*, 2nd ed. Oxford: Blackwell Scientific.

Chanarin I. 1990. *The Megaloblastic Anaemias*, 3rd ed. Boston: Blackwell Scientific.

Dagnelie PC, van Staveren WA, Hautvast JG. 1991. Stunting and nutrient deficiencies in children on alternative diets. *Acta Pediatr Scand Suppl* 374:111–118.

Darby WJ, Bridgforth EB, Le Brocquy J, Clark SL, De Oliviera JD, Kevany J, McGanity WJ, Perez C. 1958. Vitamin B_{12} requirement of adult man. *Am J Med* 25: 726–732.

Day PL, Payne LD, Dinning JS. 1950. Procarcinogenic effect of vitamin B_{12} on *p*-dimethylaminoazobenzene-fed rats. *Proc Soc Exp Biol Med* 74:854–857.

Doi K, Matsuura M, Kawara A, Tanaka T, Baba S. 1983. Influence of dietary fiber (konjac mannan) on absorption of vitamin B_{12} and vitamin E. *Tohoku J Exp Med* 141:677–681.

Donangelo CM, Trugo NM, Koury JC, Barreto Silva MI, Freitas LA, Feldheim W, Barth C. 1989. Iron, zinc, folate and vitamin B_{12} nutritional status and milk composition of low-income Brazilian mothers. *Eur J Clin Nutr* 43:253–266.

Doscherholmen A, Hagen PS. 1957. A dual mechanism of vitamin B_{12} plasma absorption. *J Clin Invest* 36:1551–1557.

Doscherholmen A, McMahon J, Ripley D. 1975. Vitamin B_{12} absorption from eggs. *Proc Soc Exp Biol Med* 149:987–990.

Doscherholmen A, McMahon J, Ripley D. 1978. Vitamin B_{12} assimilation from chicken meat. *Am J Clin Nutr* 31:825–830.

Doscherholmen A, McMahon J, Economon P. 1981. Vitamin B_{12} absorption from fish. *Proc Soc Exp Biol Med* 167:480–484.

Doscherholmen A, Silvis S, McMahon J. 1983. Dual isotope Schilling test for measuring absorption of food-bound and free vitamin B_{12} simultaneously. *Am J Clin Pathol* 80:490–495.

Draper A, Lewis J, Malhotra N, Wheeler E. 1993. The energy and nutrient intakes of different types of vegetarian: A case for supplements? *Br J Nutr* 69:3–19.

Dugois P, Amblard P, Imbert R, Bignicourt B. 1969. Acne caused by vitamin B_{12}. *Lyon Med* 221:1165–1167.

Dupre A, Albarel N, Bonafe JL, Christol B, Lassere J. 1979. Vitamin B_{12}-induced acnes. *Cutis* 24:210–211.

El Kholty S, Gueant JL, Bressler L, Djalali M, Boissel P, Gerard P, Nicolas JP. 1991. Portal and biliary phases of enterohepatic circulation of corrinoids in humans. *Gastroenterology* 101:1399–1408.

Fernandes-Costa F, Metz J. 1982. Levels of transcobalamins I, II, and III during pregnancy and in cord blood. *Am J Clin Nutr* 35:87–94.

Fernandes-Costa F, van Tonder S, Metz J. 1985. A sex difference in serum cobalamin and transcobalamin levels. *Am J Clin Nutr* 41:784–786.

Ford JE, Hutner SH. 1955. Role of vitamin B$_{12}$ in the metabolism of micro-organisms. *Vitam Horm* 13:101–136.

Foulds WS. 1968. Hydroxocobalamin in the treatment of Leber's hereditary optic atrophy. *Lancet* 1:896–897.

Foulds WS. 1969a. Cyanide induced optic neuropathy. *Ophthalmologica* 158:350–358.

Foulds WS. 1969b. The optic neuropathy of pernicious anemia. *Arch Ophthalmol* 82:427–432.

Foulds WS. 1970. The investigation and therapy of the toxic amblyopias. *Trans Ophthalmol Soc UK* 90:739–763.

Fréry N, Huel G, Leroy M, Moreau T, Savard R, Blot P, Lellouch J. 1992. Vitamin B$_{12}$ among parturients and their newborns and its relationship with birthweight. *Eur J Obstet Gynecol Reprod Biol* 45:155–163.

Gambon RC, Lentze MJ, Rossi E. 1986. Megaloblastic anaemia in one of monozygous twins breast fed by their vegetarian mother. *Eur J Pediatr* 145:570–571.

Garry PJ, Goodwin JS, Hunt WC. 1984. Folate and vitamin B$_{12}$ status in a healthy elderly population. *J Am Geriatr Soc* 32:719–726.

Georgadze GE. 1960. Effect of vitamin B$_1$ and B$_{12}$ on induction of malignant growths in hamsters. *Vopr Onkol* 6:54–58.

Grasbeck T, Nyberg W, Reizenstein P. 1958. Biliary and fecal vitamin B$_{12}$ excretion in man. An isotope study. *Proc Soc Exp Biol Med* 97:780–784.

Green R, Kinsella LJ. 1995. Current concepts in the diagnosis of cobalamin deficiency. *Neurology* 45:1435–1440.

Green R, Jacobsen DW, Van Tonder SV, Kew MC, Metz J. 1982. Absorption of biliary cobalamin in baboons following total gastrectomy. *J Lab Clin Med* 100:771–777.

Gueant JL, Champigneulle B, Gaucher P, Nicolas JP. 1990. Malabsorption of vitamin B$_{12}$ in pancreatic insufficiency of the adult and of the child. *Pancreas* 5:559–567.

Hall CA, Finkler AE. 1966. Function of transcobalamin II: A B$_{12}$ binding protein in human plasma. *Proc Soc Exp Biol Med* 123:55–58.

Hansen HA, Weinfeld A. 1962. Metabolic effects and diagnostic value of small doses of folic acid and B$_{12}$ in megaloblastic anemias. *Acta Med Scand* 172:427–443.

Healton EB, Savage DG, Brust JC, Garrett TJ, Lindenbaum J. 1991. Neurologic aspects of cobalamin deficiency. *Medicine (Baltimore)* 70:229–245.

Heinrich HC. 1964. Metabolic basis of the diagnosis and therapy of vitamin B$_{12}$ deficiency. *Semin Hematol* 1:199–249.

Hellegers A, Okuda K, Nesbitt RE Jr, Smith DW, Chow BF. 1957. Vitamin B$_{12}$ absorption in pregnancy and in the newborn. *Am J Clin Nutr* 5:327–331.

Herbert V, Jacob E, Wong KT, Scott J, Pfeffer RD. 1978. Low serum vitamin B$_{12}$ levels in patients receiving ascorbic acid in megadoses: Studies concerning the effect of ascorbate on radioisotope vitamin B$_{12}$ assay. *Am J Clin Nutr* 31:253–258.

Herzlich B, Herbert V. 1988. Depletion of serum holotranscobalamin II. An early sign of negative vitamin B$_{12}$ balance. *Lab Invest* 58:332–337.

Heyssel RM, Bozian RC, Darby WJ, Bell MC. 1966. Vitamin B_{12} turnover in man. The assimilation of vitamin B_{12} from natural foodstuff by man and estimates of minimal daily requirements. *Am J Clin Nutr* 18:176–184.

Hoey H, Linnell JC, Oberholzer VG, Laurance BM. 1982. Vitamin B_{12} deficiency in a breastfed infant of a mother with pernicious anaemia. *J R Soc Med* 75:656–658.

Houston GA, Files JC, Morrison FS. 1985. Race, age, and pernicious anemia. *South Med J* 78:69–70.

Hsing AW, Hansson L-E, McLaughlin JK, Nyren O, Blot WJ, Ekbom A, Faumeni JF. 1993. Pernicious anemia and subsequent cancer: A population-based cohort study. *Cancer* 71:745–750.

Hurwitz A, Brady DA, Schaal SE, Samloff IM, Dedon J, Ruhl CE. 1997. Gastric acidity in older adults. *J Am Med Assoc* 278:659–662.

Isaacs R, Friedman A. 1938. Standards for maximum reticulocyte percentage after intramuscular liver therapy in pernicious anemia. *Am J Med Sci* 196:718–719.

Isaacs R, Bethell FH, Riddle MC, Friedman A. 1938. Standards for red blood cell increase after liver and stomach therapy in pernicious anemia. *JAMA* 111:2291.

Jadhav M, Webb JK, Vaishnava S, Baker SJ. 1962. Vitamin B_{12} deficiency in Indian infants. *Lancet* 1962:903–907.

Jathar VS, Inamdar-Deshmukh AB, Rege DV, Satoskar RS. 1975. Vitamin B_{12} and vegetarianism in India. *Acta Haematol* 53:90–97.

Johnsen R, Bernersen B, Straume B, Forde OH, Bostad L, Burhol PG. 1991. Prevalences of endoscopic and histological findings in subjects with and without dyspepsia. *Br Med J* 302:749–752.

Johnson PR Jr, Roloff JS. 1982. Vitamin B_{12} deficiency in an infant strictly breastfed by a mother with latent pernicious anemia. *J Pediatr* 100:917–919.

Jones BP, Broomhead AF, Kwan YL, Grace CS. 1987. Incidence and clinical significance of protein-bound vitamin B_{12} malabsorption. *Eur J Haematol* 38:131–136.

Joosten E, Pelemans W, Devos P, Lesaffre E, Goossens W, Criel A, Verhaeghe R. 1993. Cobalamin absorption and serum homocysteine and methylmalonic acid in elderly subjects with low serum cobalamin. *Eur J Haematol* 51:25–30.

Joosten E, Lesaffre E, Riezler R. 1996. Are different reference intervals for methylmalonic acid and total homocysteine necessary in elderly people? *Eur J Haematol* 57:222–226.

Kalnev VR, Rachkus I, Kanopkaite SI. 1977. Influence of methylcobalamin and cyanocobalamin on the neoplastic process in rats. *Prikl Biochim Mikrobiol* 13:677.

Kano Y, Sakamoto S, Miura Y, Takaku F. 1985. Disorders of cobalamin metabolism. *Crit Rev Oncol Hematol* 3:1–34.

Karnaze DS, Carmel R. 1990. Neurologic and evoked potential abnormalities in subtle cobalamin deficiency states, including deficiency without anemia and with normal absorption of free cobalamin. *Arch Neurol* 47:1008–1012.

Kato N, Narita Y, Kamohara S. 1959. Liver vitamin B_{12} levels in chronic liver diseases. *J Vitam* 5:134–140.

Krasinski SD, Russell RM, Samloff IM, Jacob RA, Dallal GE, McGandy RB, Hartz SC. 1986. Fundic atrophic gastritis in an elderly population: Effect on hemoglobin and several serum nutritional indicators. *J Am Geriatr Soc* 34:800–806.

Kuhne T, Bubl R, Baumgartner R. 1991. Maternal vegan diet causing a serious infantile neurological disorder due to vitamin B_{12} deficiency. *Eur J Pediatr* 150:205–208.

Lindenbaum J, Healton EB, Savage DG, Brust JC, Garrett TJ, Podell ER, Marcell PD, Stabler SP, Allen RH. 1988. Neuropsychiatric disorders caused by cobalamin deficiency in the absence of anemia or macrocytosis. *N Engl J Med* 318:1720–1728.

Lindenbaum J, Savage DG, Stabler SP, Allen RH. 1990. Diagnosis of cobalamin deficiency: 2. Relative sensitivities of serum cobalamin, methylmalonic acid, and total homocysteine concentrations. *Am J Hematol* 34:99–107.

Lindenbaum J, Rosenberg IH, Wilson PW, Stabler SP, Allen RH. 1994. Prevalence of cobalamin deficiency in the Framingham elderly population. *Am J Clin Nutr* 60:2–11.

Linnell JC, Smith AD, Smith CL, Wilson J, Matthews DM. 1968. Effects of smoking on metabolism and excretion of vitamin B$_{12}$. *Br Med J* 2:215–216.

Loria A, Vaz-Pinto A, Arroyo P, Ramirez-Mateos C, Sanchez-Medal L. 1977. Nutritional anemia. 6. Fetal hepatic storage of metabolites in the second half of pregnancy. *J Pediatr* 91:569–573.

Low-Beer TS, McCarthy CF, Austad WI, Brzechwa-Ajdukiewicz A, Read AE. 1968. Serum vitamin B$_{12}$ levels and vitamin B$_{12}$ binding capacity in pregnant and non-pregnant Europeans and West Indians. *Br Med J* 4:160–161.

Luhby AL, Cooperman JM, Donnenfeld AM, Herrero JM, Teller DN, Wenig JB. 1958. Observations on transfer of vitamin B$_{12}$ from mother to fetus and newborn. *Am J Dis Child* 96:532–533.

Mangiarotti G, Canavese C, Salomone M, Thea A, Pacitti A, Gaido M, Calitri V, Pelizza D, Canavero W, Vercellone A. 1986. Hypervitaminosis B$_{12}$ in maintenance hemodialysis patients receiving massive supplementation of vitamin B$_{12}$. *Int J Artif Organs* 9:417–420.

Martin DC, Francis J, Protech J, Huff J. 1992. Time dependency of cognitive recovery with cobalamin replacement: Report of a pilot study. *J Am Geriatr Soc* 40:168–172.

McEvoy AW, Fenwick JD, Boddy K, James OF. 1982. Vitamin B$_{12}$ absorption from the gut does not decline with age in normal elderly humans. *Age Ageing* 11:180–183.

Metz J, Hart D, Harpending HC. 1971. Iron, folate, and vitamin B$_{12}$ nutrition in a hunter-gatherer people: A study of the Kung Bushmen. *Am J Clin Nutr* 24:229–242.

Miller DR, Specker BL, Ho L, Norman EJ. 1991. Vitamin B-12 status in a macrobiotic community. *Am J Clin Nutr* 53:524–529.

Miller A, Furlong D, Burrows BA, Slingerland DW. 1992. Bound vitamin B$_{12}$ absorption in patients with low serum B$_{12}$ levels. *Am J Hematol* 40:63–166.

Moelby L, Rasmussen K, Jensen MK, Pedersen KO. 1990. The relationship between clinically confirmed cobalamin deficiency and serum methylmalonic acid. *J Intern Med* 228:373–378.

Mollin DL, Ross GI. 1952. The vitamin B$_{12}$ concentrations of serum and urine of normals and of patients with megaloblastic anaemias and other diseases. *J Clin Pathol* 5:129–139.

Moss AJ, Levy AS, Kim I, Park YK. 1989. *Use of Vitamin and Mineral Supplements in the United States: Current Users, Types of Products, and Nutrients.* Advance Data, Vital and Health Statistics of the National Center for Health Statistics, No. 174. Hyattsville, MD: National Center for Health Statistics.

Muir M, Landon M. 1985. Endogenous origin of microbiologically-inactive cobalamins (cobalamin analogues) in the human fetus. *Br J Haematol* 61:303–306.

Narayanan MN, Dawson DW, Lewis MJ. 1991. Dietary deficiency of vitamin B_{12} is associated with low serum cobalamin levels in non-vegetarians. *Eur J Haematol* 47:115–118.

Naurath HJ, Joosten E, Riezler R, Stabler SP, Allen RH, Lindenbaum J. 1995. Effects of vitamin B_{12}, folate, and vitamin B_6 supplements in elderly people with normal serum vitamin concentrations. *Lancet* 346:85–89.

Nilsson-Ehle H, Jagenburg R, Landahl S, Lindstedt G, Swolin B, Westin J. 1986. Cyanocobalamin absorption in the elderly: Results for healthy subjects and for subjects with low serum cobalamin concentration. *Clin Chem* 32:1368–1371.

Norman EJ, Morrison JA. 1993. Screening elderly populations for cobalamin (vitamin B_{12}) deficiency using the urinary methylmalonic acid assay by gas chromatography mass spectrometry. *Am J Med* 94:589–594.

Ostryanina AD. 1971. Effect of vitamin B_{12} on the induction of tumors in mouse skin. *Patol Fiziol Eksperim Terapiya* 15:48–53.

Pennypacker LC, Allen RH, Kelly JP, Matthews LM, Grigsby J, Kaye K, Lindenbaum J, Stabler SP. 1992. High prevalence of cobalamin deficiency in elderly outpatients. *J Am Geriatr Soc* 40:1197–1204.

Puissant A, Vanbremeersch F, Monfort J, Lamberton J-N. 1967. A new iatrogenic dermatosis: Acne caused by vitamin B_{12}. *Bull Soc Fr Dermatol Syphiligr* 74:813–815.

Quinn K, Basu TK. 1996. Folate and vitamin B_{12} status of the elderly. *Eur J Clin Nutr* 50:340–342.

Rauma AL, Torronen R, Hanninen O, Mykkanen H. 1995. Vitamin B-12 status of long-term adherents of a strict uncooked vegan diet ("living food diet") is compromised. *J Nutr* 125:2511–2515.

Reizenstein P. 1959. Excretion of non-labeled vitamin B_{12} in man. *Acta Med Scand* 165:313–320.

Reizenstein P, Ek G, Matthews CM. 1966. Vitamin B_{12} kinetics in man. Implications on total-body B_{12} determinations, human requirements, and normal and pathological cellular B_{12} uptake. *Phys Med Biol* 11:295–306.

Robertson JA, Gallagher ND. 1983. Increased intestinal uptake of cobalamin in pregnancy does not require synthesis of new receptors. *Biochim Biophys Acta* 757:145–150.

Rogers AE. 1975. Variable effects of a lipotrobe-deficient, high-fat diet on chemical carcinogens in rats. *Cancer Res* 35:2469–2474.

Rosenberg LE, Fenton WA. 1989. Disorders of propionate and methylmalonate metabolism. In: Scriver CR, Beaudet AL, Sly WS, Valle D, eds. *The Metabolic Basis of Inherited Disease*, 6th ed. New York: McGraw-Hill. Pp. 821–844.

Rosner F, Schreiber ZA. 1972. Serum vitamin B_{12} and vitamin B_{12} binding capacity in chronic myelogenous leukemia and other disorders. *Am J Med Sci* 263:473–480.

Russell RM. 1992. Vitamin B_{12}. In: Hartz SC, Russell RM, Rosenberg IH, eds. *Nutrition in the Elderly. The Boston Nutritional Status Survey.* London: Smith-Gordon. Pp. 141–145.

Sahyoun NR, Otradovec CL, Hartz SC, Jacob RA, Peters H, Russell RM, McGandy RB. 1988. Dietary intakes and biochemical indicators of nutritional status in an elderly, institutionalized population. *Am J Clin Nutr* 47:524–533.

Savage D, Gangaidzo I, Lindenbaum J. 1994a. Vitamin B_{12} deficiency is the primary cause of megaloblastic anemia in Zimbabwe. *Br J Haematol* 86:844–850.

Savage DG, Lindenbaum J, Stabler SP, Allen RH. 1994b. Sensitivity of serum methylmalonic acid and total homocysteine determinations for diagnosing cobalamin and folate deficiencies. *Am J Med* 96:239–246.

Scarlett JD, Read H, O'Dea K. 1992. Protein-bound cobalamin absorption declines in the elderly. *Am J Hematol* 39:79–83.

Schilling RF. 1953. Intrinsic factor studies II. The effect of gastric juice on the urinary excretion of radioactivity after the oral administration of radioactive vitamin B$_{12}$. *J Lab Clin Med* 42:860–866.

Schneede J, Dagnelie PC, van Staveren WA, Vollset SE, Refsum H, Ueland PM. 1994. Methylmalonic acid and homocysteine in plasma as indicators of functional cobalamin deficiency in infants on macrobiotic diets. *Pediatr Res* 36:194–201.

Scott JM. 1997. Bioavailability of vitamin B$_{12}$. *Eur J Clin Nutr* 51 Suppl 1:S49–S53.

Scott JM, Bloomfield FJ, Stebbins R, Herbert V. 1974. Studies on derivation of transcobalamin 3 from granulocytes. Enhancement by lithium and elimination by fluoride of in vitro increments in vitamin B$_{12}$-binding capacity. *J Clin Invest* 53:228–239.

Seetharam B, Alpers DH. 1982. Absorption and transport of cobalamin (vitamin B$_{12}$). *Annu Rev Nutr* 2:343–369.

Shapiro J, Alberts HW, Welch P, Metz J. 1965. Folate and vitamin B$_{12}$ deficiency associated with lactation. *Br J Haematol* 11:498–504.

Sherertz EF. 1991. Acneiform eruption due to "megadose" vitamins B$_6$ and B$_{12}$. *Cutis* 48:119–120.

Sklar R. 1986. Nutritional vitamin B$_{12}$ deficiency in a breast-fed infant of a vegan-diet mother. *Clin Pediatr* 25:219–221.

Specker BL, Miller D, Norman EJ, Greene H, Hayes KC. 1988. Increased urinary methylmalonic acid excretion in breast-fed infants of vegetarian mothers and identification of an acceptable dietary source of vitamin B$_{12}$. *Am J Clin Nutr* 47:89–92.

Specker BL, Black A, Allen L, Morrow F. 1990. Vitamin B-12: Low milk concentrations are related to low serum concentrations in vegetarian women and to methylmalonic aciduria in their infants. *Am J Clin Nutr* 52:1073–1076.

Srikantia SG, Reddy V. 1967. Megaloblastic anaemia of infancy and vitamin B$_{12}$. *Br J Haematol* 13:949–953.

Stabler SP, Marcell PD, Podell ER, Allen RH, Savage DG, Lindenbaum J. 1988. Elevation of total homocysteine in the serum of patients with cobalamin or folate deficiency detected by capillary gas chromatography-mass spectrometry. *J Clin Invest* 81:466–474.

Stabler SP, Lindenbaum J, Allen RH. 1996. The use of homocysteine and other metabolites in the specific diagnosis of vitamin B-12 deficiency. *J Nutr* 126:1266S–1272S.

Stahlberg KG, Radner S, Norden A. 1967. Liver B$_{12}$ in subjects with and without vitamin B$_{12}$ deficiency. A quantitative and qualitative study. *Scand J Haematol* 4:312–330.

Stewart JS, Roberts PD, Hoffbrand AV. 1970. Response of dietary vitamin B$_{12}$ deficiency to physiological oral doses of cyanocobalamin. *Lancet* 2:542–545.

Sullivan LW, Herbert V. 1965. Studies on the minimum daily requirement for vitamin B$_{12}$. Hematopoietic responses to 0.1 microgram of cyanocobalamin or coenzyme B$_{12}$ and comparison of their relative potency. *N Engl J Med* 272:340–346.

Suter PM, Golner BB, Goldin BR, Morrow FD, Russell RM. 1991. Reversal of protein-bound vitamin B_{12} malabsorption with antibiotics in atrophic gastritis. *Gastroenterology* 101:1039–1045.

Teo NH, Scott JM, Neale G, Weir DG. 1980. Effect of bile on vitamin B_{12} absorption. *Br Med J* 281:831–833.

Termanini B, Gibril F, Sutliff VE, Yu F, Venzon DJ, Jensen RT. 1998. Effect of long-term gastric acid suppressive therapy on serum vitamin B_{12} levels in patients with Zollinger-Ellison syndrome. *Am J Med* 104:422–430.

Toh B-H, van Driel IR, Gleeson PA. 1997. Pernicious anemia. *N Engl J Med* 337:1441–1448.

Trugo NM, Sardinha F. 1994. Cobalamin and cobalamin-binding capacity in human milk. *Nutr Res* 14:22–33.

Tudhope GR, Swan HT, Spray GH. 1967. Patient variation in pernicious anaemia, as shown in a clinical trial of cyanocobalamin, hydroxocobalamin and cyanocobalamin-zinc tannate. *Br J Haematol* 13:216–228.

USDA (U.S. Department of Agriculture). 1997. USDA, ARS Nutrient Data Laboratory. [WWW document]. URL http://www.nal.usda.gov/fnic/foodcomp/.

van Asselt DZ, van den Broek WJ, Lamers CB, Corstens FH, Hoefnagels WH. 1996. Free and protein-bound cobalamin absorption in healthy middle-aged and older subjects. *J Am Geriatr Soc* 44:949–953.

Vaz Pinto A, Torras V, Sandoval JF, Dillman E, Mateos CR, Cordova MS. 1975. Folic acid and vitamin B_{12} determination in fetal liver. *Am J Clin Nutr* 28:1085–1086.

Vu T, Amin J, Ramos M, Flener V, Vanyo L, Tisman G. 1993. New assay for the rapid determination of plasma holotranscobalamin II levels: Preliminary evaluation in cancer patients. *Am J Hematol* 42:202–211.

WHO (World Health Organization). 1970. *Requirements of Ascorbic Acid, Vitamin D, Vitamin B_{12}, Folate, and Iron.* Report of a Joint FAO/WHO Expert Group. Technical Report Series No. 452. Geneva: WHO.

Will JJ, Mueller JF, Brodine C, Kiely CE, Friedman B, Hawkins VR, Dutra J, Vilter RN. 1959. Folic acid and vitamin B_{12} in pernicious anemia. Studies on patients treated with these substances over a ten-year period. *J Lab Clin Med* 53:22–38.

Wilson J, Matthews DM. 1966. Metabolic inter-relationships between cyanide, thiocyanate and vitamin B_{12} in smokers and non-smokers. *Clin Sci* 31:1–7.

Winawer SJ, Streiff RR, Zamcheck N. 1967. Gastric and hematological abnormalities in a vegan with nutritional vitamin B_{12} deficiency: Effect of oral vitamin B_{12}. *Gastroenterology* 53:130–135.

10

Pantothenic Acid

SUMMARY

Pantothenic acid functions as a component of coenzyme A and phosphopantetheine, which are involved in fatty acid metabolism. Pantothenic acid is widely distributed in foods; deficiency has been reported only as a result of feeding semisynthetic diets or an antagonist to the vitamin. The primary criterion used to estimate the Adequate Intake (AI) for pantothenic acid is intake adequate to replace urinary excretion. The AI for adults is 5 mg/day. There are no nationally representative estimates of the intake of pantothenic acid from food or from both food and supplements. There is not sufficient scientific evidence on which to base a Tolerable Upper Intake Level (UL) for pantothenic acid.

BACKGROUND INFORMATION

Function

Pantothenic acid is vital to the synthesis and maintenance of coenzyme A (CoA), a cofactor and acyl group carrier for many enzymatic processes, and acyl carrier protein, a component of the fatty acid synthase complex (Tahiliani and Beinlich, 1991). As such, pantothenic acid is essential to almost all forms of life. Most tissues transport pantothenic acid into cells for the synthesis of CoA.

357

Physiology of Absorption, Metabolism, and Excretion

Absorption

CoA in the diet is hydrolyzed in the intestinal lumen to dephospho-CoA, phosphopantetheine, and pantetheine, with the pantetheine subsequently hydrolyzed to pantothenic acid (Shibata et al., 1983). Pantothenic acid was the only one of these pantothenate-containing compounds absorbed by rats in studies on absorption of the various forms. Absorption is by active transport at low concentrations of the vitamin and by passive transport at higher concentrations in animal models (Fenstermacher and Rose, 1986). Because the active transport system is saturable, absorption will be less efficient at higher concentrations of intake, but the intake levels at which absorptive efficiency decreases in humans are not known. Intestinal microflora have been observed to synthesize pantothenic acid in mice (Stein and Diamond, 1989), but the contribution of bacterial synthesis to body pantothenic acid levels or fecal losses in humans has not been quantified. If microbial synthesis is substantial, balance studies in humans may have underestimated pantothenic acid absorption and requirements.

Metabolism

The synthesis of CoA from pantothenate is regulated primarily by pantothenate kinase, an enzyme that is inhibited by the pathway end products, CoA and acyl CoA. Thus CoA production does not reflect the amount of available pantothenate (Tahiliani and Beinlich, 1991). CoA, in forms such as acetyl CoA and succinyl CoA, plays an important role in the tricarboxylic acid cycle and in the synthesis of fatty acids and membrane phospholipids, amino acids, steroid hormones, vitamins A and D, porphyrin and corrin rings, and neurotransmitters. It is also required for the acetylation and acylation of proteins and the synthesis of α-tubulin (Plesofsky-Vig, 1996).

Excretion

CoA is hydrolyzed to pantothenate in a multiple-step reaction. The pantothenic acid is excreted intact in urine, where it can be measured by using a *Lactobacillus plantarum* assay or a radioimmunoassay. The amount excreted varies proportionally with dietary intake over a discrete yet wide range of intake values.

Clinical Effects of Inadequate Intake

Pantothenic acid deficiency has only been observed in individuals who were fed diets virtually devoid of pantothenic acid (Fry et al., 1976) or who were given a pantothenic acid metabolic antagonist, ω-methyl pantothenic acid (Hodges et al., 1958, 1959). The subjects exhibited various degrees of signs and symptoms, including irritability and restlessness; fatigue; apathy; malaise; sleep disturbances; gastrointestinal complaints such as nausea, vomiting, and abdominal cramps; neurobiological symptoms such as numbness, paresthesias, muscle cramps, and staggering gait; and hypoglycemia and an increased sensitivity to insulin. After 9 weeks of a semisynthetic diet devoid of pantothenic acid, blood and urine concentrations were substantially lower (Fry et al., 1976). Historically, pantothenic acid was implicated in the "burning feet" syndrome that affected prisoners of war in Asia during World War II. The condition improved after pantothenic acid supplementation but not when other B-complex vitamins were given (Glusman, 1947).

SELECTION OF INDICATORS FOR ESTIMATING THE REQUIREMENT FOR PANTOTHENIC ACID

Urinary Excretion

Urinary excretion on a typical American diet is approximately 2.6 mg/day of pantothenic acid (Tarr et al., 1981) but it is strongly dependent on intake. In a group of healthy adolescents aged 13 to 19 years, pantothenic acid intake (assessed from 4 days of food intake records) was significantly correlated with the pantothenic acid concentration in urine ($r = 0.6$) (Eissenstat et al., 1986). Total daily urinary excretion was not measured. Excretion of pantothenic acid in the urine approached zero after 11 weeks of a diet devoid of the vitamin (Hodges et al., 1958). In 10 young men, the urinary concentration of pantothenic acid fell gradually from 3.05 to 0.79 mg/day in the six men who were fed a semisynthetic diet devoid of the vitamin for 84 days (Fry et al., 1976). The other four men were supplemented with 10 mg pantothenic acid/day for a 63-day period. The excretion of the vitamin in their urine increased from 3.9 to 5.8 mg/day. In a final 7-day period, all 10 subjects were given 100 mg/day of pantothenic acid, and urinary excretion increased to approximately 60 mg/day. The authors suggested that these results implied that substantial amounts of the vitamin can be stored when intakes are high. However, it is also possible that intestinal absorption was markedly less.

Blood Levels of Pantothenic Acid

Whole Blood

Normal values for pantothenic acid in whole blood have been reported to be 1.57 to 2.66 μmol/L when care is taken to fully release pantothenate from CoA (Wittwer et al., 1989). Concentrations fell from 8.9 to 6.4 μmol/L (1.95 to 1.41 μg/mL) when six adult male prisoners were fed a diet free of pantothenic acid for 28 days (Fry et al., 1976). No further reduction was seen during the subsequent 5 weeks of depletion. In comparison, in four similar individuals supplemented with 10 mg/day of pantothenic acid, concentrations at the end of 9 weeks were not increased compared with baseline, suggesting that a value of approximately 9 μmol/L (2 μg/mL) represented normal blood concentrations of the vitamin for these subjects. In a study of 63 healthy adolescents, whole-blood concentrations and intake were significantly correlated ($r = 0.4$), but there was no correlation between whole-blood content and urinary excretion of the vitamin (Eissenstat et al., 1986).

Serum or Plasma

Concentrations in plasma are much lower than in whole blood. They do not correlate with whole-blood levels (Cohenour and Calloway, 1972) because the latter also contain CoA and other coenzymes containing pantothenic acid. Plasma concentrations are less reflective of changes in intake or status (Baker et al., 1969). Whole-blood and erythrocyte pantothenic acid concentrations are strongly correlated (Eissenstat et al. 1986).

Erythrocytes

In rats, erythrocytes were found to contain pantothenic acid, 4'-phosphopantothenic acid, and pantotheine but not CoA (Annous and Song, 1995). The correlation between erythrocyte and whole-blood concentrations of pantothenic acid in a group of 57 apparently well-nourished adolescents was 0.8, and the correlation with dietary intake was 0.4 (Eissenstat et al., 1986). Average erythrocyte concentrations were 1.5 μmol/L (334 ng/mL). Correlations between erythrocyte concentrations and both intake and urinary excretion were similar to those for whole-blood concentrations. There was no significant correlation with urinary excretion of the vitamin. Although it is theoretically possible that erythrocyte concentrations are a more accurate representation of status than whole-blood concentrations because of the contribution of serum pantothenic acid to the latter, no clear advantage of using erythrocyte values was evident in this population group. A model was developed that predicted erythrocyte pantothenic acid concentrations from intake and urinary concentrations, but it explained only 30 percent of the variance in erythrocyte concentrations, which may have resulted from errors in the estimation of intake, variability in tissue storage and utilization, and differences among subjects in the amount absorbed.

FACTORS AFFECTING THE PANTOTHENIC ACID REQUIREMENT

Bioavailability

Little information is available on the bioavailability of dietary pantothenic acid. Values of 40 to 61 percent (mean 50 percent) have been given for absorbed food-bound pantothenic acid. These values were derived by comparing urinary excretion of the vitamin after feeding a formula diet containing 8.2 mg/day of pantothenic acid (of which 6.0 mg was free crystalline pantothenic acid) with excretion after ingestion of natural foods containing 11.5 mg/day (Tarr et al., 1981). It was assumed that 100 percent of the crystalline vitamin was absorbed.

Nutrient-Nutrient Interactions

There is almost no information on the interaction between pantothenic acid and other nutrients. Koyanagi and colleagues (1969) studied subjects consuming a constant diet that contained 2.37 mg of pantothenic acid, 1.17 mg of thiamin, 0.87 mg of riboflavin, 735 IU of vitamin A, and 50 mg of ascorbic acid from food. A supplement containing either 10 mg of pantothenic acid, 5 mg of thiamin, 5 mg of riboflavin, 2,500 IU of vitamin A, or 100 mg of ascorbic acid was given to one of five subject groups (four to five subjects per group) for one week in addition to the diet. The groups receiving the pantothenic acid, thiamin, or riboflavin supplements had increases in the serum and urinary concentrations of pantothenic acid from prestudy values; the groups receiving ascorbic acid or vitamin A supplements did not. This study suggests that thiamin, and to a lesser extent riboflavin, resulted in changes in pantothenic acid metabolism and excretion. However, the levels of significance were not determined for the nutrients studied.

Oral Contraceptive Agents

Lewis and King (1980) investigated whether high-dose oral contraceptive agents affected pantothenic acid metabolism in 13 women between the ages of 19 and 24 years enrolled in a 12-day confined study. At the end of the study, blood levels and urinary pantothenic acid excretion were similar in the subjects and controls. The investigators concluded that high-dose oral contraceptive agents do not cause significant changes in the biochemical parameters of pantothenic acid.

FINDINGS BY LIFE STAGE AND GENDER GROUP

Infants Ages 0 through 12 Months

Method Used to Set the Adequate Intake

There are no functional criteria for pantothenic acid status that reflect response to dietary intake in infants. Thus, recommended intakes of pantothenic acid are based on an Adequate Intake (AI) that reflects the observed mean intake of infants fed principally with human milk.

Ages 0 through 6 Months. On the basis of a summary of recent studies in North America and the United Kingdom (Picciano, 1995), the average pantothenic acid concentration of mature human milk ranges from 2.2 to 2.5 mg/L. Values at the upper end of this range included those from women taking multivitamin supplements containing pantothenic acid. The AI is based on a reported average intake of human milk of 0.78 L/day for this age group (see Chapter 2) and an average pantothenic acid concentration of milk of 2.2 mg/L. This gives an AI for pantothenic acid of 1.7 mg/day for infants ages 0 through 6 months.

Ages 7 through 12 Months. If the reference body weight ratio method described in Chapter 2 is used to extrapolate from the AI for pantothenic acid for infants ages 0 through 6 months, the AI for pantothenic acid for older infants would be 2.2 mg/day. This is somewhat higher than the value obtained from the second method (see Chapter 2) by extrapolating down from the AI for adults to estimate a recommended intake, which results in an AI for pantothenic acid of 1.4 mg/day. The AI for pantothenic acid for older infants is set as the mean obtained from these two methods of extrapolation, 1.8 mg/day.

Pantothenic Acid AI Summary, Ages 0 through 12 Months

AI for Infants
0–6 months	**1.7 mg/day of pantothenic acid**	**≈0.2 mg/kg**
7–12 months	**1.8 mg/day of pantothenic acid**	**≈0.2 mg/kg**

Children Ages 1 through 13 Years

No data were found on which to base an Estimated Average Requirement (EAR) and thus a Recommended Dietary Allowance (RDA) for pantothenic acid for children or adolescents of any age group. Thus, AIs have been set instead.

Method Used to Set the AI

Ages 1 through 3 Years. In the absence of additional information, AIs for these age groups have been extrapolated from adult values by using the method described in Chapter 2, which gives an AI of 2 mg/day.

Ages 4 through 13 Years. In a study of 40 preschool children aged 3 to 5 years, dietary intake of pantothenic acid was measured by 3-day food records (Kerrey et al., 1968). The children were grouped by socioeconomic status. Children in the high socioeconomic group had lower reported dietary intakes of pantothenic acid than did children in the low socioeconomic group (approximately 4 and 5 mg/day, respectively). However, the mean urinary excretion of pantothenic acid was 3.36 mg/day in the high socioeconomic group compared with 1.74 mg/day in the low socioeconomic group.

Pace and colleagues (1961) studied 35 healthy girls aged 7 to 9 years during three study periods. Urinary excretion was measured while feeding controlled diets. Dietary intake ranged from 2.79 ± 0.33 (standard deviation [SD]) mg/day to 5.00 ± 0.82 mg/day. Average daily excretion of urinary pantothenic acid was 1.3 mg (47 percent of intake) when intake was 2.79 ± 0.33 mg/day and 2.7 mg (63 percent of intake) when intake was 4.49 ± 0.76 mg/day. These data indicate that intakes of 2.8 to 5.00 mg/day exceed urinary excretion.

By extrapolating the AI for children from the adult AI for pantothenic acid using the method described in Chapter 2, values of 3 and 4 mg/day are obtained for children ages 4 through 8 years and 9 through 13 years, respectively. These values are consistent with the results reviewed above.

Pantothenic Acid AI Summary, Ages 1 through 13 Years

AI for Children
1–3 years   **2 mg/day of pantothenic acid**
4–8 years   **3 mg/day of pantothenic acid**

AI for Boys
9–13 years   **4 mg/day of pantothenic acid**

AI for Girls
9–13 years   **4 mg/day of pantothenic acid**

Adolescents Ages 14 through 18 Years

Evidence Considered in Setting the AI

Eissenstat and colleagues (1986) studied 26 healthy males and 37 healthy females aged 14 to 19 and 13 to 17 years, respectively. The subjects kept 4-day dietary records. The average pantothenic acid intake was 6.3 ± 2.1 (SD) mg/day for boys and 4.1 ± 1.2 mg/day for girls. Only six subjects took pantothenic acid supplements, which provided at least 5 mg/day of additional pantothenic acid. Pantothenate in subjects' daily pooled urine samples was measured by radioimmunoassay. The average urinary pantothenate excretion for the unsupplemented subjects was 3.3 ± 1.3 (SD) mg/g of creatinine for the boys and 4.5 ± 1.9 mg/g of creatinine for the girls (approximately 5.0 mg/day for the boys and 4.2 mg/day for the girls, based on average creatinine values for adult individuals of this height) (Schneider et al., 1983).

Whole-blood and erythrocyte pantothenate were also measured. Whole-blood averages were 1.86 ± 0.47 μmol/L (411.9 ± 102.8 ng/mL) for boys and 1.57 ± 0.52 μmol/L (344.5 ± 113.6 ng/mL) for girls. Erythrocyte pantothenate concentrations averaged 1.70 ± 0.47 μmol/L (375.6 ± 104.3 ng/mL) for boys and 1.36 ± 0.42 μmol/L (301.4 ± 93.5 ng/mL) for girls. These data indicate that intakes of less than 4 mg/day supported normal concentrations of pantothenic acid (1.57 to 2.66 μmol/L) (Wittwer et al., 1989) in whole blood.

Similarly, Kathman and Kies (1984) reported that during a 4-day test period, eight boys and four girls aged 11 to 16 years had average pantothenic acid intakes of 5.6 mg/day (range 4.0 to 7.9 mg/day). These values were calculated from diet diaries and check lists. Average urinary pantothenic acid excretion was 3.74 mg/g of creatinine. However, over this 4-day period there was no statistically significant correlation between pantothenic acid intake and excretion.

By extrapolating the AI for adolescents from the adult AI for pantothenic acid using the method described in Chapter 2, a value of 5 mg/day is obtained when urinary excretion was converted to mg/day from mg/g creatinine, which is consistent with the results reviewed above.

Pantothenic Acid AI Summary, Ages 14 to 18 Years

AI for Boys
14–18 years   **5 mg/day of pantothenic acid**

AI for Girls
14–18 years   **5 mg/day of pantothenic acid**

Adults Ages 19 through 50 Years

Evidence Considered in Setting the AI

The usual pantothenic acid intake is 4 to 7 mg/day, as reported for small groups of U.S. adults and adolescents (Bull and Buss, 1982; Kathman and Kies, 1984; Srinivasan et al., 1981; Tarr et al., 1981). There is no evidence suggesting that this range of intake is inadequate. Thus, the approximate midpoint—5 mg/day—is set as the AI for adults. The adequacy of this intake is supported by the only study of the relationship between daily intake and excretion in adults (Fox and Linkswiler, 1961). Eight healthy women aged 18 to 24 years were studied to determine the urinary excretion of pantothenic acid on three levels of intake in the normal range. On self-selected diets the women consumed 3.4 to 10.3 mg/day of pantothenic acid as estimated from the tables of Zook et al. (1956) and Sarrett et al. (1946). Diets were then standardized to include 2.8 mg/day of pantothenic acid for 15 days, 7.8 mg/day for 10 days, and 12.8 mg/day for 10 days. The mean urinary excretion of pantothenic acid after consumption of these pantothenic acid intakes was stated to be 3.2, 4.5, and 5.6 mg, respectively. However, the published plot of these data suggests that excretion averaged closer to 4 and 5 mg/day at the 2.8 and 7.8 mg/day intakes, respectively, lending some uncertainty to the results. From the regression equation given relating intake to urinary excretion, a pantothenic acid intake of approximately 4 mg/day would result in a similar amount of urinary excretion of this vitamin. Because of uncertainties in the accuracy of the published values in foods used to estimate intakes, small number of subjects studied, and lack of information about the effects of intake on the efficiency of absorption and storage of the vitamin, these results can only be used to support the adequacy of the AI and not to set an EAR and RDA.

There is no information on pantothenic acid requirements of middle-age adults aged 31 through 50. The AI for younger adults, 5 mg/day, is therefore recommended for the age range 19 through 50 years. Similarly, there is no basis for determining a separate recommendation based on gender, so the AIs for men and women are the same.

Pantothenic Acid AI Summary, Ages 19 through 50 Years

AI for Men
19–30 years   **5 mg/day of pantothenic acid**
31–50 years   **5 mg/day of pantothenic acid**

AI for Women
19–30 years   **5 mg/day of pantothenic acid**
31–50 years   **5 mg/day of pantothenic acid**

Adults Ages 51 Years and Older

Evidence Considered in Setting the AI

In a study of 65 noninstitutionalized men and women aged 65 years or older (mean age 73 years), pantothenic acid intakes from food averaged 2.9 mg/1,000 kcal, or 5.9 ± 0.1 (standard error) mg/day (range 2.5 to 9.5 mg/day) (Srinivasan et al., 1981). Sixty percent of these elderly consumed supplements that increased this usual intake by 17 mg/day. The supplements did not, however, increase blood concentrations of the vitamin. Urinary pantothenic acid excretion of unsupplemented individuals averaged 6 mg/day. These data support the adequacy of the 5.9 mg/day intake from diet alone. There was no change in urinary excretion with age. Because there is no basis for expecting an increased pantothenic acid requirement in the elderly, the AI is set at 5 mg/day—the same as for younger adults.

Pantothenic Acid AI Summary, Ages 51 Years and Older

AI for Men
51–70 years   **5 mg/day of pantothenic acid**
> 70 years   **5 mg/day of pantothenic acid**

AI for Women
51–70 years   **5 mg/day of pantothenic acid**
> 70 years   **5 mg/day of pantothenic acid**

Pregnancy

Evidence Considered in Setting the AI

There is little information on pantothenic acid requirements during pregnancy. In a longitudinal study of 26 pregnant women during their third trimester and at 2 weeks and 3 months postpartum, blood pantothenate concentrations were significantly lower than those of 17 nonpregnant control women, but there was no difference in daily urinary excretion during late pregnancy compared with control subjects (Song et al., 1985). Moreover, when data for unsupplemented women measured in the third trimester and again at 2 weeks postpartum were combined, average intake exceeded excretion across the range of intakes (mean dietary intake 5.3 ± 1.7 [SD] mg/day in pregnancy, 5.9 ± 2.0 [SD] mg/day in lactation, and 2 to 11 mg/day overall). In the absence of information showing that usual intakes in the United States and Canada are inadequate to support a healthy pregnancy outcome, and rounding up from this average intake, an AI of 6 mg/day of pantothenic acid is set for pregnant women.

Pantothenic Acid AI Summary, Pregnancy

AI for Pregnancy
14–18 years   **6 mg/day of pantothenic acid**
19–30 years   **6 mg/day of pantothenic acid**
31–50 years   **6 mg/day of pantothenic acid**

Lactation

Evidence Considered in Setting the AI

The pantothenic acid content of milk appears to increase with increased intake of the vitamin. In India usual intakes of pantothenic acid were correlated with the concentration of the vitamin in human milk (Deodhar and Ramakrishnan, 1960). A similar finding was reported for 26 mothers who were nursing infants at 2 and 12 weeks postpartum (Song et al., 1984); milk pantothenate content was significantly correlated with dietary intake ($r = 0.51$) and urinary excretion ($r = 0.57$) of the vitamin and weakly correlated with blood concentrations ($r = 0.19$). The pantothenic acid content of the milk of supplemented mothers was approximately five times higher than that of the unsupplemented mothers. Blood pantothenic acid concentrations were significantly lower in lactating women at 3 months postpartum (Song et al., 1985) and at 6 weeks postpartum (Cohenour and Calloway, 1972) than for control women who had not been pregnant. Although there is no evidence that pantothenic acid intakes are inadequate to support function during lactation, on the basis of the additional secretion of the vitamin in human milk (1.7 mg/day) and the lower maternal blood concentrations reported when intakes are about 5 to 6 mg/day, an AI of 7 mg/day of pantothenic acid is recommended.

Pantothenic Acid AI Summary, Lactation

AI for Lactation
14–18 years   7 mg/day of pantothenic acid
19–30 years   7 mg/day of pantothenic acid
31–50 years   7 mg/day of pantothenic acid

INTAKE OF PANTOTHENIC ACID

Food Sources

Pantothenic acid is found both free and conjugated in virtually all plant and animal cells. To estimate the dietary intake of pantothenic acid in foods, it is necessary to convert bound pantothenic acid, for example, in coenzyme A (CoA) and fatty acid synthetase, to the free form. Various analytical methods have been used to gather information on the pantothenic acid content of foods (Orr, 1969; Schroeder, 1971; Walsh et al., 1981; Zook et al., 1956). Older data on food composition were based on microbiological assays for pantothenic acid. A high-performance liquid chromatography method was published relatively recently and has been used for the analysis of pantothenic acid in infant formulas (Romera et al., 1996). However, data on the pantothenic acid content of food regardless of method are very limited. Chicken, beef, potatoes, oat cereals, tomato products, liver, kidney, yeast, egg yolk, broccoli, and whole grains are reported to be major sources of pantothenic acid (Plesofsky-Vig, 1996; Walsh et al., 1981). Royal bee jelly and ovaries of tuna and cod have very high levels of pantothenic acid (Robinson, 1966), but refined grains, fruit products, and meats and fish with added fats or cereal extenders appear to be lower in pantothenic acid content. Freezing and canning of vegetables, fish, meat, and dairy products has been shown to decrease the pantothenic acid content of those foods (Schroeder, 1971). Processing and refining grains resulted in a 37 to 74 percent loss of pantothenic acid (Walsh et al., 1981).

Dietary Intake

The major surveys of nutrient intake used in this report (the U.S. Department of Agriculture Continuing Survey of Food Intakes by Individuals, the Third National Health and Nutrition Examination Survey, and the Boston Nutritional Status Survey) do not estimate the pantothenic acid intake from diet, largely because of the incompleteness of data on the pantothenic acid content of food. Usual daily intakes of about 4 to 7 mg have been reported quite consistently in small groups of adolescents and adults of various ages (Bull and Buss, 1982; Kathman and Kies, 1984; Srinivasan et al., 1981; Tarr et al., 1981). Data from a survey conducted in one province in Canada indicated median daily intakes of pantothenic acid from foods of approximately 5 mg for men and 4 mg for women (Santé Québec, 1995).

Intake from Supplements

Results from the 1986 National Health Interview Survey indicate that 22 percent of U.S. adults took a supplement containing pantothenic acid (Moss et al., 1989).

TOLERABLE UPPER INTAKE LEVELS

Hazard Identification

Adverse Effects

No reports of adverse effects of oral pantothenic acid in humans or animals were found. Therefore, a quantitative risk assessment cannot be performed and a Tolerable Upper Intake Level (UL) cannot be derived for pantothenic acid.

In the absence of known toxic effects by ingestion, a lowest-observed-adverse-effect level (LOAEL) and an associated no-observed-adverse-effect level (NOAEL) cannot be determined. A search of the literature revealed no evidence of toxicity associated with the intake of pantothenic acid. Vaxman et al. (1996) noted no toxic effects of 0.2 to 0.9 g/day of pantothenate combined with ascorbic acid (1 to 3 g/day) in a study of effects on wound healing. However, another study (Haslam et al., 1984) indicated that a combination of 1.2 g of calcium pantothenate, 0.6 g of pyridoxine, 3 g of niacinamide, and 3 g of ascorbic acid taken daily for 6 weeks was associated with elevations in serum transaminase levels in children. One of these doses or the combination may therefore cause hepatotoxicity, but it is not possible from this study alone to ascribe to pantothenic acid the reported adverse effect in liver function.

Special Considerations

A review of the literature failed to identify special subgroups that are distinctly susceptible to adverse effects of excess pantothenic acid intake.

Intake Assessment

Because national surveys do not provide data on the intake of pantothenic acid, a reasonable intake assessment of the 90th and 95th percentiles from U.S. or Canadian national surveys is not possible.

Risk Characterization

No adverse effects have been associated with high intakes of pantothenic acid.

RESEARCH RECOMMENDATIONS FOR PANTOTHENIC ACID

Relatively little information is available about pantothenic acid as a nutrient; priority research areas for this vitamin include the following:

• Pantothenic acid requirements of different age groups, especially infants, children, and the elderly.

• Bioavailability of pantothenic acid from different foods and mixed diets and of the extent to which synthesis by intestinal bacteria contributes to meeting the requirement.

• Use of newer methods, such as high-pressure liquid chromatography, to analyze pantothenic acid in foods. At present, pantothenic acid intakes are not calculated in national surveys such as the Third National Health and Nutrition Examination Survey because of a lack of information on the pantothenic acid content of foods.

REFERENCES

Annous KF, Song WO. 1995. Pantothenic acid uptake and metabolism by red blood cells of rats. *J Nutr* 125:2586–2593.

Baker H, Frank O, Thomson AD, Feingold S. 1969. Vitamin distribution in red blood cells, plasma, and other body fluids. *Am J Clin Nutr* 22:1469–1475.

Bull NL, Buss DH. 1982. Biotin, pantothenic acid and vitamin E in the British household food supply. *Hum Nutr Appl Nutr* 36:190–196.

Cohenour SH, Calloway DH. 1972. Blood, urine, and dietary pantothenic acid levels of pregnant teenagers. *Am J Clin Nutr* 25:512–517.

Deodhar AD, Ramakrishnan CV. 1960. Studies on human lactation. Relation between the dietary intake of lactating women and the chemical composition of milk with regard to vitamin content. *J Trop Pediatr* 6:44–70.

Eissenstat BR, Wyse BW, Hansen RG. 1986. Pantothenic acid status of adolescents. *Am J Clin Nutr* 44:931–937.

Fenstermacher DK, Rose RC. 1986. Absorption of pantothenic acid in rat and chick intestine. *Am J Physiol* 250:G155–G160.

Fox HM, Linkswiler H. 1961. Pantothenic acid excretion on three levels of intake. *J Nutr* 75:451–454.

Fry PC, Fox HM, Tao HG. 1976. Metabolic response to a pantothenic acid deficient diet in humans. *J Nutr Sci Vitaminol (Tokyo)* 22:339–346.

Glusman M. 1947. The syndrome of "burning feet" (nutritional melagia) as a manifestation of nutritional deficiency. *Am J Med* 3:211–223.

Haslam RH, Dalby JT, Rademaker AW. 1984. Effects of megavitamin therapy on children with attention deficit disorders. *Pediatrics* 74:103–111.

Hodges RE, Ohlson MA, Bean WB. 1958. Pantothenic acid deficiency in man. *J Clin Invest* 37:1642–1657.

Hodges RE, Bean WB, Ohlson MA, Bleiler R. 1959. Human pantothenic acid deficiency produced by omega-methyl pantothenic acid. *J Clin Invest* 38:1421–1425.

Kathman JV, Kies C. 1984. Pantothenic acid status of free living adolescent and young adults. *Nutr Res* 4:245–250.

Kerrey E, Crispin S, Fox HM, Kies C. 1968. Nutritional status of preschool children. I. Dietary and biochemical findings. *Am J Clin Nutr* 21:1274–1279.

Koyanagi T, Hareyama S, Kikuchi R, Takanohashi T, Oikawa K, Akazawa N. 1969. Effect of administration of thiamine, riboflavin, ascorbic acid and vitamin A to students on their pantothenic acid contents in serum and urine. *Tohoku J Exp Med* 98:357–362.

Lewis CM, King JC. 1980. Effect of oral contraceptive agents on thiamin, riboflavin, and pantothenic acid status in young women. *Am J Clin Nutr* 33:832–838.

Moss AJ, Levy AS, Kim I, Park YK. 1989. *Use of Vitamin and Mineral Supplements in the United States: Current Users, Types of Products, and Nutrients.* Advance Data, Vital and Health Statistics of the National Center for Health Statistics, No. 174. Hyattsville, MD: National Center for Health Statistics.

Orr ML. 1969. *Pantothenic Acid, Vitamin B_6 and Vitamin B_{12} in Foods.* Home Economics Research Report No. 36. Washington, DC: U.S. Department of Agriculture.

Pace JK, Stier LB, Taylor DD, Goodman PS. 1961. Metabolic patterns in preadolescent children. 5. Intake and urinary excretion of pantothenic acid and of folic acid. *J Nutr* 74:345–351.

Picciano MF. 1995. Vitamins in milk. Water-soluble vitamins in human milk. In: Jensen RG, ed. *Handbook of Milk Composition.* San Diego: Academic Press.

Plesofsky-Vig N. 1996. Pantothenic acid. In: Ziegler EE, Filer LJ Jr, eds. *Present Knowledge in Nutrition,* 7th ed. Washington, DC: ILSI Press. Pp. 236–244.

Robinson FA. 1966. *The Vitamin Co-Factors of Enzyme Systems.* Oxford: Pergamon Press.

Romera JM, Ramirez M, Gil A. 1996. Determination of pantothenic acid in infant milk formulas by high performance liquid chromatography. *J Dairy Sci* 79:523–526.

Santé Québec. 1995. *Les Québécoises et les Québécois Mangent-Ils Mieux? Rapport de l'Enquête Québécoise sur la Nutrition, 1990.* Montréal: Ministère de la Santé et des Services Sociaux, Gouvernement du Québec.

Sarrett HP, Bennett MJ, Riggs TR, Cheldelin VH. 1946. Thiamine, riboflavin, nicotinic acid, pantothenic acid and ascorbic acid content of restaurant foods. *J Nutr* 31:755.

Schneider HA, Anderson CE, Coursin DB. 1983. *Nutritional Support of Medical Practice, 2nd Edition.* Philadelphia: Harper and Row.

Schroeder HA. 1971. Losses of vitamins and trace minerals resulting from processing and preservation of foods. *Am J Clin Nutr* 24:562–573.

Shibata K, Gross CJ, Henderson LM. 1983. Hydrolysis and absorption of pantothenate and its coenzymes in the rat small intestine. *J Nutr* 113:2107–2115.

Song WO, Chan GM, Wyse BW, Hansen RG. 1984. Effect of pantothenic acid status on the content of the vitamin in human milk. *Am J Clin Nutr* 40:317–324.

Song WO, Wyse BW, Hansen RG. 1985. Pantothenic acid status of pregnant and lactating women. *J Am Diet Assoc* 85:192–198.

Srinivasan V, Christensen N, Wyse BW, Hansen RG. 1981. Pantothenic acid nutritional status in the elderly—institutionalized and noninstitutionalized. *Am J Clin Nutr* 34:1736–1742.

Stein ED, Diamond JM. 1989. Do dietary levels of pantothenic acid regulate its intestinal uptake in mice? *J Nutr* 119:1973–1983.

Tahiliani AG, Beinlich CJ. 1991. Pantothenic acid in health and disease. *Vitam Horm* 46:165–228.

Tarr JB, Tamura T, Stokstad EL. 1981. Availability of vitamin B_6 and pantothenate in an average American diet in man. *Am J Clin Nutr* 34:1328–1337.

Vaxman F, Olender S, Lambert A, Nisand G, Grenier JF. 1996. Can the wound healing process be improved by vitamin supplementation? Experimental study on humans. *Eur Surg Res* 28:306–314.

Walsh JH, Wyse BW, Hansen RG. 1981. Pantothenic acid content of 75 processed and cooked foods. *J Am Diet Assoc* 78:140–144.

Wittwer CT, Schweitzer C, Pearson J, Song WO, Windham CT, Wyse BW, Hansen RG. 1989. Enzymes for liberation of pantothenic acid in blood: Use of plasma pantetheinase. *Am J Clin Nutr* 50:1072–1078.

Zook EG, MacArthur MJ, Toepfer EW. 1956. *Pantothenic Acid in Foods.* USDA Handbook. Washington, DC: U.S. Department of Agriculture. P. 97.

11

Biotin

SUMMARY

Biotin functions as a coenzyme in bicarbonate-dependent carboxylation reactions. Values extrapolated from the data for infants and limited estimates of intake are used to set the Adequate Intake (AI) for biotin because of limited data on adult requirements. The AI for adults is 30 µg/day. There are no nationally representative estimates of the intake of biotin from food or from both food and supplements. There are not sufficient data on which to base a Tolerable Upper Intake Level (UL) for biotin.

BACKGROUND INFORMATION

After biotin's initial discovery in 1927 (Boas, 1927), it took nearly 40 years of research for it to be fully recognized as a vitamin. In mammals this colorless, water-soluble vitamin functions as a cofactor for enzymes that catalyze carboxylation retentions (Dakshinamurti, 1994).

Function

Biotin functions as a required cofactor for four carboxylases found in mammalian species, which each covalently bind the biotin moiety (Bonjour, 1991; McCormick, 1976). Of the four biotin-dependent carboxylases, three are mitochondrial (pyruvate carboxylase, methylcrotonyl-coenzyme A [CoA] carboxylase, and propionyl-CoA carboxylase) whereas the fourth (acetyl-CoA carboxylase) is found in both the mitochondria and the cytosol. An inactive form of acetyl-CoA carboxylase has been postulated to serve as storage for biotin in the mitochondria (Allred and Roman-Lopez, 1988; Allred et al., 1989; Shriver et al., 1993).

Acetyl-CoA carboxylase catalyzes the carboxylation of acetyl CoA to form malonyl CoA. Malonyl CoA then serves as a substrate for fatty acid elongation. The second biotin-dependent carboxylase, pyruvate carboxylase, catalyzes the carboxylation of pyruvate to form oxaloacetate, which serves as an intermediate in the tricarboxylic acid cycle. Oxaloacetate thus formed is converted to glucose in the liver, kidney, and other gluconeogenic tissues.

A third biotin-dependent carboxylase, β-methylcrotonyl-CoA carboxylase, is required for the degradation of leucine, a branch-chained amino acid. Low activity of this enzyme resulting from biotin deficiency leads to the production of 3-hydroxyisovaleric acid and 3-methylcrotonylglycine by an alternate pathway (Mock, 1996). Thus, elevated levels of these abnormal metabolites in urine reflect reduced activity of β-methylcrotonyl-CoA carboxylase, usually resulting from biotin deficiency.

A fourth biotin-dependent carboxylase, propionyl-CoA carboxylase, carboxylates propionyl-CoA to form D-methylmalonyl-CoA, which is racemized to the L-isomer, then undergoes isomerization to succinyl-CoA, and subsequently enters the tricarboxylic acid cycle. Reduction in activity of this carboxylase results in increased excretion of 3-hydroxypropionic acid and 3-methylcitric acid in urine (Mock, 1996).

In the normal breakdown of cellular proteins, these biotin-containing enzymes are degraded to biocytin (ε-N-biotinyl-L-lysine) or short oligopeptides containing biotin-linked lysyl residues (Mock, 1996). Biotinidase (earlier called biocytinase), a hydrolase, releases biotin from this oligopeptide for reuse (Mock, 1996).

Physiology of Absorption, Metabolism, and Excretion

Biotin exists as free biotin and in protein-bound forms in foods. The mechanism of intestinal hydrolysis of protein-bound biotin has not been well characterized, and little is known about factors that affect bioavailability. Although most dietary biotin appears to be protein bound in both meats and cereals, biotin in cereals appears to be less bioavailable (Mock, 1996). Avidin, a protein found in appreciable amounts in raw egg white, has been shown to bind biotin in the small intestine and prevent its absorption (Mock, 1996).

Biotinidase is thought to play a critical role in the release of biotin from covalent binding to protein (Wolf et al., 1984). Doses of free (unbound) biotin in the range of the estimated typical dietary intake (50 to 150 μg/day) given to individuals who have biotinidase deficiency have been shown to prevent the symptoms seen in biotinidase deficiency, indicating that biotinidase deficiency results in a relative biotin deficiency through lack of adequate digestion of protein-bound biotin, inadequate renal reabsorption, or both.

Intestinal Absorption and Microbial Synthesis

A biotin carrier located in the intestinal brush border membrane transports biotin against a sodium ion concentration gradient and is structurally specific, temperature dependent, and electroneutral; at pharmacological concentrations, diffusion predominates (Mock, 1996).

Biotin is synthesized by intestinal microflora (Bonjour, 1991). Although transporter-mediated absorption of biotin is most active in the proximal small intestine of the rat, significant absorption of biotin from the proximal colon occurs, which gives credence to the concept that biotin from microbial synthesis within the colon can contribute to meeting the human requirement. From reports of increased blood concentrations after colonic instillation of biotin, it appears that biotin is absorbed from the human colon (Innis and Allardyce, 1983; Oppel, 1948; Sorrell et al., 1971). However, Kopinski and colleagues (1989a, b) have shown that biotin synthesized by enteric flora may not be present at a location or in a form that contributes importantly to absorbed biotin.

Transport

The mechanism of biotin transport to the liver and other tissues after absorption has not been well established (Mock, 1996). Biotinidase has been identified as possibly serving as a biotin-binding protein in plasma or as a transporter protein to assist biotin's entry into the cell (Chauhan and Dakshinamurti, 1988; Wolf et al., 1985). Other studies suggest that serum biotin is more than 80 percent unbound (Hu et al., 1994; Mock and Malik, 1992; Schenker et al., 1993). An acid anion carrier with relative specificity for biotin resembling the intestinal carrier appears to mediate uptake by liver cells (Bowers-Komro and McCormick, 1985). Placental uptake of biotin and transport to the fetus have been demonstrated and appear to be specific for biotin (Hu et al., 1994; Karl and Fisher, 1992; Schenker et al., 1993); however, because the fetus does not concentrate biotin, placental transfer appears to be passive.

Metabolism and Excretion

Isolation and chemical identification of more than a dozen metabolites of biotin have established the main features of utilization in microbes and mammals (McCormick, 1976; McCormick and Wright, 1971). About half of the biotin undergoes metabolism to bisnorbiotin and biotin sulfoxide before excretion. Biotin, bisnorbiotin, and biotin sulfoxide are present in molar proportions of approximately 3:2:1 in human urine and plasma (Mock, 1996). Two additional minor metabolites, bisnorbiotin methyl ketone and biotin sulfone, were recently identified in human urine (Zempleni et al., 1997). The urinary excretion and serum concentrations of biotin and its metabolites increase roughly in the same proportion in response to either intravenous or oral administration of large doses of biotin (Mock and Heird, 1997; Zempleni et al., 1997).

Clinical Effects of Inadequate Intake

Signs of biotin deficiency in humans have been demonstrated conclusively in individuals who consume raw egg white over long periods (Baugh et al., 1968) and in total parenteral nutrition (TPN) before biotin supplementation in patients with malabsorption, including short-gut syndrome (Mock et al., 1981). The clinical findings of biotin deficiency include dermatitis, conjunctivitis, alopecia, and central nervous system abnormalities (Mock, 1996).

In adults fed raw egg white or receiving biotin-free TPN for months to years, thinning of hair, frequently with loss of hair color, was reported. Most adults with the deficiency demonstrated a red, scaly, skin rash, frequently around the eyes, nose, and mouth. Most of the adults had neurological symptoms, including depression, lethargy, hallucinations, and paresthesia of the extremities.

In infants on biotin-free TPN, symptoms of biotin deficiency begin to appear within 3 to 6 months after initiation of the TPN regimen, which is earlier than that seen in adults, probably because of the increased biotin requirement related to growth (Mock, 1996). The associated rash appears first around the mouth, eyes, and nose. The rash and the unusual distribution of facial fat observed in these infants together are called *biotin deficiency facies*. As the rash progresses, the ears and perineal orifices are affected. The resultant rash is similar in appearance to that of cutaneous candidiasis (so termed because *Candida* can usually be cultured from the lesions) and is quite similar to that seen in zinc deficiency.

Hair loss has been noted in infants after 6 to 9 months of TPN; two infants evaluated had lost all hair, including eyelashes and eyebrows, within 3 to 6 months of the onset of hair loss (Mock, 1996). In biotin-deficient infants, hypotonia, lethargy, and developmental delay, along with a peculiar withdrawn behavior, are all characteristic of a neurological disorder resulting from a lack of biotin; it is thought that the withdrawn behavior may represent the equivalent of depression seen in adults that is due to central nervous system dysfunction.

SELECTION OF INDICATORS FOR ESTIMATING THE REQUIREMENT FOR BIOTIN

The most useful information concerning indicators of the adequacy of biotin intake arises from (1) clinical observations of patients receiving biotin-free intravenous nutrition, individuals with inborn errors of metabolism, and persons who consume large amounts of raw egg white; (2) 2 studies in which biotin deficiency was experimentally induced by feeding raw egg white; and (3) fewer than 10 studies of biotin bioavailability and pharmacokinetics.

Biotin and 3-Hydroxyisovalerate Excretion

The indicators of biotin status that have been validated to the greatest extent are an abnormally decreased urinary excretion of biotin and an abnormally increased urinary excretion of 3-hydroxyisovaleric acid (NI Mock et al., 1997). The urinary excretion of biotin decreased dramatically with time in normal subjects on a raw egg white diet, reaching markedly abnormal values in 9 of 10 subjects by day 20. Bisnorbiotin excretion declined in parallel, providing evidence for regulated catabolism of biotin. By day 14 of egg white feeding, 3-hydroxyisovalerate acid excretion was abnormally increased (greater than 195 µmol/day) in all 10 subjects, providing evidence that biotin depletion decreased the activity of β-methylcrotonyl-coenzyme A (CoA) and altered leucine metabolism relatively early in biotin deficiency (NI Mock et al., 1997). Normal values are 77 to 195 µmol/day (112 ± 38 [standard deviation]); at day 10 the values for the deficient people were 272 ± 92 µmol/day. Abnormally decreased excretion of biotin, abnormally increased excretion of 3-hydroxyisovalerate acid, or both have also been reported in several overt cases of biotin deficiency (Carlson et al., 1995; Gillis et al., 1982; Kien et al., 1981; Lagier et al., 1987; Mock et al., 1981, 1985). Gender differences are not apparent in these two indicators.

Plasma Biotin

A low plasma biotin concentration is not a sensitive indicator of inadequate biotin intake. Abnormally decreased plasma biotin was absent in half the subjects fed raw egg white (NI Mock et al., 1997) and in some overt case reports of biotin deficiency (Carlson et al., 1995; Khalidi et al., 1984; Kien et al., 1981; Matsusue et al., 1985; Mock et al., 1981, 1985).

Odd-Chain Fatty Acid Composition of Plasma Lipids

Odd-chain fatty acid composition in plasma lipids may reflect biotin status (Kramer et al., 1984; Liu et al., 1994; Mock et al., 1988a, b), but the sensitivity and clinical utility of this measurement remains to be determined. The accumulation of odd-chain fatty acids is thought to result from propionyl-CoA carboxylase deficiency.

Methodology

All published studies on biotin nutriture (Zempleni et al., 1997) have used one of three basic types of assays to estimate biotin: bioassays (most studies), avidin-binding assays, or fluorescent derivative assays. Recent modifications of bioassays generally have adequate sensitivity to measure biotin in blood, urine, and foods. For example, the bioassay based on *Kloeckera brevis* has both excellent sensitivity and metabolite discrimination (Guilarte, 1985). However, the bacterial bioassays (and perhaps the eukaryotic bioassays as well) suffer interference from unrelated substances and variable growth response to biotin metabolites. The acid hydrolysis or protein digestion required to release bound biotin may also release other compounds that support bacterial growth.

There are major discrepancies among the various bioassays and avidin-binding assays concerning the true concentration of biotin in human plasma. Reported mean values range from approximately 0.5 nmol/L to more than 10 nmol/L. The avidin-based radioimmunoassay with high-performance liquid chromatography is among the best current assays (Mock, 1996).

FACTORS AFFECTING THE BIOTIN REQUIREMENT

Several factors have been identified that affect the biotin require-ment: the ingestion of large quantities of raw eggwhite, which con-tains a substance (avidin) that binds biotin; biotinidase deficiency (a genetic defect); the use of anticonvulsants that induce biotin catabolism in some individuals; and pregnancy. In the latter two conditions, the ratio of biotin metabolites to biotin in urine is in-creased (Mock and Dyken, 1997; Mock and Stadler, 1997; DM Mock et al., 1997b). Inherited biotinidase deficiency is particularly rele-vant to understanding biotin deficiency because the clinical mani-festations appear to result largely from a secondary biotin deficiency in the presence of normal dietary intakes.

FINDINGS BY LIFE STAGE AND GENDER GROUP

Infants Ages 0 through 12 Months

Method Used to Set the Adequate Intake

An Adequate Intake (AI) is used as the goal for intake by infants. The AI reflects the observed mean biotin intake of infants fed prin-cipally with human milk.

Ages 0 through 6 Months. In early and transitional human milk, the concentration of biotin metabolites is nearly twice the concentra-tion of biotin in samples (DM Mock et al., 1997a). With postpartum maturation of milk production, the biotin concentration increases but the metabolites still account for approximately one-third of total biotin at 5 weeks postpartum. In mature human milk (greater than 21 days postpartum) the concentration of biotin varies substantially (Mock et al., 1992); it exceeds the concentration in serum by one to two orders of magnitude. This suggests that there is an active biotin transport system into milk. According to Hirano and coworkers (1992), estimates of the biotin content of milk are 3.8 ± 1.2 (stan-dard deviation) µg/L as free biotin determined microbiologically and 5.2 ± 2.1 µg/L after acid hydrolysis—slightly higher than earlier estimates of 4.5 µg/L (Salmenpera et al., 1985) and 7 µg/L for total biotin from bioassays (Paul and Southgate, 1978). With greatest weight given to the recent results (Hirano et al., 1992), but with the value within the range of the two other studies (Paul and South-gate, 1978; Salmenpera et al., 1985), the biotin content of human milk was estimated to be 6 µg/L. The adequate intake for biotin for infants ages 0 through 6 months is based on the reported mean volume of milk consumed by this age group (0.78 L/day; see Chapter 2) and the estimate of the biotin concentration in human milk of 6 µg/L (0.78 L × 6 µg/L = 5 µg).

Ages 7 through 12 Months. If the reference body weight ratio method described in Chapter 2 to extrapolate up from the AI for biotin for infants ages 0 through 6 months is used, the AI for biotin for the older infants is 6 µg/day after rounding.

Biotin AI Summary, Ages 0 through 12 Months

AI for Infants

0–6 months	**5 µg/day of biotin**	**≈0.7 µg/kg**
7–12 months	**6 µg/day of biotin**	**≈0.7 µg/kg**

Children and Adolescents Ages 1 through 18 Years

Method Used to Set the AI

Evidence concerning the biotin requirement is minimal and does not justify the setting of an Estimated Average Requirement (EAR). No definitive studies demonstrate evidence of biotin deficiency in normal individuals in any age group resulting from inadequate intakes. In the absence of additional information, including data on needs of adults, AIs for children and adolescents have been extrapolated from values for infants by using the formula

$$AI_{child} = (AI_{young\ infant})\ (weight_{child}/weight_{infant})^{0.75}.$$

Biotin AI Summary, Ages 1 through 18 Years

AI for Children

1–3 years	**8 µg/day of biotin**
4–8 years	**12 µg/day of biotin**

AI for Boys

9–13 years	**20 µg/day of biotin**
14–18 years	**25 µg/day of biotin**

AI for Girls

9–13 years	**20 µg/day of biotin**
14–18 years	**25 µg/day of biotin**

Adults Ages 19 Years and Older

Method Used to Set the AI

In the absence of data on biotin deficiencies in normal individuals, a reasonable inference would be that the average current dietary intake of biotin should meet the dietary requirement. With this approach, the AI for biotin might be set at either 40 or 60 µg/day depending on the data set used (see "Dietary Intake"). Extrapolation from the AI for infants exclusively fed human milk would be expected to overestimate the requirement for adults because adults require biotin only for maintenance. The result of such an extrapolation using the formula

$$AI_{adult} = (AI_{young\ infant})\ (weight_{adult}/weight_{infant})^{0.75}$$

is 30 µg/day of biotin. Based on this very limited evidence, the AI for adults is set at 30 µg/day of biotin. This value should be adequate for maintaining normal excretion of 3-hydroxyisovaleric acid in adults (NI Mock et al., 1997). Data are not sufficient to set separate values for men and women or for the elderly.

Biotin AI Summary, Ages 19 Years and Older

AI for Men
19–30 years	30 µg/day of biotin
31–50 years	30 µg/day of biotin
51–70 years	30 µg/day of biotin
> 70 years	30 µg/day of biotin

AI for Women
19–30 years	30 µg/day of biotin
31–50 years	30 µg/day of biotin
51–70 years	30 µg/day of biotin
> 70 years	30 µg/day of biotin

Pregnancy

Evidence Considered in Setting the AI

Two recent studies (Mock and Stadler, 1997; DM Mock et al., 1997b) have raised questions, previously expressed (NRC, 1989), about the adequacy of biotin status during pregnancy. Some studies have detected low plasma concentrations of biotin (Bhagavan, 1969; Dostalova, 1984); others have not (Mock and Stadler, 1997). DM Mock and colleagues (1997b) detected increased 3-hydroxyisovaleric acid in more than half of healthy pregnant women by the third trimester, and urinary excretion of biotin was decreased in about 50 percent of the women studied. It is not known whether these changes in values are normal for pregnant women or indicate low biotin intake relative to need. However, these data are not sufficient to justify an increase in the AI to meet the needs of pregnancy except for pregnant adolescents.

Biotin AI Summary, Pregnancy

AI for Pregnancy
14–18 years	**30 µg/day of biotin**
19–30 years	**30 µg/day of biotin**
31–50 years	**30 µg/day of biotin**

Lactation

Method Used to Set the AI

To cover the amount of biotin secreted in milk, the AI is increased by 5 µg/day for lactating adolescents and women. No distinction is made for the stage of lactation or age.

Biotin AI Summary, Lactation

AI for Lactation
14–18 years	**35 µg/day of biotin**
19–30 years	**35 µg/day of biotin**
31–50 years	**35 µg/day of biotin**

Special Considerations

Persons receiving hemodialysis or peritoneal dialysis may have an increased requirement for biotin (Livaniou et al., 1987; Yatzidis et al., 1984) as would persons with genetic biotinidase deficiency (Mock, 1996).

INTAKE OF BIOTIN

Food Sources

Biotin contents have been determined for relatively few foods and are not ordinarily included in food composition tables. Although biotin is widely distributed in natural foodstuffs, its concentration varies substantially. For example, liver contains biotin at about 100 µg/100 g whereas fruits and most meats contain only about 1 µg/100 g.

Dietary Intake

The U.S. Department of Agriculture Continuing Survey of Food Intakes by Individuals, the Third National Health and Nutrition Examination Survey (NHANES III), and the Boston Nutritional Status Survey do not report biotin intake. Murphy and Calloway (1986), using food intake data from the NHANES II, estimated the mean biotin intake of young women aged 18 to 24 years to be 39.9 ± 26.9 (standard deviation) µg/day. This result is considerably lower than the estimated dietary intake of biotin in a composite Canadian diet (62 µg/day) and an actual analysis of the diet (60 µg/day) (Hoppner et al., 1978). Calculated average intakes of biotin for the British population of adults and children were similar to the U.S. estimate—33 and 35 µg/day (Bull and Buss, 1982; Lewis and Buss, 1988).

Intake from Supplements

According to the 1986 National Health Interview Survey, approximately 17 percent of U.S. adults take a supplement containing biotin (Moss et al., 1989). Specific data on intake from supplements are not available.

TOLERABLE UPPER INTAKE LEVELS

Hazard Identification

No reported adverse effects of biotin in humans or animals were found. Toxicity has not been reported in patients treated with daily doses up to 200 mg orally and up to 20 mg intravenously to treat biotin-responsive inborn errors of metabolism and acquired biotin deficiency (Mock, 1996).

Several studies reported that acute doses of biotin (10 mg/100 g body weight) in pregnant rats (at the pre- and postimplantation stages) caused inhibition of fetal and placental growth and resorption of fetuses and placentae (Paul and Duttagupta, 1975, 1976). The dose used was equivalent to a human dose of 7 g for a 70-kg person, which is considerably greater than the recommended intake. These results are not considered useful for deriving a Tolerable Upper Intake Level (UL) for human intakes because of the high doses used, mode and vehicle of administration used (subcutaneous injection administration of 0.1 mol/L of NaOH, which would itself be toxic), and lack of an adequate control group.

Dose-Response Assessment

The data on adverse effects from high biotin intake are not sufficient for a quantitative risk assessment, and a UL cannot be derived. Several studies involving high biotin intakes reported no adverse effects. Koutsikos et al. (1996) found no adverse effects after intravenous administration of 50 mg of biotin to hemodialysis patients. Roth et al. (1982) administered 10 mg/day of biotin during the ninth month of pregnancy and found no adverse effects in the mother or infant. Ramaekers et al. (1993) reported no adverse effects in a 15-year-old boy given 10 mg/day of biotin to treat multiple carboxylase deficiency resulting from an inborn error of metabolism. Colamaria et al. (1989) similarly found no adverse effects in an infant treated with 10 mg/day of biotin to reverse a syndrome consisting of lethargy, sparse scalp hair, autistic-like behavior, myoclonus, and drug-resistant seizures. Taken together, these studies indicate a possible range for intake levels in any future studies of toxic effects of biotin.

Intake Assessment and Risk Characterization

Neither an intake assessment nor a risk characterization is currently possible because national surveys do not provide data on the dietary intake of biotin.

RESEARCH RECOMMENDATIONS FOR BIOTIN

There is a serious lack of data useful for setting Estimated Average Requirements (EARs) for biotin. The understanding of the nutrition of biotin is rudimentary compared with that of some other B vitamins. Although the limited information seems to indicate that there is little cause for concern about the adequacy of biotin intake for healthy people, information on the human requirements, intake, bioavailability, toxicity, and metabolic effects of this compound is needed.

REFERENCES

Allred JB, Roman-Lopez CR. 1988. Enzymatically inactive forms of acetyl-CoA carboxylase in rat liver mitochondria. *Biochem J* 251:881–885.

Allred JB, Roman-Lopez CR, Jurin RR, McCune SA. 1989. Mitochondrial storage forms of acetyl-CoA carboxylase: Mobilization/activation accounts for increased activity of the enzyme in liver of genetically obese Zucker rats. *J Nutr* 119:478–483.

Baugh CM, Malone JH, Butterworth CE Jr. 1968. Human biotin deficiency. A case history of biotin deficiency induced by raw egg consumption in a cirrhotic patient. *Am J Clin Nutr* 21:173–182.

Bhagavan HN. 1969. Biotin content of blood during gestation. *Int Z Vitaminforsch* 39:235–237.

Boas MA. 1927. The effect of desiccation upon the nutritive properties of egg white. *Biochem J* 21:712–724.

Bonjour J-P. 1991. Biotin. In: Machlin LJ, ed. *Handbook of Vitamins.* New York: Marcel Dekker. Pp. 393–427.

Bowers-Komro DM, McCormick DB. 1985. Biotin uptake by isolated rat liver hepatocytes. *Ann NY Acad Sci* 447:350–358.

Bull NL, Buss DH. 1982. Biotin, pantothenic acid and vitamin E in the British household food supply. *Hum Nutr Appl Nutr* 36:190–196.

Carlson GL, Williams N, Barber D, Shaffer JL, Wales S, Isherwood D, Shenkin A, Irving MH. 1995. Biotin deficiency complicating long-term total parenteral nutrition in an adult patient. *Clin Nutr* 14:186–190.

Chauhan J, Dakshinamurti K. 1988. Role of human serum biotinidase as biotin-binding protein. *Biochem J* 256:265–270.

Colamaria V, Burlina AB, Gaburro D, Pajno-Ferrara F, Saudubray JM, Merino RG, Dalla Bernardina B. 1989. Biotin-responsive infantile encephalopathy: EEG-polygraphic study of a case. *Epilepsia* 30:573–578.

Dakshinamurti K. 1994. Biotin. In: Shils ME, Olson JA, Shike M, eds. *Modern Nutrition in Health and Disease.* Philadelphia: Lea & Febiger. Pp. 426–431.

Dostalova L. 1984. Vitamin status during puerperium and lactation. *Ann Nutr Metab* 28:385–408.

Gillis J, Murphy FR, Boxall LB, Pencharz PB. 1982. Biotin deficiency in a child on long-term TPN. *J Parenter Enteral Nutr* 6:308–310.

Guilarte TR. 1985. Measurement of biotin levels in human plasma using a radiometric-microbiological assay. *Nutr Rep Int* 31:1155–1163.

Hirano M, Honma K, Daimatsu T, Hayakawa K, Oizumi J, Zaima K, Kanke Y. 1992. Longitudinal variations of biotin content in human milk. *Int J Vitam Nutr Res* 62:281–282.

Hoppner K, Lampi B, Smith DC. 1978. An appraisal of the daily intakes of vitamin B_{12}, pantothenic acid and biotin from a composite Canadian diet. *Can Inst Food Sci Technol J* 11:71–74.

Hu Z-Q, Henderson GI, Mock DM, Schenker S. 1994. Biotin uptake by basolateral membrane vesicles of human placenta: Normal characteristics and role of ethanol. *Proc Soc Exp Biol Med* 206:404–408.

Innis SM, Allardyce DB. 1983. Possible biotin deficiency in adults receiving long-term total parenteral nutrition. *Am J Clin Nutr* 37:185–187.

Karl PI, Fisher SE. 1992. Biotin transport in microvillous membrane vesicles, cultured trophoblasts and isolated perfused human placenta. *Am J Physiol* 262:C302–C308.

Khalidi N, Wesley JR, Thoene JG, Whitehouse WM Jr, Baker WL. 1984. Biotin deficiency in a patient with short bowel syndrome during home parenteral nutrition. *J Parenter Enteral Nutr* 8:311–314.

Kien CL, Kohler E, Goodman SI, Berlow S, Hong R, Horowitz SP, Baker H. 1981. Biotin-responsive in vivo carboxylase deficiency in two siblings with secretory diarrhea receiving total parenteral nutrition. *J Pediatr* 99:546–550.

Kopinski JS, Leibholz J, Bryden WL. 1989a. Biotin studies in pigs. 3. Biotin absorption and synthesis. *Br J Nutr* 62:767–772.

Kopinski JS, Leibholz J, Bryden WL. 1989b. Biotin studies in pigs. 4. Biotin availability in feedstuffs for pigs and chickens. *Br J Nutr* 62:773–780.

Koutsikos D, Fourtounas C, Kapetanaki A, Agroyannis B, Tzanatos H, Rammos G, Kopelias I, Bosiolis B, Bovoleti O, Darema M, Sallum G. 1996. Oral glucose tolerance test after high-dose i.v. biotin administration in normoglucemic hemodialysis patients. *Ren Fail* 18:131–137.

Kramer TR, Briske-Anderson M, Johnson SB, Holman RT. 1984. Effects of biotin deficiency on polyunsaturated fatty acid metabolism in rats. *J Nutr* 114:2047–2052.

Lagier P, Bimar P, Seriat-Gautier S, Dejode JM, Brun T, Bimar J. 1987. Zinc and biotin deficiency during prolonged parenteral nutrition in infants. *Presse Med* 16:1795–1797.

Lewis J, Buss DH. 1988. Trace nutrients: Minerals and vitamins in the British household food supply. *Br J Nutr* 60:413–424.

Liu YY, Shigematsu Y, Bykov I, Nakai A, Kikawa Y, Fukui T, Sudo M. 1994. Abnormal fatty acid composition of lymphocytes of biotin-deficient rats. *J Nutr Sci Vitaminol* 40:283–288.

Livaniou E, Evangelatos GP, Ithakissios DS, Yatzidis H, Koutsicos DC. 1987. Serum biotin levels in patients undergoing chronic hemodialysis. *Nephron* 46:331–332.

Matsusue S, Kashihara S, Takeda H, Koisumi S. 1985. Biotin deficiency during total parenteral nutrition: Its clinical manifestation and plasma nonesterified fatty acid level. *J Parenter Enteral Nutr* 9:760–763.

McCormick DB. 1976. Biotin. In: Hegsted M, ed. *Present Knowledge in Nutrition.* Washington, DC: The Nutrition Foundation. Pp. 217–225.

McCormick DB, Wright LD. 1971. The metabolism of biotin and analogues. In: Florkin M, Stotz EH, eds. *Comprehensive Biochemistry,* Vol. 21. Amsterdam: Elsevier. Pp. 81–110.

Mock DM. 1996. Biotin. In: Ziegler EE, Filer LJ Jr, eds. *Present Knowledge in Nutrition,* 7th ed. Washington, DC: ILSI Nutrition Foundation. Pp. 220–235.

Mock DM, Dyken ME. 1997. Biotin catabolism is accelerated in adults receiving long-term therapy with anticonvulsants. *Neurology* 49:1444–1447.

Mock DM, Heird GM. 1997. Urinary biotin analogs increase in humans during chronic supplementation: The analogs are biotin metabolites. *Am J Physiol* 272:E83–E85.

Mock DM, Malik MI. 1992. Distribution of biotin in human plasma: Most of the biotin is not bound to protein. *Am J Clin Nutr* 56:427–432.

Mock DM, Stadler DD. 1997. Conflicting indicators of biotin status from a cross-sectional study of normal pregnancy. *J Am Coll Nutr* 16:252–257.

Mock DM, Delorimer AA, Liebman WM, Sweetman L, Baker H. 1981. Biotin deficiency: An unusual complication of parenteral alimentation. *N Engl J Med* 304:820–823.

Mock DM, Baswell DL, Baker H, Holman RT, Sweetman L. 1985. Biotin deficiency complicating parenteral alimentation: Diagnosis, metabolic repercussions, and treatment. *J Pediatr* 106:762–769.

Mock DM, Johnson SB, Holman RT. 1988a. Effects of biotin deficiency on serum fatty acid composition: Evidence for abnormalities in humans. *J Nutr* 118:342–348.

Mock DM, Mock NI, Johnson SB, Holman RT. 1988b. Effects of biotin deficiency on plasma and tissue fatty acid composition: Evidence for abnormalities in rats. *Pediatr Res* 24:396–403.

Mock DM, Mock NI, Langbehn SE. 1992. Biotin in human milk: Methods, location, and chemical form. *J Nutr* 122:535–545.

Mock DM, Mock NI, Stratton SL. 1997a. The concentrations of biotin metabolites in human milk. *J Pediatr* 131:456–458.

Mock DM, Stadler DD, Stratton SL, Mock NI. 1997b. Biotin status assessed longitudinally in pregnant women. *J Nutr* 127:710–716.

Mock NI, Malik MI, Stumbo PJ, Bishop WP, Mock DM. 1997. Increased urinary excretion of 3-hydroxyisovaleric acid and decreased urinary excretion of biotin are sensitive early indicators of decreased status in experimental biotin deficiency. *Am J Clin Nutr* 65:951–958.

Moss AJ, Levy AS, Kim I, Park YK. 1989. *Use of Vitamin and Mineral Supplements in the United States: Current Users, Types of Products, and Nutrients.* Advance Data, Vital and Health Statistics of the National Center for Health Statistics, No. 174. Hyattsville, MD: National Center for Health Statistics.

Murphy SP, Calloway DH. 1986. Nutrient intake of women in NHANES II, emphasizing trace minerals, fiber, and phytate. *J Am Diet Assoc* 86:1366–1372.

NRC (National Research Council). 1989. *Recommended Dietary Allowances,* 10th ed. Washington, DC: National Academy Press.

Oppel TW. 1948. Studies of biotin metabolism in man: 4. Studies of the mechanism of absorption of biotin and the effect of biotin administration on a few cases of seborrhea and other conditions. *Am J Med Sci* 215:76–83.

Paul AA, Southgate DAT. 1978. *McCance and Widdowson's The Composition of Foods.* London: Her Majesty's Stationery Office.

Paul PK, Duttagupta PN. 1975. The effect of an acute dose of biotin at the pre-implantation stage and its relation with female sex steroids in the rat. *J Nutr Sci Vitaminol (Tokyo)* 21:89–101.

Paul PK, Duttagupta PN. 1976. The effect of an acute dose of biotin at a post-implantation stage and its relation with female sex steroids in the rat. *J Nutr Sci Vitaminol (Tokyo)* 22:181–186.

Ramaekers VT, Brab M, Rau G, Heimann G. 1993. Recovery from neurological deficits following biotin treatment in a biotinidase Km variant. *Neuropediatrics* 24:98–102.

Roth KS, Yang W, Allan L, Saunders M, Gravel RA, Dakshinamurti K. 1982. Prenatal administration of biotin in biotin responsive multiple carboxylase deficiency. *Pediatr Res* 16:126–129.

Salmenpera L, Perheentupa J, Pispa JP, Siimes MA. 1985. Biotin concentrations in maternal plasma and milk during prolonged lactation. *Int J Vitam Nutr Res* 55:281–285.

Schenker S, Hu Z, Johnson RF, Yang Y, Frosto T, Elliott BD, Henderson GI, Mock DM. 1993. Human placental biotin transport: Normal characteristics and effect of ethanol. *Alcohol Clin Exp Res* 17:566–575.

Shriver BJ, Roman-Shriver C, Allred JB. 1993. Depletion and repletion of biotinyl enzymes in liver of biotin-deficient rats: Evidence of a biotin storage system. *J Nutr* 123:1140–1149.

Sorrell MF, Frank O, Thompson AD, Aquino H, Baker H. 1971. Absorption of vitamins from the large intestine in vivo. *Nutr Rep Int* 3:143–148.

Wolf B, Heard GS, McVoy JR, Raetz HM. 1984. Biotinidase deficiency: The possible role of biotinidase in the processing of dietary protein-bound biotin. *J Inherit Metab Dis* 7:121–122.

Wolf B, Grier RE, McVoy JR, Heard GS. 1985. Biotinidase deficiency: A novel vitamin recycling defect. *J Inherit Metab Dis* 8:53–58.

Yatzidis H, Koutsicos D, Agroyannis B, Papastephanidis C, Francos-Plemenos M, Delatola Z. 1984. Biotin in the management of uremic neurologic disorders. *Nephron* 36:183–186.

Zempleni J, McCormick DB, Mock DM. 1997. Identification of biotin sulfone, bis-norbiotin methyl ketone, and tetranorbiotin-1-sulfoxide in human urine. *Am J Clin Nutr* 65:508–511.

12
Choline

SUMMARY

Choline functions as a precursor for acetylcholine, phospholipids, and the methyl donor betaine. The primary criterion used to estimate the Adequate Intake (AI) for choline is the prevention of liver damage as assessed by measuring serum alanine aminotransferase levels. The AI for adults is 550 mg/day of choline for men and 425 mg/day for women. There are no nationally representative estimates of the intake of choline from food or food supplements. Choline in the diet is available as free choline or is bound as esters such as phosphocholine, glycerophosphocholine, sphingomyelin, or phosphatidylcholine. The critical adverse effect from high intake of choline is hypotension, with corroborative evidence on cholinergic side effects (e.g., sweating and diarrhea) and fishy body odor. The Tolerable Upper Intake Level (UL) for adults is 3.5 g/day.

BACKGROUND INFORMATION

Choline is a dietary component that is important for the structural integrity of cell membranes, methyl metabolism, cholinergic neurotransmission, transmembrane signaling, and lipid and cholesterol transport and metabolism. Human cells grown in culture have an absolute requirement for choline (Eagle, 1955). When cells are deprived of choline, they die by apoptosis (Albright et al., 1996; Cui et al., 1996; Holmes-McNary et al., 1997; James et al., 1997; Shin et

390

al., 1997; Zeisel et al., 1997). There is an endogenous pathway for the de novo biosynthesis of the choline moiety via the sequential methylation of phosphatidylethanolamine using S-adenosylmethionine as the methyl donor (Bremer and Greenberg, 1961) (see Figure 12-1). Thus, the demand for dietary choline is modified by metabolic methyl-exchange relationships between choline and three nutrients: methionine, folate, and vitamin B_{12} (lipotropes) (Zeisel and Blusztajn, 1994).

With this type of nutrient interdependence, designation of the essential nature of a nutrient depends on showing that de novo synthesis rates are not adequate to meet the demand for the nutrient when the other nutrients are available in amounts sufficient to sustain normal growth and function. Healthy men with normal folate and vitamin B_{12} status fed a choline-deficient diet have diminished plasma choline and phosphatidylcholine concentrations and develop liver damage (Zeisel et al., 1991). For these individuals, de novo synthesis of choline was not adequate to meet the demand for

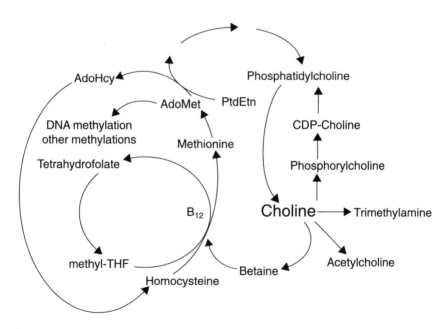

FIGURE 12-1 Choline, folate, and methionine metabolism are closely interrelated. AdoHcy = S-adenosylhomocysteine, AdoMet = S-adenosylmethionine, B_{12} = vitamin B_{12}, CDP-Choline = cytidine diphosphocholine, PtdEtn = phosphatidylethanolamine, THF = tetrahydrofolate. Reprinted with permission, from Zeisel and Blusztajn (1994). Copyright 1994 by Annual Reviews.

the nutrient. Information about women, infants, children, and older adults is not sufficient to know whether choline is needed in the diet of these groups.

Function

Choline can be acetylated, phosphorylated, oxidized, or hydrolyzed. Several comprehensive reviews of the metabolism and functions of choline have been published (Kuksis and Mookerjea, 1978; Zeisel, 1981; Zeisel and Blusztajn, 1994).

Choline accelerates the synthesis and release of acetylcholine, an important neurotransmitter involved in memory storage, muscle control, and many other functions (Cohen and Wurtman, 1975; Haubrich et al., 1974; Wecker, 1986). It is also a precursor for the synthesis of (1) phospholipids, including phosphatidylcholine (a membrane constituent important for the structure and function of membranes), for intracellular signaling (Exton, 1994; Zeisel, 1993) and hepatic export of very low-density lipoproteins (Yao and Vance, 1988, 1989); (2) sphingomyelin (another membrane constituent) for structural and signaling functions (Hannun, 1994); and (3) platelet activating factor, a potent messenger molecule (Frenkel et al., 1996). Choline is a precursor for the formation of the methyl donor betaine. Betaine is also required by renal glomerular cells, which use betaine and glycerophosphocholine as organic osmolytes to adapt to osmotic stress (Bauernschmitt and Kinne, 1993; Burg, 1995; Garcia-Perez and Burg, 1991; Grossman and Hebert, 1989).

Physiology of Absorption, Metabolism, and Excretion

Dietary choline is absorbed from the lumen of the small intestine via transporter proteins in the enterocyte (Herzberg and Lerner, 1973; Herzberg et al., 1971; Kuczler et al., 1977; Sheard and Zeisel, 1986). Before choline can be absorbed from the gut, some is metabolized by bacteria to form betaine (which may be absorbed and used as a methyl donor) and methylamines (which are not methyl donors) (Zeisel et al., 1983). No other component of the diet has been identified as competing with choline for transport by intestinal carriers. Choline is found in foods as free choline and as esterified forms such as phosphocholine, glycerophosphocholine, sphingomyelin, and phosphatidylcholine. Lecithin is a phosphatidylcholine-rich fraction prepared during commercial purification of phospholipids, and this term is often used interchangeably with phosphatidylcholine. Lecithin is often added to foods as an emulsifying agent.

Pancreatic enzymes can liberate choline from dietary phosphatidylcholine, phosphocholine, and glycerophosphocholine (Zeisel and Blusztajn, 1994). The free choline that is formed enters the portal circulation of the liver (Le Kim and Betzing, 1976) whereas phosphatidylcholine may enter via lymph in chylomicrons.

All tissues accumulate choline by diffusion and mediated transport (Zeisel, 1981). A specific carrier mechanism transports free choline across the blood-brain barrier at a rate that is proportional to the serum choline concentration. In the neonate this choline transporter has an especially high capacity (Cornford and Cornford, 1986). The rate at which the liver takes up choline is sufficient to explain the rapid disappearance of choline injected systemically (Zeisel et al., 1980c). The kidney also accumulates choline (Acara and Rennick, 1973). Some of this choline appears in the urine unchanged but most is oxidized within the kidney to form betaine (Rennick et al., 1977).

In the predominant pathway for phosphatidylcholine biosynthesis, choline is phosphorylated, converted to cytidine diphosphocholine, and then converted to phosphatidylcholine (Kennedy and Weiss, 1956; Vance, 1990) (Figure 12-1). In an alternative pathway, phosphatidylethanolamine is sequentially methylated to form phosphatidylcholine by the enzyme phosphatidylethanolamine-*N*-methyltransferase with *S*-adenosylmethionine as the methyl donor (Bremer and Greenberg, 1961; Vance and Ridgway, 1988). This is the major (perhaps only) pathway for de novo synthesis of the choline moiety in adult mammals. It is most active in the liver but has been identified in many other tissues (Blusztajn et al., 1979; Crews et al., 1981; Yang et al., 1988). Best estimates of in vivo activity of this enzyme, based on in vitro data, are that 15 to 40 percent of the phosphatidylcholine present in the liver is derived from the phosphatidylethanolamine-*N*-methyltransferase pathway, with the remainder coming from the cytidine diphosphocholine pathway (Bjornstad and Bremer, 1966; Sundler and Akesson, 1975). No estimates are available as to the relative extent of choline obtained from cell turnover. Dietary intake of phosphatidylcholine is approximately 6 to 10 g/day (Zeisel et al., 1991).

A significant portion of choline is oxidized to form betaine in the liver and kidney (Bianchi and Azzone, 1964; Weinhold and Sanders, 1973). The methyl groups of betaine can be scavenged and reused in single-carbon metabolism (Finkelstein et al., 1982) (see "Nutrient-Nutrient Interactions").

Clinical Effects of Inadequate Intake

Humans

Although choline is clearly essential to life, there is only one published study examining the effects of inadequate dietary intake in healthy men. That study reported decreased choline stores and liver damage (elevated alanine aminotransferase) when men were fed a choline-deficient diet containing adequate methionine, folate, and vitamin B_{12} for 3 weeks (Zeisel et al., 1991) (Figures 12-2 and 12-3). Another study, in which men were fed a choline- and methyl-deficient diet, reported decreased choline stores but did not report on liver function (Jacob et al., 1995). Individuals fed with total parenteral nutrition (TPN) solutions devoid of choline but adequate for methionine and folate develop fatty liver and liver damage as assessed by elevated alanine aminotransferase; in some individu-

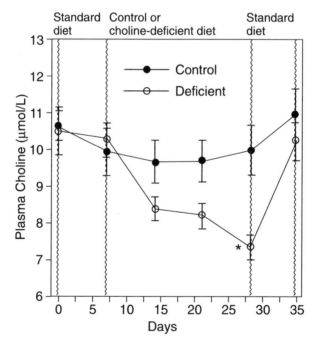

FIGURE 12-2 Plasma choline in healthy men ingesting a control (500 mg/day of choline) or choline-deficient (13 mg/day of choline) diet. *Difference from day 7 value: $p < 0.01$. Reprinted with permission, from Zeisel et al. (1991). Copyright 1991 by the Federation of American Societies for Experimental Biology.

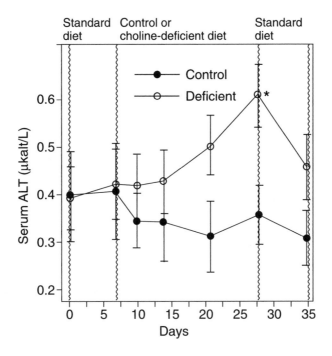

FIGURE 12-3 Serum alanine aminotransferase (ALT) activity in men ingesting a control or choline-deficient diet. Serum ALT was determined by using an automated spectrophotometric assay. Data are expressed as mean activity ± standard error of the mean. *Difference from day 7 value: $p < 0.05$. Reprinted with permission, from Zeisel et al. (1991). Copyright 1991 by the Federation of American Societies for Experimental Biology.

als, this is resolved when a source of dietary choline is provided (Buchman et al., 1992, 1993, 1995; Chawla et al., 1989; Shapira et al., 1986; Sheard et al., 1986). In a double-blind protocol, investigators administered lecithin (30 percent phosphatidylcholine) orally to patients receiving TPN twice daily for 6 weeks. At the end of this time, plasma choline had risen by more than 50 percent in the lecithin group whereas in the placebo group it had decreased by 25 percent. In the treated group, liver fat decreased by 30 percent (Buchman et al., 1992). In another small clinical study (Buchman et al., 1995), four patients who had low plasma concentrations of free choline after treatment with TPN (which contained no additional choline) were given 1 to 4 g/day of choline chloride for 6 weeks. During choline administration, plasma choline concentration increased into the normal range but decreased back to baseline when choline supplementation was discontinued. Fatty liver was resolved completely during choline supplementation but steatosis (fatty liver) recurred in one patient after 10 weeks of return to choline-free TPN. The available data support the provisional conclusion that de novo synthesis of choline is not always sufficient to meet human requirements for choline.

Animals

Supporting animal studies (in many species, such as the baboon) also found that a choline-deficient diet resulted in decreased choline stores and liver dysfunction (Hoffbauer and Zaki, 1965; Sheard et al., 1986; Tayek et al., 1990; Yao and Vance, 1990). The following animals fed a choline-deficient diet may be susceptible to developing growth retardation, renal dysfunction and hemorrhage, or bone abnormalities: baboon (Hoffbauer and Zaki, 1965), chicken (Blair et al., 1973; Ketola and Nesheim, 1974), dog (Best and Huntsman, 1932; Hershey, 1931), guinea pig (Tani et al., 1967), hamster (Handler, 1949), pig (Blair and Newsome, 1985; Fairbanks and Krider, 1945), quail (Ketola and Young, 1973), rat (Newberne and Rogers, 1986), and trout (Ketola, 1976).

SELECTION OF INDICATORS FOR ESTIMATING THE REQUIREMENT FOR CHOLINE

Markers of Liver Dysfunction

The liver is damaged when humans consume an otherwise adequate diet that is deficient in choline, resulting in elevated alanine aminotransferase levels in blood (Burt et al., 1980; Tayek et al., 1990; Zeisel et al., 1991). Fatty infiltration of liver also occurs in choline deficiency but is difficult to use as a functional marker without special liver imaging techniques (Buchman et al., 1992).

Hepatic choline and choline metabolite concentrations have been shown to decrease during choline deficiency in the rat (Zeisel et al., 1989). Phosphocholine concentration in liver is highly correlated with dietary choline intake, decreasing to 10 to 20 percent of control values after 2 weeks on a diet sufficient in methionine, folate, and vitamin B_{12} but deficient in choline (Pomfret et al., 1990). Hepatic phosphocholine concentration was most sensitive to modest dietary choline deficiency, decreasing to 10 to 20 percent of control values after 2 weeks of a deficient diet (Pomfret et al., 1990). This measurement is not easily undertaken in humans, although magnetic resonance spectroscopy does makes it possible (Cohen et al., 1995).

Plasma Concentrations

Plasma choline concentration varies in response to diet and is found in the water-soluble fraction as free choline (Buchman et al., 1993; Burt et al., 1980; Chawla et al., 1989; Sheard et al., 1986; Zeisel et al., 1991). It decreases approximately 30 percent in subjects fed a choline-deficient diet for 3 weeks (Zeisel et al., 1991). Plasma choline concentration can increase twofold after a meal high in choline content and three- or fourfold after a supplemental choline dose (Zeisel et al., 1980b). Fasting plasma choline concentrations vary from 7 to 20 µmol/L, with most subjects having concentrations of 10 µmol/L. The disadvantage of using plasma choline as a functional indicator is that these concentrations do not appear to decline below approximately 50 percent of normal, even when subjects fast for more than 1 week (Savendahl et al., 1997). Perhaps this is because membrane phospholipids, which are a large storage pool for choline, are hydrolyzed to maintain plasma choline concentration above this minimal level. Fasting plasma phosphatidylcholine concentrations (mostly as part of plasma lipoproteins) are approximately 1 to 1.5 mmol/L (Aquilonius et al., 1975; Zeisel et al., 1980b, 1991). Plasma phosphatidylcholine concentration also decreases in choline deficiency (Zeisel et al., 1991) but is also influenced by factors that change plasma lipoprotein levels.

Reduction of Risk of Chronic Disease

Dementia

Studies in rodents suggest that dietary intake of choline early in life can diminish the severity of memory deficits in aged animals (Bartus et al., 1980; Meck and Williams, 1997a, b, c). Most available human studies have used choline-containing compounds to treat rather than prevent the symptoms of dementia and therefore did not address whether dementias could be prevented. In the absence of food composition data, epidemiological studies on the association of choline intake with dementia are not available. More human studies are needed to determine whether dietary choline intake is useful in the prevention of dementia.

Cardiovascular Disease

The choline-containing phospholipid phosphatidylcholine (lecithin) has been used as a treatment to lower cholesterol concentrations because lecithin-cholesterol acyltransferase has an important role in the removal of cholesterol from tissues. In humans phosphatidyl-choline ingestion is associated with a modest reduction in plasma cholesterol (Hirsch et al., 1978; Wood and Allison, 1982; Zeisel et al., 1991). In addition, choline or betaine treatment has been used to lower high plasma homocysteine concentrations (Anonymous, 1997; Dudman et al., 1987; Wendel and Bremer, 1984; Wilcken et al., 1983, 1985), and choline-deficient rodents have elevated plasma homocysteine concentrations (Varela-Moreiras et al., 1995) (see Chapter 8, "Vascular Disease"). Wendel and Bremer (1984) report-ed that betaine treatment was more effective than folate treatment in normalizing plasma homocysteine and methionine concentra-tions of a child with homocystinuria, a genetic disease caused by 5,10-methylenetetrahydrofolate reductase deficiency (choline is the precursor for betaine, which itself is found in sugar beets and wine). Therefore, dietary choline intake might be correlated with cardio-vascular disease risk. More human studies are needed before con-clusions can be drawn about whether dietary choline intake is use-ful in preventing cardiovascular disease.

Cancer

In rodents dietary choline deficiency is associated with increased incidence of liver cancer and increased sensitivity to carcinogenic chemicals (Newberne and Rogers, 1986). The mechanisms of the carcinogenic actions of choline deficiency are not known but may be mediated by changes in protein kinase C activity (da Costa et al., 1993, 1995). There are no human data; studies in humans are needed to assess the role of dietary choline in the prevention of cancer.

FACTORS AFFECTING THE CHOLINE REQUIREMENT

Nutrient-Nutrient Interactions

Any consideration of the requirements for choline and methionine needs to include the close interrelationships with other methyl donors. Choline, methionine, and folate metabolism inter-act at the point that homocysteine is converted to methionine.

Betaine-homocysteine methyltransferase catalyzes the methylation of homocysteine using betaine as the methyl donor (see Figure 12-1) (Finkelstein et al. 1982; Mudd and Poole, 1975; Wong and Thompson, 1972). In an alternative pathway, 5-methyltetrahydrofolate-homocysteine methyltransferase regenerates methionine by using a methyl group derived de novo from the single-carbon pool (Finkelstein et al., 1982, 1988). Methionine adenosyltransferase converts methionine to *S*-adenosylmethionine (the active methylating agent for many enzymatic methylations, including the methylation of phosphatidylethanolamine to form phosphatidylcholine [Ridgway and Vance, 1988]).

Perturbing the metabolism of one of the methyl donors reveals the intermingling of these metabolic pathways. Total hepatic folate content decreased by 31 to 40 percent in rats after 2 weeks on a choline-deficient diet (Selhub et al., 1991; Varela-Moreiras et al., 1995). This effect was reversed by refeeding choline (Varela-Moreiras et al., 1995). Rats fed diets deficient in both methionine and choline for 5 weeks had hepatic folate concentrations that were half of those present in controls (Horne et al., 1989). Tetrahydrofolate deficiency in rats, induced by treatment with methotrexate (Barak and Kemmy, 1982; Barak et al., 1984; Freeman-Narrod et al., 1977; Pomfret et al., 1990; Svardal et al., 1988) or by dietary folate deficiency (Kim et al., 1994) resulted in diminished hepatic total choline, with the greatest decrease occurring in hepatic phosphocholine concentrations. During choline deficiency in rats, hepatic *S*-adenosylmethionine concentrations also decreased by as much as 50 percent (Barak et al., 1982; Poirier et al., 1977; Shivapurkar and Poirier, 1983; Zeisel et al., 1989). In rats choline deficiency for 2 weeks doubled plasma homocysteine levels (Varela-Moreiras et al., 1995). See Chapters 7 and 8 for more information on plasma homocysteine.

Gender

Males may have a higher choline requirement than do females. Female rats are less sensitive to choline deficiency than are male rats (Tessitore et al., 1995), perhaps because of females' enhanced capacity to form the choline moiety de novo. Females rats have greater phosphatidylethanolamine-*N*-methyltransferase activity in liver than do males (Arvidson, 1968; Bjornstad and Bremer, 1966; Lyman et al., 1971). Estimates of the amount of increased activity vary between 10 (Lyman et al., 1971) and 50 percent (Bjornstad and Bremer, 1966). A woman's capacity to form the choline moiety de novo may decrease after menopause (Lindblad and Schersten, 1976), because estrogens increase hepatic phosphatidylethanolamine-*N*-methyltransferase activity in rats (Drouva et al., 1986; Young, 1971).

Exercise

Strenuous physical activity in trained athletes reduced the plasma choline concentration by approximately 40 percent, from 14.1 to 8.4 µmol/L (Conlay et al., 1986). A choline supplement given to marathon runners modestly enhanced performance (Sandage et al., 1992). In 10 top-level triathletes who were given either a placebo or lecithin at 0.2 g/kg body mass 1 hour before each type of exercise, plasma choline concentrations in all the triathletes decreased on average by 16.9 percent after the bicycle exercise when placebo was taken before the race but did not do so when lecithin was given (Von Allworden et al., 1993).

Bioavailability

No estimates are available for percentage absorption of the various forms of choline in humans. The water-soluble choline-derived compounds (choline, phosphocholine, and glycerophosphocholine) are absorbed via the portal circulation whereas the lipid-soluble compounds (phosphatidylcholine and sphingomyelin) present in foods are absorbed into lymph as chylomicrons via the thoracic duct. This results in differential delivery and kinetics of distribution to tissues (Cheng et al., 1996; Zeisel et al., 1980b).

FINDINGS BY LIFE STAGE AND GENDER GROUP

Data are not sufficient for deriving an Estimated Average Requirement (EAR) for choline. The two published studies in healthy humans used male subjects only and tested a single level of choline intake. For these reasons only an Adequate Intake (AI) can be estimated. This amount will be influenced by the availability of methionine and folate in the diet. It may be influenced by gender, pregnancy, lactation, and stage of development. Although AIs are set for choline, it may be that the choline requirement can be met by endogenous synthesis at some of these stages.

To date, all studies have used choline-free diets and compared them with choline-containing diets; no intermediate levels of deficiency have been reported. Careful dose-response experiments are needed before an EAR can be derived.

Infants Ages 0 through 12 Months

Method Used to Set the AI

An AI is used as the goal for intake by infants.

Ages 0 through 6 Months. The AI reflects the observed mean intake of choline by infants consuming human milk. Thus the choline AI for young infants is based on mean intake data from infants fed human milk exclusively for their first 6 months and uses the choline concentration of milk produced by well-nourished mothers. Human milk contains 160 to 210 mg (1.5 to 2 mmol)/L of choline moiety delivered as choline, phosphocholine, glycerophosphocholine, phosphatidylcholine, and sphingomyelin (Holmes-McNary et al., 1996; Zeisel et al., 1986). The choline phospholipids sphingomyelin and phosphatidylcholine are part of the milk fat-globule membrane (Holmes-McNary et al., 1996; Zeisel et al., 1986).

Rat pups denied access to milk have lower serum choline concentrations than do their fed litter mates (Zeisel and Wurtman, 1981). Thus, milk intake contributes to the maintenance of high serum choline concentrations in the neonate. In the rat, supplemental choline is concentrated in the rat dam's milk (Garner et al., 1995; Zeisel, 1987). In women consuming a low-choline diet, milk choline content is lower than that in women consuming a more adequate diet (Zeisel et al., 1982). Consumption of either a choline-deficient or choline-supplemented diet by lactating rat dams results in significant changes in the phosphocholine concentration of their milk (Holmes-McNary et al., 1996; Zeisel et al., 1986).The concentration of total choline in human milk is 160 mg/L (1.5 mmol/L). For the mean volume of output of human milk of 0.78 L/day and the average choline content of 160 mg/L, the AI for choline is 125 mg/day (1.2 mmol/day) for infants ages 0 through 6 months. For the reference infant weight of 7 kg, this corresponds to an AI of 18 mg/kg of body weight/day (0.17 mmol/kg/day).

Ages 7 through 12 Months. If the reference body weight ratio method described in Chapter 2 to extrapolate from the AI for choline for infants ages 0 through 6 months is used, the AI for choline for the older infants would be 150 mg/day (1.4 mmol/day). The second method (see Chapter 2), extrapolating from the AI for adults, gives an AI that is essentially the same as that from extrapolating from infants. There are no data estimating choline intake from foods for this age group.

Choline AI Summary, Ages 0 through 12 Months

AI for Infants

| 0–6 months | 125 mg/day of choline | ≈18 mg/kg |
| 7–12 months | 150 mg/day of choline | ≈17 mg/kg |

Special Considerations

Although commercially available infant formulas and bovine milk both contain choline and choline-containing compounds (Holmes-McNary et al., 1996; Rohlfs et al., 1993; Zeisel et al., 1986), human milk has a significantly higher phosphocholine concentration (718 μmol/L) than does either cow milk or infant formulas. However, cow milk and cow-milk-derived infant formulas have the same glycerophosphocholine concentration as human milk (400 to 800 μmol/L) (Holmes-McNary et al., 1996) or higher (415 μmol/L) (Holmes-McNary et al., 1996). Soy-derived infant formulas have lower glycerophosphocholine concentration (115 μmol/L or less) (Holmes-McNary et al., 1996). Human milk phosphatidylcholine and sphingomyelin concentrations do not differ significantly from those in cow milk and cow-milk-derived infant formulas (200 μmol/L) (Holmes-McNary et al., 1996). Soy-derived infant formulas contain more phosphatidylcholine than do either human milk or cow-milk-derived formulas but less sphingomyelin than human milk (Holmes-McNary et al., 1996). Unesterified choline concentration in mature human milk is 30 to 80 percent lower than in either cow milk or the infant formulas (Holmes-McNary et al., 1996). The relative bioavailability of choline, phosphocholine, and glycerophosphocholine is similar in a rat model (Cheng et al., 1996) but no information is available for humans. Thus, it is not known whether these differences in milk and formula composition are clinically relevant.

Children and Adolescents Ages 1 through 18 Years

Method Used to Set the AI

No direct data on choline were found on which to base an EAR or AI for children and adolescents. In the absence of additional information, AIs for these age groups have been extrapolated from adult values by using the method described in Chapter 2.

Choline AI Summary, Ages 1 through 18 Years

AI for Children	**1–3 years**	**200 mg/day of choline**
	4–8 years	**250 mg/day of choline**
AI for Boys	**9–13 years**	**375 mg/day of choline**
	14–18 years	**550 mg/day of choline**
AI for Girls	**9–13 years**	**375 mg/day of choline**
	14–18 years	**400 mg/day of choline**

Adults Ages 19 Years and Older

Method Used to Set the AI

An intake level of 500 mg/day (4.8 mmol/day; approximately 7 mg/kg/day [0.7mmol/kg/day]) of choline base is the dose that prevented alanine aminotransferase abnormalities in healthy men (Zeisel et al., 1991). This estimate for an AI is uncertain because it is based on a single published study; it may need revision when other data become available. This estimate fits within the bracketing estimates derived from patients on total parenteral nutrition for whom approximately 2 mg/kg/day of choline moiety did not prevent a deficiency syndrome (Sheard et al., 1986) and 31 mg/kg/day of choline moiety restored normal choline status (Buchman et al., 1992, 1993). The amount estimated as adequate for men should be sufficient to prevent an increase in alanine aminotransferase but it resulted in a small decrease in plasma choline in the one study in which it was evaluated, which suggests that dietary intake normally might be slightly higher. Thus the AI is set at approximately 7 mg/kg/day or, for the reference man weighing 76 kg, at 550 mg after rounding.

To arrive at an estimate for AI for women, it is assumed that data from men can be used even though women may use choline more efficiently (see "Gender"). No experimental attempts to make healthy women choline deficient have been reported. However, women on total parenteral nutrition were just as likely as were men to develop low plasma choline concentrations and fatty liver (Buchman et al., 1995).

No experimental data are available from which to calculate an AI for life stage groups other than adults as a whole.

Choline AI Summary, Ages 19 Years and Older

The AI for choline in all forms for men in all age groups is 550 mg and for women is 425 mg. It is not known whether women have the same requirement on a body weight basis as men, but this AI is likely to be adequate on the basis of the earlier discussion on gender. Although there is some evidence that transport across the blood-brain barrier is diminished in the elderly, which suggests the possibility of a higher requirement than for younger adults (Cohen et al., 1995), no adjustment has been made in the AI for the elderly.

AI for Men	**19–30 years**	**550 mg/day of choline**
	31–50 years	**550 mg/day of choline**
	51–70 years	**550 mg/day of choline**
	> 70 years	**550 mg/day of choline**
AI for Women	**19–30 years**	**425 mg/day of choline**
	31–50 years	**425 mg/day of choline**
	51–70 years	**425 mg/day of choline**
	> 70 years	**425 mg/day of choline**

Pregnancy

Evidence Considered in Setting the AI

The need for choline is probably higher for pregnant than for nonpregnant women on the basis of animal data. Pregnancy renders female rats as vulnerable to deficiency as males (Zeisel et al., 1995). During pregnancy in humans (Welsch 1978; Welsch et al., 1981), guinea pigs (Swiery and Yudilevich, 1985; Swiery et al., 1986; Yudilevich and Sweiry, 1985), and rats (Jorswieck, 1974) large amounts of choline are delivered to the fetus through the placenta. Transport of choline from mother to fetus depletes maternal stores of choline; the choline concentration of maternal liver fell from a mean of 130 μmol/L in adult nonpregnant rats to 38 μmol/L in late pregnancy (Gwee and Sim, 1978).

Choline availability during embryogenesis and perinatal development may be especially important. In rats fed adequate diets during pregnancy, postnatally, and at weaning, 1 mmol/day of extra dietary choline results in long-lasting enhancement of spatial memory

(Meck and Williams, 1997a, b, c), altered morphology of septal neurons (Loy et al., 1991; Williams et al., 1998), and enhanced hippocampal long-term potentiation (Pyapali et al., 1998) and cholinergic neurotransmission (Cermak et al., 1998; Holler et al., 1996). The two periods of sensitivity to extra choline occur during embryonic days 12 to 17 and postnatal days 16 to 30 (Loy et al., 1991; Meck et al., 1988, 1989).

In mammals the placenta transports choline to the fetus (Welsch, 1976); choline concentration in amniotic fluid is 10-fold greater than that in maternal blood (S. Zeisel, University of North Carolina School of Public Health, unpublished observations, 1997). At birth, humans and other mammals have plasma choline concentrations that are much higher than those in adults (Zeisel et al., 1980a). It is not known whether de novo synthesis of choline increases during pregnancy.

The AI for pregnant women is greater than that for the adult by the amount needed for the fetus and placenta. Through the use of published values for the choline concentration of various adult rat tissues (Pomfret et al., 1989) and with the assumption of a body organ weight percentage as estimated by Widdowson (1963) for the human fetus, the fetal choline content can be estimated as approximately 5 mmol/kg (520 mg/kg) fetal weight. Human placental tissue has been estimated to average 1.26 ± 0.24 mmol/kg (mean $\pm$ standard error) in a small sample ($n = 7$) (Welsch, 1976); a value of approximately 2 mmol of choline per kg of placental tissue should cover almost all pregnant women. If it is thus assumed that the average choline content of fetal and placental tissue combined is approximately 3 mmol/kg (312 mg/kg), that there is no extra synthesis during pregnancy, and that there is no contribution of choline by placental or fetal synthesis, the required dietary amount of choline for the 10 kg of tissue that comprises the fetus (3 kg) and organs of pregnancy (7 kg) is 30 mmol, or 3,000 mg (10 kg tissue $\times$ 312 mg), which is approximately 11 mg/day (10 μmol/day) of additional dietary choline throughout pregnancy. This amount would be achieved by increasing the AI (after rounding) to 450 mg/day of choline for pregnancy.

Choline AI Summary, Pregnancy

The increase in the AI to support pregnancy is based on the fetal and placental accumulation of choline.

AI for Pregnancy	**14–18 years**	**450 mg/day of choline**
	19–30 years	**450 mg/day of choline**
	31–50 years	**450 mg/day of choline**

Lactation

Method Used to Set the AI

The need for choline is likely to be increased during lactation because a substantial amount of choline is secreted in human milk, and mechanisms for conserving maternal choline status have not been identified. Lactating rats are more sensitive to choline deficiency than are nonlactating rats (Zeisel et al., 1995).

The AI for women during the first 6 months of lactation should be increased above that in the nonpregnant, nonlactating woman to cover the choline that is transferred into milk. For the assumption of an average volume production of 0.78 L/day (see Chapter 2) and an average choline content of milk of 156 mg/L (1.5 mmol/L), this increase is 125 mg/day (1.2 mmol/day). This increase is based on an assumption of 100 percent efficiency. It is not known whether de novo synthesis of choline increases during lactation. Women who are breastfeeding older infants who are also eating solid foods may need slightly less because of a lower volume of milk production.

Choline AI Summary, Lactation

AI for Lactation	**14–18 years**	**550 mg/day of choline**
	19–30 years	**550 mg/day of choline**
	31–50 years	**550 mg/day of choline**

INTAKE OF CHOLINE

Food Sources

Choline is widely distributed in foods, with most of it in the form of phosphatidylcholine in membranes. Foods that are especially rich in choline compounds are milk, liver, eggs, and peanuts. It is possible to consume a diet of normal foods that delivers 1 g/day of choline (Zeisel et al., 1980b). Lecithins added during food processing may increase the average daily per capita consumption of phosphatidylcholine by 1.5 mg/kg of body weight for adults (this corresponds to 0.225 mg/kg of body weight of choline moiety) (SCOGS/ LSRO, 1979).

Dietary Intake

Choline intake is not reported in the Third National Health and Nutrition Examination Survey (Perloff et al., 1990), the Continuing Survey of Food Intake by Individuals (Perloff et al., 1990), or the Boston Nutritional Status Survey (Hartz et al., 1992), and the choline content of foods is not included in major nutrient databases. There are no reports on choline intake from Canada. Estimated average choline dietary intake in adults consuming a typical U.S. or Canadian diet (as free choline and the choline in phosphatidylcholine and other choline esters) is approximately 730 to 1,040 mg/day (7 to 10 mmol/day) (LSRO/FASEB, 1981; Zeisel, 1981). Calculations of dietary choline intake are based on estimates of the free choline and phosphatidylcholine content of foods (Engel, 1943; McIntire et al., 1944; Weihrauch and Son, 1983; Zeisel et al., 1986). Older assay procedures for choline were imprecise and did not always include glycerophosphocholine or phosphocholine content, making many of the available data unreliable. On the basis of a finding of decreased plasma choline and phosphatidylcholine concentrations when humans were switched from a diet of normal foods to a defined diet containing 500 mg/day of choline (Zeisel et al., 1991), the average dietary intake of choline probably exceeds this level in adults. Infant formulas contain approximately 240 mg/L (2.3 mmol/L) of choline in its various forms. (Holmes-McNary et al., 1996).

Intake from Supplements

Choline is available as a dietary supplement as choline chloride or choline bitartrate and as lecithin, which usually contains approximately 25 percent phosphatidylcholine or 3 to 4 percent choline by weight. In the treatment of neurological diseases, large doses (5 to 30 g) of choline and phosphatidylcholine have been administered to humans (LSRO/FASEB, 1981). There are no reliable estimates of the frequency of use or amount of these dietary supplements consumed by individuals in the United States and Canada.

TOLERABLE UPPER INTAKE LEVELS

Hazard Identification

Adverse Effects

Choline doses that are orders of magnitude greater than estimated intake from food have been associated with body odor, sweating, salivation, hypotension, and hepatotoxicity in humans (LSRO/FASEB, 1975, 1981). There are no indications in the literature that excess choline intake produces any additional adverse effects in humans. The animal data provide supportive evidence for a low degree of toxicity of choline. However, some animal studies have indicated growth suppression at high intakes (LSRO/FASEB, 1975). Because of the large doses and routes of administration used (e.g., intravenous and intraperitoneal injection), they were considered not relevant to human intakes from food and supplements (Davis, 1944; Hodge, 1945; Sahu, 1989; Sahu et al., 1986).

Body Odor, Sweating, and Salivation. High doses of choline have been associated with fishy body odor, vomiting, salivation, sweating, and gastrointestinal effects (LSRO/FASEB, 1981). These symptoms were reported in patients with tardive dyskinesia and cerebellar ataxia treated with choline chloride at 150 and 220 mg/kg of body weight/day for 2 to 6 weeks (10 and 16 g/day, respectively) (Davis et al., 1975; Growdon et al., 1977b; Lawrence et al., 1980). Studies of the production of methylamines from ingested choline suggest that fishy odor would have been observed in healthy populations (Zeisel et al., 1983). Fishy body odor results from the excretion of excessive amounts of trimethylamine, a choline metabolite, as the result of bacterial action. Lecithin, a choline-containing phospholipid, does not present a risk of fishy body odor because it generates little methylamine because the bacterial enzyme cannot cleave the ester (Zeisel et al., 1983).

Hypotension. Oral administration of 10 g/day of choline chloride (which is equivalent to 7.5 g [72 mmol] of choline alone) had a slight hypotensive effect in humans (Boyd et al., 1977). Choline could be acting by increasing vagal tone to the heart or by dilating arterioles. Although added choline increases acetylcholine release from in vitro preparations of heart (Loffelholz, 1981), changes in cardiac rate have not been observed in healthy humans treated with choline.

Hepatotoxicity. Mild hepatotoxicity was reported in patients receiving choline magnesium trisalicylate (1,500 mg twice daily for 8 days) (Cersosimo and Matthews, 1987). There is also one reported case of severe hypersensitivity hepatitis with striking tissue and peripheral eosinophilia after ingestion of choline magnesium trisalicylate (Nadkarni et al., 1992). However, it is likely that hepatotoxicity was induced by salicylate rather than by choline (Cersosimo and Matthews, 1987). Humans with and without cirrhosis have been treated with large doses of choline chloride (6 g/day for 4 weeks) with no resultant liver toxicity (Chawla et al., 1989).

Nonspecific Toxicity. Tinnitus and pruritus have been reported in patients treated with doses of 3 g/day of choline magnesium trisalicylate for 6 weeks. These side effects were transient and probably caused by salicylate (Mody et al., 1983). The salicylate effect likely accounts for many of these observations, and the others are likely unusual anomalies, such as the one case of contact dermatitis reported after dermal exposure to choline chloride (Fischer, 1984).

Identification of Sensitive Subpopulations

Individuals with trimethylaminuria (fish odor syndrome), renal disease, liver disease, depression, and Parkinson's disease may have increased susceptibility to the adverse effects of choline. Trimethylaminuria results from a rare genetic deficiency that causes excessive excretion of trimethylamine and, therefore, an increased risk of developing fishy body odor (Al-Waiz et al., 1988, 1989; Humbert et al., 1970; Shelley and Shelley, 1984). Individuals with renal or liver disease may have increased susceptibility because of increased levels of plasma choline (after ingestion of supplemental choline) compared with healthy individuals (Acara and Rennick, 1973; Acara et al., 1983; Chawla et al., 1989; Rennick et al., 1976). In rare cases, consumption of large amounts of choline has been associated with depression (Davis et al., 1979; Tamminga et al., 1976). Finally, mild and transient Parkinsonian signs (bradykinesia, tremor, and rigidity) were observed at high doses (12.7 g/day) of choline as a chloride in people with tardive dyskinesia (Gelenberg et al., 1979), which suggests that supplemental choline intake by Parkinsonian patients may exacerbate symptoms.

Summary

On the basis of considerations of causality, relevance, and the quality and completeness of the database, hypotension was selected as the critical effect in deriving a Tolerable Upper Intake Level (UL); fishy body odor was selected as the secondary consideration.

Dose-Response Assessment

Adults

Data Selection. The data used to derive the UL for choline include a single case report of hypotension and several other studies involving cholinergic effects and fishy body odor after oral administration of large choline doses.

Identification of a no-observed-adverse-effect level (NOAEL) and a lowest-observed-adverse-effect level (LOAEL). There are no adequate data demonstrating a NOAEL for excess choline intake. A LOAEL of approximately 7.5 g/day of choline can be identified from evaluation of a pilot study that reported hypotension in seven patients treated with choline for Alzheimer senile dementia (Boyd et al., 1977) and reports of fishy body odor in individuals treated with choline for tardive dyskinesia and Huntington's disease (Gelenberg et al., 1979; Growdon et al., 1977a, b; Lawrence et al., 1980). Boyd et al. (1977) treated seven older adult patients with 4 g/day of oral choline as choline chloride for 2 weeks followed by 2 weeks of choline at 7.5 g/day. At 4 g/day of choline, daily blood pressure recordings revealed no hypotension. In addition, there were no reports of nausea or diarrhea or other evidence of cholinergic effects at this dose level. At 7.5 g/day of choline, nausea, diarrhea, and a small decrease in blood pressure were reported in some patients. Other supportive data on cholinergic effects and fishy body odor after excess choline intake are summarized in Table 12-1.

Uncertainty Assessment. An uncertainty factor (UF) of 2 was selected because of the limited data regarding hypotension and the interindividual variation in response to cholinergic effects.

Derivation of a UL. A LOAEL of 7.5 g/day was divided by an UF of 2 to obtain a UL of 3.75 for adults, which was rounded down to 3.5 g/day.

Choline UL Summary, Adults

Because of the scarcity of data for any adult age group and

TABLE 12-1 Studies Reporting on Cholinergic Effects and Fishy Body Odor after Excess Choline Intake

Study	No. of Subjects	Dose	Duration (wk)	Adverse Effects
Growdon et al., 1977a[a]	20	9 g/d (wk 1); 12 g/d (wk 2)[b,c]	2	Mild cholinergic toxicity: lacrimation, blurred vision, anorexia, and diarrhea.
Growdon et al., 1977b	10	8–20 g/d[d]	2–17	Fishy body odor in all subjects; at 250–300 mg/kg/d, produced lacrimation, anorexia, vomiting, and diarrhea.
Gelenberg et al., 1979[e]	5	8–19 g/d[d]	6–8	100% with fishy body odor after several days; gastrointestinal irritation.[f]
Lawrence et al., 1980[g]	14	0.2 g/d (3 wk); 9 g/d (3 wk)[c,d]	6	At 150 mg/kg/d: 5 of 14 with fishy body odor; 12 of 15 with nausea and diarrhea.

[a] Study involved a double-blind, crossover protocol.

[b] Choline was given as a chloride or bitartrate.

[c] Doses were calculated from data in the report using a reference body weight of 61 kg. Depending on the body weights of the individuals in Lawrence et al. (1980) and Growdon et al. (1977a), the lowest-effect dose may be less than 7.5 g/d.

[d] Choline was given as a chloride.

[e] Nonblinded study; did not include a control group.

[f] Mild, transient Parkinsonian signs (bradykinesia, tremor, and rigidity) were also reported.

[g] Double-blind protocol; included control group.

because no specific physiological function might be expected to affect sensitivity to excess amounts of choline in older persons, no adjustments are proposed for the elderly.

UL for Adults   19 years and older   3.5 g/day of choline

Other Life Stage Groups

For infants, the UL was judged not determinable because of lack of data concerning adverse effects in this age group and concern about the infant's ability to handle excess amounts. The only source of intake for infants should be from food or formula to prevent high levels of intake. There are no data to suggest that during pregnancy or lactation increased susceptibility to developing cholinergic effects or fishy body odor from excess choline intake would occur. Therefore, the UL of 3.5 g/day is also set for pregnant and lactating women. The UL of 3.5 g/day for adults was adjusted for children and adolescents on the basis of relative body weight as described in Chapter 3, with the use of reference weights from Chapter 1, Table 1-2. Values have been rounded down.

Choline UL Summary, Other Life Stage Groups

UL For Infants
0–12 months   **Not possible to establish; source of**
                          intake should be formula and food only

UL for Children   **1–3 years**   1 g/day of choline
                          4–8 years   1 g/day of choline
                          9–13 years   2 g/day of choline

UL for Adolescents   **14–18 years**   3 g/day of choline

UL for Pregnancy   **14–18 years**   3 g/day of choline
                          19 years and older 3.5 g/day of choline

UL for Lactation   **14–18 years**   3 g/day of choline
                          19 years and older 3.5 g/day of choline

Special Considerations

Individuals with the following conditions may be at risk of adverse effects with choline intakes at the UL: trimethylaminuria, renal disease, liver disease, depression, and Parkinson's disease.

Intake Assessment

National surveys do not provide data on the dietary intake of choline. The UL applies to the weight of the choline moiety in the compound; for example, choline chloride contains more choline by weight than does choline bitartrate. Dietary supplements containing choline are available; however, reliable estimates of the amount of these supplements consumed in the United States and Canada are unavailable.

Risk Characterization

Because there is no information from national surveys on choline intakes or on supplement usage, the risk of adverse effects within the United States or Canada can not be characterized.

RESEARCH RECOMMENDATIONS FOR CHOLINE

High Priority Recommendations

Sufficient human data are not available for determining whether choline is essential in the human diet, how much is required if it is essential, and the public health impact of poor choline nutriture. For this reason, research that could provide such human data is assigned the highest priority:

• Examination of the effects of the use of graded levels of dietary intake of choline on parameters of health. This would include assessing plasma and tissue choline compounds and metabolites; plasma cholesterol and homocysteine concentrations; erythrocyte folate; and liver, renal, brain, and other organ function. To facilitate this process, food composition data are needed for choline, phosphocholine, glycerophosphocholine, sphingomyelin, phosphatidylcholine, and betaine and the analytic sensitivity and specificity of methods for analysis of food composition need to be validated.

• Human studies on interrelationships among requirements for choline, methionine, folate, vitamin B_6, and vitamin B_{12} to compare the homocysteine-lowering effects of combinations of these nutrients.

Other Research Areas

Two additional topics also merit attention:

• The relative effectiveness of different choline-containing compounds in the diet in promoting health and determination of the sparing effect of endogenous synthesis of choline. It will be important to conduct studies on the bioavailability of choline and choline compounds and on the rate of de novo synthesis of choline in vivo.

• Studies using increasing levels of dietary intake designed to assess toxicity for all organ systems, including heart, liver, brain and kidney; fishy body odor; and possible growth suppression in children from observational data and as determined by experimental studies in animal models.

REFERENCES

Acara M, Rennick B. 1973. Regulation of plasma choline by the renal tubule: Bidirectional transport of choline. *Am J Physiol* 225:1123–1128.

Acara M, Rennick B, LaGraff S, Schroeder ET. 1983. Effect of renal transplantation on the levels of choline in the plasma of uremic humans. *Nephron* 35:241–243.

Albright CD, Liu R, Bethea TC, da Costa KA, Salganik RI, Zeisel SH. 1996. Choline deficiency induces apoptosis in SV40-immortalized CWSV-1 rat hepatocytes in culture. *FASEB J* 10:510–516.

Al-Waiz M, Ayesh R, Mitchell SC, Idle JR, Smith RL. 1988. Trimethylaminuria ("fish-odour syndrome"): A study of an affected family. *Clin Sci* 74:231–236.

Al-Waiz M, Ayesh R, Mitchell SC, Idle JR, Smith RL. 1989. Trimethylaminuria: The detection of carriers using a trimethylamine load test. *J Inherit Metab Dis* 12:80–85.

Anonymous. 1997. Betaine for homocystinuria. *Med Lett Drugs Ther* 39:12.

Aquilonius SM, Ceder G, Lying-Tunell U, Malmlund HO, Schuberth J. 1975. The arteriovenous difference of choline across the brain of man. *Brain Res* 99:430–433.

Arvidson GA. 1968. Biosynthesis of phosphatidylcholines in rat liver. *Eur J Biochem* 5:415–421.

Barak AJ, Kemmy RJ. 1982. Methotrexate effects on hepatic betaine levels in choline-supplemented and choline-deficient rats. *Drug Nutr Interact* 1:275–278.

Barak AJ, Tuma DJ, Beckenhauer HC. 1984. Methotrexate hepatotoxicity. *J Am Coll Nutr* 3:93–96.

Bartus RT, Dean RL, Goas JA, Lippa AS. 1980. Age-related changes in passive avoidance retention: Modulation with dietary choline. *Science* 209:301–303.

Bauernschmitt HG, Kinne RK. 1993. Metabolism of the "organic osmolyte" glycerophosphorylcholine in isolated rat inner medullary collecting duct cells. I. Pathways for synthesis and degradation. *Biochim Biophys Acta* 1148:331–341.

Best CH, Huntsman ME. 1932. The effects of the components of lecithine upon deposition of fat in the liver. *J Physiol* 75:405–412.

Bianchi G, Azzone GF. 1964. Oxidation of choline in rat liver mitochondria. *J Biol Chem* 239:3947–3955.

Bjornstad P, Bremer J. 1966. In vivo studies on pathways for the biosynthesis of lecithin in the rat. *J Lipid Res* 7:38–45.

Blair R, Newsome F. 1985. Involvement of water-soluble vitamins in diseases of swine. *J Anim Sci* 60:1508–1517.

Blair R, Whitehead CC, Bannister DW, Evans AJ. 1973. Involvement of diet in fatty liver and kidney syndrome in broiler chickens. *Vet Rec* 92:118–119.

Blusztajn JK, Zeisel SH, Wurtman RJ. 1979. Synthesis of lecithin (phosphatidylcholine) from phosphatidylethanolamine in bovine brain. *Brain Res* 179:319–327.

Boyd WD, Graham-White J, Blackwood G, Glen I, McQueen J. 1977. Clinical effects of choline in Alzheimer senile dementia. *Lancet* 2:711.

Bremer J, Greenberg D. 1961. Methyl transfering enzyme system of microsomes in the biosynthesis of lecithin (phosphatidylcholine). *Biochim Biophys Acta* 46:205–216.

Buchman AL, Dubin M, Jenden D, Moukarzel A, Roch MH, Rice K, Gornbein J, Ament ME, Eckhert CD. 1992. Lecithin increases plasma free choline and decreases hepatic steatosis in long-term total parenteral nutrition patients. *Gastroenterology* 102:1363–1370.

Buchman AL, Moukarzel A, Jenden DJ, Roch M, Rice K, Ament ME. 1993. Low plasma free choline is prevalent in patients receiving long term parenteral nutrition and is associated with hepatic aminotransferase abnormalities. *Clin Nutr* 12:33–37.

Buchman AL, Dubin M, Moukarzel A, Jenden D, Roch M, Rice K, Gornbein J, Ament M. 1995. Choline deficiency: A cause of hepatic steatosis during parenteral nutrition that can be reversed with intravenous choline supplementation. *Hepatology* 22:1399–1403.

Burg MB. 1995. Molecular basis of osmotic regulation. *Am J Physiol* 268:F983–F996.

Burt ME, Hanin I, Brennan MF. 1980. Choline deficiency associated with total parenteral nutrition. *Lancet* 2:638–639.

Cermak JM, Holler T, Jackson DA, Blusztajn JK. 1998. Prenatal availability of choline modifies development of the hippocampal cholinergic system. *FASEB J* 12:349–357.

Cersosimo RJ, Matthews SJ. 1987. Hepatotoxicity associated with choline magnesium trisalicylate: Case report and review of salicylate-induced hepatotoxicity. *Drug Intell Clin Pharm* 21:621–625.

Chawla RK, Wolf DC, Kutner MH, Bonkovsky HL. 1989. Choline may be an essential nutrient in malnourished patients with cirrhosis. *Gastroenterology* 97:1514–1520.

Cheng W-L, Holmes-McNary MQ, Mar M-H, Lien EL, Zeisel SH. 1996. Bioavailability of choline and choline esters from milk in rat pups. *J Nutr Biochem* 7:457–464.

Cohen BM, Renshaw PF, Stoll AL, Wurtman RJ, Yurgelun-Todd D, Babb SM. 1995. Decreased brain choline uptake in older adults. An in vivo proton magnetic resonance spectroscopy study. *J Am Med Assoc* 274:902–907.

Cohen EL, Wurtman RJ. 1975. Brain acetylcholine: Increase after systemic choline administration. *Life Sci* 16:1095–1102.

Conlay LA, Wurtman RJ, Blusztajn K, Coviella IL, Maher TJ, Evoniuk GE. 1986. Decreased plasma choline concentrations in marathon runners. *N Engl J Med* 315:892.

Cornford EM, Cornford ME. 1986. Nutrient transport and the blood-brain barrier in developing animals. *Fed Proc* 45:2065–2072.

Crews FT, Calderini G, Battistella A, Toffano G. 1981. Age-dependent changes in the methylation of rat brain phospholipids. *Brain Res* 229:256–259.

Cui Z, Houweling M, Chen MH, Record M, Chap H, Vance DE, Tercé F. 1996. A genetic defect in phosphatidylcholine biosynthesis triggers apoptosis in Chinese hamster ovary cells. *J Biol Chem* 271:14668–14671.

da Costa KA, Cochary EF, Blusztajn JK, Garner SC, Zeisel SH. 1993. Accumulation of 1,2-*sn*-diradylglycerol with increased membrane-associated protein kinase C may be the mechanism for spontaneous hepatocarcinogenesis in choline-deficient rats. *J Biol Chem* 268:2100–2105.

da Costa KA, Garner SC, Chang J, Zeisel SH. 1995. Effects of prolonged (1 year) choline deficiency and subsequent re-feeding of choline on 1,2-*sn*-diradylglycerol, fatty acids and protein kinase C in rat liver. *Carcinogenesis* 16:327–334.

Davis JE. 1944. Depression of normal erythrocyte number by soybean lecithin or choline. *Am J Physiol* 142:65–67.

Davis KL, Berger PA, Hollister LE. 1975. Choline for tardive dyskinesia. *N Engl J Med* 293:152.

Davis KL, Hollister LE, Berger PA. 1979. Choline chloride in schizophrenia. *Am J Psychiatry* 136:1581–1584.

Drouva SV, LaPlante E, Leblanc P, Bechet JJ, Clauser H, Kordon C. 1986. Estradiol activates methylating enzyme(s) involved in the conversion of phosphatidylethanolamine to phosphatidylcholine in rat pituitary membranes. *Endocrinology* 119:2611–2622.

Dudman NP, Tyrrell PA, Wilcken DE. 1987. Homocysteinemia: Depressed plasma serine levels. *Metabolism* 36:198–201.

Eagle H. 1955. The minimum vitamin requirements of the L and HeLa cells in tissue culture, the production of specific vitamin deficiencies, and their cure. *J Exp Med* 102:595–600.

Engel RW. 1943. The choline content of animal and plant products. *J Nutr* 25:441–446.

Exton JH. 1994. Phosphatidylcholine breakdown and signal transduction. *Biochim Biophys Acta* 1212:26–42.

Fairbanks BW, Krider JL. 1945. Significance of the B vitamins in swine nutrition. *N Am Vet* 26:18–23.

Finkelstein JD, Martin JJ, Harris BJ, Kyle WE. 1982. Regulation of the betaine content of rat liver. *Arch Biochem Biophys* 218:169–173.

Finkelstein JD, Martin JJ, Harris BJ. 1988. Methionine metabolism in mammals. The methionine-sparing effect of cystine. *J Biol Chem* 263:11750–11754.

Fischer T. 1984. Contact allergy to choline chloride. *Contact Dermatitis* 10:316–317.

Freeman-Narrod M, Narrod SA, Custer RP. 1977. Chronic toxicity of methotrexate in rats: Partial to complete protection of the liver by choline. *J Natl Cancer Inst* 59:1013–1017.

Frenkel R, Muguruma K, Johnston J. 1996. The biochemical role of platelet-activating factor in reproduction. *Prog Lipid Res* 35:155–168.

Garcia-Perez A, Burg MB. 1991. Role of organic osmolytes in adaptation of renal cells to high osmolality. *J Membr Biol* 119:1–13.

Garner SC, Mar MH, Zeisel SH. 1995. Choline distribution and metabolism in pregnant rats and fetuses are influenced by the choline content of the maternal diet. *J Nutr* 125:2851–2858.

Gelenberg AJ, Doller-Wojcik J, Growdon JH. 1979. Choline and lecithin in the treatment of tardive dyskinesia: Preliminary results from a pilot study. *Am J Psychiatry* 136:772–776.

Grossman EB, Hebert SC. 1989. Renal inner medullary choline dehydrogenase activity: Characterization and modulation. *Am J Physiol* 256:F107–F112.

Growdon JH, Cohen EL, Wurtman RJ. 1977a. Huntington's disease: Clinical and chemical effects of choline administration. *Ann Neurol* 1:418–422.

Growdon JH, Hirsch MJ, Wurtman RJ, Wiener W. 1977b. Oral choline administration to patients with tardive dyskinesia. *N Engl J Med* 297:524–527.

Gwee MC, Sim MK. 1978. Free choline concentration and cephalin-*N*-methyltransferase activity in the maternal and foetal liver and placenta of pregnant rats. *Clin Exp Pharmacol Physiol* 5:649–653.

Handler P. 1949. Response of guinea pigs to diets deficient in choline. *Proc Soc Exp Biol Med* 70:70–73.

Hannun YA. 1994. The sphingomyelin cycle and the second messenger function of ceramide. *J Biol Chem* 269:3125–3128.

Hartz SC, Russell RM, Rosenberg IH. 1992. *Nutrition in the Elderly. The Boston Nutritional Status Survey.* London: Smith-Gordon.

Haubrich DR, Wedeking PW, Wang PF. 1974. Increase in tissue concentration of acetylcholine in guinea pigs in vivo induced by administration of choline. *Life Sci* 14:921–927.

Hershey JM. 1931. Substitution of lecithin for raw pancreas in the diet of depancreatized dog. *Am J Physiol* 93:657–658.

Herzberg GR, Lerner J. 1973. Intestinal absorption of choline in the chick. *Biochim Biophys Acta* 307:234–242.

Herzberg GR, Sheerin H, Lerner J. 1971. Cationic amino acid transport in chicken small intestine. *Comp Biochem Physiol* 40A:229–247.

Hirsch MJ, Growdon JH, Wurtman RJ. 1978. Relations between dietary choline or lecithin intake, serum choline levels, and various metabolic indices. *Metabolism* 27:953–960.

Hodge HC. 1945. Chronic oral toxicology of choline chloride in rats. *Proc Exp Biol Med* 58:212–215.

Hoffbauer FW, Zaki FG. 1965. Choline deficiency in baboon and rat compared. *Arch Pathol* 79:364–369.

Holler T, Cermak JM, Blusztajn JK. 1996. Dietary choline supplementation in pregnant rats increases hippocampal phospholipase D activity of the offspring. *FASEB J* 10:1653–1659.

Holmes-McNary MQ, Cheng WL, Mar MH, Fussell S, Zeisel SH. 1996. Choline and choline esters in human and rat milk and in infant formulas. *Am J Clin Nutr* 64:572–576.

Holmes-McNary MQ, Loy R, Mar MH, Albright CD, Zeisel SH. 1997. Apoptosis is induced by choline deficiency in fetal brain and in PC12 cells. *Brain Res Dev Brain Res* 101:9–16.

Horne DW, Cook RJ, Wagner C. 1989. Effect of dietary methyl group deficiency on folate metabolism in rats. *J Nutr* 119:618–621.

Humbert JA, Hammond KB, Hathaway WE. 1970. Trimethylaminuria: The fish-odor syndrome. *Lancet* 2:770–771.

Jacob RA, Pianalto FS, Henning SM, Zhang JZ, Swendseid ME. 1995. In vivo methylation capacity is not impaired in healthy men during short-term dietary folate and methyl group restriction. *J Nutr* 125:1495–1502.

James ST, Miller BT, Basnakian AG, Pogribny IP, Pogribna M, Muskhelishvili L. 1997. Apoptosis and proliferation under conditions of deoxynucleotide pool imbalance in liver of folate/methyl deficient rats. *Carcinogenesis* 18:287–293.

Jorswieck I. 1974. Proceedings: Penetration of choline through rat placenta in vivo. *Naunyn Schmiedebergs Arch Pharmakol* 282:R42.

Kennedy EP, Weiss SB. 1956. The function of cytidine coenzymes in the biosynthesis of phospholipids. *J Biol Chem* 222:193–214.

Ketola HG. 1976. Choline metabolism and nutritional requirement of lake trout (*Salvelinus namaycush*). *J Anim Sci* 43:474–477.

Ketola HG, Nesheim MC. 1974. Influence of dietary protein and methionine levels on the requirement for choline by chickens. *J Nutr* 104:1484–1489.

Ketola HG, Young RJ. 1973. The need for dietary choline by young Japanese quail. *Poult Sci* 52:2362–2363.

Kim Y-I, Miller JW, da Costa K-A, Nadeau M, Smith D, Selhub J, Zeisel SH, Mason JB. 1994. Severe folate deficiency causes secondary depletion of choline and phosphocholine in rat liver. *J Nutr* 124:2197–2203.

Kuczler FJ, Nahrwold DL, Rose RC. 1977. Choline influx across the brush border of guinea pig jejunum. *Biochim Biophys Acta* 465:131–137.

Kuksis A, Mookerjea S. 1978. Choline. *Nutr Rev* 36:201–207.

Lawrence CM, Millac P, Stout GS, Ward JW. 1980. The use of choline chloride in ataxic disorders. *J Neurol Neurosurg Psychiatry* 43:452–454.

Le Kim D, Betzing H. 1976. Intestinal absorption of polyunsaturated phosphatidylcholine in the rat. *Hoppe Seylers Z Physiol Chem* 357:1321–1331.

Lindblad L, Schersten T. 1976. Incorporation rate in vitro of choline and methylmethionine into human hepatic lecithins. *Scand J Gastroenterol* 11:587–591.

Loffelholz K. 1981. Release of acetylcholine in the isolated heart. *Am J Physiol* 240:H431–H440.

Loy R, Heyer D, Williams CL, Meck WH. 1991. Choline-induced spatial memory facilitation correlates with altered distribution and morphology of septal neurons. *Adv Exp Med Biol* 295:373–382.

LSRO/FASEB (Life Sciences Research Office/Federation of American Societies for Experimental Biology). 1975. *Evaluation of the Health Aspects of Choline Chloride and Choline Bitartrate as Food Ingredients.* Report # PB-223 845/9. Washington, DC: Department of Health, Education and Welfare.

LSRO/FASEB (Life Sciences Research Office/Federation of American Societies for Experimental Biology). 1981. *Effects of Consumption of Choline and Lecithin on Neurological and Cardiovascular Systems.* Report # PB-82-133257. Bethesda, MD: LSRO/FASEB.

Lyman RL, Sheehan G, Tinoco J. 1971. Diet and 14CH3-methionine incorporation into liver phosphatidylcholine fractions of male and female rats. *Can J Biochem* 49:71–79.

McIntire JM, Schweigert BS, Elvehjem CA. 1944. The choline and pyridoxine content of meats. *J Nutr* 28:219–223.

Meck WH, Williams CL. 1997a. Characterization of the facilitative effects of perinatal choline supplementation on timing and temporal memory. *Neuroreport* 8:2831–2835.

Meck WH, Williams CL. 1997b. Perinatal choline supplementation increases the threshold for chunking in spatial memory. *Neuroreport* 8:3053–3059.

Meck WH, Williams CL. 1997c. Simultaneous temporal processing is sensitive to prenatal choline availability in mature and aged rats. *Neuroreport* 8:3045–3051.

Meck WH, Smith RA, Williams CL. 1988. Pre- and postnatal choline supplementation produces long-term facilitation of spatial memory. *Dev Psychobiol* 21:339–353.

Meck WH, Smith RA, Williams CL. 1989. Organizational changes in cholinergic activity and enhanced visuospatial memory as a function of choline administered prenatally or postnatally or both. *Behav Neurosci* 103:1234–1241.

Mody GM, Naidoo PD, Singh TG. 1983. Clinical evaluation of choline magnesium trisalicylate in rheumatoid arthritis. *S Afr Med J* 64:195–196.

Mudd SH, Poole JR. 1975. Labile methyl balances for normal humans on various dietary regimens. *Metabolism* 24:721–735.

Nadkarni MM, Peller CA, Retig J. 1992. Eosinophilic hepatitis after ingestion of choline magnesium trisalicylate. *Am J Gastroenterol* 87:151–153.

Newberne PM, Rogers AE. 1986. Labile methyl groups and the promotion of cancer. *Annu Rev Nutr* 6:407–432.

Perloff BP, Rizek RL, Haytowitz DB, Reid PR. 1990. Dietary intake methodology. II. USDA's Nutrient Data Base for Nationwide Dietary Intake Surveys. *J Nutr* 120:1530–1534.

Poirier LA, Grantham PH, Rogers AE. 1977. The effects of a marginally lipotrope-deficient diet on the hepatic levels of *S*-adenosylmethionine and on the urinary metabolites of 2-acetylaminofluorene in rats. *Cancer Res* 37:744–748.

Pomfret EA, da Costa K-A, Schurman LL, Zeisel SH. 1989. Measurement of choline and choline metabolite concentrations using high-pressure liquid chromatography and gas chromatography-mass spectrometry. *Analy Biochem* 180:85–90.

Pomfret EA, da Costa K, Zeisel SH. 1990. Effects of choline deficiency and methotrexate treatment upon rat liver. *J Nutr Biochem* 1:533–541.

Pyapali GK, Turner DA, Williams CL, Meck WH, Swartzwelder HS. 1998. Prenatal dietary choline supplementation decreases the threshold for induction of long-term potentiation in young adult rats. *J Neurophysiol* 79:1790–1796.

Rennick B, Acara M, Hysert P, Mookerjee B. 1976. Choline loss during hemodialysis: Homeostatic control of plasma choline concentrations. *Kidney Int* 10:329–335.

Rennick B, Acara M, Glor M. 1977. Relations of renal transport rate, transport maximum, and competitor potency for tetraethylammonium and choline. *Am J Physiol* 232:F443–F447.

Ridgway ND, Vance DE. 1988. Kinetic mechanism of phosphatidylethanolamine *N*-methyltransferase. *J Biol Chem* 263:16864–16871.

Rohlfs EM, Garner SC, Mar MH, Zeisel SH. 1993. Glycerophosphocholine and phosphocholine are the major choline metabolites in rat milk. *J Nutr* 123:1762–1768.

Sahu AP. 1989. Effect of Choline and Mineral Fibres (Chrysotile Asbestos) on Guinea-pigs. Lyon, France: IARC Scientific Publications.

Sahu AP, Saxena AK, Singh KP, Shanker R. 1986. Effect of chronic choline administration in rats. *Indian J Exp Biol* 24:91–96.

Sandage BW, Sabounjian L, White R, Wurtman RJ. 1992. Choline citrate may enhance athletic performance. *Physiologist* 35:236.

Savendahl L, Mar M-H, Underwood LE, Zeisel SH. 1997. Prolonged fasting in humans results in diminished plasma choline concentrations but does not cause liver dysfunction. *Am J Clin Nutr* 66:622–625.

SCOGS/LSRO (Select Committee on GRAS Substances, Life Sciences Research Office). 1979. *Evaluation of the Health Aspects of Lecithin as a Food Ingredient.* Report # PB301405. Springfield, VA: National Technical Information Service.

Selhub J, Seyoum E, Pomfret EA, Zeisel SH. 1991. Effects of choline deficiency and methotrexate treatment upon liver folate content and distribution. *Cancer Res* 51:16–21.

Shapira G, Chawla RK, Berry CJ, Williams PJ, Roy RGB, Rudman D. 1986. Cysteine, tyrosine, choline and carnitine supplementation of patients on total parenteral nutrition. *Nutr Int* 2:334–339.

Sheard NF, Zeisel SH. 1986. An in vitro study of choline uptake by intestine from neonatal and adult rats. *Pediatr Res* 20:768–772.

Sheard NF, Tayek JA, Bistrian BR, Blackburn GL, Zeisel SH. 1986. Plasma choline concentration in humans fed parenterally. *Am J Clin Nutr* 43:219–224.

Shelley ED, Shelley WB. 1984. The fish odor syndrome. Trimethylaminuria. *JAMA* 251:253–255.

Shin OH, Mar MH, Albright CD, Citarella MT, daCosta KA, Zeisel SH. 1997. Methyl-group donors cannot prevent apoptotic death of rat hepatocytes induced by choline-deficiency. *J Cell Biochem* 64:196–208.

Shivapurkar N, Poirier LA. 1983. Tissue levels of *S*-adenosylmethionine and *S*-adenosylhomocysteine in rats fed methyl-deficient, amino acid-defined diets for one to five weeks. *Carcinogenesis* 4:1051–1057.

Sundler R, Akesson B. 1975. Regulation of phospholipid biosynthesis in isolated rat hepatocytes. Effect of different substrates. *J Biol Chem* 250:3359–3367.

Svardal AM, Ueland PM, Berge RK, Aarsland A, Aarsaether N, Lonning PE, Refsum H. 1988. Effect of methotrexate on homocysteine and other sulfur compounds in tissues of rats fed a normal or a defined, choline-deficient diet. *Cancer Chemother Pharmacol* 21:313–318.

Sweiry JH, Yudilevich DL. 1985. Characterization of choline transport at maternal and fetal interfaces of the perfused guinea-pig placenta. *J Physiol* 366:251–266.

Sweiry JH, Page KR, Dacke CG, Abramovich DR, Yudilevich DL. 1986. Evidence of saturable uptake mechanisms at maternal and fetal sides of the perfused human placenta by rapid paired-tracer dilution: Studies with calcium and choline. *J Dev Physiol* 8:435–445.

Tamminga CA, Smith RC, Chang S, Haraszti JS, Davis JM. 1976. Depression associated with oral choline. *Lancet* 2:905.

Tani H, Suzuki S, Kobayashi M, Kotake Y. 1967. The physiological role of choline in guinea pigs. *J Nutr* 92:317–324.

Tayek JA, Bistrian B, Sheard NF, Zeisel SH, Blackburn GL. 1990. Abnormal liver function in malnourished patients receiving total parenteral nutrition: A prospective randomized study. *J Am Coll Nutr* 9:76–83.

Tessitore L, Sesca E, Greco M, Pani P, Dianzani M. 1995. Sexually differentiated response to choline in choline deficiency and ethionine intoxication. *Int J Exp Pathol* 76:125–129.

Vance DE. 1990. Boehringer Mannheim Award lecture. Phosphatidylcholine metabolism: Masochistic enzymology, metabolic regulation, and lipoprotein assembly. *Biochem Cell Biol* 68:1151–1165.

Vance DE, Ridgway ND. 1988. The methylation of phosphatidylethanolamine. *Prog Lipid Res* 27:61–79.

Varela-Moreiras G, Ragel C, Perez de Miguelsanz J. 1995. Choline deficiency and methotrexate treatment induces marked but reversible changes in hepatic folate concentrations, serum homocysteine and DNA methylation rates in rats. *J Am Coll Nutr* 14:480–485.

Von Allworden HN, Horn S, Kahl J, Feldheim W. 1993. The influence of lecithin on plasma choline concentrations in triatheletes and adolescent runners during exercise. *Eur J Appl Physiol* 67:87–91.

Wecker L. 1986. Neurochemical effects of choline supplementation. *Can J Physiol Pharmacol* 64:329–333.

Weihrauch JL, Son Y-S. 1983. The phospholipid content of foods. *J Am Oil Chem Soc* 60:1971–1978.

Weinhold PA, Sanders R. 1973. The oxidation of choline by liver slices and mitochondria during liver development in the rat. *Life Sci* 13:621–629.

Welsch F. 1976. Studies on accumulation and metabolic fate of (*N*-Me3H)choline in human term placenta fragments. *Biochem Pharmacol* 25:1021–1030.

Welsch F. 1978. Choline metabolism in human term placenta—studies on de novo synthesis and the effects of some drugs on the metabolic fate of [*N*-methyl 3H]choline. *Biochem Pharmacol* 27:1251–1257.

Welsch F, Wenger WC, Stedman DB. 1981. Choline metabolism in placenta: Evidence for the biosynthesis of phosphatidylcholine in microsomes via the methylation pathway. *Placenta* 2:211–221.

Wendel U, Bremer H. 1984. Betaine in the treatment of homocystinuria due to 5,10-methylenetetrahydrofolate reductase deficiency. *Eur J Pediatr* 142:147–150.

Widdowson EM. 1963. Growth and composition of the fetus and newborn. In: Assali N, ed. *Biology of Gestation,* Vol. 2. New York: Academic Press. Pp. 1–51.

Wilcken DE, Wilcken B, Dudman NP, Tyrrell PA. 1983. Homocystinuria—the effects of betaine in the treatment of patients not responsive to pyridoxine. *N Engl J Med* 309:448–453.

Wilcken DE, Dudman NP, Tyrrell PA. 1985. Homocystinuria due to cystathionine β-synthase deficiency—the effects of betaine treatment in pyridoxine-responsive patients. *Metabolism* 34:1115–1121.

Williams CL, Meck WH, Heyer D, Loy R. 1998. Hypertrophy of basal forebrain neurons and enhanced visuospatial memory in perinatally choline-supplemented rats. *Brain Res* 794:225–238.

Wong ER, Thompson W. 1972. Choline oxidation and labile methyl groups in normal and choline-deficient rat liver. *Biochim Biophys Acta* 260:259–271.

Wood JL, Allison RG. 1982. Effects of consumption of choline and lecithin on neurological and cardiovascular systems. *Fed Proc* 41:3015–3021.

Yang EK, Blusztajn JK, Pomfret EA, Zeisel SH. 1988. Rat and human mammary tissue can synthesize choline moiety via the methylation of phosphatidylethanolamine. *Biochem J* 256:821–828.

Yao ZM, Vance DE. 1988. The active synthesis of phosphatidylcholine is required for very low density lipoprotein secretion from rat hepatocytes. *J Biol Chem* 263:2998–3004.

Yao ZM, Vance DE. 1989. Head group specificity in the requirement of phosphatidylcholine biosynthesis for very low density lipoprotein secretion from cultured hepatocytes. *J Biol Chem* 264:11373–11380.

Yao ZM, Vance DE. 1990. Reduction in VLDL, but not HDL, in plasma of rats deficient in choline. *Biochem Cell Biol* 68:552–558.

Young DL. 1971. Estradiol- and testosterone-induced alterations in phosphatidylcholine and triglyceride synthesis in hepatic endoplasmic reticulum. *J Lipid Res* 12:590–595.

Yudilevich DL, Sweiry JH. 1985. Membrane carriers and receptors at maternal and fetal sides of the placenta by single circulation paired-tracer dilution: Evidence for a choline transport system. *Contrib Gynecol Obstet* 13:158–161.

Zeisel SH. 1981. Dietary choline: Biochemistry, physiology, and pharmacology. *Annu Rev Nutr* 1:95–121.

Zeisel SH. 1987. Choline availability in the neonate. In: Dowdall MJ, Hawthorne JN, eds. *Cellular and Molecular Basis of Cholinergic Function.* Chichester, England: Horwood. Pp. 709–719.

Zeisel SH. 1993. Choline phospholipids: Signal transduction and carcinogenesis. *FASEB J* 7:551–557.

Zeisel SH, Blusztajn JK. 1994. Choline and human nutrition. *Annu Rev Nutr* 14:269–296.

Zeisel SH, Wurtman RJ. 1981. Developmental changes in rat blood choline concentration. *Biochem J* 198:565–570.

Zeisel SH, Epstein MF, Wurtman RJ. 1980a. Elevated choline concentration in neonatal plasma. *Life Sci* 26:1827–1831.

Zeisel SH, Growdon JH, Wurtman RJ, Magil SG, Logue M. 1980b. Normal plasma choline responses to ingested lecithin. *Neurology* 30:1226–1229.

Zeisel SH, Story DL, Wurtman RJ, Brunengraber H. 1980c. Uptake of free choline by isolated perfused rat liver. *Proc Natl Acad Sci USA* 77:4417–4419.

Zeisel SH, Stanbury JB, Wurtman RJ, Brigida M, Fierro BR. 1982. Choline content of mothers' milk in Ecuador and Boston. *N Engl J Med* 306:175–176.

Zeisel SH, Wishnok JS, Blusztajn JK. 1983. Formation of methylamines from ingested choline and lecithin. *J Pharmacol Exp Ther* 225:320–324.

Zeisel SH, Char D, Sheard NF. 1986. Choline, phosphatidylcholine and sphingomyelin in human and bovine milk and infant formulas. *J Nutr* 116:50–58.

Zeisel SH, Zola T, daCosta K, Pomfret EA. 1989. Effect of choline deficiency on *S*-adenosylmethionine and methionine concentrations in rat liver. *Biochem J* 259:725–729.

Zeisel SH, daCosta K-A, Franklin PD, Alexander EA, Lamont JT, Sheard NF, Beiser A. 1991. Choline, an essential nutrient for humans. *FASEB J* 5:2093–2098.

Zeisel SH, Mar M-H, Zhou Z-W, da Costa K-A. 1995. Pregnancy and lactation are associated with diminished concentrations of choline and its metabolites in rat liver. *J Nutr* 125:3049–3054.

Zeisel SH, Albright CD, Shin O-H, Mar M-H, Salganik RI, da Costa K-A. 1997. Choline deficiency selects for resistance to p53-independent apoptosis and causes tumorigenic transformation of rat hepatocytes. *Carcinogenesis* 18:731–738.

13

Uses of
Dietary Reference Intakes

OVERVIEW

In the past, Recommended Dietary Allowances (RDAs) were the primary values available to health professionals for planning and assessing the diets of individuals and groups and for making judgments about excessive intake. However, the RDAs were not ideally suited for many of these purposes (IOM, 1994). The Dietary Reference Intakes (DRIs) developed in this report—RDAs, Adequate Intakes (AIs), Tolerable Upper Intake Levels (ULs), and Estimated Average Requirements (EARs)—are a more complete set of reference values. Each type of DRI has specific uses. The transition from using RDAs alone to using all the DRIs appropriately will require time and effort by health professionals and others.

The most widespread uses of DRIs—diet assessment and planning—are described briefly in this chapter. Also included are two specific applications to the nutrients discussed in this report: using dietary folate equivalents and selecting sources of vitamin B_{12} for the elderly.

Three of the DRIs—the RDA, AI, and EAR—are set with reference to a specific criterion of adequacy. With few exceptions the criterion of adequacy chosen for each of the B vitamins is the same for each life stage and gender group. Each nutrient chapter identifies the primary criterion that defines adequacy for the specific life stage and gender group.

Dietary Reference Intakes (DRIs)

RDA (Recommended Dietary Allowance): the intake level that is sufficient to meet the nutrient requirement of nearly all (97 to 98 percent) healthy individuals in a group.

AI (Adequate Intake): a value based on observed or experimentally determined approximations of nutrient intake by a group (or groups) of healthy people—used when an RDA cannot be determined.

UL (Tolerable Upper Intake Level): the highest level of daily nutrient intake that is likely to pose no risk of adverse health effects to almost all individuals in the general population. As intake increases above the UL, the risk of adverse effects increases.

EAR (Estimated Average Requirement): a nutrient intake value that is estimated to meet the requirement of half the healthy individuals in a group.

NOTE: DRIs are expressed as intakes per day but they are meant to represent intakes averaged over time.

USING RECOMMENDED DIETARY ALLOWANCES

Nutrient Recommendations for Individuals

The Recommended Dietary Allowance (RDA) is the value to be used in guiding healthy individuals to achieve adequate nutrient intake. It is a goal for average intake over time; day-to-day variation is to be expected. RDAs are set separately for specified life stage groups and sometimes they differ for males and females. The RDAs are not intended to be used for planning diets for groups or assessing the nutrient intakes of free-living groups (Beaton, 1994). Rather, the RDAs are intended to ensure the adequacy of nutrient intake. If a healthy person meets the RDA for a nutrient, there is only a very slight chance that the intake is inadequate for this person.

The RDA is expressed as a single absolute value. For example, from Chapter 6, the RDA and thus the recommended daily intake of niacin for women ages 19 through 30 years is 16 mg/day of niacin equivalents. This would be the case for a woman in this age range weighing 50 kg (110 lb), 55 kg (121 lb), or 70 kg (154 lb) despite the differences in size and energy expenditure.

Reference weights are provided (Chapter 1, Table 1-2) to allow a calculation, when necessary, of the amount of nutrient per unit of body weight for individuals who are greatly outside the typical range of body size. These weights may also be useful for infants and children, whose average weights change substantially with age. Ordinarily, adjusting values for the nutrients included in this report based on reported energy intake has not been recommended; little justification for this practice could be found in the literature for individuals eating typical diets.

Assessing the Adequacy of Nutrient Intakes of an Individual

Although RDAs have been used to assess the adequacy of an individual's nutrient intake in the past, this practice has serious limitations if a person's intake is less than the RDA. An individual's nutrient requirement is never known with certainty; only the approximate distribution of requirements for the life stage group may be known. If the individual's intake on average meets or exceeds the RDA, there is good assurance that the intake is adequate for the specified criterion. If an individual's average intake over time is less than the RDA, it can be inferred only that there is some increased likelihood that the intake is inadequate. The likelihood increases the further the intake falls below the RDA. For a requirement that is normally distributed, when the usual intake is less than 2 standard deviations below the Estimated Average Requirement (EAR), the likelihood that the individual's requirement would be met would be small (NRC, 1986). A usual intake that is well below the RDA may indicate the need for further assessment of nutritional status by biochemical tests or clinical examination.

USING ADEQUATE INTAKES

As is true for the Recommended Dietary Allowance (RDA), healthy individuals with an average daily intake at or above the Adequate Intake (AI) are assumed to be at low risk of intake inadequate for a defined state of nutrition. This is its most important use. Similarly, if an individual's average intake over time is less than the AI, it can be inferred only that there is some likelihood that the intake is inadequate. The likelihood increases the further the intake falls below the AI. Just as with the RDA, a usual intake that is well below the AI may indicate the need for further assessment of nutritional status by biochemical tests or clinical examination.

The inherent limitations of the AI must always be taken into account in its application. The AI depends on a greater degree of judgment than is applied in estimating the Estimated Average Requirement (EAR) and subsequently deriving an RDA. Therefore, the AI provides a more imprecise basis for the planning and assessment of nutrient intakes for individuals than would an RDA. For example, when an AI is used, it is less clear at what point it would be nearly certain that the individual's requirement would not be met. The AI might deviate significantly from the RDA if it could be determined, and might be numerically higher than the RDA would be if it were known. For this reason, AIs must be applied with greater care than is the case for RDAs.

When applying an AI to the nutrition of individuals, professionals should first assess particular risk factors and other characteristics of apparently healthy individuals that are relevant to the specific nutrient or food component. In most circumstances AIs are useful only as surrogates for RDAs for individuals. In the absence of EARs and depending on how the AIs are derived, it may be possible for an AI to be taken into account when nutritionally adequate diets are planned for populations or the diets of population groups are assessed. The extent to which AIs can be used in population planning and assessment will be addressed further in subsequent reports.

USING TOLERABLE UPPER INTAKE LEVELS

The Tolerable Upper Intake Level (UL) may be used to examine the possibility of overconsumption of a nutrient. The evaluation of true status requires clinical, biochemical, and anthropometric data. If an individual's usual nutrient intake remains below the UL, it is unlikely that there would be an increased risk of adverse effects from excessive intake. The risk of adverse effects increases as intakes above the UL are continued over time. However, the intake at which a given individual will develop adverse effects as a result of taking large amounts of a nutrient cannot be known with certainty. There is no established benefit for healthy individuals from consuming amounts of nutrients that exceed the Recommended Dietary Allowance (RDA) or Adequate Intake (AI).

The UL may also be used to determine the prevalence of intakes that pose a risk of adverse effects. For the B vitamins the form of the vitamin may need to be considered when intake data are examined, and this may be impossible with current survey data (see "Specific

Applications"). For example, the UL for folate applies only to folic acid, not to folate naturally found in foods, and the UL for niacin applies only to niacin in supplements, not to the niacin in food either added or found naturally or to the niacin equivalents from its precursor, tryptophan.

USING ESTIMATED AVERAGE REQUIREMENTS

Assessing the Adequacy of Nutrient Intake of Groups

The Estimated Average Requirement (EAR) (Table 13-1) may be used to estimate the prevalence of inadequate nutrient intake for a life stage group. To do so, one determines the percentage of the

TABLE 13-1 Estimated Average Requirements for B Vitamins

Life Stage and Gender Group	Thiamin (mg/d)	Riboflavin (mg/d)	Niacin (mg/d)[a]	Vitamin B_6 (mg/d)	Folate (µg/d)[b]	Vitamin B_{12} (µg/d)
Children						
1–3 y	0.4	0.4	5	0.4	120	0.7
4–8 y	0.5	0.5	6	0.5	160	1.0
Males						
9–13 y	0.7	0.8	9	0.8	250	1.5
14–18 y	1.0	1.1	12	1.1	330	2.0
19–30 y	1.0	1.1	12	1.1	320	2.0
31–50 y	1.0	1.1	12	1.1	320	2.0
51–70 y	1.0	1.1	12	1.4	320	2.0
> 70 y	1.0	1.1	12	1.4	320	2.0
Females						
9–13 y	0.7	0.8	9	0.8	250	1.5
14–18 y	0.9	0.9	11	1.0	330	2.0
19–30 y	0.9	0.9	11	1.1	320	2.0
31–50 y	0.9	0.9	11	1.1	320	2.0
51–70 y	0.9	0.9	11	1.3	320	2.0
> 70 y	0.9	0.9	11	1.3	320	2.0
Pregnancy (all ages)	1.2	1.2	14	1.6	520	2.2
Lactation (all ages)	1.2	1.3	13	1.7	450	2.4

NOTE: Estimated Average Requirements (EARs) have not been set for infants or for pantothenic acid, biotin, or choline.

[a] As niacin equivalents. 1 mg of niacin = 60 mg of tryptophan.

[b] As dietary folate equivalents. 1 dietary folate equivalent = 1 µg food folate = 0.7 µg of folate added to food or as a supplement consumed with food = 0.5 µg of folate taken as a supplement on an empty stomach.

individuals in the group whose usual intakes are less than the EAR (Beaton, 1994) (Figure 13-1). The estimate is most accurate if the variability of intakes is at least twice as large as the variability of requirements. Before intake is compared with the EAR, methods should be used to remove the day-to-day variation in intake (Nusser et al., 1996) so that the intake data better reflect usual intakes. The adjustment narrows the intake distribution and thus gives a better estimate of the percentage of the group with intakes below the EAR (Figure 13-2).

Examples of estimating the prevalence of inadequate intake appear in Figures 13-3 and 13-4, which use adjusted data from the Third National Health and Nutrition Examination Survey. A large percentage of adults reportedly have a total folate intake less than the EAR (Figure 13-3), and this percentage is greater for women than for men. However, because the reported folate intake is considered to be substantially underestimated (partly because of methodological problems, partly because adjustment has not been made for the better bioavailability of folate in fortified cereals, and partly because these data were obtained before the fortification of cereal grains with folate was required), it is not known to what extent this

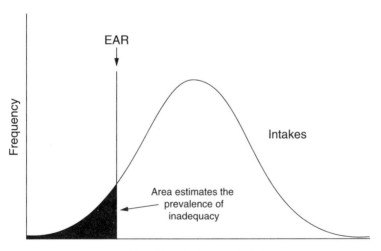

FIGURE 13-1 Estimation of the population prevalence of inadequate intakes. For this method to apply, the variance of intakes must be greater than the variance of requirements, and the requirement distribution must be symmetrical. EAR = Estimated Average Requirement.

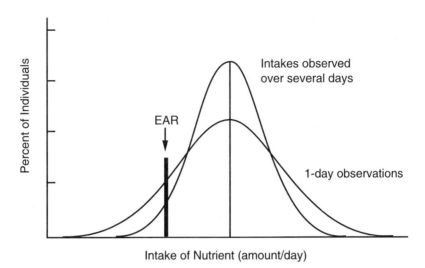

FIGURE 13-2 Effect of multiple days of observation on the apparent distribution of nutrient intake. The mean is the same for both, but the percentage of individuals with intakes less than the Estimated Average Requirement (EAR) is overestimated by the 1-day observations. Adapted from NRC (1986); originally from Hegsted (1972).

discrepancy between the EAR and intake represents a true concern. In contrast, Figure 13-4 reveals that nearly all adults have intakes of vitamin B_{12} that exceed the EAR. This graph does not provide any perspective on the problem of vitamin B_{12} absorption by many elderly, however, because the EAR assumes that adequate absorption of food-bound vitamin B_{12} occurs.

Overestimates of the prevalence of inadequate intakes could result from the underreporting of food intake; underestimates of the prevalence of inadequate intake are also possible. (See Chapter 2 for a discussion of many of the potential sources of error in self-reported dietary data.)

As suggested above, several questions need to be considered when assessing the intake of populations:

• What kinds of adjustments can be made, if any, for biases in the food intake data?
• What factors should be considered in interpreting the findings in different populations?

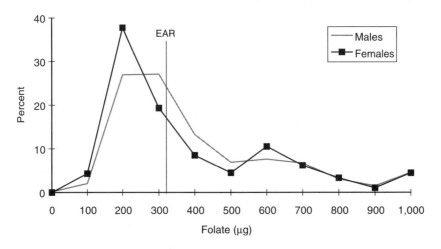

FIGURE 13-3 Distribution of reported total folate intake for men and women aged 19 years and older, Third National Health and Nutrition Examination Survey, 1988–1994. The area under each curve represents 100 percent of that population. More than 50 percent of young women have reported folate intakes (diet plus supplements) below the Estimated Average Requirement (EAR). However, these data are not adjusted for the higher bioavailability of folate as consumed in fortified foods and supplements as was done in determining the EAR. Furthermore, the reported intakes are likely to be underestimates of the actual intake because of limitations in the methods used to analyze food folate. Data have been adjusted for within-person variability using the method of Nusser et al. (1996). Folate intake values were rounded to the nearest 100 µg and all values greater than 1,000 µg were recorded as 1,000 µg. Data points are from unpublished data on percentiles of B vitamin intake from food and supplements, J.D. Wright, National Center for Health Statistics, Centers for Disease Control and Prevention, 1998.

• At what level of intake should concern be raised for a population?

Planning Nutrient Intakes of Groups

The EAR also may be used as a basis for planning or making recommendations for the nutrient intakes of free-living groups. A group mean intake that would be associated with a low prevalence of inadequate intakes can be based on the EAR and the variance of intake. This can be done by using the same principles that were used in the estimation of the prevalence of inadequate nutrient intakes above (Figure 13-5). A detailed explanation of the method is

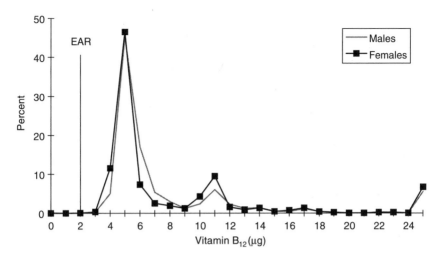

FIGURE 13-4 Distribution of reported total vitamin B_{12} intake for men and women aged 19 years and older, Third National Health and Nutrition Examination Survey, 1988–1994. Reported intakes of vitamin B_{12} (diet plus supplements) are well above the Estimated Average Requirement (EAR) for young men and women. Data were adjusted for within-person variability. Vitamin B_{12} intake values were rounded to the nearest 1.0 µg and all values greater than 25 µg were recorded as 25 µg. Data points are from unpublished data on percentiles of B vitamin intake from food and supplements, J.D. Wright, National Center for Health Statistics, Centers for Disease Control and Prevention, 1998.

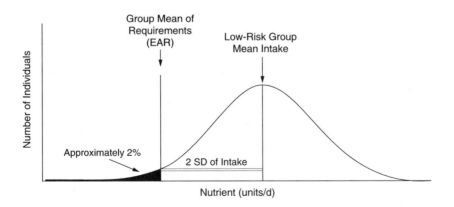

FIGURE 13-5 Derivation of a low-risk group mean intake. If the variation of intake is smaller than depicted, a lower group mean intake would also be low risk. SD = standard deviation. Adapted from WHO (1996).

presented by Beaton (1994), and an example of the application of the method is found in *Trace Elements in Human Nutrition and Health* (WHO, 1996).

Using the EAR in planning intakes of groups involves a number of key decisions and the analysis of questions such as the following:

- Should actual or ideal distributions of populations intakes be used to calculate recommendations for groups? (Actual distributions are seldom normally distributed.)
- What factors should influence the selection of the degree of risk that can be tolerated when planning for groups?
- Do the results point to a need to take concrete steps to increase the intake of those at risk or of the entire group?

It is anticipated that methods of using the EAR for assessment and planning will be addressed in future reports.

OTHER USES OF DIETARY REFERENCE INTAKES

For many years, the Recommended Dietary Allowances in the United States and the Canadian Recommended Nutrient Intakes have been used by many national and federal agencies for a variety of purposes. For example, they have been considered in setting regulations for feeding programs, setting standards for feeding in group facilities (nursing homes, school cafeterias, and correctional facilities), developing recommended intakes for the military, and setting reference values for food labels. They have been used for comparative purposes in many computer programs for nutrient analysis and by dietitians as a basis when modifying diets of patients. Guidance for using Dietary Reference Intakes (DRIs) for many purposes will be addressed in future reports. The development of this guidance is an important component of the overall DRI project.

SPECIFIC APPLICATIONS

Dietary Folate Equivalents and Folate Sources

Diet Assessment

Individuals. Currently, nutrient databases and nutrition labels do not distinguish between sources of folate (food folate and folic acid) or express the folate content of food in dietary folate equivalents

(DFEs), which takes into account the different bioavailabilities of folate sources. DFEs and types of folate are related as follows:

1 µg of DFEs = 1.0 µg of food folate
           = 0.6 µg of folate added to foods (as a fortificant or folate supplement with food)
           = 0.5 µg of folate taken as a supplement (without food).

1 µg of food folate	= 1.0 µg of DFEs
1 µg of folate added as a fortificant or as a supplement consumed with meals	= 1.7 µg of DFEs
1 µg of folate supplement taken without food	= 2.0 µg of DFEs.

When intakes of folate in the diet of an individual are assessed, it is possible to approximate the DFE intake by estimating the amount present added in fortification and the amount present naturally as food folate by using the relationship 1 µg of folate added as a fortificant = 1.7 µg of DFEs (the reciprocal of 1 µg of DFEs = 0.6 µg folate added to food).

The following four-step method is proposed to approximate DFEs when estimating the dietary intake of an individual:

1. Group foods into (a) fortified cereal grain foods and specially fortified foods and (b) all others.
2. If other current data are not available for cereal grains, assume the following levels of fortification (read the label of the product to determine whether folate has been added in amounts greater than the required fortification level; this primarily refers to cereals):

one slice of bread provides 20 µg of added folate;
one serving (about 1 cup) of cooked pasta provides 60 µg of added folate; and
one serving (about 1 cup) of cooked cereal or rice provides 60 µg of added folate.

Moderately fortified ready-to-eat cereals provide approximately 25 percent of the daily value per serving according to the product label, which is currently equivalent to 100 µg of added folate (25 percent of 400 µg). Highly fortified ready-to-eat cereals provide 100 percent of the daily value per serving, or 400 µg of added folate. Serving sizes of ready-to-eat cereals vary widely.

3. Combine the folate contributed by all the fortified cereal grains and multiply the result by 1.7 to obtain DFEs from folate added to foods.

4. Add DFEs from cereal grains to the folate content (in μg) from all other foods obtained from existing nutrient databases to obtain the total folate content in DFEs. For example, if the fortified cereal grains consumed were

8 slices of bread at 20 μg of added folate per slice (160 μg of total folate),

1 serving of moderately fortified ready-to-eat cereal (100 μg of folate), and

1 one-cup serving of pasta (60 μg of folate),

the total content would be 320 μg of added folate. The other foods in the diet—fruits, vegetables, meats, legumes, and milk products—provide 250 μg of food folate as determined by food composition data.

Therefore, total folate intake in DFEs = $(1.7 \times 320) + 250 = 794$ μg of DFEs.

Populations. If dietary folate intake has been reported for groups without adjusting for DFEs and if members of the group have consumed foods fortified with folate, the amount of available folate will be higher than reported for those group members. Adjustments can be made only at the individual level, not at the group level.

Recommendation

Because of the difference in the bioavailability of food folate and folate added to foods or taken as supplements, it is recommended that both food folate and added folate be included in tables of food composition and in reports of intake. In other words, the content or intake of naturally occurring food folate should be reported separately from that of added folate in fortified foods and supplements. This allows the computation of DFEs for the assessment of intake and for diet planning, the computation of added folate intake for comparison with the Tolerable Upper Intake Level (UL), and analysis of the relative contributions of the two forms to total intake.

Diet Planning for Individuals

As can be inferred from the example above, individuals who wish to increase their folate consumption to the Recommended Dietary Allowance (RDA) level (400 μg/day of DFEs) could readily do so by consuming fortified cereal grains. They could also increase intake of foods that are naturally high in folate, such as orange juice and cooked dried peas and beans, eat more foods that are good sources of folate, and eat fewer foods that contain none (such as soft drinks and most candy). For women capable of becoming pregnant, taking 400 μg/day of folate from fortified food, supplements, or a combination of the two is recommended to reduce the risk of neural tube defects.

One combination of fortified foods that would provide 400 μg of added folate is

2 servings of moderately fortified ready-to-eat cereal (200 μg of folate),
7 slices of bread or alternate (140 μg of folate), and
1 cup of rice or pasta (60 μg of added folate).

According to the serving sizes specified in the Food Guide Pyramid (USDA, 1992), this would be 11 servings of cereal grains— much more than that consumed by many women. It is the upper limit of the recommendation to consume 6 to 11 servings made in the Food Guide Pyramid.

Obtaining Synthetic Vitamin B_{12} from Food

It is recommended that adults over age 50 obtain their vitamin B_{12} mainly as synthetic B_{12} from fortified foods or supplements because absorption of food-bound B_{12} may be limited (see Chapter 9). The main food sources of synthetic vitamin B_{12} are fortified ready-to-eat cereals and soy-based meat substitutes (see Table 9-10). If these foods provide 25 percent of the current daily value designated on the product label, they provide 1.5 μg of synthetic B_{12} (more than half the RDA for B_{12} for older adults of 2.4 μg/day).

SUMMARY

Each reference value should be used for its intended purpose. The Recommended Dietary Allowance (RDA) and the Adequate Intake (AI) are intended for use primarily as goals for intake by individuals. Special attention to the form of folate and vitamin B_{12} is often warranted.

REFERENCES

Beaton GH. 1994. Criteria of an adequate diet. In: Shils ME, Olson JA, Shike M, eds. *Modern Nutrition in Health and Disease,* 8th ed. Philadelphia: Lea & Febiger. Pp. 1491–1505.

Hegsted DM. 1972. Problems in the use and interpretation of the Recommended Dietary Allowances. *Ecol Food Nutr* 1:255–265.

IOM (Institute of Medicine). 1994. *How Should the Recommended Dietary Allowances Be Revised?* Washington, DC: National Academy Press.

NRC (National Research Council). 1986. *Nutrient Adequacy. Assessment Using Food Consumption Surveys.* Washington, DC: National Academy Press.

Nusser SM, Carriquiry AL, Dodd KW, Fuller WA. 1996. A semiparametric transformation approach to estimating usual daily intake distributions. *J Am Stat Assoc* 91:1440–1449.

USDA (U.S. Department of Agriculture). 1992. *The Food Guide Pyramid.* Home and Garden Bulletin Number 252. Washington, DC: US Government Printing Office.

WHO (World Health Organization). 1996. *Trace Elements in Human Nutrition and Health.* Prepared in collaboration with the Food and Agriculture Organization of the United Nations and the International Atomic Energy Agency. Geneva: WHO.

14

A Research Agenda

The Standing Committee on the Scientific Evaluation of Dietary Reference Intakes was charged with developing a research agenda to provide a basis for public policy decisions related to recommended intakes of the B vitamins and choline and ways to achieve them, along with information needed to establish tolerable upper intake levels (ULs). This chapter describes the approach used to develop the research agenda, briefly summarizes gaps in knowledge, and presents a prioritized research agenda. A section at the end of each nutrient chapter (Chapters 4 through 12) presents a prioritized list of research topics for the nutrient.

APPROACH

The following approach was used to develop the research agenda:

- Gaps in knowledge were identified in
  - nutrient requirements,
  - methodological problems related to the assessment of intake of these nutrients and to the assessment of adequacy of intake,
  - relationships of nutrient intake to public health, and
  - adverse effects of nutrients.
- Data were examined to identify any major discrepancies between intake and the Estimated Average Requirements (EARs), and possible reasons for such discrepancies were considered.
- Listings of studies currently funded by the National Institutes of

437

Health were obtained and examined along with other listings by expert groups and comments submitted by experts.

• The need to protect individuals with extreme or distinct vulnerabilities resulting from genetic predisposition or disease conditions was considered.

• Expert opinion was used to weigh alternatives and set priorities.

IMPORTANT FEATURES OF STUDIES TO ESTIMATE REQUIREMENTS

Derivation of an Estimated Average Requirement (EAR) involves identification of the criterion for a particular status indicator or combination of indicators that is consistent with impaired status as defined by some clinical consequence. For many of the B vitamins, there is a dearth of information on the biochemical values that reflect abnormal function. One priority should be determination of the relationship of existing status indicators to clinical endpoints to allow their use for setting EARs. For some nutrients new clinical endpoints of impaired function need to be identified and related to status indicators.

The depletion-repletion research paradigms that are often used in studies of requirements, although not ideal, are still probably the best approach to determining vitamin requirements. However, these studies should be designed to meet three important criteria:

• An indicator of vitamin status is needed for which a cutoff point has been identified, below (or above) which vitamin status is documented to be impaired. (In the case of folate, an erythrocyte level of 300 nmol/L [140 ng/mL] fits this criterion because lower levels are associated with megaloblastic changes in blood cells. In the case of vitamin B_6 and several other B vitamins, however, there is little information relating levels of status indicators to functional sufficiency or insufficiency. Instead, the levels of indicators normally used to assess requirements are those exhibited by subjects on a baseline adequate diet—even though there is no information regarding whether this level of intake is greatly in excess of adequate, barely adequate, or deficient.) The amount needed for restoration of biochemical status indicators to baseline values is not necessarily equivalent to the requirement for the nutrient.

• The depletion and repletion periods should be sufficiently long to allow a new steady state to be reached. This can be very problematic because turnover rates of total body content for B vitamins range from less than 1 to about 3 percent per day, which suggests that long periods are needed for equilibrium. In the case of erythrocyte folate, theoretically the erythrocytes have to turn over completely (approximately 90 days). Study design should allow for examination of the effects of initial status on response to maintenance or depletion and repletion

• Intakes used in repletion regimens should bracket the expected EAR intake to assess the EAR more accurately and to allow for a measure of variance. In addition, an accurate assessment of variance requires a sufficient number of subjects.

A relatively new and increasingly popular approach to determining requirements is kinetic modeling of body pools using steady-state compartmental analyses. This approach is unlikely to supplant depletion-repletion studies because it has a number of drawbacks; for example, assumptions that cannot be tested experimentally are often needed and the numbers obtained for body pool sizes are inherently imprecise. Even if accurate assessments of body pools were possible and were obtained, such information would be useful in setting a requirement only if the size of the body pool at which functional deficiency occurs could be established.

MAJOR KNOWLEDGE GAPS

Requirements

For all the B vitamins and choline, there is a serious lack of data useful for setting Estimated Average Requirements (EARs) for children, adolescents, pregnant and lactating women, and the elderly. Studies should use graded levels of nutrient intake and a combination of response indices and should consider other points raised in the preceding section. For some of the B vitamins (e.g., folate), studies should examine whether the requirement varies substantially by trimester of pregnancy. The nutrients and life stage and gender groups for which studies of requirements appear to be priorities from a public health perspective are vitamin B_{12} requirements of the elderly and folate requirements by trimester of pregnancy. In addition, priority should be given to the identification of indicators on which to base vitamin B_6 requirements.

This short list does not imply that studies of requirements of other nutrients or age groups are not important, merely that it seems less likely that such studies will produce results that will have significant benefit on the health of the U.S. or Canadian populations. Research topics for each of the B vitamins and choline appear at the end of Chapters 4 through 12.

The understanding of the nutrition of pantothenic acid, biotin, and choline is rudimentary compared with that of the other B vitamins. Little information is available on human requirements, intake, bioavailability, toxicity, and metabolic effects of these compounds. Although choline can be formed in the human body from endogenous precursors, little is known about the relative amounts of choline derived from the diet and from endogenous synthesis. Research to date has indicated little cause for concern about the adequacy of pantothenic acid or biotin intake for healthy people; deficiency states can be produced only by actively interfering with the absorption or bacterial production of these vitamins. On the other hand, animal studies suggest that choline intake may affect long-term health.

A growing number of studies suggests that there are complex interrelationships among nutrients (e.g., vitamin B_6, folate, vitamin B_{12}, and perhaps choline, methionine, and riboflavin), but these are not well understood in relation to the maintenance of normal nutritional status and to the prevention of chronic degenerative disease. These interactions may affect the need for one or more of the nutrients.

Methodology

For some nutrients there are serious limitations in the methods available to analyze laboratory values indicative of nutrient status, to determine the nutrient content of foods, or both. These limitations have slowed progress in conducting or interpreting studies of nutrient requirements. The most serious gaps were judged to be those relating to analytical methodology for blood folate analysis and methods for the analysis of the folate content of food. A related gap, which is not strictly methodological, concerns the bioavailability of various forms of folate. Major needs include a comparison of the bioavailability of food folate from mixed diets and of folate in the form of folic acid (from supplements or fortification) consumed with food and an examination of the mechanisms by which bioavailability is altered by food matrices.

Relationships of Intake to Public Health

Developmental Disorders

For the B vitamins the developmental disorder of greatest concern is neural tube defect (NTD). Major gaps in knowledge include the mechanisms by which maternal folate sufficiency reduces the occurrence of NTD in the infant (e.g., evaluation of whether increased NTD risk is due to folate deficiency or to the mode of action of folate sufficiency [does it act on mother, embryo, or both?]); the relative efficacy of food folate, folate added to food, and folate supplements in reducing NTD risk; the process, if any, by which folate influences the embryonic process of neurulation; and the genes that are responsible for the heritability and folate-responsiveness of NTD. This latter area could include (1) linkage analyses in suitable genetically homogeneous human populations to assess the etiologic relationship between NTD and a variety of genetic alterations (including the thermolabile variant of 5,10-methylenetetrahydrofolate reductase) and in the genes responsible for NTD in the curly tail mouse; (2) investigation of whether alterations in any of these genes produce NTD when induced in mouse models, yield folate-responsive NTD in mouse models, and provide suitable markers for assessing NTD risk in human populations; and (3) identification of an animal model for common human NTDs that is responsive to relevant levels of folate.

Chronic Degenerative Disease Risk

Although interest is high and numerous studies have been conducted, there are still serious gaps in knowledge of the relationship of B vitamin intake to risk of vascular disease and other chronic degenerative diseases. With the new U.S. regulations on the fortification of cereal grains with folate, it is now possible to investigate the health effects, both positive and negative, of folate fortification on folate intake and health status by life stage and gender.

Adverse Effects

For B vitamins and choline as a group, only a few studies have been conducted that were explicitly designed to address adverse effects of chronic high intake. Thus, information on which to base Tolerable Upper Intake Levels (ULs) is extremely scanty. Because it appears that vitamin B_{12} deficiency greatly increases the potential of folate to cause adverse effects, efforts are needed to improve methods to detect and correct vitamin B_{12} deficiency before adverse hematological or neurological changes occur and to determine the prevalence of vitamin B_{12} deficiency.

THE RESEARCH AGENDA

The Standing Committee on the Scientific Evaluation of Dietary Reference Intakes agreed to assign highest priority to research that has potential to prevent or retard human disease processes and to prevent deficiencies with functional consequences. The following five areas for research were assigned the highest priority:

• Studies to provide the basic data for constructing risk curves and benefit curves across the exposures to food folate and to folate added to foods and taken as a supplement. Such studies would provide estimates of the risk of developing neural tube defects, vascular disease, and neurological complications in susceptible individuals consuming different levels of folate.

• Studies of the magnitude of effect of folate, vitamin B_6, vitamin B_{12}, and related nutrients for the prevention of vascular disease and of possible mechanisms for the influence of genetic variation.

• Studies to overcome the methodological problems in the analysis of folate. This includes the development of sensitive and specific indicators of deficiency and the development of practical, improved methods for analyzing the folate content of foods and determining its bioavailability.

• Studies to develop economical, sensitive, and specific methods for assessing the prevalence, causes, and consequences of vitamin B_{12} malabsorption and deficiency and for preventing and treating these conditions.

• Studies of how folate and related nutrients influence normal cellular differentiation and development, including embryogenesis and neoplastic transformation.

Although data are not sufficient for deriving a Tolerable Upper Intake Level (UL) for most of the B vitamins and additional research is required on the adverse effects of B vitamins and choline, it was concluded that higher priority should be given to the areas listed above because of relatively low expectation of adverse effects or toxicity.

A

Origin and Framework of the Development of Dietary Reference Intakes

This report is the second in a series of publications resulting from the comprehensive effort being undertaken by the Food and Nutrition Board's (FNB) Standing Committee on the Scientific Evaluation of Dietary Reference Intakes (DRI Committee) and its panels and subcommittees.

ORIGIN

This initiative began in June 1993, when the FNB organized a symposium and public hearing entitled "Should the Recommended Dietary Allowances Be Revised?" Shortly thereafter, to continue its collaboration with the larger nutrition community on the future of the Recommended Dietary Allowances (RDAs), the FNB took two major steps: (1) it prepared, published, and disseminated the concept paper "How Should the Recommended Dietary Allowances Be Revised?" (IOM, 1994), which invited comments regarding the proposed concept, and (2) it held several symposia at nutrition-focused professional meetings to discuss the FNB's tentative plans and to receive responses to this initial concept paper. Many aspects of the conceptual framework of the DRIs came from the United Kingdom's *Dietary Reference Values for Food Energy and Nutrients for the United Kingdom* report (COMA, 1991).

The five general conclusions presented in the FNB's 1994 concept paper are as follows:

1. Sufficient new information has accumulated to support a reassessment of the RDAs.

2. Where sufficient data for efficacy and safety exist, reduction in the risk of chronic degenerative disease is a concept that should be included in the formulation of future recommendations.

3. Upper levels of intake should be established where data exist regarding risk of toxicity.

4. Components of food of possible benefit to health, although not meeting the traditional concept of a nutrient, should be reviewed, and if adequate data exist, reference intakes should be established.

5. Serious consideration must be given to developing a new format for presenting future recommendations.

Subsequent to the symposium and the release of the concept paper, the FNB held workshops at which invited experts discussed many issues related to the development of nutrient-based reference values, and FNB members have continued to provide updates and engage in discussions at professional meetings. In addition, the FNB gave attention to the international uses of the earlier RDAs and the expectation that the scientific review of nutrient requirements should be similar for comparable populations.

Concurrently, Health Canada and Canadian scientists were reviewing the need for revision of the Recommended Nutrient Intakes (RNIs) (Health Canada, 1990). Consensus after a symposium for Canadian scientists cosponsored by the Canadian National Institute of Nutrition and Health Canada in April 1995 was that the Canadian government should pursue the extent to which involvement with the developing FNB process would be of benefit to both Canada and the United States in terms of leading toward harmonization.

On the basis of extensive input and deliberations, the FNB initiated action to provide a framework for the development and possible international harmonization of nutrient-based recommendations that would serve, where warranted, for all of North America. To this end, in December 1995 the FNB began a close collaboration with the government of Canada and appointed the Standing Committee on the Scientific Evaluation of Dietary Reference Intakes (DRI Committee). It is hoped that representatives from Mexico will join in future deliberations.

THE CHARGE TO THE COMMITTEE

In 1995 the DRI Committee was appointed by the Institute of Medicine, National Academy of Sciences, to oversee and conduct this project. To accomplish this task over a period of 5 years, the DRI Committee devised a plan involving the work of seven or more expert nutrient group panels and two overarching subcommittees (Figure A-1). The process described below for this report is expected to be used for subsequent reports.

The Panel on Folate, Other B Vitamins, and Choline, composed of experts on those nutrients, has been responsible for reviewing the scientific literature concerning the B vitamins and choline for each stage of the lifespan, considering the roles of nutrients in decreasing the risk of chronic and other diseases and conditions, and interpreting the current data on intakes in North American population groups.

The panel had additional tasks that are specifically related to this group of nutrients and are thus not necessarily part of the DRI process: an analysis of information specific to the prevention of neural tube defects, an analysis of information specific to the diagnosis and prevention of pernicious anemia and vitamin B_{12} deficiency, and the identification of a research agenda to provide a basis for public policy decisions related to recommended intakes and ways to achieve those intakes.

The panel was charged with analyzing the literature, evaluating possible criteria or indicators of adequacy, and providing substantive rationales for their choices of each criterion. By using the criterion chosen for each stage of the lifespan, the panel was to estimate the average requirement for each nutrient or food component reviewed, assuming that adequate data were available to do so. As the panel members reviewed data on Tolerable Upper Intake Levels (ULs), they also interacted with the Subcommittee on Upper Reference Levels of Nutrients, which assisted the panel in applying the risk assessment model to each B vitamin and choline. The DRI values in this report are a product of the joint efforts of the DRI Committee, the Panel on Folate, Other B Vitamins, and Choline, and the Subcommittee on Upper Reference Levels of Nutrients. The Subcommittee on Interpretation and Uses of Dietary Reference Intakes had not yet been appointed and thus did not participate in the development of this report.

446

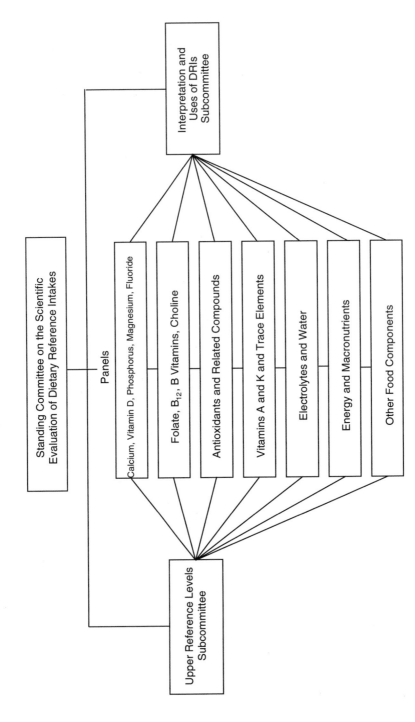

FIGURE A-1 Structure of Dietary Reference Intakes project.

REFERENCES

COMA (Committee on Medical Aspects of Food Policy). 1991. *Dietary Reference Values for Food Energy and Nutrients for the United Kingdom.* Report on Health and Social Subjects, No. 41. London: HMSO.

Health Canada. 1990. *Nutrition Recommendations. The Report of the Scientific Review Committee.* Ottawa: Canadian Government Publishing Centre.

IOM (Institute of Medicine). 1994. *How Should the Recommended Dietary Allowances Be Revised?* Washington, DC: National Academy Press.

B

Acknowledgments

The Panel on Folate, Other B Vitamins, and Choline; the Subcommittee on Upper Reference Levels of Nutrients; the Standing Committee on the Scientific Evaluation of Dietary Reference Intakes; and the Food and Nutrition Board (FNB) staff are grateful for the time and effort of the many contributors to the report and the workshops and meetings leading up to the report. Through openly sharing their considerable expertise and different outlooks, these individuals and organizations brought clarity and focus to the challenging task of setting thiamin, riboflavin, niacin, vitamin B_6, folate, vitamin B_{12}, pantothenic acid, biotin, and choline requirements and upper levels for humans. The list below mentions those individuals with whom we worked closely, but many others also deserve our heartfelt thanks. Those individuals, whose names we do not know, made important contributions to the report by offering suggestions and opinions at the many professional meetings and workshops the committee members attended.

The panel, subcommittee, and committee members, as well as the FNB staff thank the following named (as well as unnamed) individuals and organizations:

INDIVIDUALS

James Allen
Steven Bailey
Adrianne Bendich
Shirley Beresford
Barbara Bowman
Judith Brown
Alan Buchman
Ralph Carmel
Stephen Coburn
Nancy Crane
J. David Erickson
Nancy Ernst
Abhimanyu Garg
Jesse Gregory, III
John Hathcock
Mary Hediger
Jane Heinig
Jean Hines
Robert Jacob
Paul Jacques
Richard Johnston, Jr.
Jocelyn Kennedy-Stephenson
Young-In Kim
Kathleen Koehler
James Leklem
Christine Lewis
Harris Lieberman
John Lindenbaum (deceased)
Simon Manning

Bernadette Marriott
Peter Martin
Edward McCabe
Linda Meyers
James Mills
Donald Mock
William Mosher
Alanna Moshfegh
Robert Nicolosi
Godfrey Oakley
Mary Frances Picciano
K. Pietrzik
Jeanne Rader
Richard Rivlin
Killian Robinson
Sheldon Rothenberg
Robert Rucker
Howerde Sauberlich
Joan Schall
Herbert Schaumberg
Jacob Selhub
Bert Spiker
Sally Stabler
Richard Stampf
David Stumpf
Tsunenobu Tamura
Richard Troiano
Jacqueline Wright

ORGANIZATIONS

American Academy of Neurology
American Academy of Pediatrics
American College of Obstetricians and Gynecologists
American Medical Association
American Society for Clinical Nutrition
American Society for Nutrition Sciences
Canadian Paediatric Society
Council for Responsible Nutrition
Federal Advisory Steering Committee for Dietary Reference Intakes

Federation of American Societies for Experimental Biology
Health Canada
Institute of Food Technologists
Interagency Human Nutrition Research Council
Life Sciences Research Organization
National Heart, Lung, and Blood Institute

C

Système International d'Unités

In the text of this report, the Système International d'Unités (SI units) are presented for hematological and clinical chemistry values, with traditional units in parentheses. The conversion factors for hematological and clinical chemistry values are given in Table C-1.

TABLE C-1   Factors Used to Convert Between Traditional Units and the Système International d'Unités (SI Units)

Nutrient	Traditional Unit	Conversion Factor	SI Unit	Molecular Weight
Thiamin	μg/24h	0.003324468	μmol/d	300.8
Thiamin hydrochloride	μg/24h	0.00296472	μmol/d	337.3
Riboflavin	μg/dL	26.5674814	nmol/L	376.4
Flavin mononucleotide (FMN)	ng/mL	2.191540653	nmol/L	456.3
Flavin-adenine dinucleotide (FAD)	ng/mL	1.205545509	nmol/L	829.5
Niacin	mg/d	8.123476848	μmol/d	123.1
Niacinamide	mg/d	8.19000819	μmol/d	122.1
Pyridoxine hydrochloride	mg	4.86381323	μmol	205.6
Vitamin B_6	ng/mL	5.910165485	nmol/L	169.2
Pyridoxal	ng/mL	5.980861244	nmol/L	167.2
Pyridoxamine	ng/mL	5.94530321	nmol/L	168.2
Pyridoxal 5′-phosphate	ng/mL	4.046944557	nmol/L	247.1
Folic acid[a]	ng/mL	2.265518804	nmol/L	441.4
5-Methyltetrahydrofolate[b]	ng/mL	2.176752286	nmol/L	459.4
Vitamin B_{12} (cyano)	pg/mL	0.737789582	pmol/L	1,355.4
Coenzyme B_{12}	pg/mL	0.633071664	pmol/L	1,579.6
Methylcobalamin	pg/mL	0.743826242	pmol/L	1,344.4
Pantothenic acid	μg/mL	4.562043796	μmol/L	219.2
Coenzyme A	μg/mL	1.302931596	μmol/L	767.5
Biotin	ng/mL	4.093327876	nmol/L	244.3
Choline base	mg/mL	8.250825083	mmol/L	121.2
Choline chloride	mg/mL	7.163323782	mmol/L	139.6

NOTE: Conversion factors convert traditional values to SI units. To convert to SI units, multiply the traditional unit by the conversion factor. To convert to traditional units, divide the SI unit by the conversion factor.

[a] Pteroylglutamic acid.

[b] Most predominant form in serum, plasma, and erythrocytes.

D
Search Strategies

REQUIREMENTS FOR B VITAMINS AND CHOLINE

Databases Searched

- Medline, 1966 through December 1996
- Embase

Search Terms Used

The search for the nutrient name was limited to titles. Additional strategies were used to identify articles missed by this restriction.

- Nutrient Names

thiamin* or B-1
riboflavin or B-2
niacin or nicotinic acid or nicotinamide
pyridoxal or pyridoxamine or pyridoxine or B-6
cobalamin or cyanocobalamin or B-12
folic acid or folacin or folate
pantothenic acid or pantothenate
biotin
choline

• Specifications and Restrictions of Computerized Search

human only
not neoplasm?/DE (secondary search done for folate *and* variations on neoplasms in the title)
   not reviews
   not letters

Other Searches

• Medline

neural tube defects or spina bifida, prevention or etiology of atrophic gastritis and other terms related to pernicious anemia

• Federal Research in Progress (FedRIP), limited to CRISP files, which cover the Public Health Service

nutrient name in the title

• CRISP search done by NICHD

search terms: anencephaly, spina bifida, meningomyelocele, fusion failure, neural tube, folate deficiency

Additional Sources of References

• References listed in publications on various nutrient reference values from the United Kingdom, Canada, and the United States
• References listed in selected textbooks and review articles and in research articles
• References provided by panel and committee members and by outside experts
• Review of recently received journals

ADVERSE EFFECTS OF HIGH INTAKE

Databases Searched

• Medline, 1965 through December 1996
• Toxline

The following mechanisms were used: National Library of Medicine's PubMed (Medline from 1966–present); WinSPIRS, Version 1.0 (Medline from 1983–May 1997); and HealthWorld/Infotrieve (Medline and Toxline from 1966–present). The choice of which mechanism to use for a particular search involved consideration of ease of access to the database, type and extent of data on adverse effects accessed by each, and presentation of the search information.

Search Terms Used

Searches consistently allowed the inclusion of animal studies unless the number of articles retrieved was large, as was the case for folic acid. Specific searches of folate and animals are specified in the listing below. Final searches were conducted July through October 1997.

- thiamin AND [toxicity OR high dose OR dermatitis OR pruritus OR sudden infant death syndrome]
- riboflavin AND [toxicity OR safety OR high dose OR excess OR phototoxicity OR photosensitization OR mutagenicity OR DNA damage]
- niacin AND [toxicity OR gastrointestinal OR gout OR flushing OR hepatotoxicity]
- pyridoxine AND [toxicity OR peripheral neuropathy OR kidney stones OR dermatosis OR photosensitivity]
- [folic acid OR folate] AND [toxicity OR zinc OR neurotoxicity OR nephropathy] (restricted to human studies)
- [folic acid OR folate] AND [kidney AND ((humans OR toxicity)) OR ((adverse effects AND animals)) OR fruit bats OR precipitation of neuropathy]
- [vitamin B12 OR cyanocobalamin OR cobalamin] AND [toxicity OR excess OR high dose]
- pantothenic acid AND [toxicity OR excess OR high dose]
- biotin AND [toxicity OR high dose]
- choline AND [toxicity OR adverse effects OR fishy odor OR hypotension OR hepatitis OR hepatotox*]

E

Methodological Problems Associated with Laboratory Values and Food Composition Data for B Vitamins

TABLE E-1 Methodological Problems with Laboratory Values for B Vitamins

	Thiamin	Riboflavin	Niacin	B_6
Are precise, accurate methods available?	Yes	Yes	Yes	Yes
What is known about the analytic sensitivity and specificity of the methods?	Good	Good	Good	Good
Is there good agreement in results from use of different methods?	Yes	Yes	Yes	Yes, for most
Is there good agreement in results if different laboratories use the same methods?	Yes	Yes	Yes	Fair
Over time, how have changes in methods affected estimates?	Generally lower because of more specific chromatographic separation (especially high-performance liquid chromatography)			

Folate	B$_{12}$	Pantothenic Acid	Biotin	Choline
Needs improvement	Yes	Needs improvement	Being improved	Yes
Good, but incomplete assays for all forms	Fair; some metabolite interference noted	Needs improvement	Variable	Mass spectrometry specific to 5 pmol
No, see Gunter et al., 1996[a]	No, e.g., *Euglena gracilis* gives lower values than does *Lactobacillis leichmannii*	Fair, limited	No	Yes
No, see Gunter et al., 1996[a]	No, nonisotopic and radioassays do not agree closely	Fair	Fair, limited	Yes
Trends vary depending on method.	Radioassays were unreliable before 1978. Recent introduction of nonisotopic assays has led to higher results.	Little change in methods	Generally lower now	No change

continued

TABLE E-1 Continued

	Thiamin	Riboflavin	Niacin	B_6
How are problems with methods addressed in the report?	Not necessary	Earlier under- and over- estimations of flavins noted.	Questions for research	Not necessary

^a Gunter EW, Bowman BA, Caudill SP, Twite DB, Adams MJ, Sampson EJ. 1996. Results of an international round robin for serum and whole-blood folate. *Clin Chem* 42:1689–1694.

TABLE E-2 Methodologic Problems with Obtaining Food Composition Data for B Vitamins

	Thiamin	Riboflavin	Niacin	B_6
Are precise, accurate methods available?[a]	Substantial, acceptable quality	Substantial, acceptable quality	Substantial, acceptable quality	Substantial, acceptable quality
Is there good agreement in results using different methods?	Fair when allowance is made for specificity differences			
Over time, how have changes in methods affected estimates?	None noted	About the same or slightly lower	None noted	Slightly higher now

[a] Ratings for the B vitamins (but not for choline) are taken from Life Sciences Research Office/Federation of American Societies for Experimental Biology. 1995. *Third Report on Nutrition Monitoring in the United States.* Washington, DC: U.S. Government Printing Office.

[b] Quality of data was rated moot if it was considered unlikely that improved data for that food component would make a difference in the assessment of nutrition-related health status and the assignment of nutrition monitoring priority status (LSRO/FASEB, 1995).

Folate	B_{12}	Pantothenic Acid	Biotin	Choline
Detailed subsections, questions for research	Need for internal reference is stated and values given when available; questions for research.	Questions for research	Short subsection	Not necessary

Folate	B_{12}	Pantothenic Acid	Biotin	Choline
Conflicting, variable quality	Conflicting, quality moot[b]	Conflicting	Lacking, being improved	Substantial, acceptable quality
No, see Gregory (1997), Martin et al. (1992), Pfeiffer et al. (1997), Tamura et al. (1997)[c]	No, tissue methods poorly developed	—	Insufficient comparisons to assess	Yes, but very limited experience
New methods give somewhat higher results for some foods	—	—	—	Old estimates were too high, early assay not specific

[c] Gregory JF 3rd. 1997. Bioavailability of folate. *Eur J Clin Nutr* 51: S54–S59; Martin DC, Francis J, Protetch J, Huff J. 1992. Time dependency of cognitive recovery with cobalamin replacement: Report of a pilot study. *J Am Geriatr Soc* 40:168–172; Pfeiffer CM, Rogers LM, Gregory JF 3rd. 1997. Determination of folate in cereal-grain food products using trienzyme extraction and combined affinity and reversed-phase liquid chromatography. *J Agric Food Chem* 45:407–413; Tamura T, Mizuno Y, Johnston KE, Jacob RA. 1997. Food folate assay with protease, α-amylase, and folate conjugase treatments. *J Agric Food Chem* 45:135–139.

F

Dietary Intake Data from the Boston Nutritional Status Survey, 1981–1984

TABLE F-1 Dietary Intake of B Vitamins in Free-Living Older Adults (aged ≥ 60 y) Who Were Not Taking Vitamin Supplements, Boston Nutritional Status Survey, 1981–1984

Intake of B Vitamins by Gender	Dietary Intake			
	Sample Size	Mean	Standard Deviation	5th Percentile
Thiamin (mg)				
Males	168	1.5	0.5	0.9
Females	283	1.2	0.4	0.7
Riboflavin (mg)				
Males	169	2.0	0.7	1.0
Females	283	1.6	0.6	0.8
Niacin (mg)				
Males	168	21.6	6.7	12.5
Females	283	17.9	5.9	10.3
Vitamin B_6 (mg)				
Males	170	1.3	0.6	0.6
Females	281	1.1	0.4	0.5
Folate (µg)				
Males	189	273	121	119
Females	344	231	94	116
Vitamin B_{12} (µg)				
Males	171	4.7	4.4	1.5
Females	280	4.0	4.9	1.0

SOURCE: Adapted from Appendixes C.5.21–C.5.26 in Hartz SC, Russell RM, Rosenberg IH. 1992. *Nutrition in the Elderly. The Boston Nutritional Status Survey.* London: Smith-Gordon.

10th Percentile	25th Percentile	50th Percentile	75th Percentile	90th Percentile	95th Percentile
1.0	1.2	1.4	1.8	2.2	2.5
0.8	0.9	1.1	1.3	1.7	1.9
1.2	1.5	1.9	2.4	2.9	3.4
1.0	1.2	1.5	1.8	2.3	2.8
13.8	16.0	21.0	25.4	30.1	33.5
11.2	14.3	16.9	20.5	26.2	28.7
0.7	0.8	1.2	1.6	2.1	2.4
0.6	0.7	1.0	1.3	1.6	1.8
139	196	254	328	425	528
140	167	208	272	356	428
1.7	2.4	3.4	5.8	8.4	10.9
1.2	1.8	2.6	3.8	7.6	12.9

TABLE F-2 Intake of B Vitamins in Free-Living Older Adults (aged ≥ 60 y) Who Were Taking Vitamin Supplements, Boston Nutritional Status Survey, 1981–1984

	Sample Size	Mean	Standard Deviation	5th Percentile
Thiamin (mg)				
Males				
Dietary intake	69	1.5	0.5	0.9
Supplement intake	69	9.5	18.3	0.5
Total intake	69	11.0	18.3	1.7
Females				
Dietary intake	166	1.2	0.5	0.7
Supplement intake	166	11.7	24.4	0.5
Total intake	166	13.0	24.5	1.6
Riboflavin (mg)				
Males				
Dietary intake	68	2.0	0.9	1.2
Supplement intake	68	5.5	9.9	0.4
Total intake	68	7.5	9.8	1.8
Females				
Dietary intake	166	1.6	0.6	0.9
Supplement intake	166	10.0	21.8	0.5
Total intake	166	11.6	21.8	2.0
Vitamin B_6 (mg)				
Males				
Dietary intake	67	1.5	0.7	0.6
Supplement intake	67	9.8	21.3	0.3
Total intake	67	11.3	21.2	1.6
Females				
Dietary intake	168	1.2	0.6	0.4
Supplement intake	168	18.9	58.8	0.7
Total intake	168	20.1	58.9	1.7
Niacin (mg)				
Males				
Dietary intake	69	22.5	7.4	11.3
Supplement intake	69	46.7	73.5	1.2
Total intake	69	69.2	75.4	22.3
Females				
Dietary intake	166	17.9	6.2	10.0
Supplement intake	166	60.3	75.6	5.3
Total intake	166	78.2	75.8	23.7

10th Percentile	25th Percentile	50th Percentile	75th Percentile	90th Percentile	95th Percentile
1.0	1.2	1.5	1.7	2.1	2.6
0.5	1.2	2.4	10.0	20.0	55.0
1.9	2.7	4.2	11.5	22.1	55.9
0.8	0.9	1.1	1.4	1.9	2.3
1.0	1.2	3.2	10.0	30.0	55.0
1.9	2.6	4.4	11.4	31.1	56.0
1.2	1.5	2.0	2.3	2.7	3.3
0.6	1.4	1.9	7.4	10.5	15.0
2.1	3.2	4.4	9.9	13.0	17.1
1.1	1.2	1.5	1.8	2.4	2.6
1.0	1.4	2.9	10.0	20.0	34.5
2.4	2.9	5.2	11.5	21.5	35.3
0.8	1.0	1.3	1.9	2.3	2.8
0.7	2.0	2.2	5.0	31.3	50.0
2.4	3.1	4.3	6.1	31.9	51.9
0.6	0.8	1.0	1.5	2.0	2.2
0.8	2.0	2.2	5.0	50.0	55.3
2.1	2.8	4.1	7.2	51.2	57.2
13.7	18.0	21.6	25.9	29.9	39.1
6.7	16.0	20.0	51.8	100.0	100.0
25.3	35.9	48.8	74.4	121.4	137.5
10.9	13.3	16.4	22.2	25.8	28.7
11.5	16.0	30.0	100.0	120.0	160.0
26.5	31.9	48.1	111.5	132.3	178.7

continued

TABLE F-2 Continued

	Sample Size	Mean	Standard Deviation	5th Percentile
Vitamin B$_{12}$ (µg)				
Males				
Dietary intake	66	5.4	6.3	1.5
Supplement intake	66	17.4	52.0	1.5
Total intake	66	22.8	52.4	3.9
Females				
Dietary intake	169	3.7	3.7	1.1
Supplement intake	169	31.0	105.5	1.5
Total intake	169	34.7	105.8	4.1
Folate (µg)				
Males				
Dietary intake	48	296	152	140
Supplement intake	48	586	645	120
Total intake	48	882	696	312
Females				
Dietary intake	105	260	120	121
Supplement intake	105	484	517	120
Total intake	105	744	567	311

SOURCE: Adapted from Appendixes C.5.21–C.5.26 in Hartz SC, Russell RM, Rosenberg IH. 1992. *Nutrition in the Elderly. The Boston Nutritional Status Survey*. London: Smith-Gordon.

10th Percentile	25th Percentile	50th Percentile	75th Percentile	90th Percentile	95th Percentile
1.8	2.3	3.6	6.7	10.5	11.8
1.7	3.0	5.0	7.0	24.0	77.0
4.1	6.8	9.7	15.1	71.8	83.1
1.3	1.9	2.7	4.0	6.2	8.7
2.3	3.0	6.0	12.0	66.7	100.0
4.8	6.2	9.0	14.9	70.0	106.3
158	199	259	327	474	640
133	300	400	400	1,400	2,400
385	581	666	918	1,569	2,667
137	178	238	316	410	520
133	267	400	400	800	1,000
340	530	628	809	1,105	1,349

G

Dietary Intake Data from the Continuing Survey of Food Intakes by Individuals (CSFII), 1994–1995

TABLE G-1 Mean and Percentiles for Usual Intake of Thiamin (mg), CSFII, 1994–1995

Gender/Age Category[a]	N	Mean	Percentile								
			1st	5th	10th	25th	50th	75th	90th	95th	99th
0–6 mo	69	0.69	0.11	0.22	0.29	0.45	0.63	0.87	1.15	1.35	1.70
SE		0.05	0.04	0.05	0.06	0.06	0.06	0.06	0.08	0.09	0.12
7–11 mo	45	0.90	0.38	0.53	0.61	0.75	0.9	1.05	1.19	1.27	1.42
SE		0.05	0.06	0.06	0.06	0.06	0.06	0.05	0.05	0.05	0.05
1–3 y	702	1.14	0.53	0.68	0.76	0.91	1.1	1.32	1.55	1.71	2.04
SE		0.02	0.02	0.02	0.02	0.02	0.02	0.02	0.03	0.04	0.05
4–8 y	666	1.44	0.75	0.92	1.01	1.19	1.4	1.65	1.91	2.09	2.49
SE		0.03	0.03	0.02	0.02	0.02	0.02	0.03	0.05	0.06	0.09
M 9–13 y	180	1.933	0.82	1.06	1.21	1.49	1.86	2.30	2.74	3.04	3.66
SE		0.07	0.08	0.08	0.08	0.07	0.07	0.08	0.11	0.13	0.19
M 14–18 y	191	2.17	0.89	1.16	1.33	1.65	2.08	2.60	3.14	3.50	4.28
SE		0.09	0.06	0.09	0.10	0.09	0.09	0.15	0.17	0.20	0.59
M 19–30 y	328	2.03	0.92	1.16	1.3	1.58	1.95	2.39	2.84	3.14	3.78
SE		0.06	0.04	0.05	0.05	0.06	0.07	0.07	0.09	0.10	0.15
M 31–50 y	627	2.02	0.87	1.12	1.28	1.55	1.92	2.37	2.88	3.25	4.12
SE		0.06	0.04	0.03	0.03	0.04	0.05	0.08	0.13	0.17	0.30
M 51–70 y	490	1.74	0.75	0.96	1.09	1.34	1.67	2.06	2.48	2.76	3.35
SE		0.04	0.03	0.03	0.03	0.04	0.04	0.06	0.08	0.09	0.14
M 71+ y	237	1.70	0.77	0.97	1.10	1.33	1.64	2.00	2.37	2.62	3.14
SE		0.07	0.04	0.04	0.04	0.05	0.06	0.09	0.12	0.15	0.22

continued

TABLE G-1 Continued

Gender/Age Category[a]	N	Mean	Percentile								
			1st	5th	10th	25th	50th	75th	90th	95th	99th
F 9–13 y	200	1.55	0.78	0.97	1.08	1.28	1.52	1.79	2.05	2.21	2.54
SE			0.06	0.05	0.05	0.05	0.05	0.06	0.06	0.07	0.08	0.10
F 14–18 y	169	1.43	0.55	0.75	0.87	1.09	1.37	1.70	2.07	2.33	2.93
SE			0.07	0.06	0.06	0.06	0.06	0.06	0.08	0.11	0.14	0.23
F 19–30 y	302	1.27	0.52	0.80	0.80	0.99	1.22	1.49	1.78	1.99	2.45
SE			0.07	0.04	0.05	0.05	0.06	0.07	0.08	0.09	0.11	0.15
F 31–50 y	590	1.34	0.51	0.71	0.82	1.03	1.28	1.59	1.92	2.15	2.67
SE			0.03	0.03	0.03	0.03	0.03	0.03	0.04	0.06	0.07	0.12
F 51–70 y	510	1.28	0.59	0.75	0.85	1.02	1.24	1.49	1.77	1.97	2.40
SE			0.03	0.03	0.02	0.02	0.02	0.03	0.03	0.05	0.06	0.09
F 70+ y	221	1.21	0.53	0.68	0.78	0.95	1.18	1.43	1.69	1.86	2.20
SE			0.04	0.04	0.03	0.03	0.04	0.04	0.04	0.06	0.07	0.10
F pregnant	33	1.53	0.67	0.87	0.98	1.2	1.48	1.81	2.14	2.36	2.82
SE			0.39	0.47	0.42	0.42	0.39	0.38	0.40	0.46	0.51	0.65
F lactating	16	2.35	0.88	1.17	1.35	1.71	2.21	2.83	3.52	3.99	5.03
SE			0.52	0.64	0.58	0.55	0.50	0.48	0.55	0.73	0.89	1.31
All individuals	5,527	1.59	0.58	0.79	0.91	1.16	1.49	1.90	2.37	2.71	3.50
SE			0.02	0.01	0.01	0.01	0.01	0.02	0.03	0.04	0.05	0.09
All indiv +P/L	5,576	1.59	0.58	0.79	0.91	1.16	1.49	1.91	2.38	2.72	3.52
SE			0.02	0.01	0.01	0.01	0.01	0.02	0.03	0.04	0.05	0.10

[a] SE = standard error; All indiv +P/L = all individuals plus pregnant and lactating women.

SOURCE: Unpublished data on the usual intake distributions for selected B vitamins, K. Dodd and A. Carriquiry, Iowa State University, 1997. Based on transformation of CSFII data using the method of Nusser SM, Carriquiry AL, Dodd KW, Fuller WA. 1996. A semiparametric transformation approach to estimating usual daily intake distributions. J *Am Stat Assoc* 91:1440–1449.

TABLE G-2 Mean and Percentiles for Usual Intake of Riboflavin (mg), CSFII, 1994–1995

Gender/Age Category[a]	N	Mean	1st	5th	10th	25th	50th	75th	90th	95th	99th
0–6 mo	69	0.97	0.18	0.34	0.44	0.65	0.93	1.25	1.56	1.77	2.18
SE		0.07	0.06	0.07	0.07	0.08	0.08	0.08	0.08	0.08	0.14
7–11 mo	45	1.53	0.50	0.75	0.90	1.15	1.45	1.82	2.25	2.58	3.33
SE		0.17	0.11	0.10	0.10	0.10	0.14	0.21	0.32	0.41	0.64
1–3 y	702	1.66	0.73	0.96	1.10	1.33	1.62	1.94	2.27	2.50	2.98
SE		0.02	0.03	0.03	0.03	0.02	0.02	0.03	0.04	0.05	0.07
4–8 y	666	1.91	0.88	1.13	1.26	1.52	1.84	2.23	2.63	2.91	3.56
SE		0.04	0.05	0.04	0.04	0.04	0.04	0.05	0.07	0.09	0.16
M 9–13 y	180	2.45	0.93	1.26	1.47	1.85	2.35	2.95	3.56	3.97	4.81
SE		0.10	0.10	0.10	0.09	0.09	0.09	0.11	0.16	0.20	0.32
M 14–18 y	191	2.64	0.99	1.32	1.53	1.94	2.50	3.18	3.92	4.43	5.54
SE		0.11	0.06	0.07	0.07	0.08	0.10	0.14	0.19	0.23	0.32
M 19–30 y	328	2.46	1.03	1.32	1.50	1.86	2.33	2.92	3.56	4.00	4.96
SE		0.10	0.05	0.06	0.07	0.08	0.10	0.12	0.16	0.19	0.27
M 31–50 y	627	2.30	0.97	1.27	1.44	1.76	2.18	2.70	3.30	3.74	4.78
SE		0.06	0.04	0.04	0.04	0.05	0.05	0.08	0.12	0.16	0.30
M 51–70 y	490	2.04	0.85	1.09	1.24	1.54	1.94	2.44	2.98	3.35	4.18
SE		0.06	0.05	0.04	0.05	0.05	0.07	0.10	0.11	0.15	0.39
M 71+ y	237	2.05	0.83	1.09	1.26	1.56	1.97	2.45	2.96	3.30	4.02
SE		0.09	0.06	0.06	0.06	0.06	0.08	0.11	0.16	0.19	0.29
F 9–13 y	200	2.01	0.92	1.18	1.34	1.62	1.97	2.35	2.73	2.98	3.46
SE		0.08	0.08	0.07	0.07	0.07	0.08	0.09	0.11	0.13	0.17
F 14–18 y	169	1.75	0.61	0.84	0.99	1.28	1.66	2.12	2.61	2.94	3.64
SE		0.09	0.07	0.09	0.10	0.09	0.08	0.15	0.18	0.22	0.64
F 19–30 y	302	1.56	0.57	0.80	0.93	1.18	1.49	1.85	2.26	2.55	3.22
SE		0.10	0.05	0.06	0.07	0.08	0.10	0.12	0.15	0.18	0.28
F 31–50 y	590	1.58	0.64	0.85	0.96	1.18	1.50	1.89	2.30	2.58	3.26
SE		0.04	0.05	0.03	0.03	0.05	0.04	0.06	0.08	0.10	0.18
F 51–70 y	510	1.54	0.66	0.85	0.97	1.18	1.47	1.81	2.20	2.49	3.16
SE		0.04	0.03	0.03	0.03	0.03	0.04	0.05	0.08	0.11	0.20
F 70+ y	221	1.47	0.63	0.83	0.94	1.14	1.40	1.72	2.08	2.34	2.95
SE		0.05	0.04	0.04	0.04	0.04	0.05	0.06	0.09	0.13	0.26
F pregnant	33	1.77	0.51	0.83	1.01	1.35	1.74	2.16	2.55	2.79	3.26
SE		0.37	0.44	0.54	0.58	0.56	0.42	2.76	0.18	0.17	0.46
F lactating	16	2.88	0.93	1.38	1.66	2.17	2.81	3.50	4.19	4.62	5.47
SE		0.50	0.59	0.60	0.60	0.57	0.52	0.47	0.46	0.49	0.62

continued

TABLE G-2 Continued

Gender/Age Category[a]	N	Mean	1st	5th	10th	25th	50th	75th	90th	95th	99th
							Percentile				
All individuals	5,527	1.93	0.69	0.94	1.10	1.40	1.81	2.31	2.89	3.32	4.35
SE		0.03	0.01	0.01	0.01	0.02	0.02	0.03	0.05	0.07	0.11
All indiv +P/L	5,576	1.94	0.69	0.95	1.10	1.41	1.82	2.32	2.90	3.33	4.36
SE		0.03	0.01	0.01	0.01	0.02	0.02	0.03	0.05	0.06	0.11

[a] SE = standard error; All indiv +P/L = all individuals plus pregnant and lactating women.

SOURCE: Unpublished data on the usual intake distributions for selected B vitamins, K. Dodd and A. Carriquiry, Iowa State University, 1997. Based on transformation of CSFII data using the method of Nusser SM, Carriquiry AL, Dodd KW, Fuller WA. 1996. A semiparametric transformation approach to estimating usual daily intake distributions. *J Am Stat Assoc* 91:1440–1449.

TABLE G-3 Mean and Percentiles for Usual Intake of Niacin (mg), CSFII, 1994–1995

Gender/Age Category[a]	N	Mean	Percentile 1st	5th	10th	25th	50th	75th	90th	95th	99th
0–6 mo	69	8.23	1.60	2.90	3.90	5.70	7.90	10.40	13.10	14.80	17.80
SE		0.63	0.52	0.62	0.63	0.61	0.63	0.76	0.92	1.04	1.37
7–11 mo	45	11.02	4.20	5.90	7.00	8.80	10.90	13.10	15.20	16.50	18.90
SE		0.75	0.69	0.69	0.69	0.72	0.78	0.84	0.89	0.93	1.00
1–3 y	702	13.45	5.40	7.30	8.40	10.40	13.00	15.90	19.10	21.20	25.90
SE		0.30	0.26	0.27	0.28	0.29	0.31	0.35	0.42	0.50	0.75
4–8 y	666	17.60	8.20	10.50	11.90	14.20	17.10	20.40	23.90	26.20	31.40
SE		0.33	0.51	0.33	0.34	0.41	0.31	0.51	0.61	0.69	1.15
M 9–13 y	180	23.71	10.60	13.60	15.40	18.70	23.00	27.90	33.00	36.30	43.10
SE		0.81	1.06	1.05	1.04	0.95	0.83	0.86	1.01	1.13	1.70
M 14–18 y	191	28.90	12.00	15.90	18.20	22.30	27.60	34.00	41.20	46.50	58.80
SE		1.28	1.08	1.10	1.10	1.12	1.25	1.55	1.99	2.34	3.32
M 19–30 y	328	31.83	13.80	17.60	19.90	24.50	30.50	37.70	45.40	50.60	61.60
SE		1.29	0.80	0.85	0.90	1.04	1.26	1.56	1.95	2.28	3.14
M 31–50 y	627	29.39	13.80	17.50	19.60	23.40	28.30	34.10	40.50	45.10	55.30
SE		0.78	0.53	0.53	0.53	0.55	0.66	0.96	1.45	1.88	3.03
M 51–70 y	490	25.84	11.00	14.30	16.30	20.00	24.80	30.50	36.70	41.10	50.80
SE		0.87	0.51	0.53	0.55	0.63	0.80	1.08	1.49	1.83	2.80
M 71+ y	237	22.50	9.80	12.60	14.20	17.50	21.70	26.60	31.80	35.30	42.70
SE		0.84	0.49	0.53	0.56	0.63	0.77	1.06	1.54	1.95	3.00
F 9–13 y	200	18.53	8.90	11.20	12.60	15.10	18.20	21.60	25.00	27.20	31.60
SE		0.72	0.55	0.59	0.61	0.66	0.73	0.83	0.97	1.09	1.40
F 14–18 y	169	18.97	7.20	9.80	11.40	14.30	18.10	22.50	27.60	31.20	39.60
SE		1.05	0.76	0.80	0.82	0.84	0.93	1.25	1.83	2.34	3.76
F 19–30 y	302	18.17	7.10	9.50	11.00	13.80	17.50	21.80	26.20	29.10	35.20
SE		0.94	0.53	0.73	0.81	0.84	0.87	1.12	1.45	1.73	3.02
F 31–50 y	590	19.30	8.40	10.90	12.40	15.10	18.60	22.70	27.00	30.00	36.40
SE		0.55	0.57	0.51	0.44	0.47	0.61	0.66	0.89	1.07	1.43
F 51–70 y	510	18.38	8.80	11.00	12.30	14.70	17.80	21.50	25.20	27.60	32.70
SE		0.38	0.40	0.40	0.41	0.40	0.40	0.48	0.55	0.63	1.23
F 70+ y	221	17.30	7.50	9.70	11.10	13.60	16.80	20.50	24.20	26.60	31.70
SE		0.45	0.47	0.47	0.47	0.47	0.48	0.51	0.62	0.73	1.05
F pregnant	33	19.53	8.60	11.20	12.80	15.60	19.10	23.00	26.80	29.30	34.30
SE		6.97	6.74	8.28	8.97	9.31	7.86	5.30	3.95	3.81	4.60
F lactating	16	27.38	11.60	15.30	17.60	21.70	26.80	32.40	38.00	41.60	48.70
SE		4.09	3.76	3.98	4.08	4.21	4.27	4.33	4.65	5.10	6.63

continued

TABLE G-3 Continued

Gender/Age Category[a]	N	Mean	Percentile								
			1st	5th	10th	25th	50th	75th	90th	95th	99th
All individuals	5,527	22.23	7.50	10.50	12.30	15.90	20.80	26.90	33.90	38.90	50.90
SE		0.28	0.20	0.29	0.29	0.18	0.37	0.37	0.76	0.93	1.39
All indiv +P/L	5,576	22.24	7.50	10.50	12.30	15.90	20.80	26.90	33.90	38.90	50.80
SE		0.27	0.19	0.33	0.33	0.18	0.40	0.36	0.76	0.91	1.34

[a] SE = standard error; All indiv +P/L = all individuals plus pregnant and lactating women.

SOURCE: Unpublished data on the usual intake distributions for selected B vitamins, K. Dodd and A. Carriquiry, Iowa State University, 1997. Based on transformation of CSFII data using the method of Nusser SM, Carriquiry AL, Dodd KW, Fuller WA. 1996. A semiparametric transformation approach to estimating usual daily intake distributions. *J Am Stat Assoc* 91:1440–1449.

TABLE G-4 Mean and Percentiles for Usual Intake of Vitamin B_6 (mg), CSFII, 1994–1995

Gender/Age Category[a]	N	Mean	1st	5th	10th	25th	50th	75th	90th	95th	99th
0–6 mo	69	0.41	0.09	0.16	0.20	0.29	0.40	0.52	0.62	0.69	0.84
SE		0.04	0.03	0.04	0.04	0.04	0.03	0.05	0.07	0.10	0.15
7–11 mo	45	0.77	0.34	0.45	0.51	0.62	0.76	0.90	1.04	1.13	1.30
SE		0.07	0.05	0.05	0.05	0.06	0.06	0.08	0.09	0.11	0.14
1–3 y	702	1.31	0.58	0.74	0.84	1.02	1.27	1.55	1.84	2.03	2.44
SE		0.03	0.03	0.03	0.03	0.03	0.03	0.04	0.05	0.06	0.09
4–8 y	666	1.53	0.68	0.88	1.00	1.21	1.48	1.80	2.13	2.36	2.85
SE		0.04	0.03	0.04	0.04	0.04	0.03	0.04	0.06	0.08	0.19
M 9–13 y	180	2.04	0.80	1.07	1.24	1.55	1.96	2.45	2.94	3.27	3.96
SE		0.09	0.08	0.08	0.08	0.08	0.09	0.10	0.12	0.14	0.20
M 14–18 y	191	2.26	0.86	1.15	1.32	1.66	2.12	2.69	3.37	3.86	5.02
SE		0.11	0.07	0.07	0.08	0.08	0.10	0.16	0.22	0.25	0.41
M 19–30 y	328	2.43	0.94	1.25	1.44	1.81	2.31	2.91	3.57	4.01	4.98
SE		0.11	0.06	0.07	0.07	0.09	0.10	0.13	0.16	0.19	0.26
M 31–50 y	627	2.26	0.90	1.19	1.37	1.70	2.13	2.68	3.31	3.77	4.86
SE		0.07	0.04	0.04	0.04	0.05	0.06	0.08	0.13	0.18	0.32
M 51–70 y	490	2.04	0.80	1.06	1.23	1.54	1.95	2.45	2.97	3.32	4.06
SE		0.06	0.04	0.04	0.04	0.05	0.06	0.08	0.11	0.13	0.19
M 71+ y	237	1.99	0.77	1.01	1.17	1.48	1.89	2.39	2.93	3.29	4.08
SE		0.09	0.04	0.05	0.05	0.06	0.08	0.12	0.17	0.21	0.32
F 9–13 y	200	1.58	0.67	0.88	1.00	1.24	1.54	1.88	2.22	2.44	2.90
SE		0.07	0.05	0.06	0.06	0.07	0.07	0.09	0.11	0.12	0.17
F 14–18 y	169	1.50	0.54	0.75	0.87	1.10	1.41	1.80	2.25	2.57	3.31
SE		0.10	0.09	0.05	0.07	0.12	0.09	0.17	0.23	0.26	0.40
F 19–30 y	302	1.44	0.55	0.76	0.88	1.10	1.38	1.71	2.06	2.31	2.86
SE		0.08	0.04	0.05	0.05	0.06	0.07	0.11	0.14	0.17	0.25
F 31–50 y	590	1.52	0.57	0.78	0.91	1.15	1.43	1.81	2.20	2.47	3.07
SE		0.04	0.04	0.03	0.03	0.03	0.04	0.05	0.06	0.08	0.12
F 51–70 y	510	1.50	0.63	0.82	0.94	1.15	1.43	1.77	2.13	2.38	2.93
SE		0.04	0.03	0.03	0.03	0.03	0.04	0.05	0.07	0.08	0.12
F 70+ y	221	1.46	0.56	0.76	0.88	1.11	1.41	1.76	2.12	2.35	2.84
SE		0.05	0.04	0.04	0.05	0.05	0.05	0.05	0.07	0.08	0.12
F pregnant	33	1.61	0.58	0.81	0.95	1.22	1.56	1.95	2.34	2.59	3.11
SE		0.42	0.35	0.39	0.41	0.43	0.44	0.44	0.44	0.45	0.50
F lactating	16	2.56	0.98	1.29	1.49	1.88	2.41	3.08	3.81	4.31	5.41
SE		0.45	0.40	0.44	0.46	0.48	0.49	0.49	0.51	0.56	0.81

continued

TABLE G-4 Continued

Gender/Age Category[a]	N	Mean	Percentile								
			1st	5th	10th	25th	50th	75th	90th	95th	99th
All individuals	5,527	1.79	0.59	0.82	0.97	1.26	1.66	2.17	2.74	3.16	4.14
SE			0.02	0.01	0.01	0.01	0.01	0.02	0.03	0.05	0.06	0.11
All indiv +P/L	5,576	1.79	0.59	0.82	0.97	0.13	1.67	2.17	2.75	3.17	4.15
SE			0.02	0.01	0.01	0.01	0.01	0.02	0.03	0.05	0.06	0.11

[a] SE = standard error; All indiv +P/L = all individuals plus pregnant and lactating women.

SOURCE: Unpublished data on the usual intake distributions for selected B vitamins, K. Dodd and A. Carriquiry, Iowa State University, 1997. Based on transformation of CSFII data using the method of Nusser SM, Carriquiry AL, Dodd KW, Fuller WA. 1996. A semiparametric transformation approach to estimating usual daily intake distributions. *J Am Stat Assoc* 91:1440–1449.

TABLE G-5 Mean and Percentiles for Usual Intake of Folate (μg), CSFII, 1994–1995

Gender/Age Category[a]	N	Mean	1st	5th	10th	25th	50th	75th	90th	95th	99th
0–6 mo	69	80	16	29	38	56	78	101	122	137	167
SE		6	6	7	7	7	6	8	9	10	13
7–11 mo	45	129	51	69	79	99	125	154	183	202	241
SE		12	9	8	8	8	11	16	22	27	42
1–3 y	702	182	70	93	108	136	173	218	268	303	382
SE		5	4	4	5	5	5	6	8	12	37
4–8 y	666	232	95	124	141	176	221	276	335	375	461
SE		6	3	3	3	4	5	8	11	14	22
M 9–13 y	180	334	112	153	180	234	310	407	517	595	770
SE		17	10	11	12	13	16	21	31	40	65
M 14–18 y	191	321	100	140	165	217	293	392	512	600	811
SE		19	7	8	9	12	18	26	35	44	77
M 19–30 y	328	322	112	148	172	223	297	394	505	584	758
SE		18	6	6	7	10	18	24	29	36	76
M 31–50 y	627	309	106	143	167	214	281	372	484	569	783
SE		11	5	5	6	7	9	14	22	30	57
M 51–70 y	490	286	97	131	153	198	263	348	447	518	680
SE		10	6	6	6	7	8	12	18	23	37
M 71+ y	237	297	101	137	161	209	276	362	458	527	680
SE		15	6	6	7	9	13	20	32	42	68
F 9–13 y	200	254	90	123	143	184	240	308	383	435	547
SE		13	8	9	9	11	13	16	21	26	41
F 14–18 y	169	223	72	98	116	152	205	273	352	409	540
SE		16	7	6	7	8	12	19	35	57	145
F 19–30 y	302	214	71	100	118	153	200	259	326	374	488
SE		14	6	6	7	9	12	18	25	31	45
F 31–50 y	590	227	74	103	120	155	208	277	357	414	551
SE		8	4	4	5	5	7	10	16	22	41
F 51–70 y	510	224	83	111	129	163	211	270	335	379	476
SE		8	5	5	5	6	7	9	12	15	22
F 70+ y	221	224	75	105	124	161	212	273	338	383	477
SE		9	5	6	6	7	9	11	14	17	25
F pregnant	33	241	114	140	156	188	230	282	338	377	462
SE		29	16	15	15	17	23	37	57	73	113
F lactating	16	461	160	215	251	324	428	561	713	822	1,068
SE		102	41	51	57	72	95	128	172	206	294

continued

TABLE G-5 Continued

Gender/Age Category[a]	N	Mean	Percentile								
			1st	5th	10th	25th	50th	75th	90th	95th	99th
All individuals	5,527	259	80	111	131	173	235	317	414	486	663
SE		4	2	2	2	3	3	5	9	12	24
All indiv +P/L	5,576	259	80	111	131	174	236	318	415	488	667
SE		4	2	2	2	2	3	5	9	12	23

[a] SE = standard error; All indiv +P/L = all individuals plus pregnant and lactating women.

SOURCE: Unpublished data on the usual intake distributions for selected B vitamins, K. Dodd and A. Carriquiry, Iowa State University, 1997. Based on transformation of CSFII data using the method of Nusser SM, Carriquiry AL, Dodd KW, Fuller WA. 1996. A semiparametric transformation approach to estimating usual daily intake distributions. *J Am Stat Assoc* 91:1440–1449.

TABLE G-6 Mean and Percentiles for Usual Intake of Vitamin B_{12} (µg), CSFII, 1994–1995

Gender/Age Category[a]	N	Mean	Percentile								
			1st	5th	10th	25th	50th	75th	90th	95th	99th
0–6 mo	69	1.37	0.36	0.60	0.76	1.03	1.34	1.67	2.00	2.22	2.68
SE			0.11	0.10	0.12	0.12	0.12	0.11	0.12	0.17	0.21	0.31
7–11 mo	45	4.53	0.80	1.30	1.70	2.40	3.60	5.50	8.40	11.00	18.10
SE			2.27	0.16	0.21	0.29	0.62	1.36	2.82	5.25	7.50	14.10
1–3 y	702	3.36	1.44	1.85	2.08	2.52	3.11	3.92	4.95	5.73	7.63
SE			0.12	0.07	0.07	0.06	0.06	0.08	0.13	0.26	0.38	0.74
4–8 y	666	3.83	1.57	2.04	2.33	2.88	3.62	4.54	5.58	6.34	8.09
SE			0.14	0.09	0.10	0.10	0.11	0.13	0.17	0.26	0.34	0.59
M 9–13 y	180	6.23	2.00	2.90	3.40	4.40	5.90	7.50	9.40	10.90	14.50
SE			0.46	0.17	0.21	0.25	0.33	0.45	0.60	0.83	1.04	1.65
M 14–18 y	191	5.67	2.30	3.00	3.50	4.30	5.40	6.70	8.20	9.30	11.80
SE			0.25	0.18	0.20	0.21	0.22	0.24	0.31	0.43	0.55	0.86
M 19–30 y	328	5.99	2.20	2.90	3.40	4.30	5.60	7.30	9.00	10.30	13.10
SE			0.30	0.16	0.15	0.17	0.20	0.27	0.41	0.56	0.66	0.96
M 31–50 y	627	6.30	2.10	2.80	3.30	4.20	5.60	7.60	10.20	12.20	17.40
SE			0.50	0.15	0.16	0.17	0.22	0.40	0.68	1.06	1.43	2.71
M 51–70 y	490	6.92	1.80	2.40	2.80	3.60	5.20	8.10	12.80	17.10	30.10
SE			1.00	0.11	0.12	0.15	0.20	0.52	1.18	2.50	4.19	10.60
M 71+ y	237	6.25	1.80	2.40	2.80	3.70	5.10	7.40	10.80	13.80	22.90
SE			0.80	0.17	0.20	0.22	0.30	0.49	0.91	1.75	2.65	5.87
F 9–13 y	200	4.07	1.72	2.25	2.57	3.16	3.92	4.82	5.77	6.42	7.82
SE			0.21	0.14	0.16	0.17	0.19	0.22	0.25	0.28	0.31	0.40
F 14–18 y	169	4.18	1.30	1.90	2.30	2.90	3.90	5.10	6.50	7.50	9.90
SE			0.33	0.14	0.18	0.19	0.23	0.29	0.42	0.63	0.82	1.31
F 19–30 y	302	3.68	1.19	1.67	1.98	2.60	3.45	4.51	5.68	6.47	8.21
SE			0.32	0.13	0.19	0.21	0.25	0.29	0.35	0.47	0.60	1.11
F 31–50 y	590	4.24	1.27	1.74	2.02	2.55	3.41	4.88	7.29	9.51	16.09
SE			0.39	0.10	0.11	0.12	0.15	0.20	0.34	1.03	1.95	5.34
F 51–70 y	510	4.91	1.40	1.80	2.10	2.80	3.90	5.80	8.70	11.30	18.90
SE			0.65	0.09	0.09	0.11	0.24	0.56	0.92	1.48	2.21	5.38
F 70+ y	221	4.61	1.09	1.49	1.75	2.33	3.32	5.09	8.28	11.63	24.24
SE			0.84	0.08	0.10	0.10	0.12	0.27	0.71	1.89	3.47	11.00
F pregnant	33	3.83	1.40	1.94	2.27	2.90	3.70	4.62	5.55	6.16	7.39
SE			1.45	1.24	1.59	1.78	1.94	1.72	1.28	1.32	1.41	1.32
F lactating	16	6.63	1.90	2.80	3.40	4.60	6.30	8.20	10.30	11.70	14.60
SE			1.55	1.11	1.30	1.39	1.52	1.62	1.69	1.77	1.86	2.19

continued

TABLE G-6 Continued

Gender/Age Category[a]	N	Mean	Percentile								
			1st	5th	10th	25th	50th	75th	90th	95th	99th
All individuals	5,527	5.03	1.30	1.90	2.30	3.00	4.10	5.90	8.50	11.10	19.00
SE		0.16	0.04	0.04	0.04	0.05	0.09	0.16	0.37	0.68	2.01
All indiv +P/L	5,576	5.04	1.30	1.90	2.30	3.00	4.10	5.90	8.60	11.10	19.00
SE		0.16	0.04	0.04	0.05	0.06	0.09	0.16	0.37	0.66	1.97

[a] SE = standard error; All indiv +P/L = all individuals plus pregnant and lactating women.

SOURCE: Unpublished data on the usual intake distributions for selected B vitamins, K. Dodd and A. Carriquiry, Iowa State University, 1997. Based on transformation of CSFII data using the method of Nusser SM, Carriquiry AL, Dodd KW, Fuller WA. 1996. A semiparametric transformation approach to estimating usual daily intake distributions. *J Am Stat Assoc* 91:1440–1449.

H

Dietary Intake Data from the Third National Health and Nutrition Examination Survey (NHANES III), 1988–1994

TABLE H-1 Dietary Thiamin Intake (mg) of Persons Aged 6 Years and Older: Mean and Selected Percentiles, United States, 1988–1994

Gender and Age (y)	Number of Examined Persons	Mean	Standard Error of the Mean
Both, 6–8	1,506	1.60	0.01
M, 9–13	1,219	1.84	0.03
M, 14–18	908	2.09	0.04
M, 19–30	1,902	1.99	0.03
M, 31–50	2,533	1.95	0.02
M, 51–70	1,942	1.87	0.03
M, >70	1,255	1.72	0.02
F, 9–13	1,238	1.62	0.02
F, 14–18	980	1.63	0.03
F, 19–30	1,972	1.56	0.02
F, 31–50	2,988	1.52	0.02
F, 51–70	2,076	1.50	0.02
F, >70	1,368	1.49	0.01
Pregnant F, 14–55	346	1.81	0.05
Both, ≥6, unadjusted	22,233	1.72	0.01

NOTE: Data from one 24-h dietary recall have been adjusted by following methods in National Research Council. 1986. *Nutrient Adequacy. Assessment Using Food Consumption Surveys.* Washington, DC: National Academy Press; and Feinleib M, Rifkind B, Sempos C, Johnson C, Bachorik P, Lippel K, Carroll M, Ingster-Moore L, Murphy R. 1993. Methodological issues in the measurement of cardiovascular risk factors: Within-person

Selected Percentiles

5th	10th	15th	25th	50th	75th	85th	90th	95th
1.13	1.20	1.26	1.35	1.55	1.79	1.96	2.08	2.27
1.15	1.24	1.31	1.42	1.66	1.97	2.19	2.35	2.60
1.20	1.32	1.37	1.52	1.86	2.33	2.60	2.85	3.42
1.07	1.18	1.30	1.44	1.78	2.27	2.57	2.88	3.41
1.06	1.21	1.29	1.43	1.75	2.19	2.49	2.71	3.21
1.01	1.13	1.20	1.33	1.63	2.03	2.29	2.50	2.90
1.03	1.13	1.19	1.31	1.56	1.92	2.15	2.39	2.68
1.09	1.18	1.24	1.33	1.54	1.79	1.94	2.07	2.38
1.09	1.15	1.21	1.31	1.55	1.84	2.01	2.16	2.49
0.94	1.02	1.09	1.21	1.45	1.76	1.99	2.20	2.49
0.92	1.02	1.09	1.19	1.41	1.72	1.92	2.08	2.34
0.93	1.01	1.06	1.17	1.38	1.65	1.82	1.96	2.23
0.94	1.02	1.09	1.18	1.38	1.65	1.81	1.98	2.21
1.07	1.18	1.25	1.35	1.68	2.04	2.26	2.40	2.96
1.00	1.10	1.18	1.29	1.56	1.90	2.15	2.36	2.73

variability in selected serum lipid measures—results from the Third National Health and Nutrition Survey (NHANES III). *Can J Cardiol* 9:87D–88D.

SOURCE: NHANES III, 1988–1994, unpublished data, C.L. Johnson and J.D. Wright, National Center for Health Statistics, Centers for Disease Control and Prevention, 1997.

TABLE H-2 Total Thiamin Intake (mg) of Persons Aged 6 Years and Older: Mean and Selected Percentiles, United States, 1988–1994

Gender and Age (y)	Number of Examined Persons	Mean	Standard Error of the Mean
Both, 6–8	1,601	1.88	0.05
M, 9–13	1,262	2.55	0.24
M, 14–18	942	3.42	0.57
M, 19–30	1,962	4.25	0.54
M, 31–50	2,612	6.67	1.23
M, 51–70	2,033	4.60	0.47
M, >70	1,329	3.87	0.45
F, 9–13	1,282	2.12	0.16
F, 14–18	1,001	1.94	0.07
F, 19–30	2,027	3.83	0.51
F, 31–50	3,073	4.05	0.27
F, 51–70	2,148	4.60	0.49
F, >70	1,444	5.04	0.96
Pregnant F, 14–55	354	5.64	1.38
Both, ≥6, unadjusted	23,070	4.3	0.26

NOTE: Data from one 24-h dietary recall have been adjusted by following methods in National Research Council. 1986. *Nutrient Adequacy. Assessment Using Food Consumption Surveys.* Washington, DC: National Academy Press; and Feinleib M, Rifkind B, Sempos C, Johnson C, Bachorik P, Lippel K, Carroll M, Ingster-Moore L, Murphy R. 1993. Methodological issues in the measurement of cardiovascular risk factors: Within-person

Selected Percentiles

5th	10th	15th	25th	50th	75th	85th	90th	95th
0.99	1.15	1.24	1.37	1.65	2.15	2.56	2.78	3.17
1.08	1.20	1.30	1.43	1.72	2.20	2.56	2.88	3.40
1.04	1.25	1.35	1.50	1.91	2.43	2.83	3.29	4.05
0.97	1.15	1.29	1.46	1.87	2.61	3.24	3.84	6.09
0.99	1.18	1.31	1.48	1.89	2.68	3.30	3.84	6.81
0.86	1.09	1.20	1.36	1.78	2.62	3.26	3.90	7.09
0.64	1.06	1.16	1.31	1.69	2.54	3.19	3.87	5.72
1.02	1.15	1.24	1.34	1.60	1.99	2.49	2.86	3.27
1.05	1.14	1.21	1.32	1.60	1.98	2.37	2.71	3.33
0.89	1.02	1.09	1.24	1.56	2.26	2.87	3.39	5.01
0.89	1.02	1.11	1.24	1.58	2.39	3.07	3.96	8.64
0.90	1.01	1.09	1.22	1.54	2.59	3.10	3.89	10.76
0.82	0.98	1.08	1.21	1.53	2.64	3.24	4.14	10.98
1.06	1.27	1.34	1.56	2.86	4.32	4.79	5.12	6.25

variability in selected serum lipid measures—results from the Third National Health and Nutrition Survey (NHANES III). *Can J Cardiol* 9:87D–88D. Includes estimated intake from supplements.

SOURCE: NHANES III, 1988–1994, unpublished data, C.L. Johnson and J.D. Wright, National Center for Health Statistics, Centers for Disease Control and Prevention, 1997.

TABLE H-3   Dietary Riboflavin Intake (mg) of Persons
Aged 6 Years and Older: Mean and Selected Percentiles,
United States, 1988–1994

Gender and Age (y)	Number of Examined Persons	Mean	Standard Error of the Mean
Both, 6–8	1,506	2.06	0.02
M, 9–13	1,219	2.28	0.03
M, 14–18	908	2.52	0.07
M, 19–30	1,902	2.33	0.03
M, 31–50	2,533	2.27	0.03
M, 51–70	1,942	2.18	0.03
M, >70	1,255	2.03	0.03
F, 9–13	1,238	1.98	0.03
F, 14–18	980	1.85	0.04
F, 19–30	1,972	1.78	0.02
F, 31–50	2,988	1.75	0.02
F, 51–70	2,076	1.74	0.02
F, >70	1,368	1.74	0.02
Pregnant F, 14–55	346	2.19	0.08
Both, ≥6, unadjusted	22,233	2.03	0.01

NOTE: Data from one 24-h dietary recall have been adjusted by following methods in National Research Council. 1986. *Nutrient Adequacy. Assessment Using Food Consumption Surveys.* Washington, DC: National Academy Press; and Feinleib M, Rifkind B, Sempos C, Johnson C, Bachorik P, Lippel K, Carroll M, Ingster-Moore L, Murphy R. 1993. Methodological issues in the measurement of cardiovascular risk factors: Within-person

Selected Percentiles

5th	10th	15th	25th	50th	75th	85th	90th	95th
1.34	1.47	1.56	1.69	1.99	2.32	2.50	2.65	2.84
1.29	1.48	1.56	1.70	2.07	2.50	2.77	2.97	3.33
1.19	1.35	1.45	1.68	2.20	2.82	3.31	3.71	4.32
1.18	1.33	1.44	1.62	2.09	2.66	3.02	3.34	3.90
1.20	1.33	1.44	1.61	2.03	2.54	2.83	3.09	3.60
1.13	1.25	1.35	1.52	1.87	2.35	2.69	2.96	3.42
1.13	1.23	1.34	1.49	1.84	2.31	2.60	2.85	3.28
1.19	1.33	1.41	1.56	1.84	2.26	2.48	2.62	2.94
1.02	1.12	1.20	1.35	1.72	2.14	2.41	2.61	3.03
0.99	1.11	1.18	1.31	1.63	2.01	2.29	2.51	2.85
0.99	1.11	1.19	1.32	1.60	1.96	2.20	2.43	2.75
0.99	1.09	1.17	1.28	1.58	1.92	2.16	2.35	2.76
1.01	1.11	1.19	1.32	1.60	1.95	2.21	2.39	2.71
1.15	1.26	1.35	1.52	1.94	2.41	2.79	3.30	3.71
1.08	1.20	1.30	1.45	1.81	2.27	2.58	2.81	3.24

variability in selected serum lipid measures—results from the Third National Health and Nutrition Survey (NHANES III). *Can J Cardiol* 9:87D–88D.

SOURCE: NHANES III, 1988–1994, unpublished data, C.L. Johnson and J.D. Wright, National Center for Health Statistics, Centers for Disease Control and Prevention, 1997.

TABLE H-4   Total Riboflavin Intake (mg) of Persons Aged 6 Years and Older: Mean and Selected Percentiles, United States, 1988–1994

Gender and Age (y)	Number of Examined Persons	Mean	Standard Error of the Mean
Both, 6–8	1,601	2.35	0.05
M, 9–13	1,262	3.01	0.24
M, 14–18	942	3.83	0.57
M, 19–30	1,962	4.60	0.54
M, 31–50	2,612	6.93	1.23
M, 51–70	2,033	4.42	0.40
M, >70	1,329	3.61	0.36
F, 9–13	1,282	2.44	0.15
F, 14–18	1,001	2.18	0.07
F, 19–30	2,027	3.98	0.50
F, 31–50	3,073	4.24	0.27
F, 51–70	2,148	4.66	0.47
F, >70	1,444	4.70	0.68
Pregnant F, 14–55	354	6.20	1.38
Both, ≥6, unadjusted	23,070	4.5	0.26

NOTE: Data from one 24-h dietary recall have been adjusted by following methods in National Research Council. 1986. *Nutrient Adequacy. Assessment Using Food Consumption Surveys.* Washington, DC: National Academy Press; and Feinleib M, Rifkind B, Sempos C, Johnson C, Bachorik P, Lippel K, Carroll M, Ingster-Moore L, Murphy R. 1993. Methodological issues in the measurement of cardiovascular risk factors: Within-person

Selected Percentiles								
5th	10th	15th	25th	50th	75th	85th	90th	95th
1.12	1.36	1.53	1.71	2.11	2.68	3.14	3.43	3.88
1.18	1.41	1.55	1.71	2.15	2.73	3.19	3.55	4.07
1.03	1.30	1.42	1.66	2.25	2.99	3.64	4.09	5.19
1.08	1.29	1.42	1.64	2.21	3.02	3.73	4.49	6.54
1.12	1.32	1.45	1.66	2.17	3.02	3.81	4.49	7.10
0.95	1.21	1.34	1.54	2.03	2.98	3.74	4.55	6.62
0.68	1.16	1.30	1.49	1.97	2.96	3.67	4.40	6.12
1.07	1.30	1.40	1.57	1.92	2.51	3.04	3.42	3.92
0.98	1.10	1.20	1.36	1.76	2.34	2.77	3.14	3.87
0.95	1.09	1.19	1.35	1.75	2.55	3.30	3.94	5.56
0.96	1.12	1.21	1.37	1.77	2.75	3.46	4.36	8.56
0.96	1.09	1.18	1.35	1.77	2.96	3.55	4.50	8.29
0.86	1.07	1.17	1.37	1.77	3.02	3.72	4.65	10.93
1.18	1.37	1.48	1.84	3.21	5.07	5.51	6.02	7.19

variability in selected serum lipid measures—results from the Third National Health and Nutrition Survey (NHANES III). *Can J Cardiol* 9:87D–88D. Includes estimated intake from supplements.

SOURCE: NHANES III, 1988–1994, unpublished data, C.L. Johnson and J.D. Wright, National Center for Health Statistics, Centers for Disease Control and Prevention, 1997.

TABLE H-5   Dietary Niacin Intake (mg) of Persons
Aged 6 Years and Older: Mean and Selected Percentiles,
United States, 1988–1994

Gender and Age (y)	Number of Examined Persons	Mean	Standard Error of the Mean
Both, 6–8	1,506	19.28	0.14
M, 9–13	1,219	22.32	0.29
M, 14–18	908	27.01	0.55
M, 19–30	1,902	28.48	0.43
M, 31–50	2,533	27.53	0.26
M, 51–70	1,942	26.03	0.29
M, >70	1,255	23.20	0.27
F, 9–13	1,238	19.84	0.20
F, 14–18	980	20.08	0.40
F, 19–30	1,972	21.48	0.26
F, 31–50	2,988	21.62	0.24
F, 51–70	2,076	20.97	0.21
F, >70	1,368	20.28	0.18
Pregnant F, 14–55	346	23.47	0.71
Both, ≥6, unadjusted	22,233	23.56	0.13

NOTE: Data from one 24-h dietary recall have been adjusted by following methods in National Research Council. 1986. *Nutrient Adequacy. Assessment Using Food Consumption Surveys.* Washington, DC: National Academy Press; and Feinleib M, Rifkind B, Sempos C, Johnson C, Bachorik P, Lippel K, Carroll M, Ingster-Moore L, Murphy R. 1993. Methodological issues in the measurement of cardiovascular risk factors: Within-person

Selected Percentiles

5th	10th	15th	25th	50th	75th	85th	90th	95th
14.81	15.33	15.93	16.81	18.62	20.90	22.31	23.35	25.27
14.57	15.63	16.33	17.41	19.98	23.39	25.86	28.12	31.85
13.79	15.75	17.01	18.75	23.68	30.72	34.63	39.18	45.24
15.00	16.95	18.23	20.42	25.30	32.30	36.46	39.75	45.60
15.79	17.37	18.59	20.73	25.12	31.09	34.99	37.68	43.16
14.39	15.67	16.70	18.53	22.56	27.86	31.28	33.90	39.57
13.84	15.06	15.85	17.31	20.79	25.48	29.02	31.64	35.67
13.79	14.68	15.33	16.37	18.70	21.56	23.59	25.19	28.19
11.93	13.41	14.30	15.63	18.81	23.20	26.55	28.63	32.64
13.23	14.20	15.00	16.46	19.69	23.96	26.89	28.96	33.56
13.23	14.57	15.42	16.89	19.84	23.96	26.52	28.59	32.13
13.04	14.08	14.81	16.06	19.20	22.68	25.36	27.07	30.36
12.74	13.84	14.57	15.82	18.78	22.43	25.06	26.76	30.30
13.84	14.99	16.28	18.11	21.55	26.15	29.75	32.13	36.40
13.71	14.87	15.79	17.29	20.62	25.60	29.14	32.01	36.78

variability in selected serum lipid measures—results from the Third National Health and Nutrition Survey (NHANES III). *Can J Cardiol* 9:87D–88D.

SOURCE: NHANES III, 1988–1994, unpublished data, C.L. Johnson and J.D. Wright, National Center for Health Statistics, Centers for Disease Control and Prevention, 1997.

TABLE H-6   Total Niacin Intake (mg) of Persons
Aged 6 Years and Older: Mean and Selected Percentiles,
United States, 1988–1994

Gender and Age (y)	Number of Examined Persons	Mean	Standard Error of the Mean
Both, 6–8	1,601	22.78	0.56
M, 9–13	1,262	25.32	0.61
M, 14–18	942	29.17	1.01
M, 19–30	1,962	35.42	1.42
M, 31–50	2,612	39.42	1.66
M, 51–70	2,033	36.85	1.30
M, >70	1,329	35.59	2.64
F, 9–13	1,282	23.32	0.48
F, 14–18	1,001	22.55	0.64
F, 19–30	2,027	29.27	1.07
F, 31–50	3,073	34.24	1.37
F, 51–70	2,148	37.05	1.86
F, >70	1,444	39.14	6.56
Pregnant F, 14–55	354	38.68	2.13
Both, ≥6, unadjusted	23,070	33.5	0.64

NOTE: Data from one 24-h dietary recall have been adjusted by following methods in National Research Council. 1986. *Nutrient Adequacy. Assessment Using Food Consumption Surveys.* Washington, DC: National Academy Press; and Feinleib M, Rifkind B, Sempos C, Johnson C, Bachorik P, Lippel K, Carroll M, Ingster-Moore L, Murphy R. 1993. Methodological issues in the measurement of cardiovascular risk factors: Within-person

Selected Percentiles

5th	10th	15th	25th	50th	75th	85th	90th	95th
13.32	14.93	15.65	16.93	19.62	24.84	30.86	33.74	38.15
13.45	15.13	16.13	17.46	20.42	26.15	31.53	34.89	40.38
12.15	14.61	16.51	18.59	24.02	32.61	38.30	43.73	51.87
13.59	16.40	18.11	20.54	26.52	36.46	43.45	50.48	68.81
14.26	17.19	18.65	21.16	27.37	36.74	44.55	50.59	69.87
12.68	15.24	16.64	18.90	24.26	35.15	43.35	49.43	69.59
8.55	14.08	15.42	17.43	22.43	33.29	40.77	46.21	58.10
12.78	14.29	15.25	16.49	19.30	24.59	30.27	34.38	39.85
11.58	13.29	14.36	15.75	19.35	25.22	30.48	34.76	42.62
12.56	14.17	15.18	16.95	21.09	29.26	38.09	42.25	51.99
12.68	14.63	15.73	17.50	21.70	31.81	40.10	46.44	67.81
12.50	14.08	15.00	16.70	21.15	34.80	41.56	47.78	77.28
11.28	13.29	14.45	16.28	20.54	34.47	40.40	46.50	71.23
14.20	16.70	18.23	20.67	33.03	41.93	47.23	52.30	60.92

variability in selected serum lipid measures—results from the Third National Health and Nutrition Survey (NHANES III). *Can J Cardiol* 9:87D–88D. Includes estimated intake from supplements.

SOURCE: NHANES III, 1988–1994, unpublished data, C.L. Johnson and J.D. Wright, National Center for Health Statistics, Centers for Disease Control and Prevention, 1997.

TABLE H-7   Dietary Vitamin B_6 Intake (mg) of Persons Aged 6 Years and Older: Mean and Selected Percentiles, United States, 1988–1994

Gender and Age (y)	Number of Examined Persons	Mean	Standard Error of the Mean
Both, 6–8	1,506	1.63	0.01
M, 9–13	1,219	1.81	0.02
M, 14–18	908	2.18	0.05
M, 19–30	1,902	2.21	0.03
M, 31–50	2,533	2.15	0.02
M, 51–70	1,942	2.10	0.03
M, >70	1,255	1.94	0.03
F, 9–13	1,238	1.65	0.02
F, 14–18	980	1.58	0.04
F, 19–30	1,972	1.66	0.02
F, 31–50	2,988	1.66	0.02
F, 51–70	2,076	1.69	0.02
F, >70	1,368	1.72	0.02
Pregnant F, 14–55	346	1.93	0.07
Both, ≥6, unadjusted	22,233	1.87	0.01

NOTE: Data from one 24-h dietary recall have been adjusted by following methods in National Research Council. 1986. *Nutrient Adequacy. Assessment Using Food Consumption Surveys.* Washington, DC: National Academy Press; and Feinleib M, Rifkind B, Sempos C, Johnson C, Bachorik P, Lippel K, Carroll M, Ingster-Moore L, Murphy R. 1993. Methodological issues in the measurement of cardiovascular risk factors: Within-person

Selected Percentiles

5th	10th	15th	25th	50th	75th	85th	90th	95th
1.22	1.27	1.32	1.41	1.58	1.79	1.94	2.02	2.18
1.15	1.27	1.34	1.43	1.63	1.92	2.13	2.30	2.65
1.01	1.13	1.27	1.47	1.86	2.43	2.87	3.20	3.71
1.16	1.30	1.41	1.60	2.02	2.57	2.97	3.31	3.91
1.14	1.29	1.41	1.57	1.96	2.48	2.82	3.07	3.54
1.03	1.15	1.24	1.40	1.77	2.32	2.60	2.92	3.31
1.02	1.12	1.20	1.35	1.72	2.23	2.50	2.78	3.22
1.06	1.18	1.24	1.33	1.56	1.82	1.98	2.14	2.44
0.88	0.98	1.05	1.19	1.49	1.86	2.14	2.37	2.70
0.93	1.04	1.12	1.24	1.54	1.93	2.21	2.44	2.77
0.94	1.05	1.13	1.25	1.53	1.89	2.15	2.34	2.72
0.92	1.02	1.11	1.24	1.51	1.90	2.13	2.31	2.65
0.92	1.04	1.13	1.24	1.53	1.93	2.20	2.43	2.76
1.03	1.23	1.31	1.43	1.76	2.25	2.54	2.88	3.63
1.00	1.13	1.21	1.35	1.65	2.08	2.40	2.66	3.09

variability in selected serum lipid measures—results from the Third National Health and Nutrition Survey (NHANES III). *Can J Cardiol* 9:87D–88D.

SOURCE: NHANES III, 1988–1994, unpublished data, C.L. Johnson and J.D. Wright, National Center for Health Statistics, Centers for Disease Control and Prevention, 1997.

TABLE H-8   Total Vitamin B$_6$ Intake (mg) of Persons Aged 6 Years and Older: Mean and Selected Percentiles, United States, 1988–1994

Gender and Age (y)	Number of Examined Persons	Mean	Standard Error of the Mean
Both, 6–8	1,601	2.89	0.95
M, 9–13	1,262	2.54	0.24
M, 14–18	942	3.51	0.57
M, 19–30	1,962	4.52	0.54
M, 31–50	2,612	6.90	1.22
M, 51–70	2,033	5.20	0.68
M, >70	1,329	3.82	0.40
F, 9–13	1,282	2.17	0.15
F, 14–18	1,001	2.07	0.13
F, 19–30	2,027	4.25	0.52
F, 31–50	3,073	4.91	0.38
F, 51–70	2,148	5.20	0.54
F, >70	1,444	4.68	0.59
Pregnant F, 14–55	354	8.77	1.45
Both, ≥6, unadjusted	23,070	4.7	0.25

NOTE: Data from one 24-h dietary recall have been adjusted by following methods in National Research Council. 1986. *Nutrient Adequacy. Assessment Using Food Consumption Surveys.* Washington, DC: National Academy Press; and Feinleib M, Rifkind B, Sempos C, Johnson C, Bachorik P, Lippel K, Carroll M, Ingster-Moore L, Murphy R. 1993. Methodological issues in the measurement of cardiovascular risk factors: Within-person

Selected Percentiles

5th	10th	15th	25th	50th	75th	85th	90th	95th
1.04	1.23	1.31	1.43	1.67	2.15	2.59	2.80	3.44
1.04	1.23	1.32	1.43	1.69	2.14	2.59	2.90	3.59
0.88	1.08	1.22	1.45	1.89	2.63	3.20	3.63	4.98
1.05	1.26	1.40	1.62	2.13	2.98	3.85	4.51	6.81
1.06	1.28	1.42	1.63	2.16	3.07	3.84	4.61	6.75
0.93	1.13	1.23	1.42	1.97	3.03	3.95	4.68	6.66
0.68	1.05	1.16	1.36	1.87	2.92	3.91	4.41	6.25
0.92	1.14	1.23	1.35	1.62	2.06	2.56	2.96	3.62
0.84	0.97	1.06	1.19	1.53	2.07	2.55	2.94	3.95
0.90	1.03	1.13	1.30	1.68	2.51	3.53	4.29	7.80
0.92	1.06	1.16	1.32	1.73	2.80	3.72	4.74	10.88
0.89	1.03	1.14	1.29	1.75	3.17	4.01	4.79	7.50
0.81	1.00	1.12	1.29	1.77	3.20	4.10	4.74	7.23
1.03	1.31	1.41	1.67	3.72	11.25	11.92	12.32	21.01

variability in selected serum lipid measures—results from the Third National Health and Nutrition Survey (NHANES III). *Can J Cardiol* 9:87D–88D. Includes estimated intake from supplements.

SOURCE: NHANES III, 1988–1994, unpublished data, C.L. Johnson and J.D. Wright, National Center for Health Statistics, Centers for Disease Control and Prevention, 1997.

TABLE H-9   Dietary Folate Intake (µg) of Persons
Aged 6 Years and Older: Mean and Selected Percentiles,
United States, 1988–1994

Gender and Age (y)	Number of Examined Persons	Mean	Standard Error of the Mean
Both, 6–8	1,506	258	2.8
M, 9–13	1,219	278	4.0
M, 14–18	908	306	7.0
M, 19–30	1,902	313	5.1
M, 31–50	2,533	317	4.1
M, 51–70	1,942	322	4.9
M, >70	1,255	302	5.4
F, 9–13	1,238	253	3.7
F, 14–18	980	239	4.4
F, 19–30	1,972	254	3.6
F, 31–50	2,988	255	3.0
F, 51–70	2,076	269	3.6
F, >70	1,368	275	4.2
Pregnant F, 14–55	346	288	9.2
Both, ≥6, unadjusted	22,233	283	1.2

NOTE: Data from one 24-h dietary recall have been adjusted by following methods in National Research Council. 1986. *Nutrient Adequacy. Assessment Using Food Consumption Surveys.* Washington, DC: National Academy Press; and Feinleib M, Rifkind B, Sempos C, Johnson C, Bachorik P, Lippel K, Carroll M, Ingster-Moore L, Murphy R. 1993. Methodological issues in the measurement of cardiovascular risk factors: Within-person

Selected Percentiles								
5th	10th	15th	25th	50th	75th	85th	90th	95th
169	178	188	204	244	295	323	351	400
170	181	190	204	255	322	368	399	471
157	171	181	204	274	368	429	477	551
163	183	194	219	277	366	430	483	564
166	185	199	225	282	365	436	477	552
165	183	197	222	285	366	437	505	625
163	178	192	213	269	348	405	443	542
156	169	176	191	234	285	330	350	416
140	150	159	170	212	283	323	348	420
145	157	166	183	223	289	338	386	497
147	160	170	189	226	287	334	374	438
152	165	175	196	246	305	357	398	454
152	166	178	200	252	322	368	410	474
155	171	181	205	258	336	381	431	543
154	168	180	200	252	326	381	429	513

variability in selected serum lipid measures—results from the Third National Health and Nutrition Survey (NHANES III). *Can J Cardiol* 9:87D–88D.

SOURCE: NHANES III, 1988–1994, unpublished data, C.L. Johnson and J.D. Wright, National Center for Health Statistics, Centers for Disease Control and Prevention, 1997.

TABLE H-10   Total Folate Intake (µg) of Persons
Aged 6 Years and Older: Mean and Selected Percentiles,
United States, 1988–1994

Gender and Age (y)	Number of Examined Persons	Mean	Standard Error of the Mean
Both, 6–8	1,601	336	6.6
M, 9–13	1,262	337	7.8
M, 14–18	942	330	9.1
M, 19–30	1,962	407	15.1
M, 31–50	2,612	410	10.1
M, 51–70	2,033	429	11.0
M, >70	1,329	387	9.3
F, 9–13	1,282	316	7.0
F, 14–18	1,001	279	7.6
F, 19–30	2,027	369	14.5
F, 31–50	3,073	406	9.7
F, 51–70	2,148	413	10.6
F, >70	1,444	384	12.2
Pregnant F, 14–55	354	858	40.9
Both, ≥6, unadjusted	23,070	393	3.4

NOTE: Total intake estimated as intake from food and from supplements. Data from one 24-h dietary recall have been adjusted by following methods in National Research Council. 1986. *Nutrient Adequacy. Assessment Using Food Consumption Surveys.* Washington, DC: National Academy Press; and Feinleib M, Rifkind B, Sempos C, Johnson C, Bachorik P, Lippel K, Carroll M, Ingster-Moore L, Murphy R. 1993. Methodological issues in the

Selected Percentiles

5th	10th	15th	25th	50th	75th	85th	90th	95th
151	176	191	215	282	460	542	602	673
163	181	192	215	290	415	514	584	674
140	166	177	204	285	399	479	573	730
147	178	191	223	299	471	621	701	893
157	185	203	235	317	496	647	728	922
152	180	199	231	330	590	689	759	939
147	175	193	224	310	552	657	700	805
148	169	178	197	258	394	530	589	669
135	148	158	172	224	321	402	507	625
144	158	170	194	256	453	606	680	909
147	163	178	199	272	554	660	748	983
146	166	181	212	298	584	660	723	849
143	165	179	209	291	561	653	701	792
151	194	226	298	966	1,238	1,306	1,366	1,671
146	168	184	210	292	500	635	708	886

measurement of cardiovascular risk factors: Within-person variability in selected serum lipid measures—results from the Third National Health and Nutrition Survey (NHANES III). *Can J Cardiol* 9:87D–88D. Includes estimated intake from supplements.

SOURCE: NHANES III, 1988–1994, unpublished data, C.L. Johnson and J.D. Wright, National Center for Health Statistics, Centers for Disease Control and Prevention, 1997.

TABLE H-11   Dietary Vitamin B$_{12}$ Intake (µg) of Persons Aged 6 Years and Older: Mean and Selected Percentiles, United States, 1988–1994

Gender and Age (y)	Number of Examined Persons	Mean	Standard Error of the Mean
Both, 6–8	1,506	4.09	0.07
M, 9–13	1,219	5.03	0.16
M, 14–18	909	6.28	0.25
M, 19–30	1,902	5.65	0.08
M, 31–50	2,533	5.58	0.06
M, 51–70	1,942	5.43	0.04
M, >70	1,255	5.30	0.06
F, 9–13	1,238	4.30	0.11
F, 14–18	980	4.43	0.11
F, 19–30	1,972	5.00	0.05
F, 31–50	2,988	5.09	0.03
F, 51–70	2,076	5.05	0.05
F, >70	1,368	4.96	0.05
Pregnant F, 14–55	346	5.30	0.18
Both, ≥6, unadjusted	22,234	5.18	0.02

NOTE: Data from one 24-h dietary recall have been adjusted by following methods in National Research Council. 1986. *Nutrient Adequacy. Assessment Using Food Consumption Surveys.* Washington, DC: National Academy Press; and Feinleib M, Rifkind B, Sempos C, Johnson C, Bachorik P, Lippel K, Carroll M, Ingster-Moore L, Murphy R. 1993. Methodological issues in the measurement of cardiovascular risk factors: Within-person

Selected Percentiles								
5th	10th	15th	25th	50th	75th	85th	90th	95th
2.57	2.83	2.92	3.17	3.78	4.65	5.09	5.81	6.48
2.78	3.01	3.19	3.60	4.45	5.45	6.22	6.94	8.23
3.10	3.37	3.63	4.05	5.14	6.78	8.17	8.81	12.36
4.42	4.55	4.66	4.81	5.22	5.82	6.28	6.70	7.56
4.48	4.59	4.68	4.81	5.20	5.76	6.16	6.55	7.37
4.46	4.55	4.62	4.77	5.10	5.65	6.09	6.41	7.26
4.45	4.52	4.60	4.69	4.99	5.38	5.81	6.09	6.81
2.49	2.71	2.87	3.14	3.87	4.84	5.57	6.12	7.07
2.63	2.74	2.95	3.26	4.06	4.97	5.51	6.13	7.12
4.27	4.36	4.43	4.53	4.77	5.17	5.46	5.70	6.23
4.38	4.45	4.50	4.60	4.84	5.19	5.44	5.66	6.18
4.36	4.44	4.49	4.58	4.80	5.14	5.39	5.59	6.10
4.37	4.42	4.45	4.53	4.74	5.05	5.31	5.54	5.99
4.27	4.44	4.51	4.70	5.05	5.48	5.76	6.24	7.01
3.32	4.04	4.38	4.55	4.89	5.41	5.82	6.22	7.15

variability in selected serum lipid measures—results from the Third National Health and Nutrition Survey (NHANES III). *Can J Cardiol* 9:87D–88D.

SOURCE: NHANES III, 1988–1994, unpublished data, C.L. Johnson and J.D. Wright, National Center for Health Statistics, Centers for Disease Control and Prevention, 1997.

TABLE H-12   Total Vitamin B$_{12}$ Intake (µg) of Persons Aged 6 Years and Older: Mean and Selected Percentiles, United States, 1988–1994

Gender and Age (y)	Number of Examined Persons	Mean	Standard Error of the Mean
Both, 6–8	1,601	5.33	0.11
M, 9–13	1,262	8.91	2.00
M, 14–18	942	10.53	2.40
M, 19–30	1,962	12.00	3.11
M, 31–50	2,612	16.96	5.06
M, 51–70	2,033	16.43	2.96
M, >70	1,329	10.49	1.49
F, 9–13	1,282	5.52	0.19
F, 14–18	1,001	7.06	1.90
F, 19–30	2,027	9.57	1.07
F, 31–50	3,073	13.96	1.43
F, 51–70	2,148	13.51	1.51
F, >70	1,444	14.36	2.40
Pregnant F, 14–55	354	13.16	1.23
Both, ≥6, unadjusted	23,070	12.52	0.85

NOTE: Data from one 24-h dietary recall have been adjusted by following methods in National Research Council. 1986. *Nutrient Adequacy. Assessment Using Food Consumption Surveys.* Washington, DC: National Academy Press; and Feinleib M, Rifkind B, Sempos C, Johnson C, Bachorik P, Lippel K, Carroll M, Ingster-Moore L, Murphy R. 1993. Methodological issues in the measurement of cardiovascular risk factors: Within-person

Selected Percentiles

5th	10th	15th	25th	50th	75th	85th	90th	95th
2.34	2.79	2.94	3.39	4.48	7.34	8.38	8.96	9.86
2.77	3.06	3.25	3.75	4.95	6.93	8.54	9.72	12.13
2.83	3.18	3.53	4.09	5.30	8.03	9.63	12.09	18.07
4.34	4.52	4.66	4.87	5.40	7.09	10.84	13.17	23.27
4.39	4.57	4.70	4.87	5.45	7.76	11.41	14.21	34.97
4.42	4.55	4.66	4.85	5.48	9.95	11.48	14.05	23.30
4.37	4.51	4.61	4.73	5.19	10.23	11.22	13.54	19.69
2.41	2.69	2.90	3.30	4.34	6.83	8.54	9.46	11.55
2.57	2.72	3.00	3.40	4.31	5.60	6.94	8.75	10.66
4.16	4.36	4.46	4.61	5.04	8.13	10.89	13.44	18.52
4.39	4.49	4.56	4.73	5.20	10.53	11.69	16.83	36.01
4.34	4.45	4.52	4.66	5.12	10.58	11.55	16.43	35.12
4.34	4.42	4.48	4.60	4.97	10.38	11.18	13.85	34.32
4.38	4.52	4.77	5.35	9.50	16.91	17.49	18.85	36.78
3.05	4.06	4.42	4.64	5.21	8.24	11.02	13.44	26.62

variability in selected serum lipid measures—results from the Third National Health and Nutrition Survey (NHANES III). *Can J Cardiol* 9:87D–88D. Includes estimated intake from supplements.

SOURCE: NHANES III, 1988–1994, unpublished data, C.L. Johnson and J.D. Wright, National Center for Health Statistics, Centers for Disease Control and Prevention, 1997.

I

Daily Intakes of B Vitamins by Canadian Men and Women, 1990, 1993

TABLE I-1 Daily Thiamin Intake (mg) by Québec Men and Women Aged 18 Years and Older; Mean and Selected Percentiles[a]

Gender and Age (y)	Number of Examined Persons	Mean	Selected Percentiles		
			25th	50th	75th
Males, 18–34	575	1.94	1.44	1.71	2.05
Males, 35–49	175	1.77	1.28	1.57	1.88
Males, 50–64	101	1.61	1.20	1.44	1.78
Males, 65–74	185	1.62	1.16	1.42	1.82
Females, 18–34	593	1.30	1.02	1.16	1.30
Females, 35–49	209	1.17	0.91	1.05	1.25
Females, 50–64	114	1.31	0.91	1.08	1.36
Females, 65–74	166	1.14	0.80	1.02	1.24

[a] Percentiles were adjusted for intraindividual variability.

SOURCE: Santé Québec. 1995. *Les Québécoises et les Québécois Mangent-Ils Mieux? Rapport de l'Enquête Québécoise sur la Nutrition, 1990.* Montréal: Ministère de la Santé et des Services Sociaux, Gouvernement du Québec.

TABLE I-2 Daily Riboflavin Intake (mg) by Québec Men and Women Aged 18 Years and Older; Mean and Selected Percentiles[a]

Gender and Age (y)	Number of Examined Persons	Mean	Selected Percentiles		
			25th	50th	75th
Males, 18–34	575	2.50	1.83	2.29	2.78
Males, 35–49	175	2.11	1.66	1.92	2.22
Males, 50–64	101	1.80	1.52	1.79	2.09
Males, 65–74	185	1.99	1.46	1.71	2.14
Females, 18–34	593	1.64	1.23	1.51	1.88
Females, 35–49	209	1.48	1.21	1.36	1.55
Females, 50–64	114	1.45	1.16	1.37	1.52
Females, 65–74	166	1.29	1.00	1.22	1.44

[a] Percentiles were adjusted for intraindividual variability.
SOURCE: Santé Québec. 1995. *Les Québécoises et les Québécois Mangent-Ils Mieux? Rapport de l'Enquête Québécoise sur la Nutrition, 1990.* Montréal: Ministère de la Santé et des Services Sociaux, Gouvernement du Québec.

TABLE I-3 Daily Niacin Intake (mg NE) by Québec Men and Women Aged 18 Years and Older; Mean and Selected Percentiles[a]

Gender and Age (y)	Number of Examined Persons	Mean	Selected Percentiles		
			25th	50th	75th
Males, 18–34	575	50.2	43.0	46.0	48.8
Males, 35–49	175	46.4	36.2	43.5	48.1
Males, 50–64	101	41.5	35.0	39.6	45.2
Males, 65–74	185	39.5	30.4	35.4	42.6
Females, 18–34	593	32.3	27.6	29.8	32.0
Females, 35–49	209	32.4	26.8	28.9	31.1
Females, 50–64	114	30.2	25.8	28.7	31.1
Females, 65–74	166	28.6	22.5	25.3	28.7

[a] Percentiles were adjusted for intraindividual variability.
SOURCE: Santé Québec. 1995. *Les Québécoises et les Québécois Mangent-Ils Mieux? Rapport de l'Enquête Québécoise sur la Nutrition, 1990.* Montréal: Ministère de la Santé et des Services Sociaux, Gouvernement du Québec.

TABLE I-4 Daily Vitamin B$_6$ Intake (mg) by Québec Men and Women Aged 18 Years and Older; Mean and Selected Percentiles[a]

Gender and Age (y)	Number of Examined Persons	Mean	Selected Percentiles		
			25th	50th	75th
Males, 18–34	575	2.11	1.70	1.99	2.27
Males, 35–49	175	1.98	1.53	1.90	2.21
Males, 50–64	101	1.89	1.49	1.79	2.07
Males, 65–74	185	1.88	1.36	1.70	2.17
Females, 18–34	593	1.41	1.18	1.32	1.48
Females, 35–49	209	1.41	1.20	1.36	1.49
Females, 50–64	114	1.52	1.23	1.40	1.52
Females, 65–74	166	1.39	1.06	1.25	1.51

[a] Percentiles were adjusted for intraindividual variability.
SOURCE: Santé Québec. 1995. *Les Québécoises et les Québécois Mangent-Ils Mieux? Rapport de l'Enquête Québécoise sur la Nutrition, 1990.* Montréal: Ministère de la Santé et des Services Sociaux, Gouvernement du Québec.

TABLE I-5 Daily Folate Intake (μg) by Québec Men and Women Aged 18 Years and Older; Mean and Selected Percentiles[a]

Gender and Age (y)	Number of Examined Persons	Mean	Selected Percentiles		
			25th	50th	75th
Males, 18–34	575	272	199	238	281
Males, 35–49	175	246	179	226	271
Males, 50–64	101	235	177	221	266
Males, 65–74	185	254	173	215	267
Females, 18–34	593	203	142	177	221
Females, 35–49	209	189	134	168	210
Females, 50–64	114	205	145	181	218
Females, 65–74	166	180	137	166	201

[a] Percentiles were adjusted for intraindividual variability.
SOURCE: Santé Québec. 1995. *Les Québécoises et les Québécois Mangent-Ils Mieux? Rapport de l'Enquête Québécoise sur la Nutrition, 1990.* Montréal: Ministère de la Santé et des Services Sociaux, Gouvernement du Québec.

TABLE I-6 Mean Daily Vitamin B_{12} Intake (µg) by Québec Men and Women Aged 18 Years and Older[a]

Gender and Age (y)	Number of Examined Persons	Mean
Males, 18–34	575	9.15
Males, 35–49	175	6.02
Males, 50–64	101	5.02
Males, 65–74	185	7.91
Females, 18–34	593	4.40
Females, 35–49	209	3.60
Females, 50–64	114	3.68
Females, 65–74	166	3.23

[a] Intakes of vitamin B_{12} were not normally distributed; therefore, adjustment for intraindividual variability could not be performed and percentiles could not be determined.

SOURCE: Santé Québec. 1995. *Les Québécoises et les Québécois Mangent-Ils Mieux? Rapport de l'Enquête Québécoise sur la Nutrition, 1990.* Montréal: Ministère de la Santé et des Services Sociaux, Gouvernement du Québec.

TABLE I-7 Daily Pantothenic Acid Intake (mg) by Québec Men and Women Aged 18 Years and Older; Mean and Selected Percentiles[a]

Gender and Age (y)	Number of Examined Persons	Mean	Selected Percentiles		
			25th	50th	75th
Males, 18–34	575	6.16	5.02	5.74	6.49
Males, 35–49	175	5.36	4.47	5.12	5.71
Males, 50–64	101	4.94	4.32	4.85	5.46
Males, 65–74	185	5.49	4.25	4.87	5.67
Females, 18–34	593	4.21	3.43	3.95	4.59
Females, 35–49	209	3.94	3.18	3.67	4.43
Females, 50–64	114	4.14	3.21	3.88	4.60
Females, 65–74	166	3.77	2.94	3.65	4.30

[a] Percentiles were adjusted for intraindividual variability.

SOURCE: Santé Québec. 1995. *Les Québécoises et les Québécois Mangent-Ils Mieux? Rapport de l'Enquête Québécoise sur la Nutrition, 1990.* Montréal: Ministère de la Santé et des Services Sociaux, Gouvernement du Québec.

TABLE I-8   Mean Daily Nutrient Intakes by Nova Scotian Men and Women

Gender and Age (y)	Thiamin (mg)	Riboflavin (mg)	Niacin (NE)	Folate (µg)
Males, 18–34	2.1	2.4	53.6	263.4
Males, 35–49	1.7	2.0	41.4	238.3
Males, 50–64	1.6	1.9	40.2	234.3
Males, 65–74	1.6	1.8	36.3	243.0
Females, 18–34	1.2	1.5	29.0	161.7
Females, 35–49	1.1	1.4	28.9	175.7
Females, 50–64	1.1	1.4	27.9	188.9
Females, 65–74	1.1	1.2	25.3	182.2

NOTE: Tabulations are based on a total of 3,204 recalls, which includes initial 24-h recalls from 2,212 respondents and 992 replicate recalls. Intakes are based on nutrient levels derived from food sources only and do not include supplements. Intakes were calculated from the 24-h recall data by using estimated conversion factors for the reported foods and quantities consumed.

SOURCE: Nova Scotia Heart Health Program. 1993. *Report of the Nova Scotia Nutrition Survey. Nova Scotia:* Nova Scotia Department of Health, Health and Welfare Canada.

J

Options for Dealing with Uncertainties in Developing Tolerable Upper Intake Levels

Methods for dealing with uncertainties in scientific data are generally understood by working scientists and require no special discussion here except to point out that such uncertainties should be explicitly acknowledged and taken into account whenever a risk assessment is undertaken. More subtle and difficult problems are created by uncertainties associated with some of the inferences that need to be made in the absence of directly applicable data; much confusion and inconsistency can result if they are not recognized and dealt with in advance of undertaking a risk assessment.

The most significant inference uncertainties arise in risk assessments whenever attempts are made to answer the following questions (NRC, 1994):

- What set or sets of hazard and dose-response data (for a given substance) should be used to characterize risk in the population of interest?
- If animal data are to be used for risk characterization, which endpoints for adverse effects should be considered?
- If animal data are to be used for risk characterization, what measure of dose (e.g., dose per unit body weight, body surface, or dietary intake) should be used for scaling between animals and humans?
- What is the expected variability in dose response between animals and humans?

• If human data are to be used for risk characterization, which adverse effects should be used?

• What is the expected variability in dose response among members of the human population?

• How should data from subchronic exposure studies be used to estimate chronic effects?

• How should problems of differences in route of exposure within and between species be dealt with?

• How should the threshold dose be estimated for the human population?

• If a threshold in the dose-response relationship seems unlikely, how should a low-dose risk be modeled?

• What model should be chosen to represent the distribution of exposures in the population of interest when data relating to exposures are limited?

• When interspecies extrapolations are required, what should be assumed about relative rates of absorption from the gastrointestinal tract of animals and of humans?

• For which percentiles on the distribution of population exposures should risks be characterized?

At least partial, empirically based answers to some of these questions may be available for some of the nutrients under review, but in no case is scientific information likely to be sufficient to provide a highly certain answer; in many cases there will be no relevant data for the nutrient in question.

It should be recognized that for several of these questions, certain inferences have been widespread for long periods of time and, thus, it may seem unnecessary to raise these uncertainties anew. When several sets of animal toxicology data are available, for example, and data are not sufficient for identifying the set (i.e., species, strain, and adverse effects endpoint) that *best* predicts human response, it has become traditional to select that set in which toxic responses occur at lowest dose (the *most sensitive* set). In the absence of definitive empirical data applicable to a specific case, it is generally assumed that there will not be more than a 10-fold variation in response among members of the human population. In the absence of absorption data, it is generally assumed that humans will absorb the chemical at the same rate as the animal species used to model human risk. In the absence of complete understanding of biological mechanisms, it is generally assumed that, except possibly for certain carcinogens, a threshold dose must be exceeded before toxicity is expressed. These types of long-standing assumptions, which are necessary to complete a risk assessment, are recognized by risk assessors as attempts to deal with uncertainties in knowledge (NRC, 1994).

A past National Research Council (NRC) report (1983) recommended the adoption of the concepts and definitions that have been discussed in this report. The NRC committee recognized that throughout a risk assessment, data and basic knowledge will be lacking and that risk assessors will be faced with several scientifically plausible options (called *inference options* by the NRC) for dealing with questions such as those presented above. For example, several scientifically supportable options for dose scaling across species and for high-to-low dose extrapolation will exist, but there will be no ready means for identifying those that are clearly best supported. The NRC committee recommended that regulatory agencies in the United States identify the needed inference options in risk assessment and specify, through written risk assessment guidelines, the specific options that will be used for all assessments. Agencies in the United States have identified the specific models to be used to fill gaps in data and knowledge; these have come to be called *default options* (EPA, 1986).

The use of defaults to fill knowledge and data gaps in risk assessment has the advantage of ensuring consistency in approach (the same defaults are used for each assessment) and for minimizing or eliminating case-by-case manipulations of the conduct of risk assessment to meet predetermined risk management objectives. The major disadvantage of the use of defaults is the potential for displacement of scientific judgment by excessively rigid guidelines. A remedy for this disadvantage was also suggested by the NRC committee: risk assessors should be allowed to replace defaults with alternative factors in specific cases of chemicals for which relevant scientific data are available to support alternatives. The risk assessors' obligation in such cases is to provide explicit justification for any such departure. Guidelines for risk assessment issued by the U.S. Environmental Protection Agency, for example, specifically allow for such departures (EPA, 1986).

The use of preselected defaults is not the only way to deal with model uncertainties. Another option is to allow risk assessors complete freedom to pursue whatever approaches they judge applicable in specific cases. Because many of the uncertainties cannot be resolved scientifically, case-by-case judgments without some guidance on how to deal with them will lead to difficulties in achieving scientific consensus, and the results of the assessment may not be credible.

Another option for dealing with uncertainties is to allow risk assessors to develop a range of estimates based on application of both defaults and alternative inferences that, in specific cases, have some degree of scientific support. Indeed, appropriate analysis of uncertainties seem to require such a presentation of risk results. Although presenting a number of plausible risk estimates has the advantage that it would seem to more faithfully reflect the true state of scientific understanding, there are no well-established criteria for using such complex results in risk management.

The various approaches to dealing with uncertainties inherent to risk assessment are summarized in Table J-1.

As will be seen in the chapters on each nutrient, specific default assumptions for assessing nutrient risks have not been recommended. Rather, the approach calls for case-by-case judgments, with the recommendation that the basis for the choices made be explicitly

TABLE J–1 Approaches for Dealing with Uncertainties in a Risk Assessment Program

Program Model	Advantages	Disadvantages
Case-by-case judgments by experts	Flexibility; high potential to maximize use of most relevant scientific information bearing on specific issues	Potential for inconsistent treatment of different issues; difficulty in achieving consensus; need to agree on defaults
Written guidelines specifying defaults for data and model uncertainties (with allowance for departures in specific cases)	Consistent treatment of different issues; maximization of transparency of process; resolution of scientific disagreements possible by resort to defaults	Possibly difficult to justify departure or to achieve consensus among scientists that departures are justified in specific cases; danger that uncertainties will be overlooked
Presentation of full array of estimates from all scientifically plausible models by assessors	Maximization of use of scientific information; reasonably reliable portrayal of true state of scientific understanding	Highly complex characterization of risk, with no easy way to discriminate among estimates; size of required effort may not be commensurate with utility of the outcome

stated. Some general guidelines for making these choices will, however, be offered.

REFERENCES

EPA (U.S. Environmental Protection Agency). 1986. Guidelines for carcinogen risk assessment. *Fed Regist* 51:33992–34003.

NRC (National Research Council). 1983. *Risk Assessment in the Federal Government: Managing the Process.* Washington, DC: National Academy Press.

NRC (National Research Council). 1994. *Science and Judgment in Risk Assessment.* Washington, DC: National Academy Press.

K

Blood Concentrations of Folate and Vitamin B$_{12}$ from the Third National Health and Nutrition Examination Survey (NHANES III), 1988–1994

TABLE K-1 Serum Folate (ng/mL) of Persons Aged 4 Years and Older: Mean and Selected Percentiles, United States, 1988–1994

Gender and Age (y)	Number of Examined Persons	Mean	Standard Error of the Mean
Both, 4–8	3,128	11.0	0.3
M, 9–13	1,129	9.0	0.3
M, 14–18	865	6.0	0.4
M, 19–30	1,856	5.0	0.1
Males, 31–50	2,493	5.7	0.2
M, 51–70	1,987	7.4	0.2
M, >70	1,370	9.0	0.3
F, 9–13	1,149	8.2	0.3
F, 14–18	936	5.7	0.2
F, 19–30	1,915	5.8	0.2
F, 31–50	2,929	6.5	0.2
F, 51–70	2,087	8.9	0.3
F, >70	1,534	10.7	0.4
Pregnant F, 14–55	327	10.5	0.8
Both, ≥4, unadjusted	23,705	7.2	0.1

NOTE: Values have been adjusted by following recommendations in Life Sciences Research Office/Federation of American Societies for Experimental Biology. 1994. *Assessment of the Folate Methodology Used in the Third National Health and Nutrition Examination Survey (NHANES III, 1988–1994)*. Raiten DJ, Fisher KD, eds. Bethesda, MD: LSRO/ FASEB. Values are presented as received from source.

Selected Percentiles

5th	10th	15th	25th	50th	75th	85th	90th	95th
4.4	5.3	5.9	6.9	9.4	12.8	14.9	16.9	21.8
3.4	4.0	4.8	5.6	7.7	10.9	12.8	15.3	18.1
2.3	2.6	2.9	3.4	4.7	7.4	8.7	10.5	13.2
1.8	2.2	2.5	2.9	4.0	6.2	7.7	8.7	11.1
1.9	2.3	2.7	3.2	4.6	7.1	9.2	10.5	12.8
2.2	2.7	3.0	3.7	5.7	9.3	12.0	14.3	17.8
2.3	3.1	3.6	4.5	6.9	11.1	15.3	17.9	21.7
3.1	3.8	4.0	4.9	6.8	10.3	11.6	13.9	17.8
2.1	2.5	2.7	3.2	4.9	7.2	8.7	9.7	12.1
1.8	2.3	2.5	3.1	4.6	7.2	9.0	10.5	13.1
1.9	2.3	2.7	3.3	4.9	8.2	10.6	12.4	16.0
2.4	2.9	3.3	4.2	6.6	11.4	14.6	17.0	22.2
2.9	3.6	4.0	5.1	8.0	13.7	17.5	20.6	26.6
2.7[a]	3.4	3.7	4.7	8.4	13.9	16.6	18.5	20.5[a]
2.1	2.5	2.9	3.6	5.5	9.0	11.4	13.5	17.2

[a] Value does not meet standard of reliability or precision.

SOURCE: NHANES III, 1988–1994, unpublished data, C.L. Johnson and J.D. Wright, National Center for Health Statistics, Centers for Disease Control and Prevention, 1997.

TABLE K-2　Erythrocyte Folate (ng/mL) of Persons Aged
4 Years and Older: Mean and Selected Percentiles,
United States, 1988–1994

Gender and Age (y)	Number of Examined Persons	Mean	Standard Error of the Mean
Both, 4–8	3,157	222	12.6
M, 9–13	1,137	202	4.2
M, 14–18	864	164	4.4
M, 19–30	1,853	164	2.8
M, 31–50	2,484	182	3.0
M, 51–70	1,968	216	4.5
M, >70	1,266	242	7.3
F, 9–13	1,145	175	3.4
F, 14–18	934	157	4.0
F, 19–30	1,917	168	3.3
F, 31–50	2,924	194	3.4
F, 51–70	2,059	238	5.5
F, >70	1,374	259	7.3
Pregnant F, 14–55	322	261	12.5
Both, ≥4, unadjusted	23,404	197	1.2

NOTE: Values have been adjusted by following recommendations in Life Sciences Research Office/Federation of American Societies for Experimental Biology. 1994. *Assessment of the Folate Methodology Used in the Third National Health and Nutrition Examination Survey (NHANES III, 1988–1994)*. Raiten DJ, Fisher KD, eds. Bethesda, MD: LSRO/ FASEB. Values are presented as received from source.

Selected Percentiles

5th	10th	15th	25th	50th	75th	85th	90th	95th
123	140	151	170	212	258	289	312	359
102	116	129	149	190	242	280	306	342
83	94	101	116	149	197	237	261	301
84	95	107	119	151	192	219	244	300
86	102	112	127	166	217	255	282	335
93	108	120	139	188	269	318	353	425
93	113	125	149	207	305	362	419	471
87	104	113	129	167	216	239	260	288
77	89	98	115	141	190	221	240	269
77	88	97	116	149	206	242	270	315
84	97	107	125	172	239	286	324	369
95	113	128	149	206	296	356	408	462
96	116	133	160	228	325	391	432	519
114[a]	123	140	164	248	344	379	429	461[a]
87	101	113	130	175	238	284	319	380

[a] Value does not meet standard of reliability or precision.

SOURCE: NHANES III, 1988–1994, unpublished data, C.L. Johnson and J.D. Wright, National Center for Health Statistics, Centers for Disease Control and Prevention, 1997.

TABLE K-3 Serum Vitamin B_{12} (pg/mL) of Persons Aged 4 Years and Older: Mean and Selected Percentiles, United States, 1991–1994

Gender and Age (y)	Number of Examined Persons	Mean	Standard Error of the Mean
Both, 4–8	1,519	781	13.2
M, 9–13	550	620	12.7
M, 14–18	458	516	14.0
M, 19–30	891	470	8.1
M, 31–50	1,201	473	8.4
M, 51–70	937	460	23.7
M, >70	614	449	21.0
F, 9–13	595	612	13.6
F, 14–18	496	509	14.6
F, 19–30	1,049	479	42.1
F, 31–50	1,639	497	42.8
F, 51–70	1,078	504	12.7
F, >70	824	536	66.1
Pregnant F, 14–55	173	426	53.1
Both, ≥4, unadjusted	12,024	517	9.0

NOTE: Values are presented as received from source.

[a] Value does not meet standard of reliability or precision.

Selected Percentiles

5th	10th	15th	25th	50th	75th	85th	90th	95th
409	474	517	569	720	909	1,034	1,103	1,270
344	379	420	470	590	746	825	868	947
261	335	344	395	485	589	656	767	871
251	281	312	355	466	550	623	666	762
236	287	305	348	436	556	633	681	803
190	247	273	317	414	540	630	688	884
187	218	239	298	382	512	629	719	881
344	383	401	453	569	694	794	880	1,060
276	300	319	362	460	618	733	791	871
214	250	277	328	437	580	669	721	805
223	262	287	332	427	589	666	729	827
224	266	291	344	465	605	698	769	919
186	232	261	315	436	622	718	795	953
232[a]	246	255	293	421	501	549	579	631[a]
233	274	303	351	466	615	707	790	924

SOURCE: NHANES III, 1988–1994, unpublished data, C.L. Johnson and J.D. Wright, National Center for Health Statistics, Centers for Disease Control and Prevention, 1997.

TABLE K-4 Serum Homocysteine (μmol/L) of Persons Aged 12 Years and Older: Mean and Selected Percentiles, United States, 1991–1994

Gender and Age (y)	Number of Examined Persons	Mean	Standard Error of the Mean
M, 12–13	177	6.61	0.20
M, 14–18	395	8.39	0.34
M, 19–30	816	10.29	0.26
M, 31–50	1,055	9.93	0.20
M, 51–70	821	11.94	0.49
M, >70	502	13.40	0.49
F, 12–13	226	6.19	0.22
F, 14–18	448	7.08	0.19
F, 19–30	938	8.05	0.15
F, 31–50	1,446	8.36	0.16
F, 51–70	937	9.68	0.19
F, >70	681	11.83	0.36
Pregnant F, 14–55	143	5.41	0.36
Both, ≥12, unadjusted	8,585	9.51	0.09

[a] Value does not meet standard of reliability or precision.

SOURCE: NHANES III, 1988–1994, unpublished data, C.L. Johnson and J.D. Wright, National Center for Health Statistics, Centers for Disease Control and Prevention, 1997.

Selected Percentiles

5th	10th	15th	25th	50th	75th	85th	90th	95th
3.90[a]	4.40	4.80	5.30	16.50	7.70	8.59	8.61	9.51[a]
4.80	5.40	5.60	6.20	7.40	9.01	9.81	11.60	15.30
5.70	6.40	6.80	7.60	9.20	11.20	12.90	13.90	17.53
6.00	6.40	6.80	7.70	9.00	10.89	12.60	13.90	16.38
6.40	7.40	7.80	8.51	10.20	12.89	13.90	15.30	18.38
6.80	7.90	8.40	9.31	12.11	15.70	17.72	19.01	22.70
3.40[a]	3.70	4.30	4.70	5.80	7.20	7.91	8.51	9.51[a]
3.50	4.20	4.90	5.60	6.50	7.90	9.01	9.90	12.10
4.70	5.10	5.40	5.90	7.20	9.40	11.10	12.20	14.09
4.50	5.09	5.50	6.10	7.50	9.20	10.91	12.40	14.30
5.20	5.70	6.30	7.40	8.90	11.01	12.80	14.10	16.01
6.40	7.01	7.50	8.40	10.61	13.40	15.31	17.30	20.53
2.80[a]	2.80[a]	3.50	3.50	4.90	6.40	7.30	7.80[a]	9.30[a]
4.90	5.50	6.00	6.70	8.51	10.81	12.60	13.90	16.51

L

Methylenetetrahydrofolate Reductase

The T^{677} polymorphism in methylenetetrahydrofolate reductase (MTHFR) affects a large percentage of the population. Depending on the group, anywhere from 2 to 15 percent are homozygous and up to 50 percent are heterozygous for this polymorphism, which affects about one-third of alleles (Frosst et al., 1995; Goyette et al., 1994; Jacques et al., 1996). Subjects homozygous for the polymorphism have lower lymphocyte MTHFR levels, higher plasma homocysteine levels, and a higher risk for vascular disease (Bousney et al., 1995; Clarke et al., 1991; Jacques et al., 1996; Kang et al., 1991; Selhub et al., 1993, 1995). Most of the adverse effects associated with this polymorphism affect the homozygote and heterozygotes behave in most regards similarly to the wild type. Because of this, most studies of this polymorphism have compared the effect of homozygosity with a control group composed of C^{677} subjects plus heterozygotes.

The T^{677} polymorphism in MTHFR results in a less stable enzyme but does not affect the affinity of substrates for the enzyme (R. M. Matthews, personal communication). Consequently, the metabolic effects of this polymorphism are lower amounts of enzyme activity rather than abnormal enzyme activity. In subjects with poor folate status, the homozygote would be expected to display lower MTHFR activity. However, for subjects with good folate status, the higher folate levels would stabilize the protein and reduce the difference in available enzyme compared with control subjects. It is expected that metabolic and adverse effects of this polymorphism would primarily affect people with poorer folate status. The level of folate that is sufficient to stabilize the mutant enzyme to the extent that homozygotes would behave identically to controls is not known.

Theoretically, a decreased activity of MTHFR would decrease the rate of formation of methyltetrahydrofolate and consequently homocysteine remethylation. However, this might also be expected to redirect some of the one-carbon flux into other pathways of folate metabolism such as nucleotide metabolism, which could have a positive effect on nucleotide synthesis. Decreased MTHFR activity may explain a recent epidemiological study that reported that homozygotes of poor folate status had a reduced cancer risk compared to control subjects with poor folate status (Ma et al., 1997).

Folate deficiency per se would be expected to adversely affect all metabolic cycles of one-carbon metabolism. However, a metabolic defect in one enzyme may adversely affect one metabolic cycle but may promote another metabolic cycle. Metabolic defects in a single enzyme also greatly complicate the interpretation of normal measures of status. Methyltetrahydrofolate is a very poor substrate for folylpolyglutamate synthetase and has to be demethylated via the methionine synthase reaction before it can be converted to polyglutamates and retained by tissues (Cichowicz and Shane, 1987; Shane, 1989). If folate status is such that MTHFR levels decrease, the rate of formation of methyltetrahydrofolate should be reduced; this would promote folate accumulation by tissues and consequently decrease plasma folate concentrations. A lower plasma folate concentration in such a case would not represent poorer folate status, merely more effective folate accumulation by tissues. Some studies have shown lower plasma folate and increased erythrocyte folate in subjects homozygous for the T^{677} mutation (van der Put et al., 1996). Folate accumulation by fibroblasts from patients with severe defects in MTHFR is normal or increased above normal (Foo et al., 1982). Other studies, however, have reported low plasma and erythrocyte folate in homozygous subjects (Molloy et al., 1997), which is more difficult to explain in terms of our current understanding of folate metabolism. It is possible that the turnover and catabolism of folate is more rapid with nonmethylfolate. Although this could explain this last observation, there is only very limited data to support such a mechanism.

Currently, there is quite good evidence suggesting that the polymorphism has an adverse effect on homocysteine concentrations in subjects with relatively poor folate status (Selhub et al., 1993). However, there is no substantial evidence suggesting that this effect is not corrected by consuming the Recommended Dietary Allowance for folate that is presented in this report.

REFERENCES

Bousney CJ, Beresford SAA, Omenn GS, Motulsky AG. 1995. A quantitative assessment of plasma homocysteine as a risk factor for vascular disease. *J Am Med Assoc* 274:1049–1057.

Cichowicz DJ, Shane B. 1987. Mammalian folylpoly-γ-glutamate synthetase. 2. Substrate specificity and kinetic properties. *Biochem* 26:513–521.

Clarke R, Daly L, Robinson K, Naughten E, Cahalane S, Bowler B, Graham I. 1991. Hyperhomocysteinemia: An independent risk factor for vascular disease. *N Engl J Med* 324:1149–1155.

Foo SK, McSloy RM, Rousseau C, Shane B. 1982. Folate derivatives in human cells: Studies on normal and 5,10-methylenetetrahydrofolate reductase-deficient fibroblasts. *J Nutr* 112:1600–1608.

Frosst P, Blom HJ, Milos R, Goyette P, Sheppard CA, Matthews RG, Boers GJ, den Heijer M, Kluijtmans LA, van den Heuvel LP, Rozen R. 1995. A candidate genetic risk factor for vascular disease: A common mutation in methylenetetrahydrofolate reductase. *Nat Genet* 10:111–113.

Goyette P, Sumner JS, Milos R, Duncan AM, Rosenblatt DS, Matthews RG, Rozen R. 1994. Human methylenetetrahydrofolate reductase: Isolation of cDNA mapping and mutation identification. *Nat Genet* 7:551.

Jacques PF, Bostom AG, Williams RR, Ellison RC, Eckfeldt JH, Rosenberg IH, Selhub J, Rozen R. 1996. Relation between folate status, a common mutation in methylenetetrahydrofolate reductase, and plasma homocysteine concentrations. *Circulation* 93:7–9.

Kang S-S, Wong PWK, Susmano A, Sora J, Norusis M, Ruggie N. 1991. Thermolabile methylenetetrahydrofolate reductase: An inherited risk factor for coronary artery disease. *Am J Hum Genet* 48:536–545.

Ma J, Stampfer MJ, Giovannucci E, Artigas C, Hunter DJ, Fuchs C, Willett WC, Selhub J, Hennekens CH, Rozen R. 1997. Methylenetetrahydrofolate reductase polymorphism, dietary interactions, and risk of colorectal cancer. *Cancer Res* 57:1098–1102.

Molloy AM, Daly S, Mills JL, Kirke PN, Whitehead AS, Ramsbottom D, Conley MR, Weir DG, Scott JM. 1997. Thermolabile variant of 5,10-methylenetetrahydrofolate reductase associated with low red-cell folates: Implications for folate intake recommendations. *Lancet* 349:1591–1593.

Selhub J, Jacques PF, Wilson PWF, Rush D, Rosenberg IH. 1993. Vitamin status and intake as primary determinants of homocysteinemia in an elderly population. *J Am Med Assoc* 270: 2693–2698.

Selhub J, Jacques PF, Bostom AG, D'Agostino RB, Wilson PW, Belanger AJ, O'Leary DH, Wolf PA, Schaefer EJ, Rosenberg IH. 1995. Association between plasma homocysteine concentrations and extracranial carotid-artery stenosis. *N Engl J Med* 332:286–291.

Shane B. 1989. Folylpolyglutamate synthesis and role in the regulation of one carbon metabolism. *Vitam Horm* 45:263–335.

van der Put NM, van den Heuvel LP, Steegers-Theunissen RP, Trijbels FJ, Eskes TK, Mariman EC, den Heyer M, Blom HJ. 1996. Decreased methylene tetrahydrofolate reductase activity due to the 677C→T mutation in families with spina bifida offspring. *J Molec Med* 74:691–694.

M

Evidence from Animal Studies on the Etiology of Neural Tube Defects

Animal models of neural tube defect (NTD) have been examined and manipulated to elucidate the mechanisms of abnormal neurulation and to test etiologic hypotheses suggested by human epidemiological data. Neurulation in mouse and in human embryos appears to be quite similar, at least anatomically, whereas neurulation in the chick embryo differs in several ways. Neural tube closure is a complex and incompletely understood morphogenetic process. The neural plate arises from the embryonic ectoderm, a layer of epithelium held together laterally by cell adhesion molecules and basally by a basement membrane. Neurulation results from both proliferation and change in shape of these neurectoderm cells: change in cell shape results in elevation of the neural folds and cell proliferation results in their elongation. Extracellular matrix deposition apparently maintains this change in tissue form (Copp and Bernfield, 1994).

Most studies have used the mouse because it provides the best compromise of developmental similarity and ease of genetic and environmental manipulation. However, NTD phenotypes in the mouse are varied and none resembles the human in every respect. Moreover, the mechanism of action of any suspected human NTD gene will need to be investigated in animal models, such as mouse models containing induced mutations in the corresponding mouse gene or mice of appropriate genetic background that contain a mutation in the suspect gene.

Numerous mouse genetic models now exist and include spontaneous mutants (at least 13 models, 3 of which have been shown to be due to a specific gene) as well as an increasing list of induced genetic mutants in which the resultant NTD phenotype was often unexpected (Copp and Bernfield, 1994). Most of the models exhibit isolated exencephaly (an anencephaly equivalent in the mouse) and some show solely posterior NTD (Copp and Bernfield, 1994), but most are associated with malformations outside the central nervous system, growth retardation, or both. One model, the curly tail mouse, shows NTD as its sole defect (Gruneberg, 1954). Curly tail mice exhibit both anterior and posterior NTDs, which can closely resemble the common forms of human NTD (Seller and Adinolfi, 1981). As in humans, there is incomplete (60 percent) penetrance, which varies with genetic background. This suggests the presence of modifier genes, one of which has been genetically mapped (Letts et al., 1995).

NUTRITION STUDIES

Several nutritional deprivations can produce NTDs in rodents (Hurley, 1980). For example, zinc deficiency in rats results in 47 percent of offspring with NTDs (Hurley and Shrader, 1972). Severe folate deficiency, induced by dietary depletion together with gut sterilization to reduce microbial sources of folate (Walzem et al., 1983), does not yield NTDs in mice (Heid et al., 1992). However, these studies have not been done in genetic NTD models. One mouse model shows reduced neural abnormalities with huge folic acid supplements (2.5 to 3.0 mg/kg of body weight/day) (Zhao et al., 1996). Despite many other attempts, folate supplementation has not been associated with changes in NTD incidence in rodent models, including the curly tail mouse model (Seller, 1994). Because evidence suggests a genetic basis for erythrocyte folate values in humans (Mitchell et al., 1997), significant interspecies differences are conceivable. Further study is needed to determine whether mice handle folate more efficiently than humans and whether they have a lower threshold for folate sufficiency.

Supplementation with other nutrients can reduce the incidence of NTDs in some rodent models. Methionine supplements reduce the incidence of rat embryo NTD induced by homocysteine in culture (Vanaerts et al., 1994). The axial-defects mouse mutant shows a reduction in posterior NTD penetrance after parenteral methionine in the pregnant dam at the onset of neural tube closure; however, large doses of folic acid and vitamin B_{12} were without effect (Essien, 1992). Inositol administration can reduce NTD incidence in the curly tail mouse model (Greene and Copp, 1997). If curly tail is a suitable mouse model of human NTD, markers of inositol status should be evaluated with regard to human NTDs.

TERATOLOGY STUDIES

Many teratogens produce NTDs in rodents; exencephaly is especially common. Examples include ethanol, retinoic acid, vitamin A, and valproate (Sulik and Sadler, 1993). The NTD phenotype depends on the timing of administration. Early administration (before closure of the anterior neural tube at mouse embryonic day 9) most often results in exencephaly whereas later administration yields posterior defects. Valproate-induced NTDs are strain specific in the mouse (Finnell et al., 1988), highlighting the importance of genetic predisposition in NTD etiology. Although some studies show an alteration in folate levels (Hendel et al., 1984) and a protective effect of coadministered folic acid in valproate-induced NTD (Trotz et al., 1987), the metabolic mechanism of valproate teratogenesis is unclear (Nau, 1994).

In humans, carbamazepine is a commonly used antiepileptic drug that has been reported to cause NTDs at a higher-than-normal rate (Rosa, 1991). The absolute risk estimated from 21 cohort studies is approximately 1 percent (Rosa, 1991).

REFERENCES

Copp AJ, Bernfield M. 1994. Etiology and pathogenesis of human neural tube defects: Insights from mouse models. *Curr Opin Pediatr* 6:624–631.

Essien FB. 1992. Maternal methionine supplementation promotes the remediation of axial defects in *Axd* mouse neural tube mutants. *Teratology* 45:205–212.

Finnell RH, Bennett GD, Karras SB, Mohl VK. 1988. Common hierachies of susceptibility to the induction of neural tube defects in mouse embryos by valproic acid and its 4-propyl-4-pentenoic acid metabolite. *Teratology* 38:313–320.

Greene ND, Copp AJ. 1997. Inositol prevents folate-resistant neural tube defects in the mouse. *Nat Med* 3:60–66.

Gruneberg H. 1954. Genetical studies on the skeleton of the mouse. 8. Curly tail. *J Genet* 52:52–67.

Heid MK, Bills ND, Hinrichs SH, Clifford AJ. 1992. Folate deficiency alone does not produce neural tube defects in mice. *J Nutr* 122:888–894.

Hendel J, Dam M, Gram L, Winkel P, Jorgensen I. 1984. The effects of carbamazepine and valproate on folate metabolism in man. *Acta Neurol Scand* 69:226–231.

Hurley LS. 1980. *Developmental Nutrition*. Englewood Cliffs, NJ: Prentice-Hall.

Hurley LS, Shrader RE. 1972. Congenital malformations of the nervous system in zinc-deficient rats. *Int Rev Neurobiol Suppl* 1:7–51.

Letts VA, Schork NJ, Copp AJ, Bernfield M, Frankel WN. 1995. A curly-tail modifier locus, *mct1*, on mouse chromosome 17. *Genomics* 29:719–724.

Mitchell LE, Duffy DL, Duffy P, Bellingham G, Martin NG. 1997. Genetic effects on variation in red-blood-cell folate in adults: Implications for the familial aggregation of neural tube defects. *Am J Hum Genet* 60:433–438.

Nau H. 1994. Valproic acid-induced neural tube defects. In: Bock G, Marsh J, eds. *Neural Tube Defects*. Ciba Foundation Symposium 181. London: John Wiley & Sons. Pp. 144–151.

Rosa FW. 1991. Spina bifida in infants of women treated with carbamazepine during pregnancy. *N Engl J Med* 324:674–677.

Seller MJ. 1994. Vitamins, folic acid and the cause and prevention of neural tube defects. In: Bock G, Marsh J, eds. *Neural Tube Defects*. Ciba Foundation Symposium 181. London: John Wiley & Sons. Pp. 161–172.

Seller MJ, Adinolfi M. 1981. The curly tail mouse: An experimental model for human neural tube defects. *Life Sci* 29:1607–1615.

Sulik KK, Sadler TW. 1993. Postulated mechanisms underlying the development of neural tube defects. Insights from in vitro and in vivo studies. *Ann NY Acad Sci* 678:8–21.

Trotz M, Wegner C, Nau H. 1987. Valproic acid-induced neural tube defects: Reduction by folinic acid in the mouse. *Life Sci* 41:103–110.

Vanaerts LA, Blom HJ, Deabreu RA, Trijbels FJ, Eskes TK, Copius Peereboom-Stegeman JH, Noordhoek J. 1994. Prevention of neural tube defects by and toxicity of L-homocysteine in cultured postimplantation rat embryos. *Teratology* 50:348–360.

Walzem RL, Clifford CK, Clifford AJ. 1983. Folate deficiency in rats fed amino acid diets. *J Nutr* 113:421–429.

Zhao Q, Behringer RR, de Crombrugghe B. 1996. Prenatal folic acid treatment suppresses acrania and meroanencephaly in mice mutant for the *Cart1* homeobox gene. *Nat Genet* 13:275–283.

N

Estimation of the Period Covered by Vitamin B$_{12}$ Stores

As indicated in Chapter 9, it is possible to estimate the length of time body stores of vitamin B$_{12}$ held in reserve will maintain adequate nutriture when dietary intake is nonexistent as a result of adopting a vegan or similar diet very low in B$_{12}$ or when decreased absorption of vitamin B$_{12}$ occurs as a result of atrophic gastritis or pernicious anemia.

This estimate is based on being able to ascertain what the expected turnover rate will be (0.1, 0.15, and 0.2 percent have been found experimentally to occur). It has been estimated that in individuals with normal absorption and reabsorption rates for vitamin B$_{12}$, the daily turnover is 0.1 percent per day. For individuals with pernicious anemia, who cannot absorb or reabsorb vitamin B$_{12}$, the turnover rate is about 0.2 percent of the body pool or stores per day. As individuals develop various degrees of atrophic gastritis, it is possible that a turnover rate of 0.15 percent is appropriate.

To estimate how long body stores of vitamin B$_{12}$ can be depended on to maintain health, it is also necessary to know the lowest pool size of vitamin B$_{12}$ consistent with health, which could be considered to be the threshold before which signs of inadequate B$_{12}$ would begin to occur. As is indicated in Chapter 9, this has been estimated to be approximately 300 µg (Bozian et al., 1963).

By calculating the ratio of the total body stores of vitamin B$_{12}$ to this assumed threshold (or to a lower threshold if expected because of differences in individual requirements), it is possible to estimate how long body stores will meet the needs of the individual.

TABLE N-1 Table of Ratios[a] Used to Estimate the Extent of Protection from Vitamin B_{12} Stores

Vitamin B_{12} Threshold, µg	Initial Store, mg									
	1	2	3	4	5	6	7	8	9	10
350	3	6	9	11	14	17	20	23	26	29
300	3	7	10	13	17	20	23	27	30	33
250	4	8	12	16	20	24	28	32	36	40
200	5	10	15	20	25	30	35	40	45	50
150	7	13	20	27	33	40	47	53	60	67
100	10	20	30	40	50	60	70	80	90	100

[a] This represents the initial vitamin B_{12} stores divided by the level of stores at which signs of vitamin B_{12} deficiency may become evident.

Table N-1 provides the ratio of the expected stores of an individual (in milligrams) to the threshold level of stores at which signs of vitamin B_{12} deficiency may appear. Usually the threshold value is not known, but studies suggest that it may be approximately 300 µg of vitamin B_{12} for adults (Bozian et al., 1963).

Tables N-2 and N-3 give the expected length in days (Table N-2) or years (Table N-3) for a given turnover rate and ratio from Table N-1 that body stores of vitamin B_{12} will sustain health in the individual. For example, from Table N-1 the ratio for an initial store of 3 mg of vitamin B_{12} and a threshold of 300 µg of vitamin B_{12} is 10. If the turnover rate is 0.1, the store would be expected to last 2,303 days, or 6.3 years.

REFERENCE

Bozian RC, Ferguson JL, Heyssel RM, Meneely GR, Darby WJ. 1963. Evidence concerning the human requirement for vitamin B_{12}. Use of the whole body counter for determination of absorption of vitamin B_{12}. *Am J Clin Nutr* 12:117–129.

TABLE N-2   Time in Days until Vitamin B_{12} Threshold Is Reached

Ratio of Initial Stores to Threshold, from Table N-1	Turnover Rate		
	0.1	0.15	0.2
70	4,248	2,832	2,124
60	4,094	2,730	2,047
50	3,912	2,608	1,956
40	3,689	2,459	1,844
30	3,401	2,267	1,701
20	2,996	1,997	1,498
17	2,833	1,889	1,417
16	2,773	1,848	1,386
15	2,708	1,805	1,354
14	2,639	1,759	1,320
13	2,565	1,710	1,282
12	2,485	1,657	1,242
11	2,398	1,599	1,199
10	2,303	1,535	1,151
9	2,197	1,465	1,099
8	2,079	1,386	1,040
7	1,946	1,297	973
6	1,792	1,195	896
5	1,609	1,073	805
4	1,386	924	693
3	1,099	732	549

TABLE N-3   Table of Time in Years until Vitamin B_{12} Threshold Below Which Deficiency Signs May Occur Is Reached

Ratio of Initial Stores to Threshold, from Table N-1	Turnover Rate		
	0.1	0.15	0.2
70	11.6	7.8	5.8
60	11.2	7.5	5.6
50	10.7	7.1	5.4
40	10.1	6.7	5.0
30	9.3	6.2	4.7
20	8.2	5.5	4.1
17	7.8	5.2	3.9
16	7.6	5.1	3.8
15	7.4	4.9	3.7
14	7.2	4.8	3.6
13	7.0	4.7	3.5
12	6.8	4.5	3.4
11	6.6	4.4	3.3
10	6.3	4.2	3.2
9	6.0	4.0	3.0
8	5.7	3.8	2.8
7	5.3	3.6	2.7
6	4.9	3.3	2.5
5	4.4	2.9	2.2
4	3.8	2.5	1.9
3	3.0	2.0	1.5

O

Biographical Sketches

ROY M. PITKIN, M.D., (Chair), is professor emeritus at the University of California at Los Angeles (UCLA) and editor of the journal O*bstetrics & Gynecology*. Previously, he was chair of the Department of Obstetrics and Gynecology at UCLA, and before that he was the chair of the Department of Obstetrics and Gynecology at the University of Iowa. His involvement in nutrition began 25 years ago with an Academic Career Development Award in Nutrition from the National Institutes of Health and continued with his receipt of the Joseph Goldberger Award in Clinical Nutrition from the American Medical Association in 1982. He has chaired several committees of the Food and Nutrition Board involving maternal and child nutrition, including the Committee on Nutrition in Pregnancy and Lactation (1988–1992). He was elected to the Institute of Medicine in 1990.

LINDSAY H. ALLEN, Ph.D., is a professor in the Department of Nutrition and a faculty member in the Program in International Nutrition at the University of California at Davis. She received her B.Sc. in nutrition and agriculture from the University of Nottingham, England, and her Ph.D. in nutrition from the University of California at Davis. This was followed by experience at the University of California at Berkeley and a faculty position at the University of Connecticut. Dr. Allen's research focuses on the causes, consequences, and prevention of micronutrient deficiencies (vitamin B_{12}, folate, iron, zinc, and riboflavin) in developing countries and on vitamin B_{12} deficiency in the elderly in the United States. She is an associate editor of the *Journal of Nutrition* and was awarded the Kellogg International Nutrition Prize in 1997. Her other responsibilities on the Food and Nutrition Board include serving as chair of the Committee on International Nutrition and as a member of the Standing Committee on the Scientific Evaluation of Dietary Reference Intakes.

LYNN B. BAILEY, Ph.D., is a professor of nutrition in the University of Florida's Food Science and Human Nutrition Department. Before joining the faculty in 1977, Dr. Bailey completed her Ph.D. and postdoctoral training at Purdue University in the area of human nutrient requirements. Her research has focused on the estimation of folate requirements and the evaluation of folate status in different life stages, including adolescence, young adulthood, pregnancy, and postmenopause. Collaborative studies involving folate labeled with stable isotopes have provided new information related to factors affecting folate bioavailability. Many scientific journal publications and book chapters have resulted from Dr. Bailey's research, and she was editor of the book *Folate in Health and Disease*. She has served on numerous expert scientific panels, including the Food and Drug Administration's Folate Subcommittee, which addressed the fortification of cereal grain products with folic acid in an effort to reduce the risk of neural tube defects. Dr. Bailey was the recipient of a national U.S. Department of Agriculture Award for Superior Service for her research accomplishments related to estimating folate requirements.

MERTON BERNFIELD, M.D., is an internationally recognized leader in developmental biology. Dr. Bernfield studies the molecular mechanisms underlying how organs take shape during embryonic development, which is important for understanding birth defects. He developed the most suitable mutant mouse model for human neural tube defects and has conducted research on prenatal care and birth outcome in blacks and whites. Dr. Bernfield serves as the Clement A. Smith Professor of Pediatrics and professor of cell biology at Harvard Medical School. Dr. Bernfield has received various honors, including Guggenheim and Macy fellowships, and distinguished lectureships, including the Swedish Zetterstrom Lecture, the Canadian Queen Elizabeth II Lecture and the Wellcome Visiting Professorship in the Basic Medical Sciences. He is a member of the Institute of Medicine, a fellow of the American Association for the Advancement of Sciences, and a member of the American Association of Physicians. He is an editor for several basic and clinical science journals, a former member of the executive committee of the American Society for Cell Biology, and a past president of the Society for Developmental Biology.

PHILIPPE DE WALS, M.D., Ph.D., is professor in the Department of Community Health Sciences at the Faculty of Medicine, University of Sherbrooke, Québec, Canada, and has been a visiting professor at the Department of Maternal and Child Health, School of Public Health, University of North Carolina at Chapel Hill. He was born and educated in Belgium and received his degrees from the Catholic University of Louvain. Dr. De Wals's research is focused on the epidemiology of adverse reproductive outcomes and on the evaluation of health services and public health programs. He has authored over 60 peer-reviewed publications and book chapters. He is a consultant in public health policy for the Regional Health Board of Monteregie, the Ministry of Health of Québec and Health Canada. In 1990 he was awarded the Jean Van Beneden Prize for distinguished achievement in public health research.

RALPH GREEN, M.D., is professor and chair of the Department of Pathology at the University of California at Davis and pathologist-in chief at the University of California at Davis Medical Center. He received his medical degree and dissertational doctorate at the University of the Witwatersrand in Johannesburg, South Africa. Dr. Green's research interests span several aspects of nutrition as applied to the blood and the nervous system. In particular, his work has focused on vitamin B_{12}, folate, and iron metabolism. He has developed improved methods for diagnosing deficiency of these nutrients. He has also been involved in population-based programs for food folate fortification in an economically deprived rural community. Recently, he has been studying the role of disturbance in homocysteine metabolism caused by vitamin deficiencies and genetic defects and how they may relate to vascular occlusive disease. Dr. Green has served on expert panels and advisory groups convened by the Food and Drug Administration, the Centers for Disease Control and Prevention, and the Life Sciences Research Office of the Federation of American Societies for Experimental Biology to assess folate nutritional status with particular reference to fortification of the U.S. diet with folate.

DONALD B. McCORMICK, Ph.D., is a professor of biochemistry at Emory University School of Medicine, Atlanta. He received his undergraduate degree in chemistry and his Ph.D. in biochemistry from Vanderbilt University. He was the L.H. Bailey Professor of Nutritional Biochemistry at Cornell University before becoming the F.E. Callaway Professor and chairman of biochemistry at Emory University. Dr. McCormick's research is focused on cofactors, including vitamins, coenzymes, and metal ions. He has elucidated major aspects of the metabolism of riboflavin, vitamin B_6, biotin, and lipoate with regard to both conversion to functional coenzymes and catabolism. Honors and awards received include Bausch and Lomb and Westinghouse Science Scholarships, U.S. Public Health Service and Guggenheim Fellowships, Mead Johnson and Osborne and Mendel Awards from the American Institute of Nutrition (now the American Society for Nutrition Sciences [ASNS]), Wellcome Visiting Professorships, and special name lectureships. He has approximately 500 publications and has been editor for *Vitamins and Coenzymes, Vitamins and Hormones,* and *Annual Review of Nutrition.* Dr. McCormick has been a member of numerous scientific societies and committees, including chairman of the National Institute of Health's nutrition study section, President of ASNS, board member for Federation of American Societies for Experimental Biology, and vice-chairman of the Food and Nutrition Board.

ROBERT M. RUSSELL, M.D., is professor of medicine and nutrition at Tufts University and an associate director of the Jean Mayer U.S. Department of Agriculture (USDA) Human Nutrition Research Center on Aging at Tufts University in Boston. Dr. Russell received his undergraduate degree from Harvard University and his medical degree from Columbia University. He has served on national and international advisory boards including the U.S. Food and Drug Administration, National Digestive Diseases Advisory Board, U.S. Department of Agriculture Human Investigation Committee (chairman), U.S. Pharmacopeial Convention, and the American Board of Internal Medicine. He has worked on international nutrition programs in several countries, including Vietnam, Iran, Guatemala, and China. In 1991 he served on the UNICEF Consulting Team for assessing malnutrition in postwar Southern Iraq. Dr. Russell is a member of numerous professional societies, has served as a councilor to the American Society of Clinical Nutrition, and was a member of the board of directors of the American College of Nutrition. As a senior scientist at the Jean Mayer USDA Human Nutrition Research Center on Aging, Dr. Russell's primary work involves studying the effects of aging on gastrointestinal absorptive function. He is a noted expert in the area of human metabolism of vitamins.

BARRY SHANE, Ph.D., is a professor of nutrition and chair of the Department of Nutritional Sciences at the University of California at Berkeley. He was born and educated in England and received his Ph.D. in biochemistry at the University of London. This was followed by postdoctoral studies at the University of California at Berkeley and a faculty position at the Johns Hopkins University. Dr. Shane's research is focused on various areas of biochemical nutrition, including the metabolic role and interrelationships of water-soluble vitamins and the influence of genetic variation on vitamin requirements. He has authored over 100 peer-reviewed articles. He is on the editorial board of the *American Journal of Clinical Nutrition, Journal of Biological Chemistry* and *Journal of Nutritional Biochemistry* and was awarded the Mead Johnson Award by the American Society of Nutrition Sciences. He currently serves on the board of the Federation of American Societies for Experimental Biology.

STEVEN H. ZEISEL, M.D., Ph.D., is professor and chair in the Department of Nutrition and professor in the Department of Pediatrics at the University of North Carolina at Chapel Hill. He received his medical degree from Harvard University and his Ph.D. in nutrition from the Massachusetts Institute of Technology. Dr. Zeisel is the author of a medical curriculum in nutrition that is being used at more than 100 medical schools. Dr. Zeisel's research focuses on nutrient metabolism, specifically that of choline, with special emphasis on establishing human nutrient requirements and on identifying cancer-causing agents that are produced within the human body. He has authored more than 140 peer-reviewed publications and book chapters. Dr. Zeisel currently serves on the Medical Education Committee of the American Society for Clinical Nutrition and on the board of directors of the Association of Departments and Programs of Nutrition. Dr. Zeisel also is the editor-in-chief of the *Journal of Nutritional Biochemistry.*

IRWIN H. ROSENBERG, M.D., (Liaison, Subcommittee on Upper Levels), is an internationally recognized leader in nutrition science who has made important and unique contributions to the understanding of nutrition metabolism in health and disease. Dr. Rosenberg is professor of physiology, medicine, and nutrition at Tufts University School of Medicine and School of Nutrition; director of the Jean Mayer U.S. Department of Agriculture (USDA) Human Nutrition Research Center on Aging at Tufts University; and dean for nutrition sciences at Tufts University. As a clinical nutrition investigator he has helped develop a nutritional focus within the field of gastroenterology, with his primary research interest being in the area of folate metabolism. His research for the past decade has focused on nutrition and the aging process. Among his many honors are the Grace Goldsmith Award of the American College of Nutrition, Robert H. Herman Memorial Award of the American Society of Clinical Nutrition, Jonathan B. Rhoads Award of the American Society for Parenteral and Enteral Nutrition, 1994 W.O. Atwater Memorial Lectureship of the USDA, and 1996 Bristol Myers Squibb/Mead Johnson Award for Distinguished Achievement in Nutrition Research. Dr. Rosenberg was elected to the Institute of Medicine in 1994.

P
Glossary and Abbreviations

ACC	Acetyl-CoA carboxylase
ADP	Adenosine diphosphate
AI	Adequate Intake
ALT	Alanine aminotransferase
apABG	Acetamidobenzoylglutamate
ATP	Adenosine triphosphate
Bioavailability	The accessibility of a nutrient to participate in metabolic and/or physiological processes.
BMI	Body mass index; weight/height^2 (kg/m^2)
CHD	Coronary heart disease
CI	Confidence interval
CoA	Coenzyme A
CSFII	Continuing Survey of Food Intakes by Individuals, a survey conducted by the Agricultural Research Service, USDA
CV	Coefficient of variation
DFE	Dietary folate equivalent
DNA	Deoxyribonucleic acid
Dose-response assessment	The step in a risk assessment in which the relationship between nutrient intake and adverse effect (in terms of incidence and/or severity of the effect) is determined.

DRI	Dietary Reference Intake
α-EALT	Erythrocyte alanine aminotransferase
EAR	Estimated Average Requirement
α-EAST	Erythrocyte aspartate aminotransferase
EEG	Electroencephalogram
EGR	Erythrocyte glutathione reductase
EGRAC	Erythrocyte glutathione reductase activity coefficient
ETK	Erythrocyte transketolase
ETKAC	Erythrocyte transketolase activity coefficient
FAD	Flavin-adenine dinucleotide
FAO	Food and Agriculture Organization of the United Nations
FIGLU	Formiminoglutamic acid
FMN	Flavin mononucleotide
FNB	Food and Nutrition Board
Hazard identification	The step in a risk assessment, which is concerned with the collection, organization, and evaluation of all information pertaining to the toxic properties of a nutrient.
HIV	Human immunodeficiency virus
HPLC	High-performance liquid chromatography
HPV	Human papilloma virus
IM	Intramuscular
IOM	Institute of Medicine
LCAT	Lecithin-cholesterol acyltransferase
LOAEL	Lowest-observed-adverse-effect level; the lowest intake (or experimental dose) of a nutrient at which an adverse effect has been identified.
MCV	Mean cell volume
MI	Myocardial infarction
MMA	Methylmalonic acid
MTHFR	Methylenetetrahydrofolate reductase
NAD	Nicotinamide adenine dinucleotide
NADP	Nicotinamide adenine dinucleotide phosphate
NAS	National Academy of Sciences
NE	Niacin equivalent

NHANES	National Health and Nutrition Examination Survey, a survey conducted periodically by the National Center for Health Statistics, Centers for Disease Control and Prevention, U.S. Department of Health and Human Services
NHIS	National Health Interview Survey, a survey conducted periodically by the National Center for Health Statistics, Centers for Disease Control and Prevention, U.S. Department of Health and Human Services
NOAEL	No-observed-adverse-effect level; the highest intake (or experimental dose) of a nutrient at which no adverse effect has been observed.
NRC	National Research Council, the organizational arm of the National Academies
NTD	Neural tube defect
4-PA	4-Pyridoxic acid
pABG	p-Aminobenzoylglutamate
PARP	Poly-ADP-ribose polymerase
PGA	Pteroylglutamic acid
PL	Pyridoxal
PLP	Pyridoxal phosphate
PM	Pyridoxamine
PMP	Pyridoxamine phosphate
PN	Pyridoxine
PNP	Pyridoxine phosphate
RDA	Recommended Dietary Allowance
Risk	Within the context of nutrient toxicity, the probability or likelihood that some adverse effect will result from a specified excess intake of a nutrient.
Risk assessment	An organized framework for evaluating scientific information that has as its objective a characterization of the nature and likelihood of harm resulting from excess human exposure to an environmental agent (in this case, a dietary nutrient). It includes the development of both qualitative and quantitative expressions of risk. The process of risk assessment can be divided into four major steps: hazard identification, dose-response assessment, exposure assessment, and risk characterization.

Risk characterization	The final step in a risk assessment, which summarizes the conclusions from the other three steps of the risk assessment and evaluates the risk. This step also includes a characterization of the degree of scientific confidence that can be placed in the UL.
Risk management	The process by which risk assessment results are integrated with other information to make decisions about the need for, method of, and extent of risk reduction. In addition to risk assessment results, risk management considers such issues as the public health significance of the risk, technical feasibility of achieving various degrees of risk control, and economic and social costs of this control.
SD	Standard deviation
SE	Standard error
SEM	Standard error of the mean
TCI, II, III	Transcobalamin I, II, and III
TMA	Trimethylamine
TPN	Total parenteral nutrition
TPP	Thiamin pyrophosphate
UL	Tolerable Upper Intake Level
UF	Uncertainty factor; a number by which the NOAEL (or LOAEL) is divided to obtain the UL. UFs are used in risk assessments to deal with gaps in data (e.g., data uncertainties) and knowledge (e.g., model uncertainties). The size of the UF varies depending on the confidence in the data and the nature of the adverse effect.
USDA	U.S. Department of Agriculture
WHO	World Health Organization

Index

A

Abortion, spontaneous, 258

Absorption of nutrients. *See also* Bioavailability of nutrients; Malabsorption syndromes
  biotin, 376
  choline, 392-393
  folate, 1, 8-9, 198, 211, 212, 233, 256, 259
  niacin, 124-125
  pantothenic acid, 358, 366
  pregnancy and, 33, 89
  riboflavin, 88-89, 102, 115
  sensitivity considerations, 48
  thiamin, 59, 62, 81-82
  vitamin B_6, 151, 160-161
  vitamin B_{12}, 33, 306, 307-309, 313, 317, 318, 320, 322, 328, 330, 331, 333, 334, 339, 429

Acetaldehyde, 164

Acetaminophen, 212

Acne, 346

Adenosine diphosphate (ADP) ribosylation, 127, 145

Adequate Intakes (AIs). *See also specific nutrients*
  applicable population, 18
  defined, 2, 5, 17-18, 21, 424
  derivation of, 5, 21, 29-31, 33
  extrapolation from other age groups, 21, 31-33
  increasing consumption of nutrients, 14-15
  indicators used to set, 7, 10, 11, 22-23
  methods used to set, 5, 22-23, 31-33
  RDAs compared, 6, 21-22
  risk of inadequacy, 19
  uses, 2, 4, 5, 21, 425-426
  weight and height and, 26

Adolescents, 14 through 18 years. *See also* Children; Life-stage groups; Puberty/pubertal development; *individual nutrients*
  AIs, 21, 31-32
  EARs, 31-32
  lactation, 78, 112, 136-137, 143-144, 178-179, 186-187, 238-240, 280-281, 341-342, 368, 383, 406, 412
  pregnancy, 76-78, 110-111, 135-136, 143-144, 176-178, 186-187, 233-238, 280-281, 339-341, 367, 382-383, 404-406, 412
  RDAs, 32-33
  ULs, 13, 56

541

N

S

FOOD AND NUTRITION BOARD, INSTITUTE OF MEDICINE–NATIONAL ACADEMY OF SCIENCES
DIETARY REFERENCE INTAKES:
RECOMMENDED INTAKES FOR INDIVIDUALS

Life Stage Group	Calcium (mg/d)	Phosphorus (mg/d)	Magnesium (mg/d)	Vitamin D (µg/d)[a,b]	Fluoride (mg/d)	Thiamin (mg/d)
Infants						
0–6 mo	210*	100*	30*	5*	0.01*	0.2*
7–12 mo	270*	275*	75*	5*	0.5*	0.3*
Children						
1–3 y	500*	**460**	**80**	5*	0.7*	**0.5**
4–8 y	800*	**500**	**130**	5*	1*	**0.6**
Males						
9–13 y	1,300*	**1,250**	**240**	5*	2*	**0.9**
14–18 y	1,300*	**1,250**	**410**	5*	3*	**1.2**
19–30 y	1,000*	**700**	**400**	5*	4*	**1.2**
31–50 y	1,000*	**700**	**420**	5*	4*	**1.2**
51–70 y	1,200*	**700**	**420**	10*	4*	**1.2**
> 70 y	1,200*	**700**	**420**	15*	4*	**1.2**
Females						
9–13 y	1,300*	**1,250**	**240**	5*	2*	**0.9**
14–18 y	1,300*	**1,250**	**360**	5*	3*	**1.0**
19–30 y	1,000*	**700**	**310**	5*	3*	**1.1**
31–50 y	1,000*	**700**	**320**	5*	3*	**1.1**
51–70 y	1,200*	**700**	**320**	10*	3*	**1.1**
> 70 y	1,200*	**700**	**320**	15*	3*	**1.1**
Pregnancy						
≤ 18 y	1,300*	**1,250**	**400**	5*	3*	**1.4**
19–30 y	1,000*	**700**	**350**	5*	3*	**1.4**
31–50 y	1,000*	**700**	**360**	5*	3*	**1.4**
Lactation						
≤ 18 y	1,300*	**1,250**	**360**	5*	3*	**1.4**
19–30 y	1,000*	**700**	**310**	5*	3*	**1.4**
31–50 y	1,000*	**700**	**320**	5*	3*	**1.4**

NOTE: This table presents Recommended Dietary Allowances (RDAs) in **bold type** and Adequate Intakes (AIs) in ordinary type followed by an asterisk (*). RDAs and AIs may both be used as goals for individual intake. RDAs are set to meet the needs of almost all (97 to 98 percent) individuals in a group. For healthy breastfed infants, the AI is the mean intake. The AI for other life-stage and gender groups is believed to cover needs of all individuals in the group, but lack of data or uncertainty in the data prevent being able to specify with confidence the percentage of individuals covered by this intake.

[a] As cholecalciferol. 1 µg cholecalciferol = 40 IU vitamin D.

[b] In the absence of adequate exposure to sunlight.

[c] As niacin equivalents (NE). 1 mg of niacin = 60 mg of tryptophan; 0–6 months = preformed niacin (not NE).

[d] As dietary folate equivalents (DFE). 1 DFE = 1 µg food folate = 0.6 µg of folic acid from fortified food or as a supplement consumed with food = 0.5 µg of a supplement taken on an empty stomach.

Riboflavin (mg/d)	Niacin (mg/d)[c]	Vitamin B_6 (mg/d)	Folate (µg/d)[d]	Vitamin B_{12} (µg/d)	Pantothenic Acid (mg/d)	Biotin (µg/d)	Choline[e] (mg/d)
0.3*	2*	0.1*	65*	0.4*	1.7*	5*	125*
0.4*	4*	0.3*	80*	0.5*	1.8*	6*	150*
0.5	6	0.5	150	0.9	2*	8*	200*
0.6	8	0.6	200	1.2	3*	12*	250*
0.9	12	1.0	300	1.8	4*	20*	375*
1.3	16	1.3	400	2.4	5*	25*	550*
1.3	16	1.3	400	2.4	5*	30*	550*
1.3	16	1.3	400	2.4	5*	30*	550*
1.3	16	1.7	400	2.4[f]	5*	30*	550*
1.3	16	1.7	400	2.4[f]	5*	30*	550*
0.9	12	1.0	300	1.8	4*	20*	375*
1.0	14	1.2	400[g]	2.4	5*	25*	400*
1.1	14	1.3	400[g]	2.4	5*	30*	425*
1.1	14	1.3	400[g]	2.4	5*	30*	425*
1.1	14	1.5	400	2.4[f]	5*	30*	425*
1.1	14	1.5	400	2.4[f]	5*	30*	425*
1.4	18	1.9	600[h]	2.6	6*	30*	450*
1.4	18	1.9	600[h]	2.6	6*	30*	450*
1.4	18	1.9	600[h]	2.6	6*	30*	450*
1.6	17	2.0	500	2.8	7*	35*	550*
1.6	17	2.0	500	2.8	7*	35*	550*
1.6	17	2.0	500	2.8	7*	35*	550*

[e] Although AIs have been set for choline, there are few data to assess whether a dietary supply of choline is needed at all stages of the life cycle, and it may be that the choline requirement can be met by endogenous synthesis at some of these stages.

[f] Because 10 to 30 percent of older people may malabsorb food-bound B_{12}, it is advisable for those older than 50 years to meet their RDA mainly by consuming foods fortified with B_{12} or a supplement containing B_{12}.

[g] In view of evidence linking folate intake with neural tube defects in the fetus, it is recommended that all women capable of becoming pregnant consume 400 µg from supplements or fortified foods in addition to intake of food folate from a varied diet.

[h] It is assumed that women will continue consuming 400 µg from supplements or fortified food until their pregnancy is confirmed and they enter prenatal care, which ordinarily occurs after the end of the periconceptional period—the critical time for formation of the neural tube.